Streamflow Generation Processes

Benchmark Papers in Hydrology, 1

Streamflow Generation Processes

Selection, Introduction and Commentary by

Keith J. Beven
University of Lancaster, UK

International Association of Hydrological Sciences / Association Internationale des Sciences Hydrologique

Published by the International Association of Hydrological Sciences 2006
IAHS Press, Centre for Ecology and Hydrology, Wallingford, Oxfordshire OX10 8BB, UK

IAHS Benchmark Papers in Hydrology Series no. 1
ISBN 1-901502-53-8

British Library Cataloguing-in-Publication Data.
A catalogue record for this book is available from the British Library.

IAHS Publications are available from:
IAHS Press, Centre for Ecology and Hydrology, Wallingford, Oxfordshire OX10 8BB, UK
tel: +44 1491 692442; fax: +44 1491 692448; email: jilly@iahs.demon.co.uk

Printed in the UK by 4Word, Bristol

Foreword to the Benchmark Papers in Hydrology Series

Hydrology is a young science. Following the seminal work by Henry Darcy in the mid-19th Century and Robert Horton's influential studies in the first half of the 20th Century, knowledge generation in the hydrological sciences exploded through the second half of the 20th Century with new discoveries in each of the components of the hydrological cycle. Despite scientific and technological advances, the field of hydrology has been highly fragmented across the engineering–science interface, between basic and applied studies, mathematical and descriptive work, field and laboratory investigations. While hydrological science is now taught in most of the world's major universities, the teachers and students may be found just as commonly in civil engineering, as in physical geography, forestry, environmental engineering, geology, soil science, or agricultural engineering. Our key journals, textbooks, conference proceedings and monographs are likewise dispersed and scattered across myriad outlets. While Internet searching and Pdf versions of more recent published material make "keeping up" easier, there is a body of important early work that is often ignored by young researchers. As a result, we stand at the beginning of the 21st Century in danger of memory loss of our young science.

The objective of the *Benchmark Papers in Hydrology* series is to define and create a world repository of the benchmark papers in hydrology for education, research and practice. In so doing, it is hoped that this series will counter memory loss in our young and highly interdisciplinary science by chronicling the development of ideas leading to current theory in hydrology. Indeed, there are many examples where experiments, ideas, etc. are being "re-invented" because the key papers are not known to new generations of students, scientists and practitioners. This hydrological foundation is what the *Benchmark Papers in Hydrology* series aims to define, publish and distil: the key papers that contributed to the current thinking of the sub-fields of the hydrological sciences. As such, the benchmark papers are hydrological exemplars. They are the best of the contributions that our field has to offer.

The idea of the *Benchmark Papers in Hydrology* series resembles the very successful series *Benchmark Papers in Geology*, published in the 1970s and 1980s by Dowden, Hutchinson and Ross Inc. and then van Nostrand Reinhold Co. That series yielded over 80 volumes and is seen today as the definitive collection in geology, and is held by libraries around the world. Given the International Association of Hydrological Science's (IAHS) lead role in international hydrology, it is appropriate that it has initiated the *Benchmark Papers in Hydrology* series. This major effort for diffusion of the most important ideas in hydrology to libraries, institutes and individuals worldwide is consistent with the IAHS mission.

The *Benchmark Papers in Hydrology* series interprets the hydrological sciences in the broadest sense. Each volume tackles a specific topic and is edited by a recognized scientific authority. The volume editor makes the initial selection of papers and prepares the Introduction and Commentary to the papers that are finally included in that volume. The series Editorial Board approves the published selection of papers. Articles are reproduced in whole (but sometimes in part) from the key papers of the 20th Century.

Jeffrey J. McDonnell
Series Editor

Editorial Board

Acknowledgements

The various publishers who granted permission to reproduce copyright material in this volume are thanked, and also the authors of those papers. Special gratitude is expressed to Elsevier Science Publishers for their generous assistance.

Several papers published in *Hydrological Sciences Journal* or the IAHS Red Book Series (IAHS Proceedings and Reports) are reprinted here and we are grateful to their authors too.

Cate Gardner at IAHS Press, and Marloes Bakker at Oregon State University, are thanked for their huge contributions to this volume. Sarah Anne Graham is thanked for the cover artwork.

Contents

Introduction

One of the things that makes hydrology an interesting and important field of study is the way in which it overlaps both pure scientific interests and practical water management and engineering, to support life and civilization. From a scientific point of view, the way in which water flows on hillslopes is of interest in understanding the processes controlling:

- flood runoff,
- the transport of nutrients and sediments to rivers and lakes,
- the spatial patterns of water availability to plants,
- the potential for hillslope failures, and
- controls on residence times of water in a catchment.

From the water management point of view, the way in which water flows on hillslopes is of interest in controlling:

- the water yield of catchment areas and its prediction for the design of flood protection measures, reservoirs for drought management, culverts and bridges,
- water quality and its prediction for the design of water treatment and effluent works,
- soil water status and its prediction for the design of agricultural drainage works and slope failures, and the design of protection measures.

In both cases, the list of topics of interest can easily be extended.

This dual interest is apparent throughout the history of hydrology, with many of the advances in hydrological concepts and theory being instigated by engineering needs as well as the study of water flows being an intrinsic and fascinating part of the science of the environment. Many celebrated hydrologists have embodied this duality, not the least being Robert E. Horton (1875–1945), Consulting Hydraulic Engineer of Voorheesville, New York, who practiced as an engineer but who had wide scientific interests (becoming, for example, President of the American Meteorological Society, 1938–1939), carried out many laboratory and field experiments (including running an experimental catchment), and perhaps did more to invigorate the scientific study of hydrology in the first half of the 20th Century than anyone else.

The area of streamflow generation has particular resonance in this duality. Except in considering issues of slope stability (which are not considered in this volume), the processes of water flux are not immediately of interest to the engineer who can determine much of the information that is needed for water management from measurements at the catchment scale (e.g. Leach *et al.*, 1933; Horton, 1935). Certainly catchments are made up of hillslopes, but the engineer is primarily interested in the net effect of all the flow processes in determining the water yield of a reservoir or intake for potable supply on a river, or the design of a flood embankment. And yet, it is readily perceived that there is a lot of science to be done at the hillslope scale that might help in both the understanding and prediction of water flows and water quality, particularly in areas where no streamgauges exist. In addition, since Horton's time, the availability of powerful computers has resulted in the ability to implement distributed models of catchments, and such models by definition represent (in more or less detail) the nature of the hillslopes in that catchment. The early studies of hillslope hydrology and streamflow generation, however, were instigated by a desire for understanding as to how the hydrology of hillslopes worked. The use of that understanding (or not) in the early implementations of distributed models came later (e.g. Huggins & Monke, 1968; Freeze, 1971, 1972; Smith & Woolhiser, 1971), though note that in the pre-computer age, Ross (1921), Zoch (1934) and others, including Horton (1938), had already used distributed calculations of runoff generation and routing.

Making a quick list of classic papers that have influenced my thinking on streamflow generation produced a total of over 100, without listing any more recent than 15 years old!! I have therefore found it necessary to limit the scope of the list to fit within the remit I was given, in the hope that other volumes planned in this series will cover some of the important areas relevant to streamflow generation that could not be included here, such as: the hydraulics of overland flow; interception by vegetation and throughfall; local controls on infiltration and soil water content profiles; snow and snowmelt processes; channel heads; hillslope hydrology and hillslope failures; remote sensing in hydrology; and, especially, models of hillslope hydrology.

The papers are very much those that I consider to have shaped perceptions of hillslope processes, rather than those that have provided, for example, modelling techniques. I have also invited suggestions from others about papers that have influenced their own thinking but the final choice was mine and necessarily reflects my perceptual model of streamflow generation as predominantly influenced by my experience in humid temperate regions.

I have also taken the liberty of including some papers that should have been on the list of benchmark papers of a wider range of hydrologists, and which are relevant to our understanding today but have suffered the ignominy of being largely forgotten (although these are also, of course, very much a personal choice). There are also several that are really representative of a larger body of publications. For example, I have chosen Weyman (1970, Paper 17), for reasons explained in the commentary accompanying that paper; others have suggested that Weyman (1973) was the benchmark paper from the work on the East Twin hillslopes.

In the period up to the 1970s it was also much more common for people to publish work in their own language. The choice here reflects my own reading and is dominated by papers in English (with one notable exception, Paper 4). However, I did not manage to identify any papers in, for example, Russian or German, that might have had a significant influence on the development of current thought on streamflow generation. I would be very happy to receive suggestions from those who do read other languages, for future reference.

There are a variety of different perceptual models of streamflow generation in the literature (see below). An obvious way of organizing a collection of benchmark papers on streamflow generation would be along classical lines of these different perceptual models of hillslope responses. That was certainly my first choice. However, further thought suggested that it is necessary to counter the impression that the different mechanisms are mutually exclusive. This is not the case. They may well occur in the same catchment at different locations on the hillslopes or at different times at the same location on the hillslopes (e.g. Beven, 1989, 2001b, 2002a). Thus I have chosen not to classify the papers in any way, but merely to present them in historical sequence. This will also allow the reader to appreciate that there were different ideas around in the same period. The emphasis is on the scientific understanding of how water flows on hillslopes, though there are reflections of the science/management duality in many of the selected papers. One paper on the effect of agricultural field drainage on streamflow generation (Paper 19) has been included as a reminder that in many catchments, man has had an important effect on the hydrological response.

There are already a variety of good reviews of streamflow generation available. The chapters by Dick Chorley and Tom Dunne in the book on *Hillslope Hydrology* edited by Mike Kirkby (1978) provide definitive summaries of studies and issues even after 25 years. Chorley (1978) is also the only published source I know of information about the extremely interesting PhD study of the small (24 ha) Whitehall catchment, Georgia (USA), by Tischendorf (1969) (though the thesis, supervised by John Hewlett, is still available on microfiche). The book edited by Sopper & Lull (1967), arising from a US National Science Foundation Advanced Science Seminar held at Pennsylvania State University in 1965 and with contributions by many of the important hydrologists of the era, was a major impetus to studies of hillslope hydrology and streamflow generation. A variety of topics are covered in the book on *Process Studies in Hillslope Hydrology* edited by Malcolm Anderson and Tim Burt (1990). There is a good review of processes in arid areas in Yair & Lavee (1985). Mike Bonell (1993, 1998, 2005) has reviewed runoff generation in forest catchments, including tropical regions; Al Freeze (1974), Roy Ward (1984), Andy Pearce and others (1986), and Jeff McDonnell (1990) cover responses of small catchments in humid areas; I have reviewed the interflow concept (Beven, 1989) and runoff generation in arid and semiarid areas (Beven, 2002a) and, with Peter Germann, the role of macropores in hillslope hydrology (Beven & Germann, 1982). There are also some papers that are too long to reproduce here; these include the monograph on surface runoff produced by Horton (1935), the summary of field experiments on overland flow hydraulics of Emmett (1970), the complete series of papers on surface hydrology by Zaslavsky & Sinai (1981), and the summaries of studies on pipeflows by Gilman & Newson (1980) and Jones (1981).

The development of ideas in streamflow generation

Horton's ideas that surface flow dominates the storm response of catchments (Paper 1) became the mainstream hydrological theory for decades, even though it was not really an appropriate model for his own experimental catchment (Beven, 2004a) or even an adequate representation of his ideas about catchment responses, which were generally much more sophisticated than they are usually represented now (see, for example, Paper 2). It is also worth noting that, in the same period, Charles R. Hursh (then Director of the USDA Forest Service Coweeta Hydrological Laboratory) was reporting results from catchments that clearly were not dominated by surface runoff (Hursh, 1936, 1944, Paper 3), though not in a way that was nearly so easy to apply in engineering calculations. Indeed, most of the concepts that underlie understanding of hillslope hydrology and streamflow generation today can be found in papers before 1945, including saturated contributing areas, macropore flows, air pressure effects, "translatory wave" effects and variability of soil properties (Beven, 2004b,c). We have simply forgotten that the scientists of the time were also serious observers of nature, equally perceptive in their responses to their observations and also, it must be said, using essentially the same measurement techniques as today (with the important exception of the use of geochemical fingerprinting of water sources).

What they did not have was such a wide range of theory, and the reason why Horton is remembered primarily for his ideas on infiltration excess (or Hortonian) overland flow is that in his last classic paper of 1945 he managed to present a fully developed theoretical analysis of hillslope erosional development based on surface runoff generation concepts. This paper was a real attempt to apply scientific principles of surface flow and transport to geomorphology in a way that had not been seen before—but based on surface runoff mechanisms that did not seem to be generally applicable. So a range of field studies was instigated, for the most part by geomorphologists and geographers interested in how landforms worked, to see if Horton's ideas were supported in the field (e.g. Schumm, 1956) or to see what happened on other types of hillslopes where surface runoff was not the dominant mechanism. There was a wave of such studies and data analyses at a variety of sites in the 1950s and 1960s, in part stimulated by the aims of the International Hydrological Decade of UNESCO from 1965–1974 (see Donoso, 2004). Benchmark papers by Tsukamoto (Paper 5), Hewlett & Hibbert (Paper 6), Betson (Paper 8), Dunne & Black (Paper 15), Weyman (Paper 17), Harr (Paper 22), and others resulted in a variety of perceptual models for the mechanisms of streamflow generation on catchments, including variable and dynamic contributing areas, subsurface stormflows, preferential flows, dynamic stream networks and perched water tables (see the review by Dunne, 1978).

However, these primarily field-based studies, while extending the range of perceived response mechanisms beyond the "Hortonian" model, did not result in a new body of quantitative theory. One reason was the lack of computational tools—analytical theory, such as that of Henderson & Wooding (1964), had obvious limitations in predicting the response of catchments over extended periods of time. The availability of hydrological models running on digital computers in the 1960s changed that. While early models were based on lumped storage elements, following the Freeze & Harlan (1969) blueprint, there were many models developed to simulate distributed surface and subsurface flows. It is interesting that this paper did not really stimulate field studies to verify the concepts in the same way as Horton's (1945) paper. With some (few) notable examples, it seems almost to have led to a separation of the field and modelling communities in hydrology, something we are only just beginning to recover from (see below).

Grossly generalizing, this seems to have been because field experiments were showing that the world was very complicated, with an increasing appreciation of the role of preferential flow pathways, and the use of geochemical separations of the hydrograph to show that in many catchments, storm runoff was made up of water that had been stored in the catchment prior to an event. The hydrological models, when calibrated to a discharge record, did not have to worry about either of these complications. There have been attempts to model the effects of preferential flow pathways (but none particularly successful); there have been even fewer attempts to model transport on hillslopes in a distributed way. What is more worrying, however, is that these complications are also left out of most hydrological text books. There is an attraction when teaching students to present a subject in a formal framework to demonstrate understanding. Looking at most hydrological texts that present the theory of hydrology, that means the Darcy-Richards equation for subsurface flows, the St Venant or kinematic equations for surface runoff, and the Penman-Monteith combination theory for evapotranspiration. These could be a good starting point, but all of them are wrong if applied to the complexity of real hillslopes. I would suggest that this volume of papers is a good way of extending the concepts of students to embrace some of that complexity.

The difficulties of studying streamflow generation

There are, of course, good reasons for why it has been so difficult to bring hydrological observation and hydrological theory into compatibility. There are some real difficulties in studying streamflow generation.

First, find your hillslope. Ideally the hillslope should be "representative" of those within a catchment area—but what does representative mean in this context? What is representative in terms of length, slope, and shape in a region? Should it be a straight hillslope, a convergent (and concave?) hollow hillslope or a divergent (and convex?) spur hillslope. In many catchments, straight and divergent hillslopes make up the bulk of the area, but the convergent slopes leading down to channel heads might be dominant in the storm response of the catchment. Dunne & Black (Paper 15) chose to study all three types of slopes, by dint of digging a long trench and separating out the flow collectors, but even they did not have replicates to check on how representative their slope segments might be. Indeed, their mapping of saturated areas within their catchment revealed that controls other than topography were important in controlling contributing areas. More recent work has revealed important local variability in the nature of responses (Huff *et al.*, Paper 29; Woods & Rowe, 1996; Western & Grayson, 1998; Freer *et al.*, 2002).

Then, you must find a way of measuring the hillslope inputs and outputs. This is also much easier said than done. There have been few field-scale experiments on hillslopes where the flow system has been artificially bounded (but see Paper 6, and more recently the Hydrohill experiments in China reported by Kendall *et al.*, 2001). Controlled boundary conditions are normally a requirement of good scientific investigation, but this is difficult in the field, particularly for the lower boundary where assumptions of an "impermeable bedrock" may be hopeful rather than realistic. Only very small rates of leakage through the base of a slope might make a significant difference to the saturated zone response (see the analysis of Ahuja & Ross, 1983). If the lower boundary is to be sealed, of course, this will then require that the soil be disturbed and replaced, causing further problems in reproducing the effects of soil structure (see Hewlett & Hibbert, Paper 6; Kendall *et al.*, 2001).

Inputs to the ground surface can be quite variable, particularly as a result of throughfall and stemflow processes on vegetated slopes and, while surface outflow collectors are not too difficult to install (though sealing can be a problem), the digging of a trench to collect the subsurface outflows from a slope segment is not an ideal measurement technique. It is only a useful technique where the depth to an impermeable layer is not more than about 2 m, and it will tend to induce a larger upslope contributing area during wet conditions than dry conditions, the changes being proportionally greater for short trenches (Atkinson, 1978). Also, a trench may not collect the flows from deeper flow pathways that lie below the base of the trench, even if it is dug to an apparently impermeable layer. Even for a long trench, the upslope area contributing to the trench might be difficult to determine, especially in cases where the fluxes into the trench might be dominated by a few preferential flow lines associated with bedrock fractures, depressions in a bedrock surface, or macropores. There is direct or indirect evidence of such fluxes in many studies of hillslope hydrology (e.g. Papers 17, 20 and 26). An interesting paper on "percolines" by Bunting (1961), that could also be considered a benchmark paper, has not been included here partly because of length and partly because it is more concerned with the geomorphological development of hillslopes.

It might be better to infer hillslope streamflow generation from incremental discharges in a stream channel (e.g. Paper 29), though this is not currently feasible on a continuous basis without installing expensive gauging structures to obtain accurate estimates of the timing and quantity of discharge. Also, this approach does not isolate the contributions from either side of the channel where the slopes might have quite different characteristics, nor does it separate surface and subsurface flow contributions.

A further problem is in trying to assess the spatial responses of a hillslope. Rainfall inputs to the ground surface as throughfall and stemflow, water tables, soil moisture storage and surface runoff, can only be measured at discrete points on the hillslope and are known to be highly spatially variable so that it is difficult to obtain enough measurements. Areas of saturation or surface runoff at the soil surface can be mapped by different techniques (though I can vouch from personal experience that this can be very labour intensive, Beven, 1978) and saturation at the surface, or even lateral flow collected from a topsoil layer, does not always imply that the whole soil is saturated in the case of perched water tables.

It would also be useful to determine the <u>sources</u> of hillslope outflows, or at least the proportion of the storm hydrograph from a hillslope made up of event water and pre-event water stored on the hillslope prior to the event. Hydrograph separations using chemical characteristics to label different sources of runoff can be used to do this (e.g. Paper 27), but samples of different water sources (e.g. throughfall, soil water and deeper groundwater) can only be taken at discrete points in the catchment and can be quite variable in their characteristics (though it is common to make the simplifying assumption that the streamflow prior to an event after a period of recession is an integration of waters from deeper layers and therefore characteristic of the "groundwater" component). This means that any such separations will be subject to some uncertainty (see Bazemore *et al.*, 1994; Genereux, 1995; Joerin *et al.*, 2002) but, even so, this information might still be useful in the calibration of streamflow generation models (see Seibert & McDonnell, 2002).

It is evident that any understanding developed from the measurement of hillslope streamflow generation is subject to the limitations of current measurement techniques, both in terms of the limited sampling that is possible at both hillslope and local scales, and the lack of techniques for measuring subsurface fluxes directly. Future work really needs a direct measurement of water fluxes so that scale-dependent model representations of hillslope responses can be developed (see discussion in Beven, 2002b).

Streamflow generation and hydrological models

As noted previously, trying to model hillslope processes has been useful in the development of streamflow generation because of the way in which conceptual ideas must, in the end, be confronted by the complexities and limitations of real world data. It has to be said that this confrontation has been largely avoided in hydrology by resorting to model calibration (albeit for very good reasons), such that the issue of <u>how</u> to model hillslope hydrology and streamflow generation, and particularly flow pathways and tracer breakthroughs, has not yet been resolved (see Beven, 2001b, 2002b; Loague *et al.*, 2005). It was not until the later 1960s that sufficient computer power became available to model quantitatively the processes of streamflow generation in any detail. The initial models concentrated on modelling the infiltration excess runoff using either grid square (Huggins & Monke, 1968) or hillslope segment (Smith & Woolhiser, 1971) discretizations of the terrain. Using one-dimensional (1-D) hillslope segments allowed the use of numerical approximations to the kinematic wave equation, analytical solutions for which had been applied to both surface and subsurface flow on hillslopes by Henderson & Wooding (1964).

In 1969, however, Freeze & Harlan outlined their blueprint for a fully distributed and integrated solution of the surface and subsurface flow equations. These nonlinear differential equations (Richards' equation for subsurface flows and the St Venant equation for surface flows) could not be solved analytically for any useful cases, so it was necessary to resort to numerical solutions using discretizations of both time and space. Al Freeze implemented 2-D and 3-D solutions based on finite difference solutions, while later Neuman (1973) and Beven (1975) independently implemented 2-D vertical slice finite element solutions, that allowed more realistic representations of hillslopes convex/concave profiles and sloping soil horizons. Even though Freeze was working at the IBM Thomas J. Watson research centre at the time, with access to the largest computers available anywhere, there were still significant constraints on what could be implemented and it was necessary to use coupled 1- and 2-D averaged approximations to the full 3-D equations, and relatively coarse spatial discretizations and short simulation periods, to effect solutions.

One useful simplification was to use only a single hillslope segment, initially of constant width. This was used by Al Freeze, first with some hypothetical hillslopes, to study the sensitivity of different types of hydrological response to parameters such as the hydraulic conductivity of the soil (Freeze, 1972), following the publication of the work of Dunne & Black (Paper 15). The simulation results were interesting, in that they appeared to suggest that only a relatively restricted range of conditions would result in soil saturation to the surface and saturation excess overland flow. This work was later extended to convergent and divergent slopes by Beven (1975, 1977), demonstrating the importance of slope convergence in producing dynamic saturated areas, and how a catchment could be represented by a collection of variable width hillslopes (though initial attempts to reproduce the behaviour of a small catchment were poor and were not published until Beven, 2001a).

These results were based on single storm simulations (I remember having a celebration in the lab when my model finally took less than 12 hours of computer time to simulate a 12-hour storm) and we probably did not have an adequate appreciation of the effect of initial conditions at the time. This was one of the issues highlighted when Stephenson & Freeze (1974) tried to apply the Freeze model to data collected on a hillslope at Reynolds Creek, Idaho. They noted the difficulty of trying to obtain a match to all the available field data (in part because of the limited number of runs that could be made). They were also remarkably perspicacious in noting that, having obtained a reasonable fit, model validation would also be difficult because of the dependence of results on uncertain initial and boundary conditions during any validation period (in fact, they may also not have had a correct representation of the processes on the slope, see Unnikrishna *et al.*, 1995). A greater appreciation of the long time-scale effects of initial conditions came later in the fully 3-D simulations of a Darcian hillslope hollow by Binley & Beven (1992).

Another implementation of the Freeze & Harlan (1969) blueprint was the SHE model (Système Hydrologique Européen), started in 1977 and finally published in Abbott *et al.* (1986), and with application to the Wye catchment in the UK by Bathurst (1986), though it has only been quite recently that computer power has allowed fully 3-D saturated/unsaturated subsurface models to be coupled to surface models. There are now a variety of models linking different solutions of the Darcian subsurface flow equations with the St Venant or kinematic wave equations for surface runoff (Beven, 2001b). These distributed models can, in principle, allow different parameter values in every element of the model but in practical applications will usually assume that whole zones of the domain have homogeneous parameters.

Al Freeze was also among the first to look at the effects of heterogeneity of soil properties on hillslope responses, initially for a saturated 1-D Darcian system (Freeze, 1975). He did so, only to "*get Jim Wallis off my back*" (Freeze, 2004). Jim Wallis also worked at IBM and was well-known for his work in stochastic hydrology. The work was later extended by Smith & Hebbert (1979), looking at the effects of spatial variability of infiltration on runoff, and Freeze (1980) who used a simplified model of subsurface and surface responses on a 2-D hillslope with stochastic rainfalls and hydraulic conductivities. Freeze (1980) concluded: "*The simulations carried out in this study have placed the author in some awe of the delicate hydrologic balance on a hillslope. If one fixes the mean hydraulic conductivity of a hillslope, then there is only a very narrow range of topographic slopes that can lead to runoff generated by the Dunne mechanism. If one fixes the topographic slope of a hillslope, then there is only a very narrow range of hydraulic conductivities that will lead to a water table that is high enough to allow the Dunne mechanism to be operative in a given climatic regime. The fact that the Dunne mechanism is so common in nature in spite of these theoretical limitations on its occurrence infers a very close relationship between climate, hydraulic conductivity, and the development of geomorphic landforms.*" (p.406–407).

The work was extended by Binley *et al.* (1989) using a 3-D model with the aim of testing whether the stochastically heterogeneous hillslopes could be represented by a homogeneous soil with equivalent effective values of the parameters. These were again, however, hypothetical simulations with even the full heterogeneous model assuming that the soil parameters were homogeneous in each model element. Also, initial conditions were assumed prior to each storm. The real difficulty comes in applying such stochastic models to real hillslopes. As noted by Beven (1981), commenting on the Freeze (1980) paper: "*A real catchment is not a collection of multiple samples from an underlying stochastic process; it is rather a single realisation of such a process. In fact, the real stochastic process may be nonstationary in space and time but must reflect the integrative nature of the geomorphological, pedological and biological history of the catchment under natural conditions.*" (p.431).

The problem is that, while the hillslopes of a catchment necessarily reflect a history of events, we need to characterize those hillslopes, with all their variability, at a particular point in time. How can this be done with the limited resources and measurement techniques available? It cannot be done by direct measurement of parameter values everywhere, and it seems unlikely that a theory of hillslope development will help much because of the lack of knowledge of boundary conditions in the past. Thus, it is not so much the stochastic nature of hillslopes that is the problem, but the uncertainty in characterizing particular hillslopes. This is why it has been so difficult, and why there are so few papers that really attempt to link field studies of streamflow generation with model predictions. There would seem to have been little real advance in such characterization since these studies, despite the availability of new remote sensing and geophysical techniques. Indeed, it can be questioned whether we are actually using the correct equations to describe locally Darcian processes on heterogeneous hillslopes because of local nonlinearities in the unsaturated zone (see Beven, 1989, 2002b,c).

The problem is, in fact, worse because the Darcian hillslope models simply do not take account of all the processes involved in streamflow generation. They may include representations of evapotranspiration and interception but they do not generally take any account of the interactions between the vegetation and the resulting heterogeneity of inputs to the ground, nor of roots and the structural or macroporosity in the subsurface due to cracking, root channels and soil fauna, or rock fractures (e.g. Beven & Germann, 1982). Other (nonstationary) processes such as hydrophobicity and soil crusting (e.g. Burch *et al.*, 1989) are also usually neglected. Such neglect in the face of such complexity is perfectly understandable (if only to think about all the additional parameters that would be required to represent such processes) but is not based on any firm understanding that such processes can be neglected relative to Darcian subsurface fluxes. That type of guidance is not yet available. There are <u>no</u> benchmark papers that can be included here to validate the assumptions of Darcian models on the basis of field measurements (except in the case of repacked soil, e.g. the Scholl & Hibbert, 1973, lysimeter). It should therefore be expected that hillslope responses can only be predicted with significant uncertainties. It remains to be seen how far those uncertainties might be constrained in the future by the development of new understanding or measurement techniques. The long history of trying to model the field data set for the R5 catchment at Coshocton, Ohio, is an interesting case study in this respect (see the commentary on Paper 28).

What is there still to learn about streamflow generation?

In any era of research it is difficult to assess what there might still be to add to our perceptual model of streamflow generation. We can, after all, now <u>perceive</u> the potential complexity of response of any hillslope in the field quite easily, including all the small-scale heterogeneities of inputs, soil characteristics, preferential flows and immobile storage, crusting and other surface effects, surface irregularities in flow directions and channelling, effects of drainage and ditching, surface and subsurface (perched or profile saturated) dynamic contributing areas, and geological structures, fractures and deeper flow pathways. Even the recent suggestions of a "double paradox" of hillslope responses (fast response of pre-event water in contributing to the storm hydrograph and relatively constant geochemical characteristics of the slow component, see Kirchner, 2003) is not really a paradox with respect to this complex perceptual model if the difference between the (complex) propagation of pressure waves, including the displacement of stored old water into macropores, and actual velocities of flow is appreciated (e.g. Beven, 1989, the commentary on Paper 31, and Bishop *et al.*, 2004). The difficulty is going beyond the perceptual model to <u>quantify</u> the response of individual hillslopes, and give guidance about what should be the dominant mechanisms of response (and consequent parameterizations) in different situations.

McDonnell & Woods (2004) suggest that one strategy to address this problem is to start with a classification of catchment (or hillslope) responses. Classification: "*is a means to identify which might be the most important controls on water fluxes and pathways from the full range of processes and mechanisms present on the globe …… a globally-agreed classification system would provide an important organising principle, complementing the concept of the hydrological cycle and the principle of mass conservation. It would provide help with both modelling and experimental approaches to hydrology, by providing guidance on the similarities and differences between basins*" (p.3). Their arguments can be traced back to the "awe" of Al Freeze, that the hillslope response system and catchment geomorphology may be finely tuned to climate forcing in producing different modes of hydrological response. It may, indeed, be possible to summarize the dominant modes of catchment response using simple parameterizations, such as the recent innovations in identifying catchment nonlinearities of Young (2003). However, this requires adequate measurements of both inputs and outputs to the system that are much easier to assess at the small-catchment scale than at the hillslope scale, and it may prove more difficult to relate those summary parameterizations to quantifiable hillslope or catchment characteristics (see, for example, Sefton & Howarth, 1998, in the UK, and Post & Jakeman, 1999, in Australia).

Beven (2000, 2004d) has suggested an alternative approach based on "uniqueness-of-place" ideas. While accepting that there may be procedures developed to allow the estimation of the response of hillslopes or small catchments on the basis of only prior knowledge of hillslope characteristics, it seems to be inevitable that the uniqueness of individual hillslopes (as single realizations of some complex developmental process and boundary conditions) will mean that any such prior estimates will be subject to significant uncertainties. Thus, characterization must take such uncertainty into account within the framework of a learning process as new data become available. New data will allow the rejection of some model representations and, hopefully, the consequent constraint on uncertainty in prediction of responses at that site (e.g. Seibert & McDonnell, 2002; Uhlenbrook & Leibundgut, 2002; Franks *et al.*, 1998; Lamb *et al.*, 1998; Blazkova *et al.*, 2003; Freer *et al.*, 2004). For cases in which all the prior model representations are rejected, it might be necessary to implement new parameterizations that will be consistent with the available data.

This approach focuses attention on the utility of data in testing the performance of models (as hypotheses about how the hillslope is working) and in constraining prediction uncertainties. There is, however, little guidance about what type of data might be most useful in this respect. Hydrologists have a gap in available measurement techniques between the integrative measurements of discharges and tracers, and the point measurements of local storages and parameter values. At the hillslope scale, even discharges have been difficult to measure, since, as noted above, troughs and trenches cannot be considered reliable in assessing total fluxes from a slope, while the few studies that have assessed incremental inputs down stream channels have suggested large variability in the lateral inputs from different hillslopes (e.g. Paper 29). Remote sensing and geophysics can give some spatially distributed indexes of behaviour but these techniques have proven difficult to relate directly to hydrological variables and water fluxes.

This leaves a closure problem for hillslope hydrology. Reggiani *et al.* (1999, 2000) have shown how balance equations for mass, energy and momentum can be written for any hillslope element at any scale. These equations involve fluxes at the boundaries of the element. Closure of the balance equations requires a representation of those fluxes in terms of the internal state variables for the element—and, possibly, adjacent elements. It is known that these fluxes will have nonlinear (and probably hysteretic) dependence on the heterogeneity of storage and physical characteristics within the elements. The difficulty is that without direct measurement of the fluxes and internal state variables at the hillslope scales of interest, it will be very difficult to develop any theory for those dependencies.

At the hillslope scale, the closure problem is difficult, and we do not have adequate measurement techniques to characterize fluxes or storages in a hillslope element. Any closure will, therefore, depend on implying fluxes or storages from other forms of measurement (e.g. stream discharge, point water table measurements) and will be necessarily more uncertain. Thus it is unlikely that real advances in streamflow generation and hillslope hydrology (and hillslope models) will be made with the existing range of measurement techniques. New measurement techniques are needed that give more useful spatially distributed information on responses at feasible costs.

At present there is no guidance available about what measurement techniques might have most value in resolving the closure problem at the hillslope element scale. Fluxes can only (currently) be measured by collecting flow in trenches that change the nature of the flows. For spatial patterns of state variables, most remote sensing techniques do not penetrate far enough to provide useful hydrological information except in exceptional circumstances, while the assimilation of uncertain surface information (e.g. the radar-derived surface soil moisture data used by Houser *et al.*, 1998, and van Loon *et al.*, 2002) may not provide sufficient constraint on subsurface responses to constrain prediction uncertainties for storm outputs. Repeated gravity anomaly measurements are being used to estimate total water storages but at scales of tens of kilometres (e.g. Rodell *et al.*, 2004). Other geophysical techniques, such as repeated ground penetrating radar might also prove to be useful in the future (at least in some locations) to obtain better estimates of changes in storage over time at hillslope scales, but the inference of fluxes from such measurements is likely to remain highly uncertain for the foreseeable future.

An alternative is to take current local methods of measurement and greatly increase the number of sites in order to have better spatial coverage. Methods for recording shallow water table changes by capacitance probes have become much cheaper, as have multiplexed TDR and capacitance measurements for soil moisture. GPS has made it possible to repeat soil moisture measurements using roving probes to obtain patterns of surface soil moisture over an entire small catchment (e.g. Western & Grayson, 1998; Wilson *et al.*, 2003). It will remain difficult, however, to have a sufficient number of sites and sufficient time resolution to fully characterize the 3-D responses of even a single hillslope, especially one in which local preferential flows are important (e.g. Freer *et al.*, 2002).

Ideally, we would like to be able to measure the vertical profile of averaged downslope surface and subsurface fluxes directly at the hillslope scale, integrating over the detail of local matrix and preferential flow pathways, and providing information on differences in such fluxes between different hillslopes. While some local subsurface flux measurement techniques have been proposed, it is difficult to see these being implemented at hillslope scales. An alternative is to infer the bulk responses of different hillslopes from the type of incremental stream discharge measurements noted above. This, however, requires the development of an accurate, continuous flow measurement in an arbitrary channel shape across a range of discharges, since estimates of incremental discharge will depend strongly on the accuracy of the basic discharge measurements. It is a matter of critical importance to the future of hydrology that such methods should be developed and a programme should be proposed to involve physicists in the development of such techniques.

The most promising techniques for gaining new understanding about how hillslopes work is in the geochemical characterization of different water sources in stream discharges, extending the line of work started by Pinder & Jones (Paper 14) and Sklash & Farvolden (Paper 27). As noted already, the assumptions of such analyses mean that there are uncertainties about the interpretation of such geochemical characterizations, but this approach has already been very important in changing ideas about the role of subsurface flows in runoff generation.

The benchmark papers in this volume extend to the mid-1980s. It is more difficult to say what papers published since then will prove to be benchmarks in due course. Even in this volume, I have made some difficult choices in reducing the number of papers, and I am sure that there will be some disagreement about those choices. It will be very interesting, in years to come, for someone to take on the task of reviewing the last couple of decades to extend the collection. Some trends are already apparent. The integration of water quantity and geochemical characteristics is proving useful in providing insights to runoff water sources (e.g. Kirchner *et al.*, 2001). There is a better appreciation of the spatial organization of catchments, particularly the way in which different hillslope forms link to the channel through a riparian area (e.g. McGlynn & Seibert, 2003). There is more interaction between field hydrologists and modellers (but still not enough). Addressing the ungauged catchment problem has been given new impetus by the Prediction in Ungauged Basins (PUB) initiative of IAHS (Sivapalan, 2003; Wagener *et al.*, 2004). We know that for both gauged and ungauged catchments we have to deal with heterogeneity of catchment characteristics and residence times but do not quite know how to go about that characterization. We still do not have an adequate way of assessing the dynamics and importance of preferential flows in the field (for a recent review see Uchida *et al.*, 2001). We are still constrained by the lack of adequate measurement techniques for studying hillslope processes (see discussion about the closure problem above). We have, still, no general guidelines about the value of different types of measurement in the field. We can but hope that the benchmark papers of the future will include the breakthrough papers that will describe the measurement techniques that allow the accurate inference of the different fluxes of the water balance at the hillslope scale.

Acknowledgements

I am grateful to all those who contributed suggestions for benchmark papers to my list, including: Bruno Ambroise, Tim Burt, Nick Chappell, Jim Freer, Francesc Gallart, Jeff McDonnell, John Quinton, Mark Robinson, Jan Seibert, Roy Sidle, Taro Uchida, Stefan Uhlenbrook and Ross Woods. Others helped provide information about authors, including Wayne Swank, Robin Hodgkinson, Mike Bonell, Kevin McGuire and Jamie Shanley.

Jeff McDonnell did a good job in his editorial role; his own extensive knowledge of the field and detailed comments helped sharpen up a number of points in the commentaries. My final choice of papers is perhaps a little more idiosyncratic than he was expecting, but I hope the reader will find this volume none the less interesting for that.

References

Ahuija, L. R. & Ross, J. D. (1983) Effect of subsoil conductivity and thickness on interflow pathways, rates and source areas for chemicals in a sloping layered soil with seepage face. *J. Hydrol.* **64**, 189–204.

Abbott, M. B., Bathurst, J. C., Cunge, J. A., O'Connell, P. E. & Rasmussen, J. (1986) An introduction to the European Hydrological System – Système Hydrologique Européen, SHE. 1. History and philosophy of a physically-based distributed modelling system. *J. Hydrol.* **87**, 45–59.

Anderson, M. G. & Burt, T. P. (1990) *Process Studies in Hillslope Hydrology*. Wiley, Chichester, UK.

Atkinson, T. C. (1978) Techniques for measuring subsurface flow on hillslopes. In: *Hillslope Hydrology* (ed. by M. J. Kirkby), 73–120. Wiley, Chichester, UK.

Bathurst, J. C. (1986) Physically-based distributed modelling of an upland catchment using the Système Hydrologique Européen. *J. Hydrol.* **87**, 79–102.

Bazemore, D. E., Eshleman, K. N. & Hollenbeck, K. (1994) The role of soil water in stormflow generation in a forested headwater catchment: synthesis of natural tracer and hydrometric evidence. *J. Hydrol.* **162**, 47–75.

Betson, R. P. (1964) What is watershed runoff? *J. Geophys. Res.* **69**, 1541–1552.

Beven, K. J. (1975) A deterministic spatially distributed model of catchment hydrology. PhD Thesis, University of East Anglia, Norwich, UK.

Beven, K. J. (1977) Hillslope hydrographs by the finite element method. *Earth Surface Processes* **2**, 13–28.

Beven, K. J. (1978) The hydrological response of headwater and slideslopes areas. *Hydrol. Sci. Bull.* **23**(4), 419–437.

Beven, K. J. (1981) Comment on "A stochastic-conceptual analysis of rainfall–runoff processes on a hillslope" by R. Allen Freeze. *Water Resour. Res.* **17**, 431–432.

Beven, K. J. (1989) Interflow. In: *Unsaturated Flow in Hydrological Modelling* (ed. by H. J. Morel-Seytoux) (Proc. NATO ARW), 191–219. Kluwer, Dordrecht, The Netherlands.

Beven, K. J. (2001a) How far can we go in distributed hydrological modelling? *Hydrol. Earth. Syst. Sci.* **5**(1), 1–12.

Beven, K. J. (2001b) *Rainfall–Runoff Modelling – The Primer*. Wiley, Chichester, UK.

Beven, K. J. (2002a) Runoff production in semi-arid areas. In: *Dryland Rivers* (ed. by M. J. Kirkby & L. Bull), 57–105. Wiley, Chichester, UK.

Beven, K. J. (2002b) Towards an alternative blueprint for a physically-based digitally simulated hydrologic response modelling system. *Hydrol. Processes* **16**(2), 189–206.

Beven, K. J. (2002c) Towards a coherent philosophy for environmental modelling. *Proc. Roy. Soc. London A* **458**, 2465–2484.

Beven, K. J. (2004a) Surface runoff at the Horton Hydrologic Laboratory (or not?). *J. Hydrol.* **293**, 219–234.

Beven, K. J. (2004b) Robert Horton's perceptual model of infiltration. *Hydrol. Processes* **18**, 3447–3460.

Beven, K. J. (2004c) Robert Horton and abrupt rises of groundwater. *Hydrol. Processes* **18**, 3687–3696.

Beven, K. J. (2004d) Towards environmental models of everywhere: advances in modelling and data assimilation. In: *Hydrology: Science and Practice for the 21st Century*, vol. 1 (ed. by B. Webb, N. Arnell, C. Onof, N. McIntyre, R. Gurney & C. Kirby), 244–250. British Hydrological Society, UK.

Beven, K. J. & Germann, P. (1982) Macropores and water flow in soils. *Water Resour. Res.* **18**(5), 1311–1325.

Binley, A. M. & Beven, K. J. (1992) Three-dimensional modelling of hillslope hydrology. *Hydrol. Processes* **6**, 347–359.

Binley, A. M., Beven, K. J. & Elgy, J. (1989) A physically-based model of heterogeneous hillslopes. II. Effective hydraulic conductivities. *Water Resour. Res.* **25**(6), 1227–1233.

Bishop, K. H., Seibert, J., Köhler, S. & Laudon, H. (2004) Resolving the double paradox of rapidly mobilised old water with highly variable responses in runoff chemistry. *Hydrol. Processes* **18**, 185–189.

Blazkova, S., Beven, K., Tacheci, P. & Kulasova, A. (2002) Testing the distributed water table predictions of TOPMODEL (allowing for uncertainty in model calibration): the death of TOPMODEL? *Water Resour. Res.* **38**(11), doi: 10.1029/2001WR000912.

Bonell, M. (1993) Progress in the understanding of runoff generation dynamics in the forests. *J. Hydrol.* **150**, 217–275.

Bonell, M. (1998) Selected challenges in runoff generation research in forests from the hillslope to headwater drainage basin scale. *J. Am. Water Resour. Assoc.* **34**(4), 765–785.

Bonell, M. (2005) Runoff generation in tropical forests. In: *Forests, Water and People in the Humid Tropics* (ed. by M. Bonell & L. A. Bruijnzeel), 314–406. UNESCO International Hydrology Series, Cambridge University Press, Cambridge, UK.

Bunting, B. T. (1961) The role of seepage moisture in soil formation, slope development and stream initiation. *Am. J. Sci.* **259**, 503–518.

Burch, G. J., Moore, I. D. & Burns, J. (1989) Soil hydrophobic effects on infiltration and catchment runoff. *Hydrol. Processes* **3**, 211–222.

Chorley, R. J. (1978) The hillslope hydrological cycle. In: *Hillslope Hydrology* (ed. by M. J. Kirkby), 1–42. Wiley, Chichester, UK.

Donoso, M. C. (2004) History of the International Hydrological Programme of UNESCO. In: *Water Resources and Environmental History* (ed. by J. R. Rogers, G. O. Brown & J. D. Garbrect) (EWRI Annual Meeting, 2004). doi: 10.1061/40738(140)21.

Dunne, T. (1978) Field studies of hillslope flow processes. In: *Hillslope Hydrology* (ed. by M. J. Kirkby), 227–294. Wiley, Chichester, UK.

Emmett, W. W. (1970) The hydraulics of overland flow on hillslopes. *USGS Prof. Paper 662-A*. Washington, DC, USA.

Franks, S. W., Gineste, Ph., Beven, K. J. & Merot, Ph. (1998) On constraining the predictions of a distributed model: the incorporation of fuzzy estimates of saturated areas into the calibration process. *Water Resour. Res.* **34**, 787–797.

Freer, J., McDonnell, J. J., Beven, K. J., Peters, N. E., Burns, D., Hooper, R. P., Aulenbach, B. & Kendal, C. (2002) The role of bedrock topography on subsurface stormflow. *Water Resour. Res.* **38**(12), doi: 10.1029/2001WR000872.

Freer, J., McMillan, H., McDonnell, J. J. & Beven, K. J. (2004) Constraining dynamic TOPMODEL responses for imprecise water table information using fuzzy rule based performance measures. *J. Hydrol.* **291**, 254–277.

Freeze, R. A. (1971) Three-dimensional, transient, saturated-unsaturated flow in a groundwater basin. *Water Resour. Res.* **7**(2), 347–366.

Freeze, R. A. (1972) Role of subsurface flow in generating surface runoff. 2. Upstream source areas. *Water Resour. Res.* **8**(5), 1272–1283.

Freeze, R. A. (1974) Streamflow generation. *Reviews of Geophysics and Space Physics* **12**, 627–647

Freeze, R. A. (1975) A stochastic-conceptual analysis of one-dimensional groundwater flow in a nonuniform homogeneous media. *Water Resour. Res.* **11**, 725–741.

Freeze, R. A. (1980) A stochastic-conceptual analysis of rainfall–runoff processes on a hillslope. *Water Resour. Res.* **16**, 391–408.

Freeze, R. A. (2004) The role of stochastic hydrogeological modeling in real-world engineering applications. *Stoch. Environ. Res. Risk Assess.* **18**(4), 286–289.

Freeze, R. A. & Harlan, R. L. (1969) Blueprint for a physically-based digitally simulated hydrologic response model. *J. Hydrol.* **9**, 237–258.

Genereux, D. (1998) Quantifying uncertainty in tracer-based hydrograph separations. *Water Resour. Res.* **34**, 915–919.

Gilman, K. & Newson, M. (1980) *Soil Pipes and Pipeflow – A Hydrological Study in Upland Wales*. British Geomorphology Research Group, Research Monograph 1, Geo Books, Norwich, UK.

Harr, R. D. (1977) Water flux in soil and subsoil on a steep forested slope. *J. Hydrol.* **33**, 37–58.

Henderson, F. & Wooding, R. (1964) Overland flow and groundwater flow from a steady rainfall of finite duration. *J. Geophys. Res.* **69**, 1531–1540.

Hewlett, J. D. & Hibbert, A. R. (1967) Factors affecting the response of small watersheds to precipitation in humid areas. In: *Proc. Int. Symp. on Forest Hydrology* (ed. by W. W. Sopper & H. W. Lull), 275–290. Pergamon, Oxford, UK.

Horton, R. E. (1935) *Surface Runoff Phenomena. Part 1. Analysis of the Hydrograph*. Horton Hydrologic Laboratory Publication 101. Edward Bros, Ann Arbor, Michigan, USA.

Horton, R. E. (1936) Maximum groundwater levels. *Trans. Am. Geophys. Un.* **17**(2), 344–357.

Horton, R. E. (1938) The interpretation and application of runoff experiments with reference to soil erosion. *Soil Sci. Soc. Am. Proc.* **3**, 340–349.

Horton, R. E. (1945) Erosional development of streams and their drainage basins: hydrophysical to quantitative morphology. *Bull. Geol. Soc. Am.* **56**, 275–370.

Houser, P. R., Shuttleworth, W. J., Famiglietti, J. S., Gupta, H. V., Syed, K. H. & Goodrich, D. C. (1998) Integration of soil moisture remote sensing and hydrologic modeling using data assimilation. *Water Resour. Res.* **34**(12), 3405–3420.

Huggins, L. F. & Monke, E. J. (1968) A mathematical model for simulating the hydrologic response of a watershed. *Water Resour. Res.* **4**, 529–539.

Hursh, C. R. (1936) Storm water and adsorption. *Trans. Am. Geophys. Un.* **17**, 301–302.

Hursh, C. R. (1944) Report of the committee on subsurface flow. *Trans. Am. Geophys. Un.* Part V, 743–746.

Joerin, C., Beven, K. J., Iorgulescu, I. & Musy, A. (2002) Uncertainty in hydrograph separations based on geochemical mixing models. *J. Hydrol.* **255**, 90–106.

Jones, J. A. A. (1981) *The Nature of Soil Piping: A Review of Research*. British Geomorphology Research Group, Monograph 3. Geo Books, Norwich, UK.

Kendall, C., McDonnell, J. J. & Gu, W. (2001) A look inside 'black box' hydrograph separation models: a study at the Hydrohill catchment. *Hydrol. Processes* **15**, 1877–1902.

Kirchner, J. W. (2003) A double paradox in catchment hydrology and geochemistry. *Hydrol. Processes* **17**, 871–874.

Kirchner, J. W., Feng, X. & Neal, C. (2001) Catchment-scale advection and dispersion as a mechanism for fractal scaling in stream tracer concentrations. *J. Hydrol.* **254**, 82–101.

Kirkby, M. J. (ed.) (1978) *Hillslope Hydrology*. Wiley, Chichester, UK.

Lamb, R., Beven, K. J. & Myrabø, S. (1998) Use of spatially distributed water table observations to constrain uncertainty in a rainfall–runoff model. *Adv. Water Resour.* **22**(4), 305–317.

Leach, H. R., Cook, H. L. & Horton, R. E. (1933) Storm-flow prediction. *Trans. Am. Geophys. Un.* **14**, 436–446

Loague, K., Heppner, C. S., Abrams, R. H., Carr, A. E., VanderKwaak, J. E. & Ebel, B. A. (2005) Further testing of the Integrated Hydrology Model (InHM): event-based simulations for a small rangeland catchment located near Chickasha, Oklahoma. *Hydrol. Processes* **19**(7), 1373–1398.

McDonnell, J. J. (1990) A rationale of old water discharge through macropores in a steep humid catchment. *Water Resour. Res.* **26**, 2821–2832.

McDonnell, J. J. & Woods, R. A. (2004) On the need for catchment classification. *J. Hydrol.* **299**, 2–3.

McGlynn, B. L. & Seibert, J. (2003) Distributed assessment of contributing area and riparian buffering along stream networks. *Water Resour. Res.* **39**(4), 1082, doi: 10.1029/2002WR001521.

Neuman, S. P. (1973) Saturated–unsaturated seepage by finite elements. *J. Hydraul. Div. ASCE* **99**, 2233–2250.

Pearce, A. J., Stewart, M. K. & Sklash, M. G. (1986) Storm runoff generation in humid headwater catchments. 1. Where does the water come from? *Water Resour. Res.* **22**, 1263–1272.

Post, D. A. & Jakeman, A. J. (1999) Predicting the daily streamflow of ungauged catchments in SE Australia by regionalising the parameters of a lumped conceptual rainfall-runoff model. *Ecol. Model.* **123,** 91–104.

Reggiani, P., Hassanizadeh, S. M., Sivapalan, M. & Gray, W. G. (1999) A unifying framework for watershed thermodynamics: constitutive relationships. *Adv. Water Resour.* **23**, 15–39.

Reggiani, P., Sivapalan, M. & Hassanizadeh, S. M. (2000) Conservation equations governing hillslope responses: exploring the physical basis of water balance. *Water Resour. Res.* **36**(7), 1845–1863.

Rodell, M., Famiglietti, J. S., Chen, J., Seneviratne, S. I., Viterbo, P., Holl, S. & Wilson, C. R. (2003) Basin scale estimates of evapotranspiration using GRACE and other observations. *Geophys. Res. Letters* **31**(20), art. no. L20504 OCT 26 2004.

Ross, C. N. (1921) The calculation of flood discharge by the use of time contour plan isochrones. *Trans. Instn Engnrs Aus.* **2**, 85–92.

Scholl, D. G. & Hibbert, A. R. (1973) Unsaturated flow properties used to predict outflow and evapotranspiration from a sloping lysimeter. *Water Resour. Res.* **9**(6), 1645–1655.

Schumm, S. A. (1956) Evolution of drainage systems and slopes in badlands at Perth Amboy, New Jersey. *Bull. Geol. Soc. Am.* **67**, 597–646.

Sefton, C. E. M. & Howarth, S. M. (1998) Relationships between dynamic response characteristics and physical descriptors of catchments in England and Wales. *J. Hydrol.* **211**, 1–16.

Seibert, J. & McDonnell, J. J. (2002) On the dialog between experimentalist and modeler in catchment hydrology: use of soft information for multi-criteria model calibration. *Water Resour. Res.* **38**(11), doi: 10.1029/2001WR000978.

Sivapalan. M. (2003) Prediction in ungauged basins: a grand challenge for theoretical hydrology. *Hydrol. Processes* **17**(15), 3163–3170.

Smith, R. E. & Hebbert, R. H. B. (1979) A Monte Carlo analysis of the hydrologic effects of the spatial variability of infiltration. *Water Resour. Res.* **15**, 419–429.

Smith, R. E. & Woolhiser, D. A. (1971) Overland flow on an infiltrating surface. *Water Resour. Res.* **7**(4), 899–913.

Sopper, W. E. & Lull, H. W. (eds) (1967) *International Symposium on Forest Hydrology*. Pergamon, Oxford, UK.

Stephenson, G. R. & Freeze, R. A. (1974) Mathematical simulation of subsurface flow contributions to snowmelt runoff, Reynolds Creek, Idaho. *Water Resour. Res.* **10**(2), 284–298.

Tischendorf, W. G. (1969) Tracing stormflow to varying source area in small forested watershed in the Southeastern Piedmont. PhD Dissertation. University of Georgia, Athens, USA.

Tsukamoto, Y. (1961) An experiment on subsurface flow. *J. Japanese Soc. Forestry* **45**, 186–190.

Uchida, T., Kosugi, K. & Mizuyama, T. (2001) Effects of pipeflow on hydrological processes and its relation to landslides: a review of pipeflow studies in forested headwater catchments. *Hydrol. Processes* **15**(11), 2151–2174.

Uhlenbrook S. & Leibundgut, C. (2002) Process-oriented catchment modelling and multiple-response validation. *Hydrol. Processes* **16**(2), 423–440.

Unnikrishna, P. V., McDonnell, J. J., Tarboton, D. G. & Kendall, C. (1995) Isotopic analysis of hydrologic processes in a small semiarid catchment. In: *Tracer Technologies for Hydrological Systems* (ed. by Ch. Leibundgut), 295–304. IAHS Publ. 229. IAHS Press, Wallingford, UK.

van Loon, E. E. & Troch, P. A. (2002) Tikhonov regularization as a tool for assimilating soil moisture data in distributed hydrological models. *Hydrol. Processes* **16**(2), 531–556.

Wagener, T., Sivapalan, M., McDonnell, J., Hooper, R., Lakshmi, V., Liang, X. & Kumar, P. (2004) Predictions in ungauged basins as a catalyst for multidisciplinary hydrology. *Eos Trans. AGU* **85**(44), 451, doi: 10.1029/2004EO440003.

Ward, R. C. (1984) On the response to precipitation of headwater streams in humid areas. *J. Hydrol.* **74**, 171–189.

Western, A. W. & Grayson, R. B. (1998) The Tarrawarra data set: soil moisture patterns, soil characteristics, and hydrological flux measurements. *Water Resour. Res.* **34**(10), 2765–2768.

Weyman, D. R. (1970) Throughflow on hillslopes and its relation to the stream hydrograph. *Bull. Int. Assoc. Hydrol. Sci.* **15**, 25–33.

Wilson, D. J., Western, A. W., Grayson, R. B., Berg, A. A., Lear, M. S., Rodell, M., Famiglietti, J. S., Woods, R. A. & McMahon, T. A. (2003) Spatial distribution of soil moisture over 6 and 30 cm depth, Mahurangi River catchment, New Zealand. *J. Hydrol.* **276**(1–4), 254–274.

Woods, R. A. & Rowe, L. K. (1996) The changing spatial variability of subsurface flow across a hillslope. *NZ J. Hydrol.* **35**, 51–86.

Yair, A. & Lavee, H. (1985) Runoff generation in arid and semi-arid zones. In: *Hydrological Forecasting* (ed. by M. G. Anderson & T. P. Burt), 183–220. Wiley, Chichester, UK.

Young, P. C. (2003) Top-down and data-based mechanistic modelling of rainfall-flow dynamics at the catchment scale. *Hydrol. Processes* **17**, 2195–2217.

Zaslavsky, D. & Sinai, G. (1981) Surface hydrology (5 parts). *J. Hydraul. Div. Am. Soc. Civ. Engrs* **107**, 1–93.

Zoch, R. T. (1934) On the relation between rainfall and streamflow. *Monthly Weather Rev.* **62**(9), 315–322.

Commentary on Paper 1

Horton, R. E. (1933) The role of infiltration in the hydrologic cycle.

Robert Elmer Horton was not the first scientific hydrologist interested in hillslope runoff processes and streamflow generation but was certainly the most important of the 20th Century. Horton was born in 1875 in Palmer, Michigan, USA, and he died in Voorheesville, New York, where he had his home and Hydrologic Laboratory, in 1945. Close to Voorheesville there is now a waterfall named after him (Schaefer, 1991). He moved there in 1916 and worked as a Consulting Hydraulic Engineer, including work for various US Government agencies. Early in his career he worked for the US Geological Survey carrying out experimental hydraulic work on weir designs in a large experimental flume (Horton, 1906). In 1932 he was awarded an honorary doctorate from his former college, Albany College, Michigan, whose library also houses a collection of his published papers. An accessible account of his life was provided by Hall (1987). Horton's scientific papers, all 94 boxes of them, are saved in the US National Archives (see Beven, 2004a,b,c). His work is notable for his creative theorizing and careful, and painstaking, measurements (for example, there is a letter in his papers in which he notes eight discrepancies in figures published by the US Weather Bureau compared with those he had supplied from the Hydrologic Laboratory).

This, Horton's most cited paper, was published in the *Transactions of the American Geophysical Union* in 1933. Interestingly, it immediately follows another paper on which he was a co-author, entitled: *Storm-flow prediction* (Leach *et al.*, 1933). The latter includes some hydrographs from the La Grange Brook, the catchment at the Horton Hydrologic Laboratory. These papers, between them, established a set of principles of storm hydrograph analysis that Horton later expanded into a full monograph (Horton, 1935), and which continue to be used to the present day. They were summarized by Leach *et al.* (1933) as: "*Storm-runoff, which is here under consideration, is very nearly equal to the part of the rain which falls at intensities exceeding the infiltration-capacity of the soil – in other words, it is the difference between total rainfall and infiltration*" (p.441).

Horton (1933) deals with definitions, methods for separating storm runoff from groundwater flow, and rainfall–intensity distributions in relation to the infiltration-capacity of the soil. Infiltration-capacity, however, is only ever treated as a catchment average value, as if it was constant in both space and time. Horton realized that, in <u>analysis</u> of hydrographs in relation to rainfall intensities, it was really only possible to derive such an average value; and from this followed the idea that Horton's concept of streamflow generation was that infiltration excess overland flow would occur everywhere on a catchment (as later queried by the partial area concepts of Betson, Paper 8, and dynamic contributing concepts of Dunne & Black, Paper 15).

This is, however, much too simplistic. Horton (1938) also showed how, <u>in prediction</u>, different infiltration capacities on different shapes of hillslope segments could be taken into account, including areas where no surface runoff might be produced. He also demonstrated how the intersection of a water table with the surface might also lead to the generation of rapid surface responses (Paper 2). He even illustrates this, taking the variable width of a hillslope segment into account (see the discussion of Beven, 2004c), crediting LeRoy K. Sherman (who published the "unitgraph" concept in 1932) with introducing the concept of a unit width hillslope segment analysis.

Horton (1933), as a primary source for "Hortonian" concepts of runoff production, is also sometimes cited as a source for the Horton infiltration equation. This is wrong (though I have to admit to also making this mistake in the past). The infiltration equation was not introduced in published form until Horton (1939, 1940) as:

$$f = f_c + (f_o - f_e)\,e^{-K_f t}$$

"*where f = infiltration-capacity, inches per hour, at time t, in hours, f_o = infiltration-capacity at time $t = 0$; f_c = minimum constant infiltration-capacity; K_f is constant for a given curve*" (Horton, 1939, p.697). Horton (1940) notes that "*this was originally given as an empirical equation. It can, however, be derived from the simple assumption that the processes involved in the reduction of f as rain continues are of the nature of exhaustion processes. These processes include rain-packing, inwashing, breaking down of the crumb-structure of the soil, the swelling of colloids and, in cases where they occur, the closing of sun-checks*" (p.399). He was also aware of the effects of macropores on infiltration to depth and the escape of air to the surface. Horton explains his concepts of infiltration in much more detail in a monograph that can be found in Box 61 of his archived papers, but which seems to have remained unfinished and unpublished when he died (Beven, 2004b).

Fortunately, he did finish the extended discussion about the role of surface runoff and erosion on the development of hillslopes in a paper (Horton, 1945) that proved hugely influential in geomorphology (see Roberts & Cobb, 1973). This paper is also available in the volume of *Benchmark Papers in Geology* edited by Cuchlaine King (1976). It was the last paper Horton submitted before his death (although a few more papers on meteorological topics were completed by Richard van Vliet and published posthumously) and served to reinforce the Hortonian surface runoff concepts of streamflow generation. It is true that Horton used these concepts widely in his consultancy work, and in the analysis of data from the La Grange basin, where surface runoff would not have been seen on the slopes in most storms (Beven, 2004a). They influenced the analyses of generations of later hydrologists and computer models of streamflow generation (including Smith & Woolhiser, 1971; and continuing to the present, see Downer *et al.*, 2002). Horton's full concepts were, however, more flexible and more subtle, and perhaps he deserves even greater recognition for his very early appreciation of some of the complexities of hillslope hydrological responses.

References

Beven, K. J. (2004a) Surface runoff at the Horton Hydrologic Laboratory (or not?). *J. Hydrol.* **293**, 219–234.

Beven, K. J. (2004b) Robert Horton's perceptual model of infiltration processes. *Hydrol. Processes* **18**, 3447–3460.

Beven, K. J. (2004c) Robert Horton and abrupt rises of groundwater. Paper submitted to *Hydrol. Processes* **18**, 3687–3696.

Downer C. W., Ogden, F. L., Martin, W. D. & Harmon, R. S. (2002) Theory, development, and applicability of the surface water hydrologic model CASC2D. *Hydrol. Processes* **16**(2), 255–275.

Hall, F. R. (1987) Contributions of Robert E. Horton. In: *AGU Monograph on the History of Geophysics*, vol. 3, 113–117. American Geophysical Union, Washington, USA.

Horton, R. E. (1906) Weir experiments, coefficients and formulas. *US Geological Survey Water Supply and Irrigation Paper no. 180*, Washington, DC, USA.

Horton, R. E. (1933) The role of infiltration in the hydrologic cycle. *Trans. Am. Geophys. Un.* **14**, 446–460.

Horton, R. E. (1935) *Surface Runoff Phenomena. Part 1. Analysis of the Hydrograph.* Horton Hydrological Laboratory, Publication 101. Edward Bros, Ann Arbor, Michigan, USA.

Horton, R. E. (1936) Maximum groundwater levels. *Trans. Am. Geophys. Un.* **17**(2), 344–357.

Horton, R. E. (1938) The interpretation and application of runoff experiments with reference to soil erosion. *Soil Sci. Soc. Am. Proc.* **3**, 340–349.

Horton, R. E. (1939) Analysis of runoff-plat experiments with varying infiltration-capacity. *Trans. Am. Geophys. Soc.* **20**, 693–711.

Horton, R. E. (1940) An approach towards physical interpretation of infiltration-capacity. *Proc. Soil Sci. Soc. Am.* **5**, 399–417.

Horton, R. E. (1945) Erosional development of streams and their drainage basins: hydrophysical to quantitative morphology. *Bull. Geol. Soc. Am.* **56**, 275–370.

King, C. A. M. (1976) *Landforms and Geomorphology*. Benchmark Papers in Geology no. 28. Wiley, Chichester, UK

Leach, H. R., Cook, H. L. & Horton, R. E. (1933) Storm-flow prediction. *Trans. Am. Geophys. Un.* **14**, 436–446.

Roberts, M. C. & Cobb, D. A. (1973) Horton's 1945 study – a cartobibliography. *Bull. Geol. Soc. Am.* **84**, 2733–2736.

Schaefer, V. (1991) Horton Falls named for pioneer hydrologist. *Trans. Am. Geophys. Un.* **72**(18), 203.

Sherman, L. K. (1932) Streamflow from rainfall by unit-graph method. *Engineering News Record* **108**, 501–505.

Smith, R. E. & Woolhiser, D. A. (1971) Overland flow on an infiltrating surface. *Water Resour. Res.* **7**(4), 899–913.

THE RÔLE OF INFILTRATION IN THE HYDROLOGIC CYCLE

Robert E. Horton

Introduction--Definitions

For some years the author has used the term "infiltration" to describe the process involved where water soaks into or is absorbed by the soil. Absorption, imbibition, and percolation are often used in much the same sense. It seems better to confine the use of "percolation" to the free downward flow by gravity of water in the zone of aeration--a process for which a distinctive term is needed. "Absorption" includes the entrance of air as well as water, both liquid and vapor, into the soil (see Patten and Gallagher, Absorption of vapors and gases by soils, U.S. Dep. Agric., Bul. 51, Bur. Soils, Wash., 1908; also Charles H. Lee, On absorption and transpiration, Trans. Amer. Geophys. Union, 1932, pp. 288-298). "Infiltration" is limited to water in the liquid form and is more accurately descriptive of the physical processes by which rain enters the soil. "Water-penetration" is also sometimes used as if synonymous with infiltration. Its use should be restricted to the depth below soil-surface reached by the given surface infiltration.

"Infiltration-capacity" will be used to describe the maximum rate at which rain can be absorbed by a given soil when in a given condition. Infiltration is (a) the sole source of soil-moisture available to supply the transpiration stream and sustain the growth of vegetation, and (b) the sole source of ground-water supply of wells, springs, and streams.

The surface of a permeable soil acts like a diverting dam and head-gate in a stream, where the head-gate can be opened to a certain width only or closed so as to still leave a fixed opening. With varying rates of flow in the stream, all the flow is diverted up to the then capacity of the opening. Similarly, with varying rain-intensity, all of the rain is absorbed for intensities not exceeding the infiltration-capacity, while for excess rainfall there is a constant rate of absorption as long as the infiltration-capacity is unchanged. As in the case of the dam and head-gate, there is usually some pondage which remains to be disposed of after the supply to the stream is cut off, so in the case of infiltration, surface-detention remains after rain ends. Infiltration divides rainfall into two parts, which thereafter pursue different courses through the hydrologic cycle. One part goes via overland flow and stream-channels to the sea as surface-runoff; the other goes initially into the soil and thence through ground-water flow again to the stream or else is returned to the air by evaporative processes. The soil therefore acts as a separating surface and the author believes that various hydrologic problems are simplified by starting at this surface and pursuing the subsequent course of each part of the rainfall as so divided, separately. This has not hitherto, in general, been undertaken.

The following terms require definition. (Quantities used in this paper are inches depth per hour unless otherwise stated.)

Infiltration-capacity may be defined as the maximum rate at which a given soil can absorb rainfall when the soil is in a specified condition. Infiltration-capacity, in general, is designated f. As subsequently shown, for a given soil there is a fairly definite maximum infiltration-capacity, designated f_g and a more definite minimum infiltration-capacity, designated f_ℓ.

Ground-rainfall may be defined as the part of the rainfall which actually reaches the ground, or it equals the total rainfall minus interception by vegetation. It is designated P_g.

Rainfall-excess (P_e) is the part of the rainfall which falls at intensities exceeding the infiltration-capacity. For example, if the rain-intensity is 1.5 inches per hour for a 30-minute interval and the infiltration-capacity is 0.8 inch per hour, then the rainfall-excess is $0.7 \times 0.5 = 0.35$ inch.

Surface-detention is that part of the rain which remains on the ground-surface during rain and either runs off or is absorbed by infiltration after rain ends. A part of this surface-detention is usually contained in small pockets or depressions in the ground-surface and none of this runs off. This portion is designated detention-storage.

Field moisture-capacity (designated fmc or m_c) is used in the sense of the quantity of water which can be permanently retained in the soil in opposition to the downward pull of gravity. It may be expressed in per cent of dry weight or in inches depth for a given depth of soil.

Field moisture-deficiency (designated fmd or m_d) for any soil and time is the quantity of water, similarly expressed, which would be required to restore the soil moisture-content to field moisture-capacity.

The term ground-water accretion is used to describe the total quantity of water added to the water-table in any given time-interval. It is synonymous with the term re-charge as used in western United States and ultimate percolation as used in England.

Soil-moisture accretion is that part of infiltration which is retained in the zone of soil-moisture and which does not pass on downward to the water-table.

The storm-interval is the natural hydrologic time-unit. It may be defined as the interval from the beginning of a rain through the rain-period and the subsequent dry or rainless period to the beginning of the next subsequent rain-period.

Rise is used, for want of a better name, to define the period during and following rainfall from the time when the hydrograph first departs from the normal depletion-curve until it again becomes coincident with the normal depletion-curve. All of the surface-runoff due to the storm takes place in this interval. A rise consists of a period of increasing flow, which may result either from increased ground-water flow, from surface-runoff, or both. This culminates in the crest or peak of the rise, which is followed by a recession period.

Concordant flow is a condition of a stream such that if the rate of ground-water inflow to the stream remained unchanged, there would be no change in the stage of or rate of outflow from any given reach of the stream.

Normal depletion-curve

In the determination of total infiltration from runoff-records it is necessary to subdivide total runoff into two principal components: Ground-water flow and surface-runoff. This may be accomplished by the use of a "normal depletion-curve." Consider a period of little or no rain. Stream-flow is then derived from storage. If there is no artificial regulation and no lakes or marshes on the drainage-basin, the stream-flow is derived either from ground-water or from waters in transit in stream-channels. At lower stages of the stream and when flood-waters are not passing down-stream, the water in transit in stream-channels is itself derived from ground-water. For practical purposes and for the sake of brevity the flow of a stream at such times may be referred to as "ground-water flow."

After the passage of a flood or a rise due to surface-runoff, a stream settles down to what may be called a condition of concordant flow. The flow is variable in a hydraulic sense in that the same quantity does not pass each cross-section but the conditions are also such that if the inflow from ground-water remained unchanged there would be no change in depth or quantity of water stored in any reach of the stream.

A curve representing a segment of a hydrograph under conditions of concordant flow may be called a normal depletion-curve. It is also a time-discharge rate relation-curve. Since the area under the depletion-curve between two dates equals the total runoff and also equals the total draft from storage for the given period, it is easy to construct a second or storage-discharge rate relation-curve. If ground-water level-records are available this may be in the form of a ground-water level and discharge-rate curve, or virtually a rating curve for ground-water flow. For streams where the ground-water level underneath the drainage-basin is at a depth beyond the reach of direct abstraction by plants or vegetation, the normal depletion-curves in different years are often nearly identical throughout their common range within the limit of errors of observation and excluding the effects of barometric changes, etc., on ground-water flow.

The normal depletion-curves may differ to a considerable extent in different years or seasons in cases where there is a direct abstraction of ground-water from the water-table by vegetation or evaporation. Even in such cases the difference between summer season depletion-curves in different years is often so small that for practical purposes in separating ground-water and surface-runoff, an average normal depletion-curve can be used. [The author first noticed the characteristics of normal depletion-curves during a long drought which occurred in the eastern United States in the spring of 1903 and published depletion-curves for this drought-period in the Annual Report of the State Engineer and Surveyor of New York, 1903, v. 2, p. 16. Such curves were foreshadowed by the "storage curves" used by Vermeule (Geol. Sur. N.J., v. 3, Water supply, 1894).] Owing to interruption by rain and surface-runoff it is generally impossible to obtain a complete normal depletion-curve directly from a hydrograph. It can be shown that for a simple phreatic drainage-basin (with no direct abstraction from the water-table by transpiration or evaporation within the soil) the equation of the normal depletion-curve is

$$q = q_o e^{-ct}$$

where q is the ground-water flow-rate at the time t, and t is time elapsed subsequent to a date when the flow was q_o (Robert E. Horton, Discussion on yield of drainage-areas, J. New England Water Works Assn., v. 28, No. 4, Dec., 1914, pp. 538-539). [This particular formula was derived by the author from theoretical considerations in 1904 and applied to the depletion-curves of several streams. Its applicability to the springs of Cerilly was shown by Maillet (Eaux souterraines, 1905). In 1921 D. Halton Thomson (Trans. Inst. Water Eng.) showed that this type of expression accurately represents the law of decrease of ground-water levels in chalk regions of West Sussex, England, during rainless periods. Maillet states that this formula as applied to ground-water was first derived by him in 1903 and was independently derived by Boussinesq the same year.] Where a drainage-basin contains many phreatic sub-basins of different characteristics the normal depletion-curve for ground-water flow from the entire drainage-basin can in general be represented by an equation of the form

$$q = q_o e^{-ct^n}$$

in which n is a constant exponent. For a drainage-basin in which the gravitational capacity of the soil decreases proceeding downward, as is not uncommonly the case, the normal depletion-curve can in general be represented by an equation similar to one or other of those above given but consisting of the sum of two exponential terms. By deriving the equation of the normal depletion-curve for a drainage-basin the curve can be extended so as to show the rate at which ground-water would be supplied to the stream: (a) During a drought-period of any duration and (b) during periods when surface-runoff is taking place if there was no coincident accretion to the ground-water table.

By shifting the normal depletion-curve horizontally across a hydrograph, segments of the hydrograph which coincide with the normal depletion-curve may be noted and these represent periods during which the runoff is derived entirely from ground-water flow. Connecting the points where the hydrograph departs from and rises above the normal depletion-curve, the accretion to ground-water and the ground-water flow during periods of surface-runoff can also be determined. In this way the ground-water runoff of the stream can be separated from the total runoff, thus making it possible by simple subtraction to determine the surface-runoff Y_s, since $Y_s = Y - Y_g$, where Y and Y_g are the total and ground-water runoff, respectively.

The many interesting and important properties of normal depletion-curves and their applications cannot be given in detail here. [Reference is made to the following papers: Samuel Hall, Stream-flow and percolation-water, Trans., Inst. Water Eng., v. 23, pp. 92-127, 1918; D. Halton Thomson, Hydrological conditions in the chalk at Compton, West Sussex, Trans., Inst. Water Eng., London, 1921. According to Gravelius (Flusskunde) the separation of ground-water from surface-runoff by drawing a line across the base of the hydrograph of a stream-rise has been practiced in France for more than 50 years. It seems necessary to point out that the method of separating ground-water flow from surface-runoff, used by Houk (Rainfall and runoff in the Miami Valley, Tech. Reports, Miami Conservancy Dist., Part 8, 1921) and by Meinzer and Stearns (W.S.P. 597-B) is incorrect and leads to underestimates of ground-water runoff, especially during wet months.] The author expects, however, to publish a full account of ground-water depletion-curves and their application in the near future.

Rain-intensity duration-curve

The rain-intensity duration-curve provides a good starting point for consideration of the rôle of infiltration in hydrology. Such a curve may be prepared either for an individual storm or for a season, a year, or other period in the same manner as the familiar runoff duration-curve. For a single storm, rain intensities for hourly or shorter intervals should be used and must be derived from the record of a recording rain-gage.

The lower diagram on Figure 1 shows the actual rain-intensity distribution in a typical summer storm on the drainage-basin of Ralston Creek at Iowa City, Iowa. The total rainfall of 1.96 inches produced a runoff of 0.67 inch. There was no initial ground-water flow and the area under the runoff-curve extended to zero represents the total surface-runoff and is also equal, when suitable scale-factor is applied, to the portion of the rainfall shown by the shaded area. The shaded area on the rainfall-diagram represents that part of the rain which fell at intensities exceeding 0.55 inch per hour, which in turn was the infiltration-capacity of the drainage-basin during this storm. Inset A of Figure 1 shows the same rainfall plotted as a rainfall-intensity duration-diagram. If a horizontal line corresponding to some particular intensity is drawn across this diagram, the area between the intensity-curve and this line is the rainfall-excess for the given intensity. The line marked "rainfall-excess" on inset B of Figure 1 was derived in this way. From this diagram it is possible to read off directly the rainfall-excess corresponding to any given infiltration-rate. The total infiltration equals the total rainfall minus the rainfall excess, and a line of total infiltration has been added to Diagram B.

If the infiltration-capacity of a drainage-basin is known it is possible by an analysis of the rainfall-excess in individual storms, in the manner described, to determine, approximately at least, the surface-runoff which will take place directly from rainfall-data.

Also having given the runoff and the rain-intensity distribution it is possible to determine from these data the infiltration-capacity of a drainage-basin. For large areas it is, of course, necessary to determine the rainfall-excess for several stations.

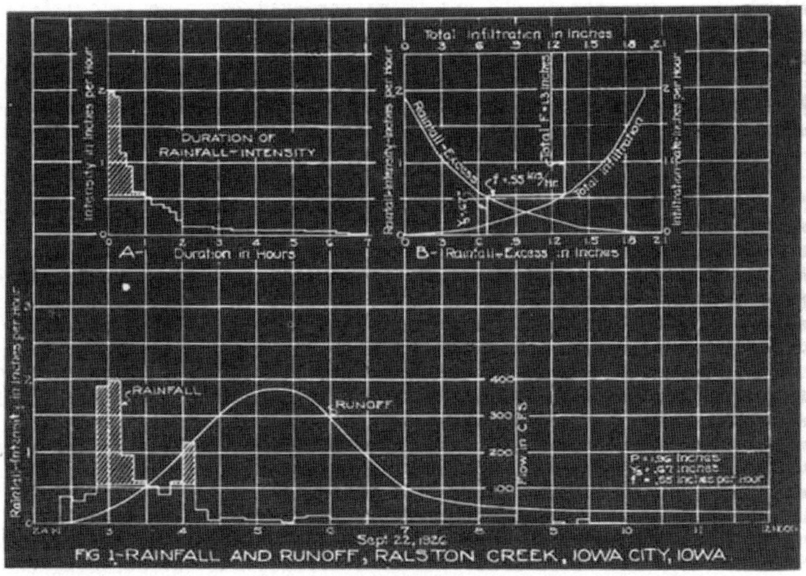

FIG 1-RAINFALL AND RUNOFF, RALSTON CREEK, IOWA CITY, IOWA

Fig. 1

Infiltration-capacity

Infiltration-capacity of a soil is closely related to but not necessarily and, in general, not actually identical with the transmission-capacity as used, for example, in Slichter's formula. The infiltration-capacity of a specific soil can be determined: (a) From laboratory experiments, using artificial rainfall; (b) from runoff-plat experiments, with natural rainfall; (c) from rainfall- and runoff-records for small drainage-basins with homogeneous soils; and (d) the average equivalent infiltration-capacity for a drainage-basin where the soil-type varies areally can be determined from runoff and rainfall-intensity records.

Few data have been published giving directly the infiltration-capacity of soils. Experimental and correlative researches for the determination of infiltration-capacity of soils have been carried out from time to time at the author's laboratory for a number of years. Only the briefest outline of the results can be given here. In the author's experiments, water was applied in a manner simulating rainfall to pans of soil of different types, under fully controlled conditions.

Various series of experiments by other investigators, made primarily for the determination of runoff-characteristics, have been analyzed for the determination of infiltration-capacity. These include laboratory experiments with artificial rainfall, made by Houk, by W. W. Horner, and by Duley and Hayes. Infiltration-capacities have also been determined from various series of runoff-plat experiments and from runoff-data for small drainage-basins, in particular, Ralston Creek, at the University of Iowa.

In discussing infiltration-capacity, two cases must be considered: (a) Natural soils, such as those of forest-covered areas, natural meadows, brushlands, and desert areas and, in general, all areas where the soil has remained fallow for a year or more; and (b) soils which have recently been cultivated.

Considering first the case of a natural soil, here the infiltration-capacity passes through a fairly definite cycle for each storm-interval. Starting with a maximum value when rain begins, the infiltration-capacity decreases rapidly at first as the result of the operation of some or all of the following processes: (1) Packing of the soil-surface by rain; (2) swelling of the soil, thus closing sun-checks and other openings; and (3) inwashing of fine materials to the soil-surface openings.

These effects are confined to a thin layer at the soil-surface. Infiltration-capacity during prolonged rain is, therefore, generally less than the gravitational transmission-capacity within the soil-mass. This is the principal reason why soils free to drain are seldom if ever fully saturated during rain, however intense or prolonged. Another reason is the necessity for escape of air as fast as the water enters the soil. This reduces the pore-space available for water within the soil.

The change from maximum to minimum infiltration-capacity during rain at once accounts for increased surface-runoff from initially wet soils or in a succession of storms, an effect commonly but erroneously attributed to saturation of the soil.

In the case of natural soils the infiltration-capacity may drop from one inch per hour or more at the beginning of rain to one-half inch per hour or less within an interval ranging commonly from one hour to two or three hours.

Thereafter, if the rain continues or the soil remains thoroughly wetted, a further slow decrease in infiltration-capacity takes place for a time, especially in the case of soils containing large percentages of colloidal material. When this colloidal readjustment becomes complete, the infiltration-capacity reaches a stable minimum value. This may not be materially less than the value attained shortly after rain begins.

After rain ends, restoration of the infiltration-capacity begins. Wind-action and differential temperatures close to the soil-surface aid in reopening the soil-pores, shrinkage of colloids takes place, perforations of earthworms and insects are restored, and the infiltration-capacity returns to its maximum value usually within a period of a day or less for sandy soils, although several days may be required for clays and fine-textured soils. Apparently the maximum and minimum infiltration-capacities for a given soil have fairly definite values at a given season of the year.

The range of variation of infiltration-capacity depends on the type of soil as well as its condition. Because of its large absorption-area per unit of surface-area and because of the temporary storage of rainfall in sun-checks, a sun-checked clay soil can initially absorb rain as fast as it falls. Similar conditions pertain in case of a newly plowed or cultivated field. In both instances the high initial infiltration-capacity is partly due to storage of water which is held in sun-checks, furrows, or small depressions until absorbed. Neglecting these exceptional cases, the range of infiltration-capacity between the maximum and minimum values is usually of the order of one inch in the case of loam soils, the range decreasing as the soil-texture becomes coarser. In the case of sandy soils which do not pack or sun-check and in which there are few earthworm or other artificial perforations, infiltration-capacity is apparently nearly constant or the range of variation is slight.

There is also a marked seasonal variation in infiltration-capacity. Apparently this is in part related to temperature but also in part results from the fact that earthworm and other perforations which freely admit water or permit the escape of air from the soil are present only during the summer season.

Table 1 shows the infiltration-capacity of the loess soil comprised in the drainage-basin of Ralston Creek at the University of Iowa, Iowa City, Iowa, derived from the analysis of 56 storms. In the case of storms designated "Class A" the soil-surface was initially dry, as there had been at least two preceding days without rain. In case of storms designated "Class B," rain of at least moderate intensity had occurred within an interval of not more than one day. It will be noted that the average infiltration-capacity during summer storms for initially dry soil was 1.25 inches per hour, while for Class B storms the average is 0.62 inch per hour or practically one-half as much. It will also be noted that the average infiltration-capacity in Class A storms varied in different years. The variation in different years is due in part to the effect of cultivation. About 20 per cent of the area was devoted to the growth of cereal crops; most of the remainder was grassland, including meadow and pasture.

An analysis was also made of ten winter storms which occurred during the months November to April. These gave an average infiltration-capacity of 0.12 inch per hour. This illustrates the marked difference in the ability of a soil to absorb rain under winter and summer conditions.

Table 1--Average values of infiltration-capacity f from analysis of individual storms, annual and eight-year means, Ralston Creek drainage-basin, Iowa

Year	Class A		Class B		All storms	
	Average f	No. of storms	Average f	No. of storms	Average f	No. of storms
(1)	(2)	(3)	(4)	(5)	(6)	(7)
			Summer storms, May-October, inclusive			
1925	1.55	3	0.53	2	1.14	5
1926	0.98	4	0.86	3	0.93	7
1927	1.43	3	0.28	5	0.71	8
1928	0.98	6	0.36	4	0.74	10
1929	1.65	3	0.97	3	1.31	6
1930	1.57	3	0.44	1	1.29	4
1931	1.07	2	1.92	1	1.35	3
1932	1.09	3			1.09	3
Average,[a] 1925-1932, inclusive	1.25	27	0.62	19	0.99	46
			Winter storms, November-April, inclusive			
Average,[a] 1925-1932, inclusive	....	..	....	..	0.12	10
			All storms			
Average,[a] 1925-1932, inclusive	....	..	....	..	0.84	56

a) Average of all storms--not average of yearly means.

Table 2--Summary of mean monthly values of infiltration-capacity, Ralston Creek at Iowa City, Iowa

Month	Average values of f, inches per hour					
	Class A		Class B		All storms	
	f	No. of values	f	No. of values	f	No. of values
(1)	(2)	(3)	(4)	(5)	(6)	(7)
February	0.17	2	....	..	0.17	2
March	0.12	3	0.03	1	0.10	4
April	0.12	1	0.10	1	0.11	2
May	1.30	3	0.24	2	0.87	5
June	1.57	9	0.81	8	1.21	17
July	1.36	4	0.32	3	0.91	7
August	0.75	5	0.47	3	0.65	8
September	1.11	6	1.92	1	1.23	7
October	....	..	0.25	2	0.25	2
November	0.11	2	....	..	0.11	2
All values	....	..	....	..	0.84	56

Note--Class A: f for storms following two or more rainless days.
Class B: f for storms following one or less than one rainless day.

Seasonal effects, also the effect of cultivation, are further revealed by the data given in Table 2, which shows the average infiltration-capacity for storms occurring in different months of the year. For Class A storms, which represent approximately maximum infiltration-capacity, the value of f is of the same order for the months May, June, July, and September, but drops to a lower value in August, in which month tillage-operations on farms are generally at a minimum, as the season of corn-cultivation is over and the sowing of fall grain has not begun.

The other experimental data above referred to give results of the same general character. It is expected that these data will be published in full in the near future.

In view of the large variations of infiltration-capacity, even for the same soil when in different conditions, the question naturally arises as to the practicability of using infiltration-capacity as an independent variable in hydrologic studies. As already noted, infiltration-capacity changes rather quickly from its maximum to its minimum value during heavy rain, and after a rain its maximum value is restored, usually before the next succeeding heavy rain.

Actual variations in infiltration-capacity are not as great as they appear to be. Owing to the method of calculation of f, especially for large areas involving a deduction for ground-water, errors in the determination of either the rainfall or runoff may combine so as to produce a considerably greater error in the calculated infiltration-capacity.

The data thus far analyzed indicate that if there has been no recent rain, f is close to its maximum value during short intense storms of the summer thunder-storm type. This would be true, for example, for storms of less than one hour total duration of rainfall. For longer storms of the same type the value of f is about the average of the maximum and minimum values. For long storms of the cyclonic type and for all storms where there has been at least a moderate fall of rain within one day preceding, the value of f is close to the minimum value. In other words, if two heavy rains occur not more than a day apart, the ground being dry antecedent to the first rain, and if both rains are of short duration, the value of f will be close to the maximum value for the first rain and close to the minimum value for the second rain. If the maximum and minimum infiltration-capacities of the soil are known, a study of the rainfall-record will usually show the value of f applicable to a given storm.

A simple method of determining the range and approximate values of f for a small drainage-area consists in noting the intensities of the heaviest rains which do not produce surface-runoff--this will indicate f_g; and also the lightest rains which produce surface-runoff--this will indicate f_ℓ.

Relation of infiltration to runoff

There are three ways of viewing runoff:

(1) As a fraction or percentage of rainfall. This is the usual concept when applied to average conditions. The average runoff can be determined by multiplying the rainfall by the ratio of runoff to rainfall or runoff-coefficient. There is, however, only an indirect relation between the average runoff-coefficient for permeable areas and the runoff-coefficient for particular storms. The average runoff may include a large percentage of ground-water flow, while runoff for an individual storm is mainly or wholly surface-runoff. Since surface-runoff is approximately equal to the rainfall-excess, and this is zero for rain-intensities less than the infiltration-capacity, it follows that the runoff-coefficient is not a constant, neither is it directly a function of the amount of rainfall. It is a direct function of the rainfall-excess intensity for rains of sufficient intensity to produce a rainfall-excess.

(2) Runoff may be considered as the difference between rainfall and water-losses. The water-losses include interception, transpiration, and evaporation from the soil. Runoff is the residuum left after these evaporative processes have taken their toll. This is the basis of the water-loss method of estimating runoff, largely used in England. It directly reflects the effect of the physical processes involved but makes no distinction between ground-water and surface-runoff and does not directly take storage-effects into account. It is therefore, applicable primarily to average conditions and not to individual storms.

(3) The infiltration-theory—Runoff from a permeable area generally consists of two parts: (a) Surface-runoff; (b) ground-water runoff. Neglecting interception by vegetation, surface-runoff is that part of the rainfall which is not absorbed by the soil by infiltration. If the soil has an infiltration-capacity f, expressed in inches depth absorbed per hour, then when the rain-intensity i is less than f the rain is all absorbed and there is no surface-runoff. It may be said as a first approximation that if i is greater than f, surface-runoff will occur at the rate (i-f).

Many attempts have been made to correlate annual seasonal or monthly runoff with rainfall-amount, or with rainfall-amount and evaporation or temperature, it being assumed that at a given locality temperature is the principal independent variable controlling evaporation-rate. The results have sometimes been expressed in terms of a method for computing runoff or extending runoff-records, such as the methods of Vermeule and Meyer. More often they have been expressed as runoff-formulas, of which there are many. Neither of these methods of procedure has met with any marked success.

An analysis of the relation of runoff for the summer season, May to October, inclusive, to rainfall, evaporation, and to rainfall and evaporation together, has been made for West Branch Delaware River for the 25-year period, 1904-29, for the purpose of determining the degree of correlation existing between these quantities. Briefly stated, the results indicate that the best which can possibly be done, even where the coefficients and relationships are derived directly from a long runoff-record, is to duplicate the record for individual seasons with an average error of about ten per cent but with much larger actual errors in the estimated runoff in individual seasons.

The imperfect correlation between total rainfall and evaporation, on the one hand, and runoff, on the other hand, is illustrated by Table 3, showing various data for West Branch Delaware River for the summer season, May to October, inclusive. The ground-water flow was determined by the use of a normal depletion-curve and the total runoff separated into its two elements: Ground-water runoff (Y_g), Column 4, and surface-runoff (Y_s), Column 5. Column 6 shows the ratio of seasonal evaporative capacity computed by the author's formula (Robert E. Horton, A new evaporation formula developed, Eng. News-Rec., Apr. 26, 1917, pp. 196-199). The apparent water losses, L', Column 7, represent the difference between rainfall and total runoff, without correction for gain or loss of storage. It will be noted that there are large differences in total runoff and surface-runoff, particularly the latter, in different years having approximately the same rainfall. For example, in 1909 we have $P = 15.59$, $Y = 6.78$, and $Y_s = 3.42$, while for 1913 we have $P = 16.20$, $Y = 2.40$, and $Y_s = 0.94$. These differences are not wholly accountable to variations in evaporation. While in most years excessive evaporation is accompanied by water-losses larger than the normal, this is not always true and, in any event, departures of evaporation from the normal are usually small compared with variations in either total or surface-runoff with the same amount of rainfall.

The author has long felt that there was some independent variable or variables other than total rainfall and evaporation-rate which should be taken into account. It will be noted that the infiltration-theory of runoff includes two new variables, namely, rainfall-excess and infiltration-capacity. Just how far the introduction of these new variables will serve to bridge the gap between observed and computed runoff remains to be determined. In accordance with this theory, total runoff consists of two parts:

(1) Surface-runoff, which is dependent on rainfall-amount, rain-intensity, and infiltration-capacity and is practically independent of evaporation-rate.

(2) Ground-water runoff. This is dependent on (a) total infiltration and hence indirectly on the same factors which control surface-runoff and is also dependent on (b) vegetational activity and evaporation, which in part determine the water losses, and on (c) the complex interrelations between infiltration-capacity, field moisture-capacity, vegetational activity, and accretion to the water-table. These relationships will be subsequently discussed.

Referring again to Table 3, Column 9 gives the ratio of the apparent water-losses to total infiltration. Attention is particularly directed to the relative constancy of this ratio. West Branch Delaware River drainage-basin is in a large de-

TABLE 3 - SEASONAL RAINFALL, RUNOFF, AND WATER-LOSSES
WEST BRANCH OF DELAWARE RIVER AT HANCOCK AND HALE EDDY, N.Y.
SUMMER SEASON - MAY TO OCTOBER INCL.
(All quantities except ratios are in Inches Depth)

Year	P	Y	Y_g	Y_s	E_r	L′	F = P−Y_s	L′/F	f [(a)]
(1)	(2)	(3)	(4)	(5)	(6)	(7)	(8)	(9)	(10)
1905	23.29	6.06	2.97	3.09	0.932	17.23	20.20	0.853	0.88
06	23.08	7.11	4.21	2.90	1.01	15.97	20.18	.791	.79
07	20.30	6.83	3.98	2.85	0.928	13.47	17.45	.772	.98
08	16.71	5.32	2.95	2.37	1.12	11.39	14.34	.794	.80
09	15.59	6.78	3.36	3.42	1.05	8.81	12.17	.724	.34
1910	15.66	4.56	2.90	1.66	1.13	11.10	14.00	.793	.68
11	20.65	5.42	3.09	2.33	1.06	15.23	18.32	.831	1.05
12	18.71	4.43	3.10	1.33	1.04	14.28	17.38	.822	1.09
13	16.20	2.40	1.46	0.94	1.08	13.80	15.26	.904	1.03
14	20.90	5.67	3.55	2.12	1.04	15.23	18.78	.811	.68
1915	24.90	8.89	5.12	3.45	0.967	16.01	21.45	.746	1.14
16	23.19	7.87	5.68	2.19	1.08	15.32	21.00	.730	1.14
17	32.14	13.41	7.46	5.95	0.953	18.73	26.19	.715	.72
18	28.60	8.62	6.40	2.22	1.08	19.98	26.38	.758	.82
19	21.18	5.25	4.00	1.25	1.00	15.93	19.93	.799	1.18
1920	27.40	8.89	5.94	2.95	1.00	18.51	24.45	.757	1.10
21	23.18	4.69	3.68	1.01	1.11	18.49	22.17	.834	.73
22	26.03	8.28	5.35	2.93	1.01	17.75	23.10	.768	1.18
23	21.85	5.08	3.17	1.91	1.00	16.77	19.94	.841	1.00
24	24.53	10.40	5.46	4.94	0.835	14.13	19.59	.721	.62
1925	23.02	8.14	4.82	3.32	0.852	14.88	19.70	.755	.82
26	20.37	5.88	3.78	2.10	0.960	14.49	18.27	.793	1.00
27	26.69	11.56	5.39	6.17	0.812	15.13	20.52	.737	.82
28	25.35	13.06	7.54	5.51	0.880	12.29	19.84	.619	.63
1929	22.39	7.18	4.96	2.22	1.01	15.21	20.17	.754	1.20
Mean	22.48	7.27	4.42	2.85		15.20	19.63	0.777	0.90

(a) This is apparent infiltration-capacity in inches per day as deduced from daily rainfall-records. The actual infiltration-capacity is greater in the ratio of 24 to the number of hours per day of rainfall-excess duration.

gree covered with natural forest vegetation. The meaning of the constancy of this ratio seems to be that the natural vegetation of a region tends to develop to such an extent that it can utilize the largest possible proportion of the available soil-moisture supplied by infiltration. The constancy of this ratio also suggests its utility as a factor for use in estimating water-losses from drainage-areas and, furthermore, it is apparent that the total infiltration represents the maximum amount of available or "effective rainfall" in relation to the growth of vegetation.

Phenomena of a stream-rise

The segment of a hydrograph from the end of a period of concordant flow when a rise begins and continuing through to the beginning of the next period of concordant flow may be designated a "rise." In reality, of course, it involves both a rise and a recession. A rise of a perennial stream begins as a result of rain of an intensity exceeding the infiltration-capacity f.

It can be shown that on a permeable drainage-area direct surface-runoff to the stream after rain ends takes place only from a restricted belt bordering the stream. From portions of the drainage-basin more remote from streams all of the overland flow or water in transit as surface-detention when rain ends enters the soil as infiltration and none of it reaches the stream.

As a result of this condition, direct surface-runoff into stream-channels ends very shortly after the end of the last period of rain-intensity exceeding infiltration-capacity. There is thereafter an interval of variable flow, its duration ranging from a few hours for very small areas to many days for large areas, after which the stream settles down again to a condition of concordant or nearly concordant flow. The flow is then represented by the normal depletion-curve. This continues until the beginning of the next rise. A hydrograph consists of a succession of alternating rises and intervals of concordant flow or flow derived from normal depletion of storage.

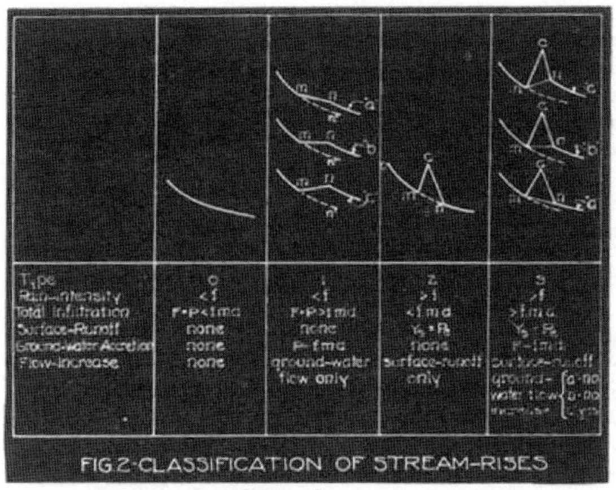

FIG.2-CLASSIFICATION OF STREAM-RISES

Fig. 2

Ground-water flow, of course, continues during the interval of the rise and follows the normal depletion-curve unless or until accretion to the water-table takes place. After accretion ends, ground-water flow in accordance with the normal depletion-curve is resumed, though perhaps at a higher level. Accretion to the water-table does not, however, occur during a rise unless the infiltration is in excess of field moisture-deficiency between the water-table and the soil-surface. These principles make it possible to classify rises appearing on a hydrograph in terms of rain-intensity, infiltration-capacity, and field moisture-deficiency.

The author's classification of rises is shown on Figure 2.

Type 0 is so designated because nothing happens as far as the stream is concerned. For this type the rain-intensity is less than the infiltration-capacity. There is, therefore, no surface-runoff. The total infiltration is less than the field moisture-deficiency and there is, therefore, no accretion to ground-water. The normal depletion-curve continues its downward course uninterrupted. There is, therefore, no rise in the stream. These phenomena are characteristic of light rains occurring during generally dry weather, particularly after long droughts when the soil has the maximum infiltration-capacity f_g. Type 0 is, however, something more than a gesture since soil moisture-accretion takes place. Soil moisture-accretion effects are cumulative and the occurrence of conditions of Type 0 may hasten the time when a real rise in the stream will occur.

Type 1--Again the rain-intensity is less than infiltration-capacity and no surface-runoff occurs. The total infiltration is greater than the field moisture-deficiency and some accretion to the water-table takes place, accompanied either by an increase in ground-water flow or a slowing down of the ground-water depletion-rate. These small irregularities in a hydrograph look like the effects of observational errors or barometric changes but they frequently result from rain. They are typical effects of light rain in the spring and of somewhat heavier rain of low intensity in the summer and fall.

Three different cases occur under Type 1, as shown on the diagram. In each case accretion to the water-table takes place during the interval denoted by mn. Normal ground-water depletion interrupted at m is resumed at n, while n' shows the corresponding stage had there been no ground-water accretion. In case (a) the rate of accretion is less than the rate of normal ground-water depletion. The depletion, therefore, continues but at a reduced rate. In case (b) the accretion and depletion rates are equal and the ground-water flow-rate remains constant for a time. In case (c) the rate of ground-water accretion exceeds the rate of normal depletion and there is a rise of the water-table and an increase in the ground-water outflow-rate.

Type 2--Here the rain-intensity exceeds the infiltration-capacity and surface-runoff occurs, but the total infiltration is less than the initial field moisture-deficiency and there is no accretion to the ground-water and hence no change in ground-water flow. The normal depletion continues during the rise and the ground-water regimen is resumed at n. The stream falls after the rise to a lower stage than pertained when the rise began. This is a growing season or midsummer type and commonly occurs when the field moisture-deficiency is large enough so that the field moisture-capacity is not fully restored by infiltration. Such rises are typical of the effect of short, sharp showers of the thunder-storm type.

Type 3--Again the rain-intensity exceeds the infiltration-capacity and surface-runoff occurs. In this type of rise, the total infiltration exceeds the field moisture-deficiency and accretion to the water-table takes place. The point n at which the rise ends is the point at which the recession-side cn of the discharge-graph coincides with the normal depletion-curve.

There are three cases under Type 3 identical with those for Type 1-rises and dependent on the rate of ground-water accretion. Normal depletion-flow is resumed at the end of a rise of Type 3 at a higher stage than for a rise of Type 2, other things equal, but the stage at the end of the rise may or may not be higher than the initial stage. In cases (a) or (b), Type 3, the stage at which the normal depletion-flow is resumed will not be higher than the initial stage; in case (c) it will be higher. Whether a given rise is of Type 2 or Type 3 can be determined by extending the normal depletion-curve underneath the rise. If the recession-side of the graph returns to this curve as extended, the rise is of Type 2. If the normal depletion-curve at the end of the rise is at a higher level than the extended curve under the graph, the rise is of Type 3.

Compound rises occur where,as the result of a second intense rainfall, a second rise begins before the completion of the recession-phase of the first rise. There are, however, only three fundamental cases: Types 1, 2, and 3, as may readily be seen. A compound rise may, of course, be decidedly complex since any of the three types may be partly superposed on any other. This gives six cases for a compound rise, Type 1 on 2, 1 on 3, 2 on 3, 3 on 2, 3 on 1, and 2 on 1. Also each of the three types may succeed itself.

As regards its relation to ground-water accretion, a compound rise may always be resolved into one of the three fundamental types, since a rise of any one of these types always involves a reduction in field moisture-deficiency. A storm which produces a rise of a given type will, if compounded with a second rise, generally produce a compound rise of an order equal to or higher than that of either of its two components. Thus a rise of Types 1 and 2 combined will generally be of Type 3.

It may be noted that in order to bring out the various relationships described on a hydrograph a fairly open time-scale is required, say one inch equals five or ten days, together with not too small a discharge-scale.

Classification of stream-rises in the manner above described not only affords a definite basis for separation of ground-water flow from surface-runoff on a hydrograph but it affords a convenient starting point for determining what is going on within the non-saturated zone of the soil during and following rain.

Interrelations of infiltration, vegetation, soil-moisture, and ground-water

There are four hydrologic factors which limit and determine the adaptability of the soil of a given region to the growth of vegetation: (1) Rainfall-amount, duration, intensity, and seasonal distribution; (2) infiltration-capacity of the soil; (3) field moisture-capacity of the soil; and (4) depth to water-table, if one exists.

Infiltration-capacity limits the amount of water which can enter the soil. The excess is rejected as surface-runoff. The field moisture-capacity limits the part of the infiltration that can be retained in the non-saturated zone of the soil and is available to supply vegetation therefrom. If at a given time when infiltration occurs, there is a certain aggregate field moisture-deficiency, measured in inches depth of water, the excess of infiltration over field moisture-deficiency, if any, is rejected and percolates downward to the water-table. In general, the part of the ground-rainfall which remains in the soil available to supply soil-moisture requirements is that which remains after the rejection (1) of rainfall-excess at the soil-surface and (2) of excess of infiltration over field moisture-deficiency within the soil.

Depth to ground-water operates in three ways: (a) The water-table if too near the surface inhibits the growth of mesophytic vegetation, although hydrophytic plants may thrive, as on swampy areas; (b) if within the reach of plant-roots, the water-table may provide an important and even the larger part of the moisture-supply used by vegetation; and (c) if at a depth below the reach of plant-roots and evaporative uplift, the presence of a water-table has no effect on vegetation.

There is a close interrelation between infiltration-capacity, field moisture-capacity, and depth to water-table which does not seem to have been pointed out.

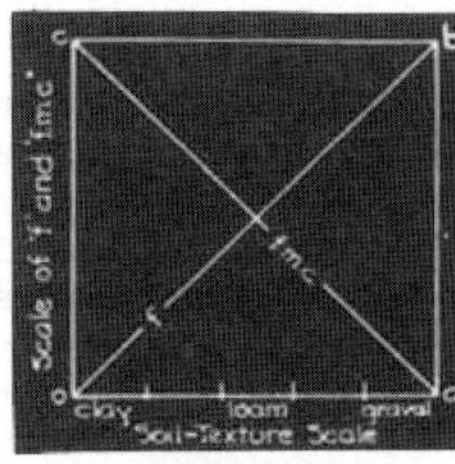

Fig. 3

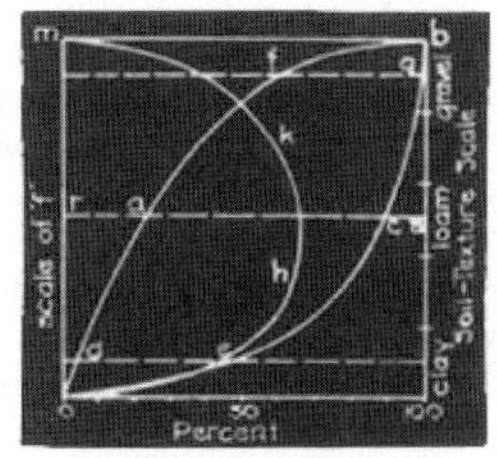

Fig. 4

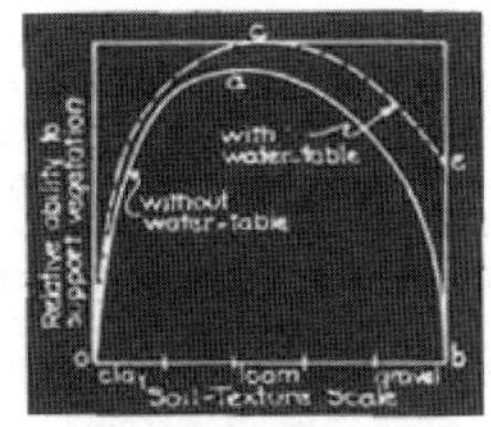

Fig. 5

Fig. 3--Relation infiltration-capacity and field moisture-capacity to soil-texture (schematic)

Fig. 4--Infiltration oab and accretion to water-table ocb, per cent of total (schematic)

Fig. 5--Relative ability of soils to support vegetation with and without a water-table (schematic)

Infiltration-capacity and field moisture-capacity are in a large measure complementary functions of soil-texture, the former increasing, the latter decreasing as one proceeds along the scale of soil-texture from clay and silt to sand and gravel. This fact and some of its consequences are illustrated by Figure 3. The base-scale oa represents soil-texture. The diagonal lines ob and ac may be taken to represent infiltration-capacity and field moisture-capacity, respectively.

For fine soil-textures near the left-hand end of the texture-scale, the infiltration-capacity f is small and the field moisture-capacity fmc is large. Plants may suffer because not enough water enters the soil to support luxuriant growth. For these conditions also there is a large rejection of rainfall at the surface and but little rejection within the soil, so that the final residuum available to supply the water-table is small; hence ground-water accretion is limited and a perennial water-table may not exist.

For coarse-textured soils, corresponding to the right-hand portion of the texture-scale, f is large and fmc is small. There is little surface-rejection of water but a large rejection of infiltration within the soil, so that, again, plants may suffer for want of soil-moisture, but there is abundant accretion of moisture to the water-table and ground-water supplies from such soils are generally plentiful.

In the central portion of the diagram there is both a moderately large infiltration-capacity and field moisture-capacity, a condition most favorable to plant-growth, bearing in mind that rain occurs in quanta, not continuously, while the demands of the plant for moisture are incessant and if the moisture-supply is interrupted, wilting, retarded growth, and perhaps death of the plant results.

It is not to be assumed that the relations of infiltration-capacity and field moisture-capacity to soil-texture are actually linear, as indicated by the lines oa and bc. Quite certainly these quantities are not linear functions of the average size of the soil-grains. This, however, does not affect the validity of the diagram for the purpose for which it is here used.

For a given rainfall-régime, as infiltration-capacity increases, the total infiltration increases, at first rapidly, then more slowly as the maximum rain-intensity is approached. This is illustrated by the line oab, Figure 4. The base-scale in this case shows infiltration expressed as a percentage of total rainfall.

As the field moisture-capacity of the soil decreases, that is, as the infiltration-capacity increases, the proportion of infiltration which is rejected within the soil increases. The total rejection within the soil equals the total accretion to the water-table and can be determined from ground-water flow and runoff-records. The relations of the accretion to the water-table or total rejection of moisture within the soil to total infiltration and infiltration-capacity are somewhat as shown by the line ocb, Figure 4. Consider, for example, a loam soil having an infiltration-capacity as indicated by the horizontal line ac. The segment ra of this line is rainfall-excess rejected at the surface and which produces surface-runoff. The segment cs is rejected excess of infiltration over field moisture-deficiency and goes to the water-table. The remainder, represented by the intercept ac, represents the percentage of total rainfall which will be retained within the soil moisture-zone, presuming that vegetation is growing on the soil to the maximum extent which the soil can support. Similarly, the intercepts de and fg represent moisture available for vegetation in clay and sand. A series of intercepts similar to those described and corresponding to different infiltration-capacities has been taken off and plotted as shown by the line ohkm, Figure 4. This shows, as far as soil-moisture alone is concerned, the relative ability of soils of different infiltration-capacities to support vegetation from soil-moisture alone. The maximum point on this curve corresponds to a soil of medium texture. It is not possible at this time to convert this directly into a diagram expressing ability of a soil to support vegetation in terms of soil-texture. However, for illustrative purposes it may be assumed that infiltration-capacity and field moisture-capacity of the soil are linear functions of the texture or fineness. The result shown by the line oab, Figure 5, is then obtained (this corresponds to line ohkm, Fig. 4). This curve, derived solely from hydrological considerations, is in good agreement with practical experience. The curve shown is not symmetrical, and probably is not in fact. Heavy impervious soils because of their large field moisture-capacity can, with equal rainfall, tide vegetation over periods of drought better than coarse-textured soils. It is to be noted that this discussion does not take into consideration other requirements of vegetation than those for soil-moisture.

For a given rainfall-regime, that soil will best support vegetation which permits the largest percentage of the rainfall to be absorbed as infiltration and at the same time permits the smallest percentage of infiltration to pass downward to the water-table. For such a soil the surface-runoff will be relatively small, the water losses a máximum; the total runoff will not be large but will be somewhat regulated by ground-water flow; ground-water supplies will be of moderate amount but generally perennial.

Turning next to a consideration of the height of water-table, in a given phreatic basin with given rainfall characteristics this is controlled jointly by two factors, namely, (a) the amount of accretion to the water-table, and (b) the transmission-capacity of the soil. With the outlet level or sole of the aquifer fixed by topographic conditions, the average height of water-table at a given point within the phreatic basin will be such that, at the existing height, the average rate of outflow equals the average rate of accretion. If there is abstraction of water directly from the water-table by vegetation or evaporation, this is to be added to the outflow.

Considering first a case where there is no such abstraction, the rate of outflow is proportional to the product of depth of ground-water in the aquifer, the slope of the water-table and the transmission-capacity of the soil.

The amount of accretion to the water-table for a given soil and rainfall-regime is doubly a function of the infiltration-capacity, since, as f increases, fmc decreases and an increasing portion of the total infiltration passes downward to the water-table; hence the average rate and the total amount of accretion increase rapidly, other things equal, as the soil-texture becomes increasingly coarser. Transmission-capacity is also a rapidly increasing function of increasing size of soil-grains. Thus proceeding from finer to coarser-textured soils the total accretion to the water-table increases, thus tending to maintain a higher ground-water level. At the same time the transmission-capacity of the soil increases and this tends toward a lowering of the water-table. The combined action of these two opposing factors tends toward the maintenance of a more or less invariant mean ground-water level under widely differing soil-conditions. The effect of their joint operation is revealed by the fact that the surface water-table in well-drained humid regions is generally at a depth varying from 10 to 30 feet, in soils ranging from those of the finest to those of the coarsest textures. Ability of a given region to supply ground-water is, however, more or less proportional to the infiltration-capacity alone.

Starting with the finest-textured soils, as the coarseness of the soil increases, ability to support vegetation and to supply ground-water both increase, the ability to support vegetation reaches a maximum for a soil of medium texture and thereafter decreases, while ability to supply ground-water continues to increase.

Thus far it has been assumed that vegetation is supplied wholly from the zone of aeration or that there is no direct abstraction from the water-table by plants. If the water-table is at a suitable but moderate depth, plants, including both native vegetation and cultivated crops, may and often do derive a part of their moisture-supply by direct abstraction from the water-table. For this condition the relation between soil-moisture and ability to support vegetation is considerably modified as compared with the case already considered, where the moisture-supply of vegetation is derived wholly from the zone of aeration. Assuming—as is apparently the case—that the plant derives moisture from the soil above the water-table and by abstraction directly from the water-table, respectively, in proportion to the facility with which it can get water from each, then the relation between soil-textures and ability of the soil to support vegetation will be somewhat as indicated by the dashed line oce, on Figure 5. The two curves on Figure 5 indicate, qualitatively, at least, the relative ability of a given soil to support vegetation according as there is or ic not a water-table at a height within the reach of plant-roots. The purpose of this diagram is to show that the range of soil-textures through which mesophytic vegetation can thrive is, in general, considerably increased by the presence of a water-table of suitable depth.

Conclusions

Infiltration-capacity is an important physical characteristic of the soil. For a given soil the infiltration-capacity varies between a maximum value when the soil is dry and a minimum value after wetting and packing. The infiltration-capacity is generally close to the maximum during short storms following dry periods and close to the minimum during prolonged wet periods.

The infiltration-capacity of a drainage-basin can be determined from runoff and rain-intensity data and, conversely, if infiltration-capacity is known, in conjunction with rain-intensity data, the surface-runoff and total infiltration to the soil can be determined.

For small areas in and adjacent to cities, where rain-intensity records are available and estimates of storm-runoff are required in connection with sewer-design, the use of this method for estimating surface-runoff appears more rational than the so-called "rational method."

It is shown that rainfall-excess and infiltration-capacity are apparently two additional variables which need to be included in order to obtain a close correlation between runoff and the factors by which it is controlled.

Stream-rises are classified on the basis of infiltration-capacity and field moisture-capacity, and a definite procedure is established for separating ground-water flow from surface-runoff in streams.

There are close interrelations between infiltration-capacity, field moisture-capacity, growth of vegetation, and ground-water levels. It appears that natural vegetation generally develops to the maximum extent to which it can be continually supplied with moisture with a given rainfall-régime.

Owing to the fact that the demands of vegetation for moisture are incessant, whereas infiltration occurs intermittently and by quanta, that soil is best able to support vegetation which has a large infiltration-capacity and also a large field moisture-capacity.

Infiltration-capacity and field moisture-capacity are complementary, the first increasing and the second decreasing as the soil-texture ranges from fine to coarse. The best combination for plant-growth, from the viewpoint of hydrological considerations alone, is therefore a medium-textured soil within the class designated "loam."

There is a close correlation between total infiltration and the water-losses or water utilized directly or indirectly by vegetation, and total infiltration appears to be the best available basis of estimating the "effective rainfall" in relation to vegetation.

Voorheesville, New York

Commentary on Paper 2

Horton, R. E. (1936) Maximum groundwater levels.

Horton (1936) is not a highly cited paper. Indeed, it has been largely forgotten perhaps because Horton's reputation was based on concepts of surface, rather than subsurface, responses. It was only on re-reading it relatively recently that I came to appreciate its importance, so I have decided to include it here as a way of trying to avoid it being left out (by default) of the history of streamflow generation. There may well be many other papers that have been undeservedly forgotten but hopefully it will soon be possible for someone to write the history of hydrology in the 20th Century to update the book by Biswas (1972), which did not go beyond the 19th Century.

In fact, the origins of this paper lie firmly in Horton's ideas about the infiltration-capacity of the soil surface. He realized that if rainfall continues indefinitely, but if infiltration is limited by the final infiltration-capacity of the soil surface, then recharge to groundwater would be limited by this infiltration-capacity, and if recharge is limited then the water table could only reach some maximum level. It was therefore possible, knowing the infiltration-capacity of the soil, to carry out a theoretical analysis under steady flow conditions to calculate this maximum level.

Horton does this for both horizontal and sloping beds and for straight and variable width (convergent or divergent) hillslopes. He noted that the maximum water table might intersect the ground surface, or a gulley, in which case there would be a sudden increase in the production of surface runoff. In doing so, he anticipates the topographic index of Kirkby (1975), the wetness index analysis of O'Loughlin (1981), and the dynamic contributing area model of Beven & Kirkby (1979) by some 40 years. The paper therefore deserves a place in the historical pantheon of hillslope response analyses.

However, correspondence between Horton and LeRoy K. Sherman and Howard L. Cook reveals that Horton still considered that rapid subsurface responses were rare (see Beven, 2004). He also mentions a theoretical analysis in which Howard Cook "*worked out under my direction the profile equations for underground back-water using Poiseuille's law instead of Manning. The resulting equations show similar conditions of alternative flow* [depths] *for groundwater……There is, for any given aquifer and rate of outflow therefrom a minimum or critical level at the outlet such that the rate of groundwater outflow* <u>*is in no way affected*</u> *by water level in the stream channel as long as the stream does not rise above this critical level.*" (letter from Horton to Sherman of 1942). It is not clear what Horton means by "alternate depths" here. The theoretical analysis would be certainly related to that presented in Paper 2 (though he does not give the "complicated" equations for the sloping bed case). It is true that the maximum groundwater profile will not be greatly affected by the downstream river level for cases in which a seepage face develops. This does not, however, seem to be directly analogous to the alternate (sub-critical and super-critical) depth analogue in a steady state river flow. A possible interpretation might be that because of the draw-down effect of a seepage face, the gradient of the water table will lead to different <u>discharges</u> at the same depth of flow on either side of a peak water table depth (with subsurface discharge continuing to increase downslope under steady recharge conditions). But this would not seem to be a good analogy, so perhaps Horton was not always right in his interpretations despite his note in a post-script to the letter to Sherman that "*I again emphasize that you cannot leave the hydraulic laws and principles out of this question. Everyone else except myself seems to have done so*".

References

Beven, K. J. (2004) Robert E. Horton and abrupt rises of ground water, *Hydrol. Processes* **18**, 3687–3696.

Beven, K. J. & Kirkby, M. J. (1979) A physically-based, variable contributing area model of basin hydrology. *Hydrol. Sci. Bull.* **24**(1), 43–69.

Biswas, A. K. (1972) *History of Hydrology*. North-Holland Publishing Company: Amsterdam, The Netherlands.

Kirkby, M. J. (1975) Hydrograph modelling strategies. In: *Processes in Physical and Human Geography* (ed. by R. Peel, M. Chisholm & P. Haggett), 69–90. Heinemann, London, UK.

O'Loughlin, E. M. (1981) Saturation regions in catchments and their relation to soil and topographic properties. *J. Hydrol.* **53**, 229–246.

MAXIMUM GROUND-WATER LEVELS

Robert E. Horton

Introduction and synopsis

Aquifers are living, moving things. It is the application of hydraulic and hydrologic principles to geologic facts that enables one to interpret their behavior. The difficulties of this treatment which would result from variation of the phenomena with time are largely avoided by limiting the discussion to equilibrium-conditions. Geology and hydrology, respectively, bear much the same relation to ground-water phenomena that anatomy and physiology bear to biology.

In the absence of rainfall the flow of perennial streams without surface-storage in their basins is derived either from ground-water or channel-storage. It can readily be shown that channel-storage becomes exhausted in a relatively short time compared with that required to exhaust ground-water storage. The rate of ground-water inflow to a stream during a given dry period increases as the initial volume of available ground-water storage increases. Hence the question of the maximum amount of ground-water storage which can be retained by a given drainage-basin becomes immediately important. Hydrologic literature is filled with discussions of the economic value of ground-water storage but one does not find a single title dealing specifically with the question of the maximum or limiting volume of ground-water storage.

The expression "maximum ground-water levels" may be used in two senses: (1) With reference to the limiting height to which it is possible for the ground-water to rise in a given well; (2) with reference to the highest observed stage in any given period or, in the generic sense, to the highest annual stages over a series of years. It has not hitherto, so far as the author is aware, been pointed out that there is a natural maximum or limiting ground-water stage and, pari passu, a maximum or limiting volume of ground-water storage in every aquifer.

[C. C. Vermeule seems to have sensed the idea that there was a maximum volume of ground-water storage which could accumulate in a given drainage-basin. In "Geological survey of New Jersey" (v. 3, Water Supply, 1894, p. 83), he states: "The ground-water flow for full ground-water seems to range close to two inches per month and I have assumed this in every case." An examination of curves used by Vermeule for computation of ground-water flow indicates that for Atlantic Coast streams he found the highest rate of ground-water flow--usually in April--to be about two inches depth of runoff per month in most cases, running up to 2.5 inches in case of some streams with deep, sandy drainage-basins. Since the rate of ground-water flow is nearly--and in many cases exactly--in proportion to the volume of ground-water storage, it follows necessarily that there must be a maximum volume of ground-water storage and corresponding maximum or limiting ground-water stages in these drainage-basins.]

Every crest or peak on a phreatograph corresponds to a point at which the inflow to and outflow from the aquifer are equal. That this must be so is readily seen from the storage-equation

$$\text{Inflow} = \text{outflow} + \text{gain or} - \text{loss of storage}$$

As long as the inflow-rate is greater than the outflow-rate, the storage increases and the ground-water level at a given well continues to rise. If the outflow-rate exceeds the inflow-rate, the storage and stage decrease. The crest is a point of change from a rising to a falling stage. Hence at the crest the stage passes through a point where the inflow- and outflow-rates are equal.

Designating the infiltration-capacity of the soil by f and the transmission-capacity of the aquifer by K_t, the ratio (f/K_t) enters as an important factor in the analytical treatment of a great variety of ground-water problems. This ratio deserves a name and it will here be designated the "transmission-ratio," since it is the ratio of the rate at which water can enter or be transmitted into the soil from rain or melting snow, to the rate at which it can flow out of the soil under unit-head.

Since the infiltration-capacity limits the rate of inflow, the limiting,maximum ground-water stage will be attained if and only if the ground-water has risen to such a stage that the ground-water outflow-rate equals the infiltration-capacity. As subsequently shown, this stage may never be reached in a given well or aquifer because rain-intensities exceeding infiltration-capacity are seldom of sufficient duration to bring the outflow-rate up to equality with f. In such cases a maximum stage or crest-stage occurs, less than the limiting stage. In such cases the crest-stage occurs when rain ends or when the rain-intensity falls below the infiltration-rate; the majority of ground-water crests are of this type.

For purposes of this discussion, surface-aquifers may be classified as (a) simple, (b) multiple, (c) compound.

In a "simple" aquifer the ground-water flows toward a single main outlet.

In many drainage-basins, especially in regions overlain with permeable glacial deposits, ground-water is continuous everywhere but each stream-valley receives ground-water from different slopes. There are no impervious divides between these sub-aquifers and such a system may be described as a "multiple" aquifer. The simple aquifers which make up a multiple aquifer may be called "sub-aquifers."

The flood-plain of a stream may contain an aquifer which is fed in part by direct infiltration and in part by ground-water drainage from adjacent hill-slope aquifers. This may be described as a "compound" aquifer. Pomperaug Valley, Connecticut, affords an excellent illustration of a compound aquifer.

While much of the discussion applies to other conditions, the treatment of this subject is primarily intended to apply to (a) surface or non-artesian aquifers, (b) aquifers in continuous porous media--not to aquifers in rock-fissures.

Following the analytical discussion of the subject, several long ground-water records are presented and examined to see if they afford evidence of behavior in accordance with the principles outlined.

Theory of limiting maximum ground-water stages

For the purpose of discussion of limiting ground-water stages, the case of a simple aquifer with an impervious horizontal sole will first be considered.

Let α be the rate of ground-water accretion, f the infiltration-capacity, and K_t the transmission-capacity of the soil, all in the same units, and with other notation as shown on Figure 1. Then for a strip of unit-width running from the stream or outlet to the phreatic divide, and for equilibrium-conditions, the inflow upstream from any point x must equal the flow past x or, measuring inflow and outflow both in terms of depth per unit of time on the entire phreatic drainage-area

$$\alpha[(\ell_0 - x)/\ell_0] = K_t(h/\ell_0)S \quad (1)$$

But the slope S = (dh/dx). This is positive since the slope is upward in the direction in which x is measured. This gives

$$(\alpha/K_t)(\ell_0 - x)\,dx = h\,dh \quad (2)$$

and

$$(\alpha/K_t)(\ell_0 x - x^2/2) = h^2/2 + \text{constant} \quad (3)$$

where x = 0, $h = h_0$ and constant = $-(h_0^2/2)$. This gives

$$h = \sqrt{h_0^2 + 2(\alpha/K_t)(\ell_0 x - x^2/2)} \quad (4)$$

This is the equation of the ground-water profile for the assumed conditions with any rate of accretion α either equal to or less than the infiltration-capacity. [This equation is general and applies also where α is negative or where there is uniform abstraction of water from the water-table by vegetation. The equation has many applications which will not be discussed here.]

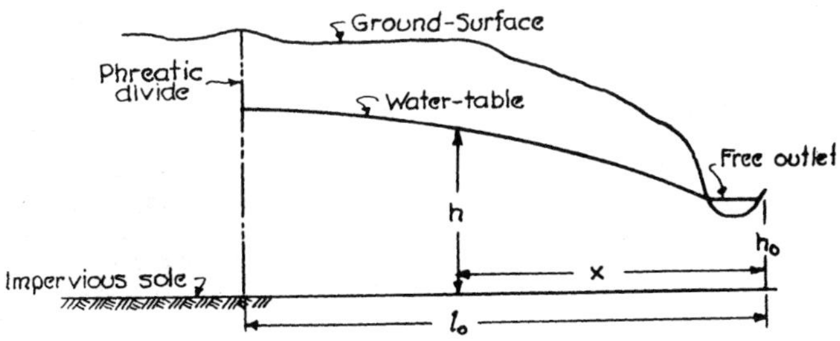

FIG. 1 - PROFILE OF SUB-AQUIFER WITH HORIZONTAL SOLE

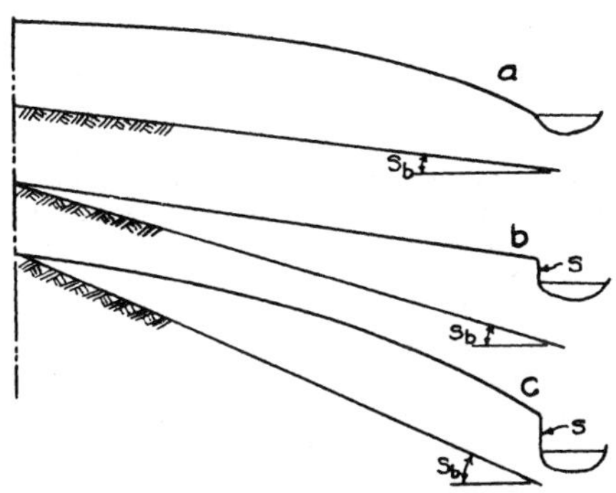

FIG. 2- GROUND-WATER EQUILIBRIUM-PROFILES, INCLINED AQUIFERS (SCHEMATIC)
(a) $S_b < 2\sqrt{\frac{f}{K_t}}$ (b) $S_b = 2\sqrt{\frac{f}{K_t}}$ (c) $S_b => 2\sqrt{\frac{f}{K_t}}$
Cases (b) and (c) are shown with vertical seep-face S at the outlet

Let h_g be the limiting maximum stage for a given infiltration-capacity f. Then

$$h_g = \sqrt{h_0^2 + 2(f/K_t)(\ell_0 x - x^2/2)} \quad (5)$$

This is the equation of a parabola with the vertex at the point p, Figure 1, that is, at the intersection of the water-table and the phreatic divide.

It has been noticed that the profiles of surface-aquifers such, for example, as those between tile-drains, are not always precisely parabolic. One reason is that they are not taken under equilibrium-conditions. Since ground-water rises as long as inflow exceeds outflow, and falls whenever outflow exceeds inflow, it follows that at every maximum point on a phreatograph a condition exists temporarily where inflow and outflow are equal, that is, an equilibrium-condition, for which the formula applies.

Effect of various factors on limiting maximum ground-water stages

The manner in which the limiting maximum stage is affected by changes in the controlling variables is most easily seen by converting formula (5) into an expression for extreme range of ground-water fluctuation. Since the outlet level h_0 represents the minimum possible stage where there is only a single outlet, it follows that

$$\text{Range} = h_g - h_0 = \sqrt{h_0^2 + (f/K_t)(2\,\ell_0 x - x^2)} - h_0 \quad (6)$$

The equation was derived with reference to a strip of unit-width and hence applies strictly only to an aquifer having a rectangular catchment-area. Catchment-areas of simple aquifers are as varied in form as those of topographic drainage-basins, which they generally resemble.

If w_a is the average width of the phreatic drainage-basin upstream from a given point x, and w its width at x, then the flow past x bears the ratio (w_a/w) to that which would result for a rectangular phreatic basin of width w. It can readily be seen that for an irregular-shaped phreatic basin the flow past any point x is the same as it would be for a rectangular basin of width w but with an infiltration-capacity $(w_a/w)f$ instead of f. Making this correction, equations (5) and (6) can be written in the form

$$h_g = \sqrt{h_0^2 + (w_a/w)(f/K_t)(2\,\ell_0 x - x^2)} \quad (7)$$

and

$$\text{Range} = \sqrt{(w_a/w)(f/K_t)}\ \sqrt{2\,\ell_0 x - x^2} \quad (8)$$

Equation (8) shows that, other things equal:

(1) The limiting stage and range of fluctuation at a given well vary in proportion to the square root of the width-ratio and also in proportion to the square root of the transmission-ratio. If either ratio is doubled, the limiting stage and range are increased 41 per cent. If halved, the stage and range are decreased 29 per cent. If both are doubled, the range is doubled. If both are halved, the range is halved, and if one is doubled and the other halved, there is no change in the range or maximum stage.

(2) For a well at a fixed distance from a stream or outlet, the range and limiting stage increase with the length ℓ_0 or the length of underground flow of the water approaching the well.

The relative values of the range for different ratios ℓ_0 to x, computed by formula (8), are as follows:

ℓ_0/x =	1	2	3	4	5	6	7	8	9	10
Range =	1.0	1.73	2.24	2.64	3.00	3.16	3.60	3.87	4.12	4.36

Wells at a given distance from the outlet in a long narrow aquifer have correspondingly larger ranges and higher maximum stages than similar wells in a shorter, wider aquifer of equal surface area.

(3) For an aquifer of a given length ℓ_0, the range and limiting stage increase with the distance of the well from the outlet or with the ratio (x/ℓ_0) as follows:

x/ℓ_0 =	0.0	0.1	0.2	0.4	0.6	0.8	1.0
Range =	0.0	0.19	0.36	0.64	0.84	0.96	1.0

For wells near the outlet the range increases rapidly with distance from the outlet. For wells near the phreatic divide, increase of range with distance from the outlet is relatively small.

Stratification

In this connection, stratification of an aquifer may be considered as any degree or arrangement of horizontal variations or diversity in the water-bearing medium which affects the hydraulic conditions. In addition to true stratification, this includes the important case of gradual variation, usually a decrease in such factors as porosity and transmission-constant proceeding downward from the surface. Extensive areas of Hudson River shale in eastern and central New York afford an excellent example. This rock weathers rapidly when exposed to the air or frost, but at a slight depth below surface it is wholly impervious. There is water in the fractured surface, decreasing in amount until at a depth, usually of a few feet, no water is found. Similar conditions exist in other classes of rocks and soils.

Let K_1 be the transmission-capacity of the aquifer at a height of one foot above the impervious level. Then at a height h the transmission-capacity is K_1h and the average below the height h is $K_1h/2$.

Considering a horizontal surface with the origin at the impervious level and at the outlet-end of the aquifer, then for equilibrium of inflow and outflow

$$(\ell_0 - x)\,f = (K_1h^2/2)\,(dh/dx)$$

Integration gives

$$(2f/K_1)\,(\ell_0 x - x^2/2) = h^3/3 + \text{constant} \tag{9}$$

When x = 0, h = h_0, therefore constant = $-\,h_0^3/3$. Inserting $(w_a/w)f$ in place of f to allow for form of the aquifer, the equation for the maximum limiting stage becomes

$$h_g = \sqrt[3]{h_0^3 + 3(w_af/wK_1)\,(2\,\ell_0 x - x^2)} \tag{10}$$

This is the equation of a parabola of two-thirds power. The profile-equation in other respects is identical with that of a homogeneous aquifer. The discussion of limiting ground-water stages and range, based on the equation for homogeneous aquifers, is therefore applicable to stratified aquifers of this type except that the maximum stages will not be the same in the two cases. It is also possible to obtain an equation for limiting stage applicable to an aquifer consisting of two or more horizontal strata with fixed lines of demarcation between them. Because of the great variety of conditions which this situation presents, it will not be discussed here.

Aquifers with inclined soles

The differential equation of the equilibrium-profile for an aquifer with inclined sole, typical of many hillside aquifers, is easily stated and has been integrated and studied at the Horton Hydrologic Laboratory. The resulting profile-equation is complicated and does not as readily show the effect of changes in individual factors as that given above. The profile-equation for horizontal aquifers has therefore been used for purposes of discussion.

While the conditions which fix limiting stages of hill-slope aquifers are essentially the same as for aquifers with horizontal soles, there are certain important differences. For horizontal aquifers both the limiting stage and the range increase proceeding upstream from the outlet. For sloping aquifers there are three principal cases and several sub-cases, dependent on the relations between the variables. The principal cases are outlined in Figure 2.

(1) Slope of sole $S_b < 2\sqrt{f/K_t}$--Underground back-water extends from the outlet to the divide and the depth at the divide is not zero. The maximum stage increases but the range generally decreases proceeding upstream from the outlet.

(2) Transition-stage, slope $S_b = 2\sqrt{f/K_t}$--It can readily be seen that for an aquifer with a sloping sole and uniform infiltration-rate, there is a certain constant slope for which the inflow upstream from a given well will equal the flow past the well. The profile of the water-table is nearly a straight line, extending from the sole at the divide to the outlet or to a minimum level adjacent to the outlet. In this case the limiting ground-water stage increases and the range decreases, both uniformly, proceeding upstream.

(3) Steep aquifers, slope $S_b > 2\sqrt{f/K_t}$--The depth at the divide is zero, and for a constant level at the outlet the stage increases proceeding upstream. The range is zero at the outlet and also at the divide and has a maximum at some intermediate point.

In some of these cases there is a minimum level at which the ground-water flow can arrive at the outlet. If the actual water-level in the outlet channel is below this minimum, there will be a "seep-surface" at the outlet, that is, ground-water will enter the stream at a level above the stream-level.

Surface and gully-outlets, bornes

In addition to the geologic and physical factors which fix the limiting ground-water stage in a given aquifer or well, there are certain topographic conditions to be considered. So far it has been assumed that the aquifer has only one outlet. Theoretically the water-table in an aquifer with a horizontal sole would intersect the ground-water if the aquifer extended back far enough from the outlet. For aquifers with inclined soles there are cases of this kind and also cases where the water-table would never intersect the ground-surface.

Swamp aquifers--In case of many swampy areas or in case of undissected flat or sloping areas with impervious floors at a slight distance below the surface, conditions of equilibrium may lead to a limiting ground-water stage h_g above the level of the soil-surface over part of the tributary area. On such areas the soil may be continually flooded or the ground-water may rise to the surface only at times. Swampy areas with deep muck-deposits underlain by impervious beds, such as the Everglades of Florida, provide excellent examples of aquifers of this type. As long as the surface is flooded, infiltration takes place at a constant rate equal to the infiltration-capacity and there is flow of ground-water to the outlet at the same constant rate. Rain tends to produce surface-runoff at an intensity approaching the rain-intensity as a limit, since in this case the ground-water outflow equals the infiltration-rate. Large reduction of runoff-intensity below rain-intensity is, however, usually provided by surface-storage. In case of swamp aquifers the maximum limiting stage is, of course, the ground-surface.

Ephemeral outlets, bournes--If the surface of the terrain is serrated by gullies, shallow relative to the main outlet of the aquifer, the conditions may be such that before reaching the limiting stage the water-table will begin to intersect the gullies. These will provide ephemeral or auxiliary outlets for ground-water. In general, the ground-water above the level of the bottom of the gully will flow a distance equal to one-half the spacing between gullies to reach an outlet. The great reduction in length of underground flow as compared with that required to reach the main outlet will limit the maximum ground-water stage to a height a little above that of the gullies. Conditions of this kind are fairly extensive. They are of particular interest because of the state of affairs to which they give rise in connection with flood-runoff. Until the water-table reaches the level of the auxiliary outlet, rain, however intense, cannot produce a surface runoff-rate exceeding the difference between the rain-intensity and infiltration-capacity. Rainfall not exceeding the infiltration-capacity, however prolonged, will therefore produce no surface-runoff. After the water-table rises to a point where gullies are intersected, and the limiting ground-water stage is attained, ground-water outflow-rate will equal the infiltration-capacity and, except for surface-storage, the runoff-intensity from then on will equal the rain-intensity. Thus it may happen that long-continued rains will produce little or no surface-runoff and no increase in ground-water flow until the field-moisture-deficiency becomes zero; then, somewhat abruptly, rain of no greater intensity will produce a relatively high runoff, substantially the same as if the infiltration-rate had been reduced to zero. Such high runoff-intensities are commonly but inaccurately ascribed to the "soil becoming saturated." Actually the soil is not saturated above the water-table but the ground-water inflow and outflow have become equal, so that the total runoff-intensity approaches the rain-intensity as a limit.

If it happens that the water-table intersects gullies when the water-level is considerably below its limiting stage, the condition known in the chalk-downs of England as a bourne may result. A bourne is a stream fed by ground-water where the head of the stream or source of the flow travels up and down the stream-valley with the fluctuations of ground-water level. I have noticed this condition in certain shallow stream-valleys on the south slope of Long Island.

Local variation of ground-water regimen

It is sometimes found that in a group of adjacent wells the phreatographs are all closely similar except that fluctuation and range of stages in some wells is greater than in others. This is commonly attributed to local differences in permeability of the material comprising the aquifer, it being assumed that the wells showing the greatest range are in the least permeable beds. It is true, as formula (8) shows, that if permeability alone varied, then, since the range varies inversely as the square root of the permeability, it would be greatest for wells supplied by the least permeable aquifers. This, I think, is not generally the correct explanation. It will be noted that the range varies as the square root of the transmission-ratio. Permeability and infiltration-capacity, though not usually identical, are closely related. They may be practically identical in sandy soils containing little colloidal material. In general, if the surface of the soil and water-bearing beds are of the same or similar material, a change in transmission-capacity is accompanied by a similar and nearly proportional change of infiltration-capacity; the transmission-ratio may be changed but little and hence the fluctuations and range for a well may not in general vary to any great extent with change of transmission-capacity of the water-bearing beds.

There are other factors quantitatively adequate to explain differences in range of adjacent wells or wells in adjacent aquifers.

Summarized briefly, the factors which fix the maximum limiting stages and range for a given well and aquifer are: (1) Length of aquifer; (2) distance from well to outlet; (3) shape, as expressed in the width-ratio (w_a/w); (4) transmission-ratio (f/K_t); (5) slope of the sole, if not horizontal; (6) homogeneity and stratification; (7) topographic controls, gully-intersections, etc.

These factors are of more or less equal importance. Differences in any one or in any combination may produce considerable variation in range of one well as compared with other adjacent wells.

The effects of the first four factors are rigorously taken into account in equation (8) for homogeneous aquifers with horizontal soles. The others can be taken into account in similar analyses.

Keeping in mind the distinction between limiting ground-water stage and other maximum or crest-stages, it will be seen that the analysis thus far given does not show if or when the limiting maximum stage will be attained.

Conditions which determine times and frequency of occurrence of limiting ground-water stages

The limiting stage may be attained at frequent intervals; for example, it may be attained, as a rule, once each year, or it may be rarely attained or never attained. The frequency with which it will be attained depends on certain factors in addition to the seven described which govern its height. These are: (1) Initial ground-water storage; (2) initial ground-water level and field-moisture-deficiency; (3) duration of rainfall-excess. Since the limiting stage can only be attained at times when there is rainfall-excess over the infiltration-capacity, the criterion which determines the minimum duration t_e of rainfall-excess which will permit the ground-water to reach the limiting stage is

$$ft_e \geqq fmd + Q_g - Q_o$$

where fmd is the field-moisture-deficiency at the beginning of any given period of rainfall-excess and Q_o and Q_g are the volumes of ground-water storage at the beginning of rainfall-excess and at the limiting ground-water stage, respectively. It will be seen that, other things equal, rainfall-duration and, more specifically, duration of rainfall-excess, is the principal controlling factor. Ground-water maximum stages occur at all seasons of the year but in general, particularly in middle latitudes, maxima in the spring months, usually just after the frost has gone out of the ground, tend to predominate.

Using the water-year beginning March 1, the 20-year record at the farm well at the Horton Hydrologic Laboratory shows that the maximum stage has occurred: Fifty per cent of the years in the spring months, March to May; 25 per cent of the years in the winter months, December to February; 15 per cent of the years in the fall months, September to November; 10 per cent of the years in the summer months, June to August.

Records of ground-water level in the chalk-downs in Sussex, England, compiled by D. Halton Thomson, and extending back continuously to 1836, show midwinter maximum levels strongly predominating. The excellent records of the Italian Hydrometric Service show midwinter maximum ground-water levels predominating at some stations, spring and early summer maxima at other stations. Several long ground-water-level records in Michigan show spring maxima strongly predominating, these usually occurring about the time the frost goes out of the ground or shortly after

There are four factors which operate together to bring about the occurrence of ground-water maxima in the early spring: (1) Seasonal variation of rainfall-duration; (2) storage-deficiency; (3) melting of snow; (4) frost-removal. The manner in which these factors operate to produce maximum and limiting ground-water stages does not seem hitherto to have been described.

(A) Rainfall-duration--In many localities, including northern United States, rainfall-duration per shower or storm is much greater, and rain-intensity in inches per hour much less, in fall, winter, and spring than during midsummer months. Analysis of hourly rain-intensity records for ten years at representative regular United States Weather Bureau stations shows, in general, winter rainfall-duration per shower or storm one and one-half to two times as great as in summer, and summer average rain-intensity two to three times as great as for winter conditions. For a given infiltration-capacity the duration of rainfall-excess is therefore generally much greater in fall, winter, and spring than in midsummer and the chances of attainment of the limiting ground-water stage are correspondingly greater than in summer, provided the infiltration-capacity is the same. For rains sufficiently long-continued to produce limiting ground-water stages, the infiltration-capacity is usually at or near its minimum value for the given soil. If the ground is not frozen, this minimum value is nearly constant throughout the year except as it is affected by temperature. If, owing to recent cultivation of the soil, the infiltration-capacity tends to be much higher than the minimum, then the limiting stage and the volume of ground-water storage which must be replenished before the limiting stage can be reached are both increased. Nevertheless, high ground-water stages are sometimes attained in midsummer, as has happened twice at the author's well.

(B) Field-moisture-deficiency--Another reason why ground-water maxima are less frequent in summer than in early spring is because there is no abstraction of moisture from the soil by vegetation during the winter season; consequently the field-moisture-deficiency is usually less in winter than in summer--in fact, it is often relatively slight, especially if it has been replenished by fall rains. If the ground has been frozen for a long period, the ground-water storage-deficiency below the limiting level may, however, be nearly or quite as great at the end of the winter as in a summer dry-period. It appears, therefore, that as between the two factors, rainfall-duration and storage-deficiency, the former is the more important in relation to fixing the time of occurrence of ground-water maxima and limiting stages.

(C) Frost-removal--The author believes, although data for quantitative proof are not available, that an important factor in producing maximum ground-water stages in the spring is the rapid flushing of large volumes of water to the water-table when the frost goes out of the ground. This phenomenon has occasionally been noticed in connection with lysimeter-records. The author has observed two or three inches of water passing downward through a 3-foot depth lysimeter within a day or two after the frost went out of the ground.

The process involved seems to be as follows. The frost melts from the surface downward and, in the case of loams and heavier soils containing colloid material, most of the melt-water remains in the soil. It is true that, while there is usually some infiltration through frozen soils, yet, except in case of sandy soils, the infiltration-rate through frozen soil is relatively small--the rate is apparently measured in terms of thousandths rather than tenths of an inch. Expansion of ice in freezing produces a degree of porosity and a transmission-capacity at the time the frost goes out greatly in excess of the normal minimum values of these quantities. This condition persists until the soil has become settled. When the frost goes out there is in the soil a column of water representing a fraction of the volume considerably in excess of the normal gravitational voids, all of which passes immediately downward to the water-table and often produces the maximum ground-water stage of the year. Incidentally this seems to be one of those wise provisions of nature, so often observed. This same volume of water, if it entered the water-table gradually over a period of two or three weeks while the soil-frost is melting, might be drained away as fast as it entered, leaving the water-table at about the same level as at the beginning of the melting period. Instead, all this water is flushed quickly into the water-table with little loss, just before the beginning of the growing season, and therefore provides the maximum available volume of ground-water storage for the subsequent dry period.

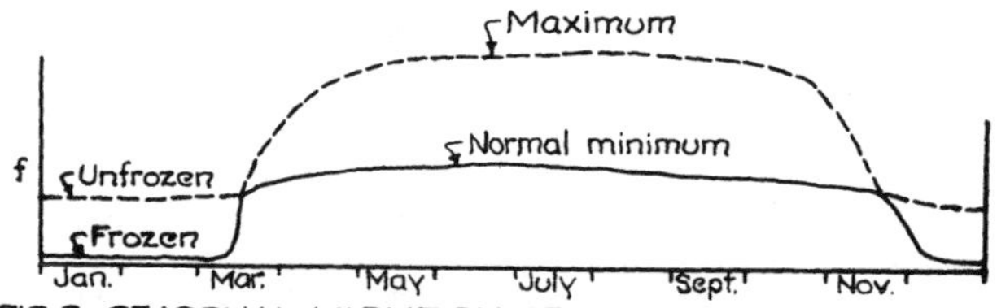

FIG. 3 - SEASONAL VARIATION OF INFILTRATION-CAPACITY, MEDIUM LOAM-SOIL (SCHEMATIC)

(D) Snow-cover--The complex phenomena taking place in the contact-zone between the soil-surface and a layer of melting snow are of great importance hydrologically but cannot be fully discussed here. It may be noted in passing that in case of heavy soils, where the process just described takes place, water derived from melting snow may contribute largely to the volume of water flushed into the water-table when the frost goes out, thus providing a high spring ground-water level in the absence of coincident rainfall. In case of sandy soils with low field-moisture-capacity, infiltration takes place at a relatively high rate even when the ground is frozen. Melting snow may provide a continuous supply of water for absorption by the soil of much longer duration than is ever provided by rainfall alone. For sandy regions in middle latitudes the occurrence of maximum ground-water stages in the early spring is likely to be due to this cause. There is, however, a limitation to the rate at which water can be supplied to the soil by this process, since snow can only melt at a rate dependent on the temperature-excess over 32° F.

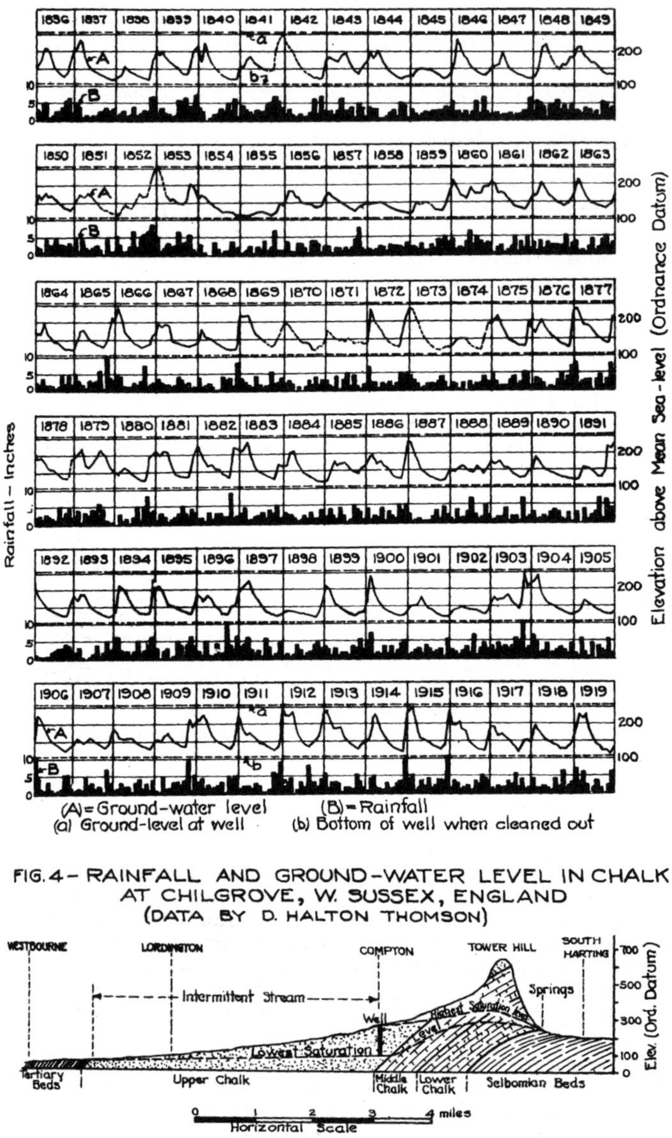

FIG. 4 - RAINFALL AND GROUND-WATER LEVEL IN CHALK AT CHILGROVE, W. SUSSEX, ENGLAND (DATA BY D. HALTON THOMSON)

FIG. 5 - GENERAL SECTION ACROSS SOUTH DOWNS ALONG COMPTON AND EMS VALLEYS (DATA BY D. HALTON THOMSON)

Seasonal variation of ground-water range

For loams and heavier colloidal soils there is a somewhat irregular seasonal cycle of variation of infiltration-capacity. The transmission-capacity of the underlying aquifer remains constant except as affected by the variation of ground-water temperature, which is usually small.

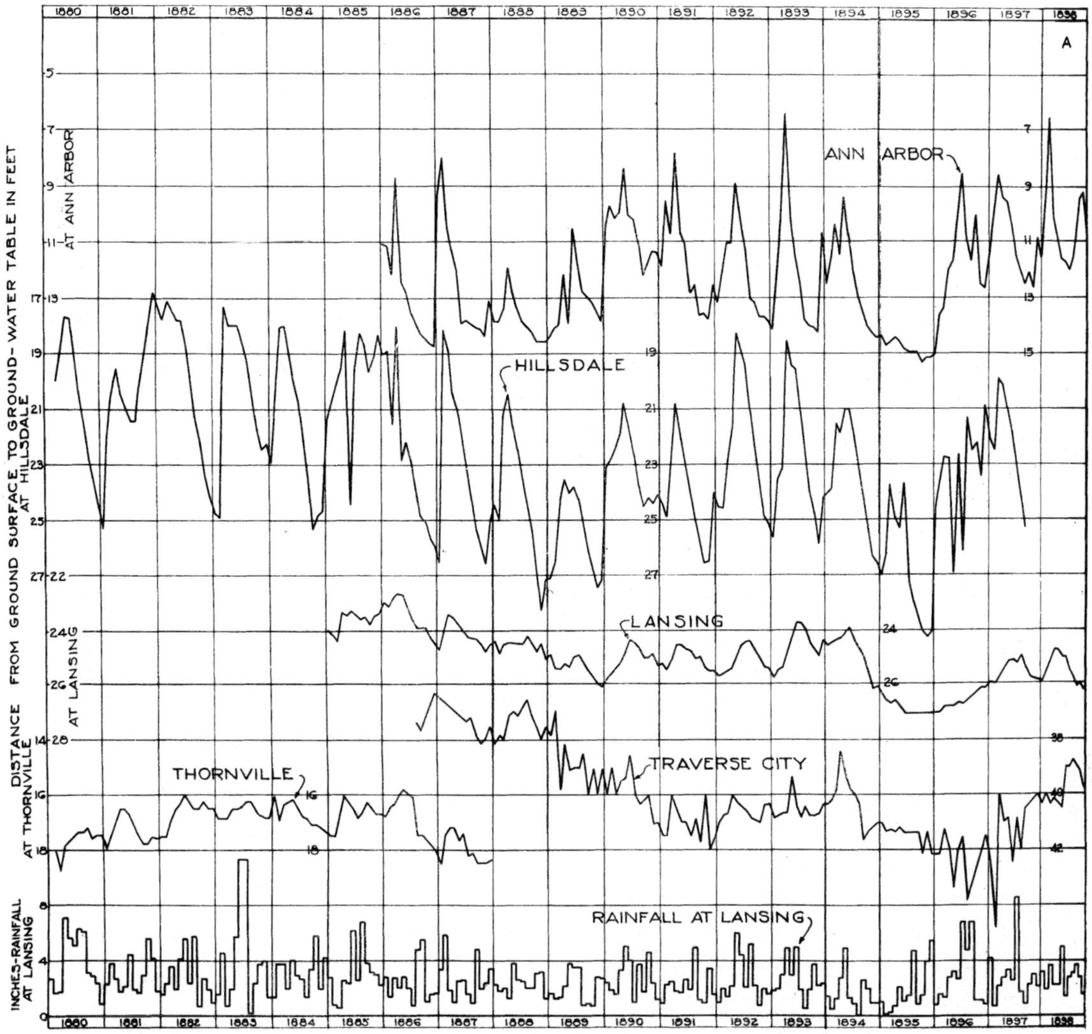

FIG.6-GROUND-WATER LEVEL RECORDS IN MICHIGAN
(DATA BY HENRY B. BAKER)

The relations of winter and summer infiltration-capacities, transmission-capacities, and transmission-ratios for a typical fertile loam soil are illustrated schematically by Figure 3. The upper of the two lines for infiltration-capacity in summer represents a condition with the soil newly cultivated, sun-checked or with numerous earthworm, insect, and root perforations. The lower line represents the minimum infiltration-capacity for a moist soil, packed by rain, and with the larger openings closed by swelling of colloids or inwashing of fine materials. The minimum infiltration-capacity seems to be nearly a constant and definite characteristic of the soil-type. Also the minimum line represents conditions usually prevailing during long-continued rains and hence corresponds to the infiltration-capacity at times when ground-water maximum stages are or may be attained. For summer conditions the minimum value of f is likely to be greater than the transmission-constant K_t because the soil-temperature is higher than the ground-water temperature. For example, in case of a sandy soil, where the transmission-ratio is unity at equal temperatures, and in a region with mean annual temperature of 50° F and ground-water temperature constant at 50° F, the transmission-ratio will be increased 15 per cent under mid-summer conditions with a ground-surface temperature of 70° F and decreased 15 per cent for winter conditions with a ground-surface temperature of 32° F.

Since the transmission-ratio is usually greater in midsummer than in the remainder of the year, it is evidently possible, other things permitting, for higher ground-water stages to be attained in midsummer than in the fall, winter, or spring.

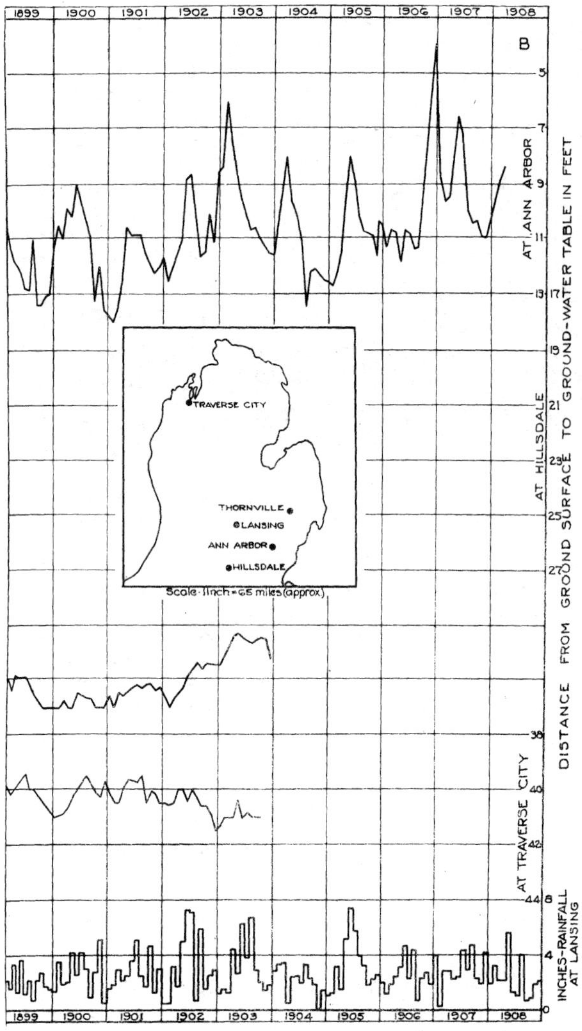

The question naturally arises: Why are the maximum ground-water levels in mid-latitudes usually attained in the early spring or late fall? The answer is derived from a consideration of the seasonal relations of rain-intensity and duration already given. A rain-intensity > f and of sufficient duration to provide ground-water equilibrium or the limiting seasonal ground-water stage may be a common occurrence in fall and spring but only a rare occurrence in midsummer. Quantitative treatment of this subject by the use of curves of rainfall-intensity versus duration for different seasons of the year can easily be carried out. Space-limitations forbid full treatment here but it is evident that, while the limiting maximum ground-water stage for spring conditions is likely to be reached in the majority of years, the higher limiting maximum stage for midsummer will be reached only at rare intervals, if ever. Figures 4, 6, 8, and 9 serve well to illustrate these conditions.

Too much precision should not be expected in relation to the limiting maximum ground-water level at a given well. Even where the well-level is independent of the outlet-level, as it usually is in case of hillside aquifers not too close to streams, and even where it is controlled solely by the minimum infiltration-capacity, there is a variation in the latter, outside the frost-season, of at least 10 to 20 per cent due to temperature and other causes. There will be a corresponding variation in the limiting maximum ground-water stage of 5 to 10 per cent. If, for example, the highest observed stage at a given well is 20 feet below ground, the aquifer is 20 feet deep and stages 18 to 20 feet occur in some years, it may be fairly assumed that any stage less than about 20 feet below ground represents the limiting maximum stage for the conditions pertaining at the time it occurs.

Ground-water levels in the chalk-downs, Sussex, England--The longest continuous ground-water-level record known to the author is that compiled and published by D. Halton Thomson for a well in the chalk at Chilgrove, West Sussex, England [The effect of rainfall on the saturation-level in the chalk at Chilgrove, West Sussex, from 1836 to 1919, British Rainfall, 1919; also Hydrological conditions in the chalk at Compton, W. Sussex, Inst. Water Eng., 1921, (second paper) 1931]. The record extends back with short intermissions to 1834 and so covers more than a century. The record to 1919 is shown graphically on Figure 4. This is an excellent example of a well which overflows at the ground-surface. It will be noted that the maximum level is very definitely fixed but is only reached in occasional years. During dry years the ground-water level may remain continuously low for several years, as in 1887-91. In order that there may be ground-water accretion, the field-moisture-deficiency must first be eliminated. In most of the years shown on Figure 4 there was some rise of the ground-water level in the spring, showing that in spite of the long-continued drought, there was no field-moisture-deficiency at the beginning of the growing season. An important point in this connection is that the effects of a prolonged drought on the ground-water supply to streams is often much more prolonged than the effect of the drought on vegetation. As a rule, accretion to the water-table occurs and there is no field-moisture-deficiency in humid climates at the beginning of the growing season in years of drought. Vegetation suffers because of subsequent deficiency in rainfall. When this

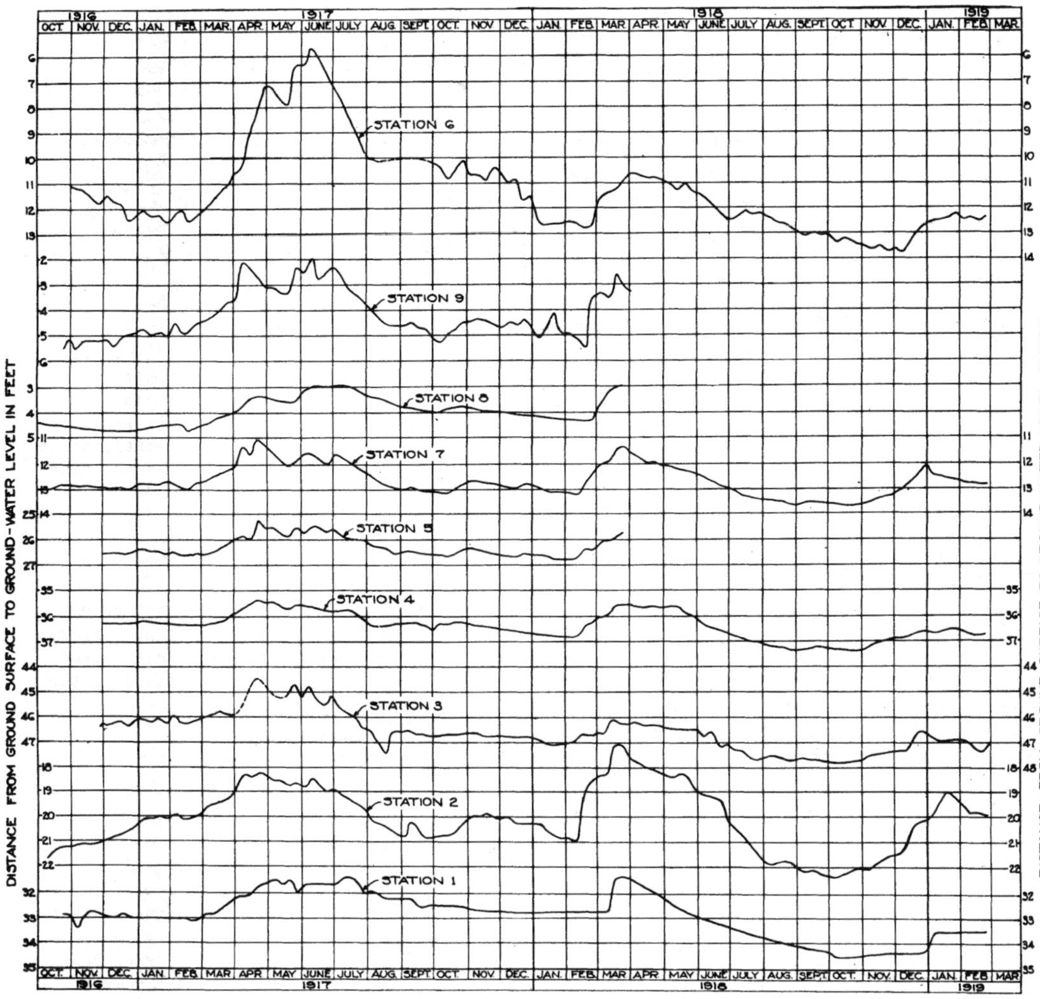

FIG. 7-GROUND-WATER LEVEL RECORDS, HURON RIVER DRAINAGE-BASIN, MICHIGAN
(DATA BY C.O. WISLER)

ends, normal soil-moisture conditions are almost invariably restored inside of a year, whereas several years may elapse before the water-table rises to its maximum or even to its normal level.

D. Halton Thomson has also kept and published an excellent record of ground-water levels in the chalk at Compton, West Sussex, beginning 1903. As large-scale phreatographs of this record have been published, it is not reproduced here. The Compton well is an excellent example of an aquifer with a bourne-outlet. Referring to the cross-section, Figure 5, it will be seen that the head of the bourne migrates throughout a range of more than four miles. This record is remarkable because of the wide range of fluctuation of the water-table. The extreme variation in height of the water-table at the Compton well is 126.2 feet. It appears that in this case the limiting stage is about 220 feet (Ordnance datum), but this stage is only attained at rare intervals.

Ground-water levels in Michigan--Dr. Henry B. Baker, long-time secretary of the State Board of Health of Michigan, believing that there was some relation between water-level in wells and the occurrence of typhoid fever of the virulent rural type, common in the middle west of his time, established well-records at several locations in southern Michigan, beginning in 1880. Some of the records were continued to 1908. Readings were taken in each well once a month, about the middle of the month. The longer and better records are shown graphically on Figure 6. In spite of the approximation due to single monthly readings, there is a close agreement between the curves for Ann Arbor and Hillsdale, although these places are 60 miles apart and in different drainage-basins. Corresponding stages are reflected, with greatly reduced range, in the records

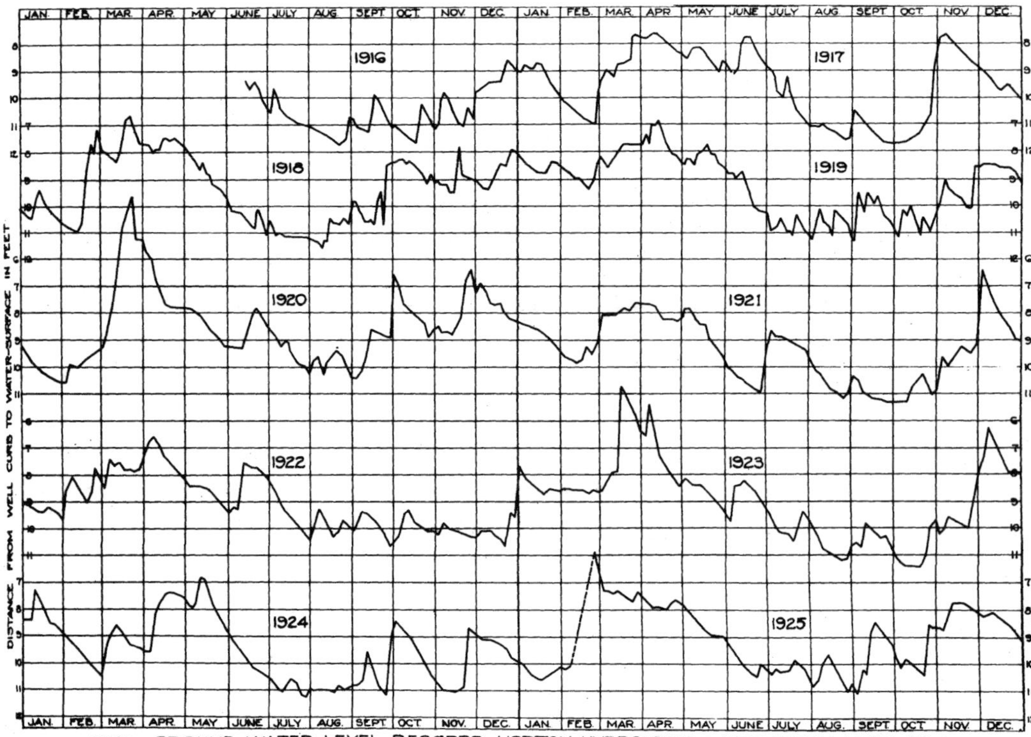

FIG.8a – GROUND-WATER LEVEL RECORDS, HORTON HYDROLOGIC LABORATORY, WELL NO. 1

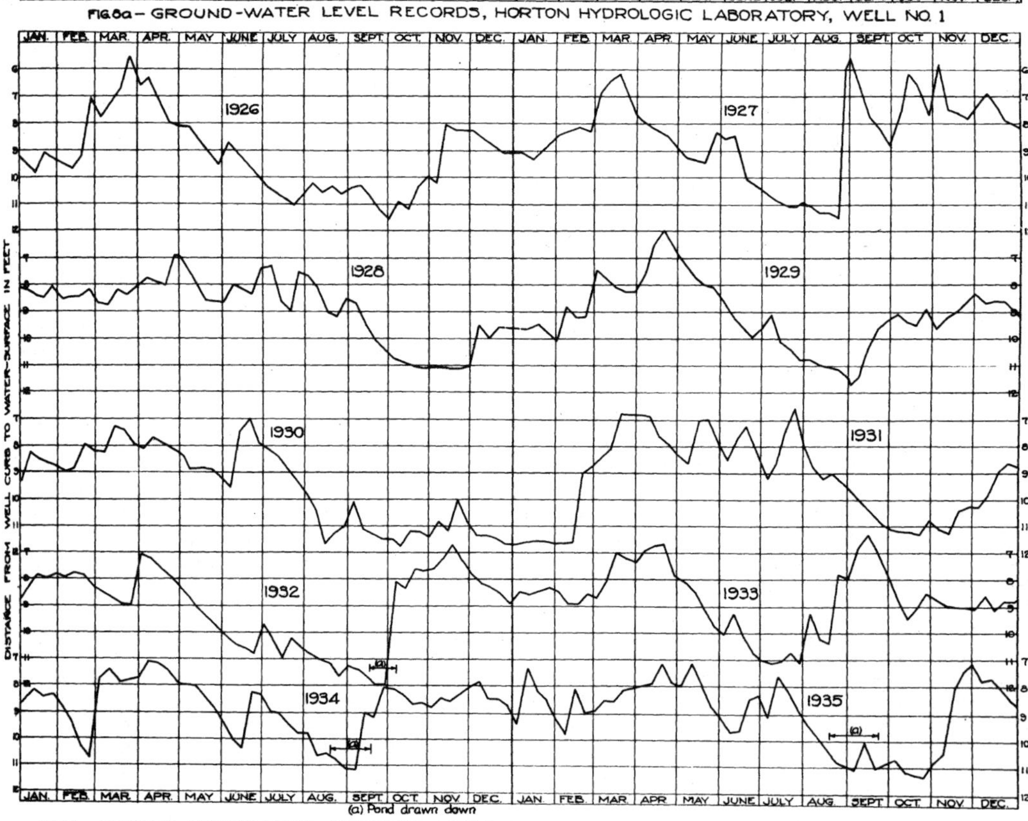

FIG.8b – GROUND-WATER LEVEL RECORDS, HORTON HYDROLOGIC LABORATORY, WELL NO. 1

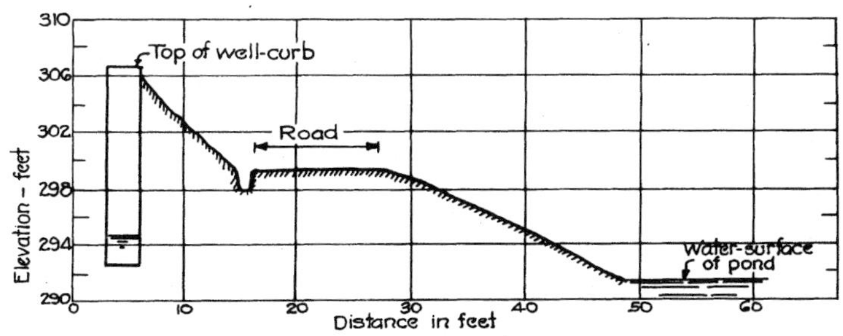

FIG.9— PROFILE OF GROUND-SURFACE ~ WELL NO.1 TO POND
HORTON HYDROLOGIC LABORATORY

for Lansing, Traverse City, and Thornville. All of the wells are located in regions of rather coarse-textured glacial sands, and just which of the reasons capable of producing the variations in range have operated to produce the marked distinction between the upper two and the lower three wells cannot be definitely stated.

Maximum stages of all wells most usually occur in the spring, with minimum in the late fall, although the highest rainfall is in general in the midsummer months. There are exceptions where the annual maximum occurred in midsummer, as in 1900.

The rock floor of lower Michigan is saucer-shaped and there is in places, in the southern part of the State, large artesian leakage into local aquifers. This may have the effect of equalizing the regimen of wells in these aquifers. As a rule, the region is poorly drained, so that the normal range of wells may vary widely as between those close to outlets and those more remote.

Huron River Basin--Figure 7 shows the results of ground-water-level observations taken continuously at different stations in the Huron River Drainage-Basin in southern Michigan. The data have been furnished by Professor C. O. Wisler. Here again we find close similarity of regimen as between different wells but a considerable difference in range, indicating the operation of some of the causes affecting range already described. These records are largely free from saw-tooth irregularities, common to some wells, and the phreatographs show evidence of the operation of some natural smoothing process. The Huron River Drainage-Basin contains numerous lakes, covering probably ten per cent or more of the area. [Well No. 1--Dexter Quadrangle; well on highway 1/4 mile east of east branch Mill Creek and 3/4 mile south of railway; also 3 miles southeast of Lima. Well No. 2--Stockbridge Quadrangle; 1/2 mile northeast from Sylvan, 5 miles southwest from Chelsea. Well No. 3--Dexter Quadrangle; well at first house north of end of north-south road, 1 mile east of small lake 5 miles northwest from Dexter. Well No. 4--Stockbridge Quadrangle; 1 mile east from Gregory. Well No. 5--Dexter Quadrangle; well at third house from south end of north-south road, 2 miles northeast from Pinckney. Well No. 6--Dexter Quadrangle; well at first house south from north end of cross-road, 1-1/2 miles southwest from Horseshoe Lake. Well No. 7--South Lyon Quadrangle; 1/2 mile east from southwest corner Oakland County, 2 miles south and a little west from South Lyon. Well No. 8--Milford Quadrangle; at house at south end of north-south road, 1/2 mile east from Ohio Mills school, 3/4 mile south from Beach Lake. Well No. 9--Milford Quadrangle; well at Wixon village. Well No. 10--Milford Quadrangle; well 1-1/4 miles east from Highland village.] Some of these have outlets to the streams, others have no surface outlets, but the surfaces of nearly all are continuous with the adjacent water-tables. Thus, as distinguished from ordinary aquifers, there is accretion to the ground-water in this region every time it rains, to an extent substantially equal to the amount of rain falling on the lake-surfaces. Thus the ground-water level rises and falls much more gradually than in cases where accretion to the water-table can take place only when and after the field-moisture-deficiency is reduced to zero.

Ground-water records at Horton Hydrologic Laboratory--Figure 8 shows a ground-water-level record covering approximately 20 years at the farm well at the Horton Hydrologic Laboratory. Readings are taken weekly or oftener. About five gallons of water per day are drawn from this well, the amount being kept practically constant. The well is in gravel underlain by Hudson River shale and is about 50 feet from the edge of a mill-pond, in which the water-level is nearly constant except during spring high-water. On a few occasions the pond has been drawn down temporarily, as noted on the diagram. This record affords an excellent example of a well where the maximum level is limited by an auxiliary outlet. When the water in the well rises to a certain elevation, water begins to appear in the ditch at the foot of the slope below the well, as shown on Figure 9. On one or two occasions abnormally high levels have been recorded in this well in the early spring as the result of LaGrange Brook overflowing upstream from the well and flooding part of the area tributary to the well. Aside from this the maximum stage, as limited by the ephemeral outlet, is about five feet below ground.

Voorheesville, New York

Commentary on Paper 3

Hoover, M. D. & Hursh, C. R. (1943) Influence of topography and soil-depth on runoff from forest land.

At the same time that Horton was pursuing his ideas on the dominance of surface runoff in forming the storm hydrograph in New England, Charles R. Hursh was working as Director of the Coweeta Hydrologic Laboratory in North Carolina, USA. At Coweeta there are deeply weathered piedmont soils under a forest cover and surface runoff is seen only rarely on slopes away from the actual channel network. Hursh was also Chairman of the Sub-committee on Subsurface-Flow of the Hydrology Section of the American Geophysical Union. He originated the term subsurface-stormflow (Hursh, 1936) and provided an early review of the role of subsurface-flow in the storm hydrograph in Hursh (1944).

In his 1944 paper, Hursh cites Horton's (1936) paper on maximum water table levels (Paper 2) but also notes the importance of soil horizons in downslope water movement and suggests that "*another factor limiting the rise of the water-table is the tremendous increase in lateral transmission-rate as the water-table approaches the soil-surface*". He therefore anticipates the "transmissivity feedback" concept (see Rohde, 1989; Bishop *et al.*, 1990, 2004). Hursh & Fletcher (1942), in a paper on the soil profile as a natural reservoir, also discuss the importance of the non-capillary pore space in the movement and storage of gravitational water in the soil: "*The exact nature of this macro-pore space occurring in different horizons of the soil profile has yet to be described. It includes all large underground channels formed from decayed roots, fractured rock, insect and animal burrows, and larger spaces that may exist. It also includes macro-pore spaces formed through the complex structural patterns created by the aggregation of soil particles in the presence of organic materials. In the upper horizons of natural soils these biological openings and structural patterns built up from lattice-like aggregates are far more important in determining non-capillary porosity than the single grain particle size. Root channels and animal burrows are of particular significance on the detention storage and draining of gravitational water. A single earthworm burrow may be far more important in draining through a block of heavy soil than the entire cross-sectional area of the pore space. In like manner, it is conceivable that a few continuous void spaces may give rise to rapid discharge of groundwater through a soil profile which, when viewed as a uniformly pervious medium, would be expected to transmit water slowly*" (p.485). As already noted, the importance of macropores in infiltration was also appreciated by Horton (Beven, 2004).

The recognition of the importance of subsurface responses at Coweeta also influenced a later Director of the Hydrologic Laboratory, John D. Hewlett (see Paper 6 and commentary).

Charles Hursh retired from the Coweeta Laboratory in 1954 and died in 1988 (see *http://www.srs.fs.usda.gov/history/hursh_bio.htm*). Summaries of hydrological research at the Coweeta Laboratory can be found in Swank & Crossley (1988) and Swank *et al.* (2001).

Marvin Hoover later worked as Chief of the Division of Watershed Management Research at the Rocky Mountain Forest and Range Experiment Station, Fort Collins, Colorado, and as a member of the graduate faculty of Colorado State University. In the 1970s he retired to Wenatchee, Washington, and he died in 2003 (*http://wa.audubon.org/new/audubon/default.cfm?pageID=163)*

References

Beven, K. J. (2004) Robert Horton's perceptual model of infiltration processes. *Hydrol. Processes* **18**, 3447–3460.

Bishop, K. H., Grip, H. & O'Neill, A. (1990) The origins of acid runoff in a hillslope during storm events. *J. Hydrol.* **116**, 35–61.

Bishop, K. H., Seibert, J., Köhler, S. & Laudon, H. (2004) Resolving the double paradox of rapidly mobilised old water with highly variable responses in runoff chemistry. *Hydrol. Processes* **18**, 185–189.

Hursh, C. R. (1936) Storm water and adsorption. *Trans. Am. Geophys. Un.* **17**, 301–302.

Hursh, C. R. (1944) Surbsurface flow. *Trans. Am. Geophys. Un.* **25**(5), 743–746.

Hursh, C. R. & Fletcher, P. W. (1942) The soil profile as a natural reservoir. *Proc. Soil Sci. Soc. Am.* **7**, 480–486.

Rohde, A. (1989) On the generation of stream runoff in till soils. *Nordic Hydrol.* **20**(1), 1–8.

Swank, W. T. & Crossley, D. A. (eds) (1988) *Forest Hydrology and Ecology at Coweeta.* Ecological Studies vol. 66. Springer-Verlag, New York, USA.

Swank, W. T., Meyer, J. L. & Crossley, D. A. (2001) Long term ecological research: Coweeta history and perspectives. In: *Holistic Science: The Evolution of the Georgia Institute of Ecology (1940–2000)* (ed. by G. W. Barrett & T. L. Barrett), 143–163. Sheridan Books, Ann Arbor, Michigan, USA.

INFLUENCE OF TOPOGRAPHY AND SOIL-DEPTH ON RUNOFF FROM FOREST LAND

M. D. Hoover and C. R. Hursh

Introduction

It has been recognized that topography, size, shape, vegetation, and soil-profile differ for individual drainage-areas. These inherent properties of watersheds together with climatic factors are the principal causes for variations in the hydrologic characteristics of different drainage-areas. These properties account for the differences in unit-tributary contributions to the runoff of the larger streams. Few watersheds are uniform within themselves as regards topography, vegetation, and soil-profile, but small watersheds show less variability in these properties and are, therefore, most useful in the study of contributions of tributaries to runoff of larger streams. They are also more useful in the study of the complete water-cycle, and of the changes within the water-cycle resulting from changes in conditions of vegetation and soil brought about through different land-use practices.

Watershed-studies requiring measurements within the water-cycle are now in progress on the Coweeta Experimental Forest in Western North Carolina. The Forest of 4,200 acres ranges in elevation from 2,200 to 5,200 feet, and contains 30 small drainages with continuous flow. Average annual rainfall ranges from 70 to 80 inches depending on elevation and is distributed throughout the year. Mean annual temperature is 55°F at the elevation of 2,200 feet. Natural growth is composed of deciduous forest with abundant shrubs and minor vegetation. Topography is favorable to the establishment of small independent experimental watersheds.

Results to date indicate that watersheds at the Coweeta Experimental Forest may be roughly classified on the basis of elevation. Watersheds lying largely below an elevation of 3,000 feet are characterized by a soil-profile common to the intermountain areas of the Appalachian Region and Piedmont Plateau. The soil-profile shows deep sub-soil horizons. Shallowest soils occur on the upper slopes, but even here depths of soil of four to six feet are common.

Watersheds above 3,000 feet in elevation are typically more steep. On the slopes, depths of soil are generally less than two feet. However, at the foot of these slopes, and along stream-courses, there is an accumulation of a large amount of angular rock-talus and finer material.

These differences in the soil-profile between the lower and higher elevations markedly influence runoff. Characteristics of storage of the deep soil-profiles found on the Coweeta Experimental Forest have been previously discussed [see 1 of "References" at end of paper]. Under natural forest-conditions, infiltration-values are well in excess of occurring rainfall-intensities. The soil-profiles store large quantities of water both as retained soil-moisture and as ground-water. Stream-flow from such areas is subject to a large degree of regulation. A characteristic winter storage of ground-water takes place which sustains ground-water flow well into the growing season. Because no surface storm-flow other than channel-precipitation occurs, peak-rates are comparatively low.

On the higher watersheds, the shallow soils on the slopes have a lower total storage-capacity which is compensated for in part by the high unit-storage of the talus-deposits. Although the latter generally cover only a small portion of any drainage, they contribute considerable storage-capacity. When the soil on these watersheds is wet to field-capacity, water moves down the slope and is stored in the talus-fill. Outflow from this storage occurs at rapid rates because of steep slopes and large voids in the fill-material. This outflow represents a dynamic form of subsurface storm-water and contributes appreciably to the hydrograph during the course of the storm. It is the form of water that accounts for the high unit-rates of storm-discharge that frequently have been reported from mountain forest land.

Higher rates of storm-discharge may be recorded at gaging-stations on the major streams even though the gages are at lower elevations, but these rates are produced by discharge from the tributaries located at high elevations. These upper tributaries not only receive more rainfall, but have less capacity to retain or store rainfall as it occurs. This is illustrated by the peak-discharges recorded at stations Nos. 8 and 9, which gage the two major streams on the Coweeta Experimental Forest (Fig. 1). Total drainage-areas above these gages are 1,877 and 1,788

Fig. 1--The Coweeta Experimental Forest photographed from 12,000 feet [outside boundaries of the forest and of the drainage-areas discussed, are shown in white (photo by Tennessee Valley Authority)]

acres, respectively. For the general storm of December 27-29, 1942, the peak-discharge at gage No. 8 was 56 second-feet per square mile, and at gage No. 9, 55 second-feet per square mile. Tributary-drainages that are at the lower elevation, produced peaks of only 23-31 second-feet per square mile. Working with a hydrograph produced by this general heavy storm from gage No. 8, it might appear logical to assign the peak to the whole drainage-area. Actually, a proportionally much greater rate of discharge occurred from the tributaries at higher elevations. To assign a high unit-discharge based on the entire drainage above gage No. 8 to the lower watersheds would lead to a very erroneous analysis of the hydrologic characteristics of tributaries within different ranges of elevation.

Presentation of Data

Figure 1 shows the Coweeta Experimental Forest photographed from 12,000 feet. Watersheds considered in this discussion are numbered. Watersheds 37, 28, and 22 are representative of the areas of higher elevation, whereas watersheds 1, 7, 10, and 14 represent areas of lower elevation. Differences in runoff from the two classes of drainages are illustrated by the storm of December 27-29, 1942. Rainfall of 6.78 to 10.85 inches produced a uniform pattern over the entire forest. Intensities were remarkedly uniform with an average of 0.15 inch per hour. At no time did the intensity exceed 0.46 inch per hour. These rates greatly eliminated the factor of intensity as being an important factor in producing surface-runoff insomuch as all of the drainage-areas are in natural forest cover, and have surface infiltration-rates far in excess of the maximum intensities of the storm. Antecedent rainfall had been sufficient to satisfy soil-moisture deficiencies and to cause a general rise of the ground-water level. Data for drainage-area and runoff are shown on Table 1.

Table 1--Drainage-area data, storm of December 27-29, 1942, Coweeta Experimental Forest

No.	Area	Elevation		Precipitation			Runoff			
		Minimum	Maximum	Amt.	Maximum	Average	Total	Peak	Prior	Base mdt. Jan. 2
	acres	ft	ft	in	in/hr	in/hr	in[a]	csm[b]	csm[b]	csm[b]
1	40	2,315	3,241	6.78	0.38	0.13	1.60	22.82	1.15	3.62
7	145	2,370	3,510	6.84	0.38	0.13	1.77	24.31	1.36	4.13
10	212	2,436	3,801	7.52	0.38	0.13	2.20	28.36	1.80	5.87
14	152	2,318	3,125	7.03	0.41	0.13	2.33	31.53	1.87	5.64
8	1877	2,292	5,249	8.40	0.46	0.14	4.22	56.43	2.04	6.81
22	88	2,780	4,089	8.37	0.46	0.16	5.04	68.27	2.96	9.62
28	350	3,162	5,097	10.84	0.46	0.16	5.33	83.34	3.29	10.91
37	100	3,390	5,240	10.85	0.42	0.15	8.78	167.36	3.82	10.44

[a]Above prior base, beginning of storm to midnight of January 2, 1943. [b]Second-feet per square mile.

Figure 2 shows the hydrographs produced by this storm for the areas mentioned, together with precipitation-data at intensity-gage No. 4, situated 3,350 feet above sea-level. Precipitation-data at other intensity-gages follow the same pattern as gage No. 4. It will be seen that the discharges for the watersheds with the deeper soil-profiles are all of the same magnitude, illustrating that they are all from the same type of storage. Peak-rates from such areas range from 22 to 31 second-feet per square mile on the tributary drainage-areas. These rates may be compared with peaks of 67 to 167 second-feet per square mile for the higher watersheds, illustrating high outflow rates from temporary storage in the talus-fills. Although the latter peaks are relatively high, the hydrographs show very definite regulation of runoff by detention-storage. Watershed 37, for which the peak was 167 second-feet per square mile, produced a discharge for the six-hour period--from $18^{h}\,00^{m}$ to midnight of December 29--amounting to 0.72 area-inch at a mean rate of 77.57 second-feet per square mile. This period was well after outflow of channel-precipitation was complete. Runoff for the following day, December 30, amounted to 1.32 area-inches, a mean daily flow of 35.3 second-feet per square inch; for December 31, 0.73 inch, a mean daily flow of 19.7. This amount of water appears to have been detained in the same manner as storm-water in a detention-reservoir. The storage and comparatively rapid release of this amount of water appears to be associated with the talus-deposits on the lower slopes and in the bottoms of the ravines.

Characteristic longitudinal stream-profiles for watersheds at different elevations within the belt of high rainfall of the Southern Appalachian Mountain Region are shown in Figure 3, in which

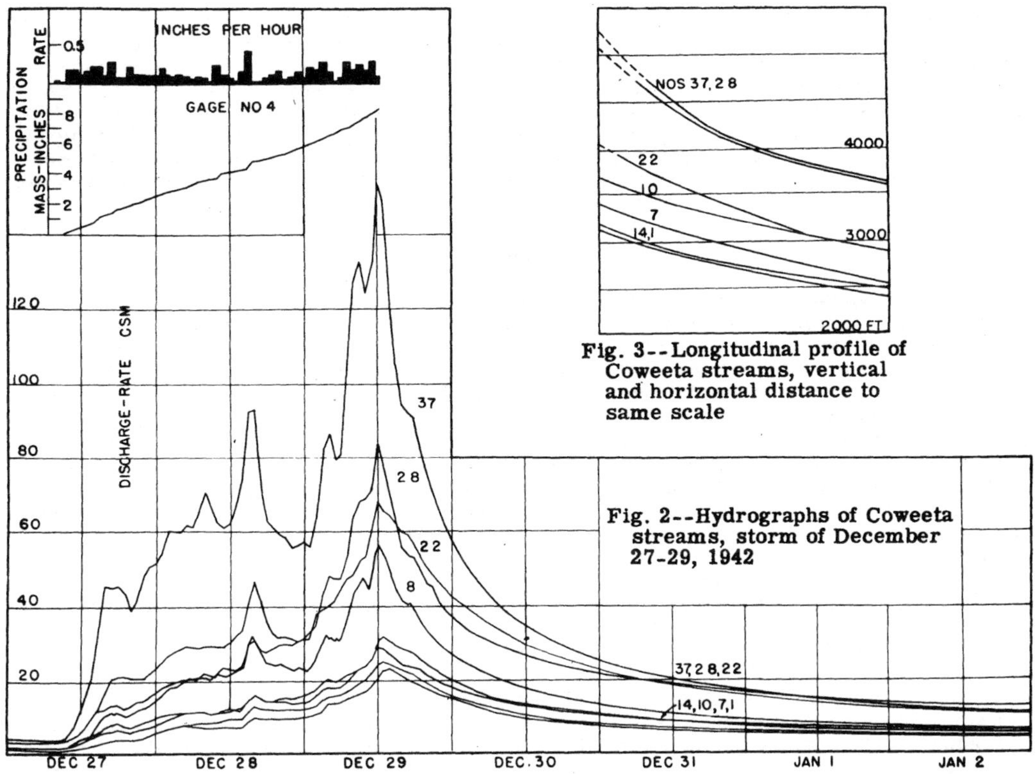

Fig. 3--Longitudinal profile of Coweeta streams, vertical and horizontal distance to same scale

Fig. 2--Hydrographs of Coweeta streams, storm of December 27-29, 1942

the lower stream-reaches for the larger areas are not shown. A very rapid increase in slope takes place in the upper portion of the profile of the higher watersheds. At the base of this slope occur the beds of talus-material that appear to exhibit a definite detention-reservoir effect during storm-periods. The profile of watersheds within the lower ranges of elevation do not exhibit these steep slopes.

In the Southern Appalachians, the amount of rainfall increases with elevation. As a result, watersheds at higher elevations generally have more moisture present and consequently operate with less opportunity for storage than drainages at lower elevations. This may have been an important factor in the storm of December 27-29, but it is not the only factor to be considered. Differences in runoff shown in Figure 2 were apparently also closely related to the soil-profile on the two classes of areas, together with the associated difference in opportunity for storage.

Discussion

By far the greatest amount of land-area in the Southern Appalachian Region is less than 3,000 feet above sea-level. Under natural forest conditions, this land generally contributes only small amounts of storm-runoff to stream-tributaries. When depleted and misused for agriculture and grazing, however, both opportunities for infiltration and storage are greatly reduced, and drainage-areas with these same soil-profiles may produce measured peaks of 800 to 1200 second-feet per square mile for watersheds of 10 to 20 acres.

Most of the land-areas below 3,000 feet in the Southern Appalachian Region are pastured or cultivated, or because of their accessibility have been heavily cut over for wood products, and hence there is very little undisturbed forest-cover. At the higher elevations, however, which are in effect the headwater-areas for the streams, there is much less farming. Particularly the steep slopes have not been used for agriculture because of the shallow soils and difficult cultural technics. In addition, the accumulations of talus, or rock-bars as they are locally called, cannot easily be logged, and consequently trees are more likely to be present, giving the appearance of a heavy forest-cover. Because vegetation has been little disturbed, runoff-characteristics of streams draining such areas have been considered to represent the effects of a forest-cover. However, on many of these headwater-areas, the geologic factors are expressed in the thin soils and deep talus-slopes in the ravines may be far more important than effects of vegetation. Vegetation actually does have a tremendous influence on the thin soil-profiles on such areas, being directly responsible for the organic material which protects the mineral soil on extremely steep slopes. Without this organic development, it is conceivable that no soil whatever would remain on the steep slopes in the Southern Appalachians and that the talus-fills would be rendered less efficient in producing the controlling effect they now exert. On forested watersheds where peak-discharges are now high, the principal consideration is not the present magnitude of the peaks, but rather what would be the actual runoff-rates if no vegetation were present. In making a comparison of the effect of different land-use types upon runoff, it is obviously necessary that the studies under consideration all be carried out on watersheds of the same general hydrologic characteristics with regard to the nature of the soil-profile.

The remarkable similarity of the unit-discharge rates from watersheds at the lower elevation shown by Figure 2 indicates that these watersheds, although differing in size, shape, and aspect, are controlled principally by the uniform soil-profile type and associated natural forest-cover. Soil and topographic conditions similar to those at the lower elevations of the Coweeta Experimental Forest may be found throughout the Southern Appalachian Mountains and Piedmont Plateau. Preliminary observations indicate that wherever the same general conditions of soil and vegetation-cover exist, there occur similar conditions of high infiltration and storage of rainfall during prolonged storms, such as have been recorded at the Coweeta Experimental Forest. This suggests that the results of studies obtained on the Coweeta Forest at an elevation below 3,000 feet may have a wide application to the land-use problems throughout the deep, red-clay soils of the Southeastern and Middle Atlantic States. The results obtained from watersheds above 3,000 feet apply principally to steeper mountain ranges of the Southern Appalachian Region.

Summary

This paper is a report of tributary-contributions from small drainage-areas to the runoff of larger streams within the Coweeta Experimental Forest. Storm-runoff is discussed for the storm of December 27-29, 1942. This storm--6.78 to 10.85 inches--produced the same general rainfall-pattern over the entire forest. Average intensities were quite low--0.15 inch per hour with maximum 0.46 inch per hour. Small tributary drainage-areas--40 to 212 acres in size--with elevation principally below 3,000 feet, produced maximum peaks of 22 to 32 second-feet per square mile. For drainage-areas of similar size, but with elevation well above 3,000 feet, peaks of 68 to 167 second-feet per square mile were recorded. Difference in peaks is assigned in part to greater rainfall at higher levels, but it is also related to soil-depth, topography, and hydrologic characteristics associated with the different elevations.

References

[1] C. R. HURSH and E. F. BRATER, Separating storm-hydrographs from small drainage-areas into surface- and subsurface-flow, Trans. Amer. Geophys. Union, pp. 863-871, 1941.

Appalachian Forest Experiment Station,
Asheville, North Carolina

Commentary on Paper 4

Cappus, P. (1960) Bassin expérimental d'Alrance: études des lois de l'écoulement. Application au calcul et à la prévision des débits.

This is another paper that was, for many years, forgotten by mainstream hydrology, even in France. Because it is in French, it is doubtful if many English speaking hydrologists knew of it at all until the 1990s when Bruno Ambroise noted its significance (Ambroise *et al.*, 1996; Ambroise, 2004). Pierre Cappus was an engineer with La Direction des Études et Recherches de l'Électricité de France (EDF) based at the Centre de Recherches et d'Etudes de Chatou. EDF continues to be one of the major hydrological research organizations in France because of the major investments in hydrolelectric power (known as La Houille Blanche, literally "white oil", the title of the journal in which this paper appears). The interest of EDF is primarily in flow forecasting, including the formation and melt of snow packs in the Alps and flow routing in major rivers, and the yield of catchment areas for dam design. The study of Cappus was carried out within this last theme, looking at the variation of runoff coefficients for a small experimental catchment (3.15 km^2, Fig. 1), in the Alrance basin in the Massif Central of France, in the Department of Aveyron. Cappus also worked on the spatial variability of rainfall (Cappus, 1958). The basin had an annual average rainfall of 1200 mm with about 15% provided in the form of snow. The catchment was operated, starting in 1950, for more than 20 years (Dosseur & Guilbot, 1974).

The data are handled at the catchment scale. The way in which daily recession coefficients (*coefficient journalier de décroissance)* vary with season and temperature is demonstrated (Figs 2 and 3). A form of hydrograph separation is carried out (Fig. 4) and definitions for storm flow (*ruissellement*) and subsurface baseflow (*l'écoulement souterrain*) are given (p.498), though Cappus notes that the separation is necessarily somewhat arbitrary. The main part of the paper is then concerned with the analysis of the variation of the runoff volumes and runoff coefficients. The important section is his "New theory of runoff" which starts on p.503, where Cappus differentiates between a "runoff area" (*zone de ruissellement*) and an "infiltration area". He defines the runoff area as consisting of impermeable areas (roads, paved tracks, soils compacted due to the effects of man or stock, rock surfaces, etc.) but also areas of permeable soil that are saturated. He notes that any rainfall falling on these areas will become runoff, either by surface routes (*ruissellement superficiel*) or near-surface pathways (*ruissellement hypodermique*).

Cappus also notes (p.504) that the runoff area should be expected to vary with the mean depth of the water table in the catchment (see Fig. 6) but, since the water table responds only slowly during any particular rainfall event, a good approximation to the runoff area might be obtained as a function of the initial groundwater discharge (it remains the best indicator of the antecedent state of a catchment to this day). In addition, Cappus notes that, while a traditional approach to runoff prediction based on infiltration theory might be applicable locally at a point, it is not applicable at the basin scale because the degree of saturation of the soil varies greatly in the basin. He suggests that infiltration rates on the infiltration area in this catchment are normally sufficient to absorb all the incident rainfall in all but rare storms, and that the runoff coefficient is much more dependent on the contributing area than on the rainfall intensity. Thus, runoff generation is almost a matter of all or nothing (*tout ou rien*) depending on whether the soil is saturated or not (there is some similarity here with the suggestion of Horton (1936, see Paper 2), on what happens when the maximum water table intersects the surface). This is, therefore, one of the first statements of a dynamic contributing area concept of streamflow generation (see also Paper 15).

Cappus also differentiates between baseflow (*le débit souterrain de base*), that would have happened in the absence of a storm, and subsurface stormflow (*le débit souterrain de crue*) (Fig. 10 and p.509). Analysis of the surface runoff and subsurface stormflow for some 30 storms suggested that the surface runoff contributing area varies linearly with initial discharge, while the total runoff coefficient increases nonlinearly to near 100% for an initial baseflow of 50 L s^{-1} km^{-2} in the absence of evaporation (Table 3, p.512). In his conclusions to his study of subsurface stormflow, Cappus notes that the subsurface stormflow discharge seems to be independent of the retention of water by the vegetation cover and (something he finds surprising) independent of the water content in the unsaturated zone ("*dans les couches superficielles qui ne sont pas atteintes par la nappe*", p.513).

A second part of the paper (not reproduced here) is concerned with the calculation and forecasting of discharges (pp.515–518). This is then followed by a report of a discussion session (pp.518–520) that contains some interesting and very modern comments on the problems of hydrograph separation, finding functional representations of processes and the generality of forms of analysis. Jean Rodier (later President of IAHS, 1971–1975) briefly reports a study on a basin also permeable and heterogeneous but where the water table is generally at a depth of 10–12 m, where baseflow discharges might be zero, and where storms were often more intense. He had found that, in these conditions, rainfall intensity was an important control on runoff generation. Jean Tixeront (later President of the Continental Erosion Commission of IAHS) notes the difficulty and rather arbitrary separation of subsurface stormflow from surface runoff. He also notes, however, personal experience on a basin underlain by mica-schists similar to the Alrance, that the depth to the water table is very variable and may consist of "*pseudo-nappes*" in a depth of soil and subsoil (perched water tables rather than true groundwater). Depth to the water table will generally be smaller in the valley bottoms and it is here that the water table responds more quickly to form saturated areas.

Pierre Guillot (later Director of Hydrometeorology at Électricité de France) raised the problem of separating successive events. M. Ferry notes the difficulties of trying to find hydrological laws and suggests that stochastic relationships might be more useful, especially in basins without the monitoring available in the Alrance. He also notes the discussion in the Commission for the Study of Flood Discharges that in using unit hydrograph predictions, estimating the volume of runoff is more important than the form of the unit hydrograph, for which a triangle might suffice (which predates similar discussions on the use of the geomorphological unit hydrograph a couple of decades later; see Cordova & Rodriguez-Iturbe, 1983, Beven, 1986). Finally, there was some discussion as to whether the runoff coefficient relationships provided by Snyder in the USA might be applicable for these types of basin in France, and whether well observations would be necessary to apply the methods elsewhere.

Pierre Cappus left the field of hydrology in 1963, though he continued to work for EDF. He died in 2002. Later measurements on the Alrance were carried out by staff of the Centre de Bassins experimentaux du Massif Central, directed from 1966 to 1970 by Hubert Dosseur who had produced his PhD thesis on the Alrance (Dosseur, 1964). The centre was closed in 1970.

References

Ambroise, B. (2004) Variable 'active' versus 'contributing' areas or periods: a necessary distinction. *Hydrol. Processes* **18**, 1149–1155.

Ambroise B., Beven, K. J. & Freer, J. (1996) Towards a generalisation of the TOPMODEL concepts: topographic indices of hydrological similarity. *Water Resources Res.* **32**(7), 2135–2145.

Beven, K. J. (1986) Runoff production and flood frequency in catchments of order *n*: an alternative approach. In: *Scale Problems in Hydrology* (ed. by V. K. Gupta, I. Rodriguez-Iturbe & E. F. Wood), 107–131. Reidel, Dordrecht, The Netherlands.

Cappus, P. (1958) Répartition des précipitations sur un bassin versant de faible superficie. In: *Land Erosion, Instruments, Precipitation* (vol. 1, Proc. IAHS General Assembly, Toronto, September 1957), 515–528. IAHS Publ. 43. IAHS Press, Wallingford, UK.

Cordova, J. R. & Rodriguez-Iturbe, I. (1983) Geomorphoclimatic estimation of extreme flow probabilities. *J. Hydrol.* **65**, 159–173.

Dosseur, H. (1964) Bassin expérimental d'Alrance: contribution à la définition des charactéristiques d'état du bassin pour la prévision hydrologique. PhD Thesis. Faculté des Sciences de Montpellier, Montpellier, France.

Dosseur, H. & Guilbot, A. (1974) Bassin expérimental d'Alrance. *Cahiers ORSTOM, Série Hydrologie* **11**(2), 127–132.

BASSIN EXPÉRIMENTAL D'ALRANCE

Étude des lois de l'écoulement Application au calcul et à la prévision des débits

THE ALRANCE EXPERIMENTAL BASIN

Investigation of the laws of flow Application to the computation and prediction of discharge

PAR P. CAPPUS

INGÉNIEUR A LA DIRECTION DES ÉTUDES ET RECHERCHES D'ÉLECTRICITÉ DE FRANCE

I. Etude du tarissement : forme des courbes de tarissement, influence de l'évapotranspiration.
Etude des crues : le ruissellement.

- *volume du ruissellement: variations du coefficient de ruissellement avec le niveau moyen de la nappe phréatique, la valeur du débit souterrain, l'humidité des couches superficielles, le montant de la pluie, l'intensité de la pluie, le développement de la végétation, l'évapotranspiration; nouvelle théorie du ruissellement;*
- *forme de l'hydrogramme de ruissellement : variations des éléments caractéristiques de la forme de l'hydrogramme avec la durée de la pluie; comparaison des résultats expérimentaux et des résultats obtenus par application de la méthode de l'hydrogramme unitaire.*
- *l'écoulement souterrain de crue : maximum du débit souterrain de crue, décroissance du débit souterrain de crue; coefficient d'écoulement souterrain; nouvelle théorie de l'écoulement.*

II. Application au calcul et à la prévision des débits instantanés, des débits moyens journaliers et des débits moyens mensuels. Précision des calculs.
Extension des résultats obtenus à d'autres bassins versants.

I. Base flow recession: shape of base flow recession curves, the effect of evapotranspiration.

Floods: Runoff:

- *the amount of runoff: variation of the runoff factor with the mean level of the water table, with ground water discharge, with soil humidity, with the amount of rainfall, with the growth of vegetation and with evapotranspiration; a new runoff theory.*
- *the shape of the runoff hydrograph: variation of the shape of hydrograph components with the duration of rainfall; comparison of experimental results with results of unit hydrograph analysis.*
- *Ground water storm flow:*
  — ground water peak flow, recession of ground water flow; ground water flow factor; new flow theory.

II. Application to the computation and prediction of instantaneous discharge, mean daily discharge and mean monthly discharge. Accuracy of computations.
Extension of results to other river basins.

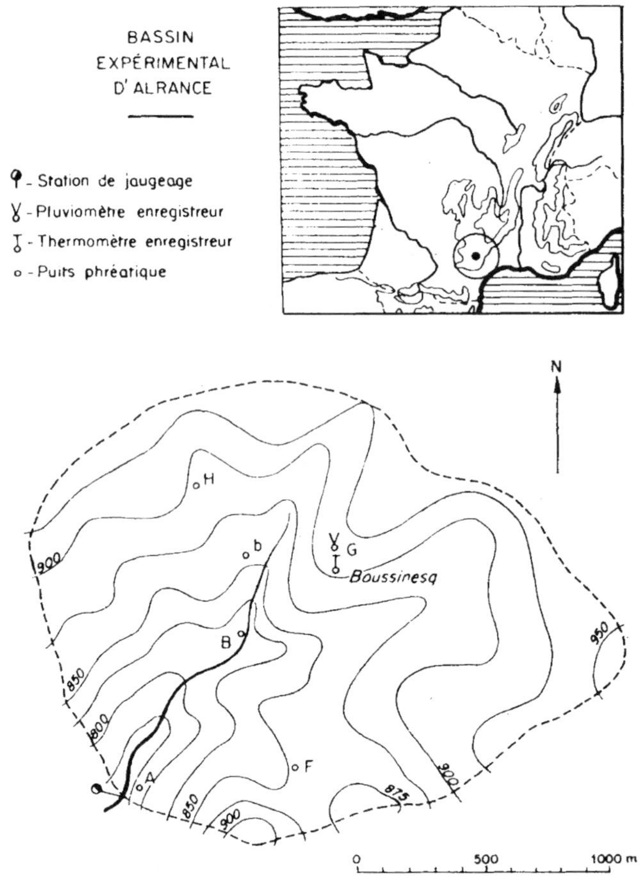

FIG. 1

Jusqu'à présent, les hydrologues ont surtout étudié les facteurs de l'écoulement ou au contraire procédé à des études directes des débits qui font très peu intervenir les facteurs de l'écoulement.

Aussi connaît-on très mal les « lois de l'écoulement » — relations entre l'écoulement et ses divers facteurs — alors que ces relations traduisent vraiment les phénomènes de l'écoulement et devraient par conséquent constituer l'objet essentiel des études hydrologiques. Et de ce fait, nous ne connaissons pas de cas où l'on ait pu apporter une solution au problème du calcul des débits d'un cours d'eau à partir de données météorologiques.

Nous avons résolu ce problème dans le cas particulier du Bassin expérimental d'Alrance : nous avons montré que l'écoulement de ce bassin obéit à quelques lois simples qui rendent compte cependant de l'influence de tous les facteurs possibles de l'écoulement sur les caractéristiques de l'hydrogramme et permettent de calculer les débits de ce bassin avec une bonne précision en ne faisant intervenir qu'un nombre réduit de facteurs.

La présente étude a pour objet, d'abord d'exposer les résultats obtenus dans l'étude des lois de l'écoulement du Bassin expérimental, ensuite de proposer une explication de ces résultats et enfin d'examiner dans quelle mesure et à quelles conditions ils peuvent rester valables sur d'autres bassins versants.

Le Bassin expérimental d'Alrance (latitude : $49^G04'$; longitude $0^G40'$ Est) est constitué par le bassin supérieur de la Saignerie, petit affluent de la rivière Alrance qui se jette elle-même dans le Tarn. Il s'étend sur 3,15 km² de terrain vallonné dont l'altitude varie de 770 à 960 mètres (voir carte de la fig. 1); l'altitude moyenne est d'environ 875 mètres et la pente moyenne de l'ordre de 10 à 15 %. Le terrain est constitué de schistes imperméables décomposés en surface; la couche de décomposition, épaisse en général de moins de 2 mètres, est très perméable. Le bassin est couvert de cultures variées, de prairies, de landes et de quelques bois (10 % de la superficie totale). Malgré sa latitude relativement méridionale, le Bassin expérimental est soumis à un climat tempéré assez froid en raison de son altitude (température moyenne : 8 °C). Le régime des pluies est nettement océanique, mais subit cependant quelques influences méditerranéennes. Le total annuel des précipitations est en moyenne de l'ordre de 1 200 mm, dont 15 % en moyenne sous forme de neige.

PREMIÈRE PARTIE

LES LOIS DE L'ÉCOULEMENT

Nous distinguons :

— *Les périodes de tarissement,* pendant lesquelles le débit décroît de façon très lente et régulière (chapitre I);

— *Les périodes de crue,* pendant lesquelles l'hydrogramme des débits présente des intumescences de formes diverses (chapitre II).

Le début de la période de crue est évidemment le moment où le débit commence à croître au lieu de continuer à décroître comme en période de tarissement; il est en général très net. La fin de la période de crue est, par définition, le moment où la décrue reprend l'allure très lente qui caractérise les périodes de tarissement : comme la pente de l'hydrogramme de décrue varie de façon très progressive, l'instant de la fin de la période de crue ne peut pas être déterminé exactement.

CHAPITRE I

LE TARISSEMENT

Les hydrogrammes de tarissement du Bassin expérimental d'Alrance, comme ceux de la plupart des bassins versants, ont l'allure d'une courbe exponentielle asymptote à l'axe des temps : il est donc normal d'étudier le rapport des débits de tarissement mesurés à intervalles réguliers, rapport qui serait constant dans le cas d'une exponentielle. On remarque aussi que le débit de tarissement diminue plus vite en saison chaude qu'en saison froide, de toute évidence à cause d'une évapotranspiration plus intense; on étudiera donc l'influence, sur l'hydrogramme de tarissement, du pouvoir évaporant de l'atmosphère, repéré par la température, et du développement de la végétation.

a) Coefficient journalier D_1 de décroissance du débit de tarissement.

$$D_1 = \frac{Q_{j+1}}{Q_j}$$

(Q_j et Q_{j+1} : débits de tarissement mesurés à 0 h les jours j et $j+1$).

1. CAS OU $Q_j < 30$ l/s/km² (condition réalisée en moyenne pendant plus de 350 jours par an) :

L'examen de 334 valeurs de D_1 (toutes les saisons des années 1951 à 1957) montre que :

— D_1 *est fonction décroissante de la température moyenne* θ_j *du jour j.*

— D_1 *est indépendant de la valeur* Q_j *du débit de tarissement* (fig. 2) (la courbe de tarissement est donc, à température constante, une exponentielle asymptote à l'axe des temps).

— D_1 *est indépendant de la saison* (autrement dit du développement de la végétation) (fig. 3).

D_1 est d'environ 0,970 pour $\theta_j = 0°$
D_1 — 0,965 pour $\theta_j = 5°$
D^1 — 0,955 pour $\theta_j = 10°$
D_1 — 0,930 pour $\theta_j = 15°$
D_1 — 0,890 pour $\theta_j = 20°$
D_1 — 0,825 pour $\theta_j = 25°$

2. CAS OU $Q_j > 40$ l/s/km² (condition réalisée en moyenne pendant moins de 15 jours par an).

Pendant les années 1951 à 1957, nous n'avons pas observé de période de tarissement correspondant à des débits supérieurs à 40 l/s/km² : en effet, bien que les plus forts débits souterrains observés atteignent 200 l/s/km², la décroissance du débit souterrain ne devient pas régulière avant qu'il ait repris des valeurs de l'ordre de 40 l/s/km².

b) Interprétation.

Il est facile d'expliquer qu'en l'absence d'évaporation, le coefficient D_1 soit à peu près indépendant de la valeur du débit de tarissement, au moins dans un certain intervalle de variation de ce débit : la courbe de tarissement est alors une exponentielle asymptote à l'axe des temps et le volume de la réserve phréatique est proportionnel au débit de tarissement (loi de Maillet).

D'autre part, il est normal que, toutes choses égales par ailleurs, la décroissance du débit de tarissement soit d'autant plus rapide que la température est plus élevée.

Sur le Bassin expérimental, la majeure partie de l'évaporation des eaux de la nappe se fait à partir des zones où la nappe affleure à la surface du sol et où, par conséquent, le terrain est saturé. L'évaporation d'un terrain saturé, comme celle d'une surface d'eau libre, est très sensible aux variations du pouvoir évaporant de l'atmosphère : on comprend ainsi que les variations du coefficient D_1 suivent au jour le jour celles de

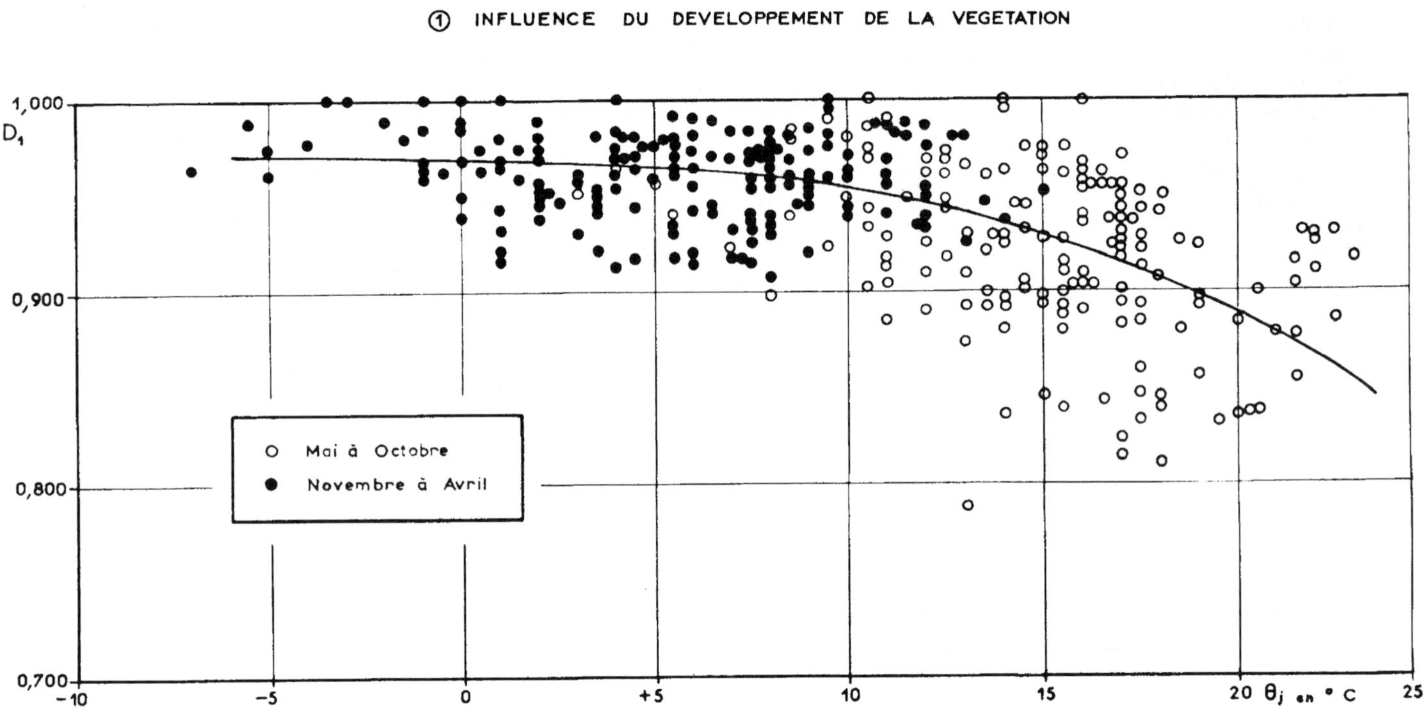

FIG. 2

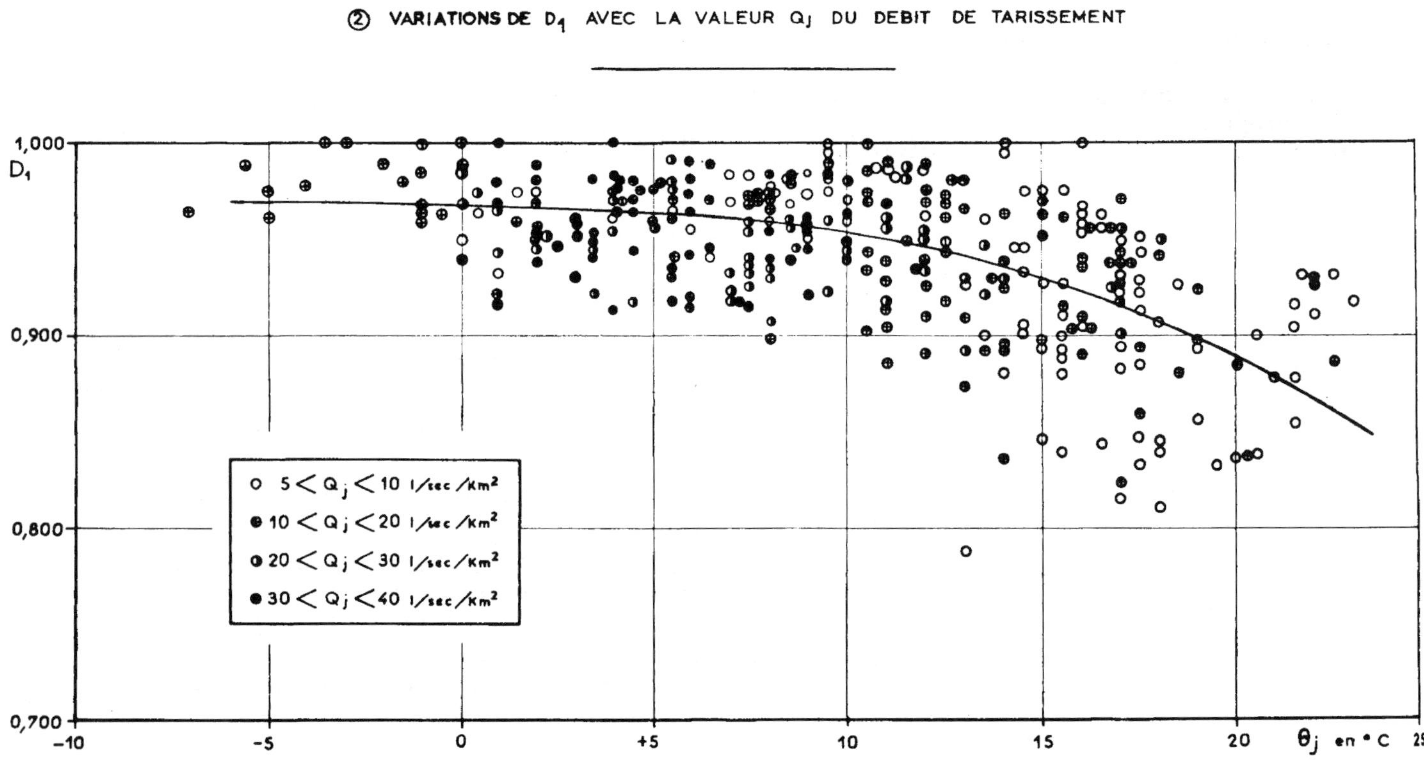

FIG. 3

$D_1 = (Q_j + 1)/Q_j$. Coefficient journalier de décroissance du débit de tarissement en fonction de Θ_j température moyenne journalière.

($Q_j < 40$ l/s/km²)

la température (en saison chaude, on distingue même nettement une diminution plus rapide du débit de tarissement aux heures les plus chaudes de la journée). On sait aussi que l'évapotranspiration à partir d'un terrain saturé varie assez peu avec le développement de la végétation : d'où l'indépendance de D_1 vis-à-vis de la saison.

Nous verrons ultérieurement que, sur le Bassin expérimental d'Alrance, l'étendue des zones saturées où affleure la nappe varie avec le niveau moyen de la nappe sur le bassin, ou, ce qui revient au même, avec le débit de tarissement; nous verrons même que la superficie de ces zones saturées est, en première approximation, proportionnelle au débit de tarissement : d'où, pour une température donnée, une diminution du débit de tarissement proportionnelle au débit de tarissement et par conséquent un coefficient D_1 indépendant du débit de tarissement.

CHAPITRE II

LES CRUES

A) FORME DES HYDROGRAMMES DE CRUE. DÉFINITION ET SÉPARATION DES DIFFÉRENTS GENRES D'ÉCOULEMENTS. — NOTATIONS

Forme des hydrogrammes de crue.

Nous avons convenu d'appeler « crues » les intumescences que présente la courbe des débits pendant les périodes pluvieuses; ces intumescences ont des formes et des dimensions variées; mais on découvre aisément entre elles des analogies de forme : la forme la plus typique est celle de l'hydrogramme de la crue produite par une pluie uniforme; dans le cas général, l'hydrogramme peut être décomposé en hydrogrammes partiels correspondant chacun à une pluie à peu près uniforme.

Dans tous les cas on appelle « *débit de base* » le débit que l'on aurait observé si la crue considérée n'avait pas eu lieu et nous appelons « *débit de crue* » le débit qui est en réalité venu s'ajouter au débit de base.

Dans le cas d'une pluie uniforme, l'hydrogramme de crue comprend (fig. 4) :

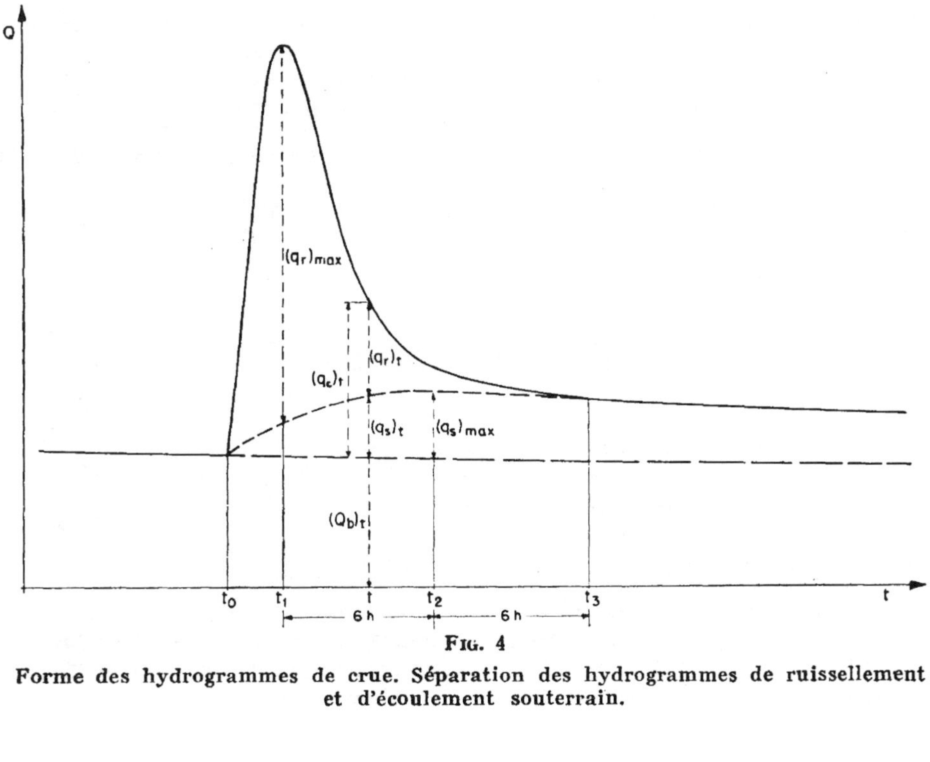

FIG. 4

Forme des hydrogrammes de crue. Séparation des hydrogrammes de ruissellement et d'écoulement souterrain.

1. Une courbe décroissante, à peu près rectiligne, et de très forte pente, que, suivant l'usage, nous appelons « *courbe de concentration* » et qui vient s'articuler brusquement à la courbe des débits de base dès le début de la pluie;
2. Une pointe plus ou moins arrondie, dite « *pointe de crue* », qui se situe approximativement à la fin de la pluie;
3. Une courbe décroissante, dite « *courbe de décrue* », dont la pente présente les variations suivantes :
   - Pendant les trois au quatre premières heures qui suivent la pointe de crue, la pente est très forte, mais cependant inférieure à celle de la courbe de concentration;
   - Entre six et douze heures après la pointe de crue, la pente subit une variation assez brusque et devient nettement moins forte;
   - Environ douze heures après la pointe de crue, la pente a repris une valeur relativement faible, un peu plus forte cependant que la pente de la courbe des débits de base au même moment;
   - Ensuite la pente ne varie plus beaucoup, mais jusqu'à la fin de la période de la crue, elle reste constamment un peu plus forte que celle de la courbe des débits de base, de sorte que l'hydrogramme réel se rapproche progressivement de l'hydrogramme de base.

Définition des différents genres d'écoulement. Séparation des hydrogrammes de ruissellement et d'écoulement souterrain.

a) LE RUISSELLEMENT :

La variation relativement brusque de la pente de la courbe de décrue, six à douze heures après la pointe de crue, correspond sans doute à la fin d'un écoulement rapide qui commence dès le début de la crue et que nous convenons d'appeler « ruissellement ».

Le ruissellement ainsi défini comprend évidemment le ruissellement superficiel mais aussi, vu les temps de parcours des eaux qui ruissellent en surface, le ruissellement hypodermique : aucune particularité de l'hydrogramme ne permet de séparer ces deux types de ruissellement dans le cas du Bassin expérimental d'Alrance.

b) L'ÉCOULEMENT SOUTERRAIN :

On n'observe pas, sur le Bassin expérimental d'Alrance, de crues sans ruissellement qui donneraient directement les caractéristiques des hydrogrammes de « crue souterraine ». Nous sommes donc obligés *de fixer de façon assez arbitraire certaines caractéristiques du tracé de l'hydrogramme souterrain :*

- *Début coïncidant avec le début de la crue* (certaines parties de la nappe sont en effet en rapport direct avec le cours d'eau);
- *Maximum six heures après la pointe de crue* (il est logique de supposer qu'il existe, pour l'écoulement souterrain comme pour le ruissellement, un décalage à peu près constant entre la fin de la pluie et le maximum du débit; nous *admettons* que ce décalage est de six heures pour l'écoulement souterrain : cette hypothèse sera vérifiée *a posteriori* par le fait qu'elle conduit à des résultats satisfaisants et cohérents aussi bien dans l'étude de l'écoulement souterrain que dans celle du ruissellement);
- *Contact avec l'hydrogramme total environ douze heures après la pointe de crue* (c'est le moment où le ruissellement se termine complètement).

Théoriquement, ces trois conditions ne définissent le tracé de l'hydrogramme souterrain que de façon approchée; mais cette approximation s'est révélée suffisante pour notre étude.

B) LES CRUES ÉTUDIÉES

Caractéristiques générales.

Notre étude porte sur *226 crues* assez bien réparties entre le *1er mai 1951 et le 31 octobre 1957*. Ces 226 crues représentent la totalité des crues survenues durant cette période de six ans et demi, à l'exception seulement :

- Des crues de fonte de neige;
- Des crues inférieures à 5 l/s/km² (ces crues sont négligeables ou bien on peut admettre qu'elles suivent les mêmes lois que les crues plus fortes).

Débits de pointe.

La plupart des crues étudiées sont petites ou moyennes ($Q_{max} < 500$ l/s/km²); quelques crues sont cependant assez fortes (500 à 1 000 l/s/km$_2$) (tableau I).

TABLEAU I

Répartition des crues étudiées d'après la valeur Q_{max} du débit de pointe total ou d'après le montant P de la pluie ou d'après la valeur initiale $(Q_s)_{t_0}$ du débit souterrain

Q_{max} en l/s/km²	*n*	P en mm	*n*	$(Q_s)_{t_0}$ en l/s/km²	*n*
5 à 100 (1)	172	2 à 10 (2)	106	2,5 DC E	10
100 à 200	24	10 à 20	84	(355 j)	
200 à 300	17	20 à 30	23	4 DC 11	9
300 à 400	7	30 à 40	8		23
400 à 500	3	40 à 50	4	7 DC 9	
500 à 600	1	50 à 60	1		35
600 à 700	1			14 DC 6	54
700 à 800				25 DC 3	
800 à 900				45 DC 1	40
900 à 1 000	1				21
(1) 5 à 20	6			80 DC M	
				(10 j)	18
10 à 20	34			150 DC	
				(1 j)	16
20 à 50	82	(2) 2 à 5	35		
50 à 100	50	5 à 10	71		

D'après les résultats obtenus sur d'autres bassins versants du Massif Central, on peut évaluer très approximativement les débits de pointe des crues quinquennale et décennale du Bassin expérimental à 700 et 900 l/s/km².

Ainsi, *si l'on cherche suivant l'usage à évaluer la probabilité des crues par la seule considération de leur débit de pointe, on peut dire qu'en gros l'échantillon des 226 crues étudiées est surtout représentatif des crues inférieures à la crue quinquennale et fournit des indications sur les crues comprises entre la crue quinquennale et la crue décennale.*

AUTRES CARACTÉRISTIQUES :

Nous montrerons ultérieurement que les caractéristiques d'une crue ne dépendent que des caractéristiques de la pluie et de l'état initial du terrain; nous montrerons aussi que l'état du terrain peut être caractérisé, du point de vue de la production des crues, par le niveau moyen de la nappe phréatique sur le bassin versant, ou, ce qui revient au même, par la valeur du débit souterrain Q_s.

Nous avons vérifié (tableau I) :

— Que, compte tenu de leur durée, les pluies envisagées représentent toutes les précipitations possibles sur le Bassin expérimental jusqu'aux averses de fréquence à peu près décennale;

— Que les états initiaux du terrain envisagés représentent tous les états d'humidité possibles, à l'exception seulement d'états extrêmes de très faible probabilité.

Nous avons d'ailleurs vérifié que les valeurs envisagées des autres facteurs éventuels de l'écoulement couvrent pratiquement tout l'intervalle des variations possibles de ces facteurs en période de crue d'origine pluvieuse.

C) RUISSELLEMENT

I. — Volume du ruissellement.

Ordre de grandeur du volume de ruissellement :

La lame d'eau ruisselée (ruissellement superficiel + ruissellement hypodermique) est en général très faible :

— Inférieure à 1 mm en période de basses et moyennes eaux, même pour des précipitations importantes, de l'ordre par exemple de 30 à 40 mm;

— Atteignant seulement 5 à 10 mm pour les plus grandes crues étudiées, qui sont des crues de fréquence à peu près décennale.

Méthode d'étude du ruissellement. Théories du ruissellement :

La « théorie classique » du ruissellement veut que le ruissellement instantané sur un bassin versant homogène soit la différence entre l'intensité de la pluie et la capacité d'infiltration f dans le terrain des couches superficielles, f étant variable avec l'humidité de ces couches.

L'application de cette théorie au Bassin expérimental d'Alrance n'a donné aucun résultat.

Par contre, nous obtenons des résultats intéressants en étudiant le coefficient de ruissellement et nous proposons une nouvelle théorie du ruissellement qui permet d'expliquer ces résultats.

Etude du coefficient de ruissellement :

$$C_r = \frac{V_r}{P}$$

(V_r, volume du ruissellement, et P, montant de la pluie, exprimés en mm de hauteur d'eau).

a) ORDRE DE GRANDEUR DU COEFFICIENT DE RUISSELLEMENT :

Le coefficient de ruissellement (superficiel+hypodermique) est en général très faible :

— En moyenne (d'après résultats du paragraphe *b* 1 ci-dessus) :

Pendant :

6 mois par an, C_r est inférieur à 2,5 %

9 mois par an, C_r est inférieur à 5 %

11 mois par an, C_r est inférieur à 10 %

364 jours par an, C_r est inférieur à 20 %

— Pour des crues à peu près décennales, C_r atteint à peine 30 %.

b) INFLUENCE SUR C_r DU NIVEAU DE LA NAPPE PHRÉATIQUE :

Le coefficient de ruissellement doit *a priori* dépendre de nombreux facteurs — humidité du sol, caractéristiques de la pluie, état de la végétation, évaporation — et même de toutes les valeurs prises par certains de ces facteurs pendant la durée du ruissellement.

La faiblesse du coefficient de ruissellement prouve que la pluie tombant sur le bassin est presque entièrement absorbée par le terrain, retenue par la végétation ou évaporée pendant la durée du ruissellement; nous verrons ultérieurement que la rétention par la végétation et l'évaporation des eaux ruisselées sont pratiquement négligeables sur le Bassin expérimental : le ruissellement y est donc avant tout fonction de l'infiltration, qui est elle-même fonction de l'humidité du sol dans toutes les couches où elle se produit.

Or, sur la majeure partie du Bassin expérimental, la nappe phréatique est très voisine de la surface du sol (la distance de la nappe à la surface du sol varie en général de 0 à 2 m, sauf dans quelques poches peu étendues où elle peut atteindre plusieurs mètres); l'humidité des couches superficielles où se fait l'infiltration est donc surtout fonction du niveau de la nappe.

On peut ainsi prévoir que *l'infiltration, et par conséquent le ruissellement, doivent être fonction principalement du niveau de la nappe et accessoirement de l'humidité du sol dans les couches superficielles qui ne sont pas atteintes par la nappe.*

Le niveau moyen de la nappe sur l'ensemble du Bassin expérimental peut être représenté par l'un des deux paramètres suivants :

— La moyenne des niveaux mesurés dans plusieurs puits;

— Le débit souterrain (on constate en effet expérimentalement que le débit souterrain et le niveau moyen de la nappe sont assez étroitement liés; ce résultat tient au fait que la répartition des pluies sur le bassin est relativement peu variable, de sorte que la surface de la nappe s'élève ou s'abaisse à peu près toujours de la même façon).

1. VARIATIONS DE C_r EN FONCTION DU DÉBIT SOUTERRAIN $(Q_s)_{t_0}$ MESURÉ A L'INSTANT INITIAL DE LA CRUE :

Le coefficient de ruissellement C_r relatif à une pluie P dépend *a priori* de toutes les valeurs prises par le débit souterrain pendant la durée de la pluie. Dans un but de simplification, nous étudions les variations de C_r *en fonction de la seule valeur* $(Q_s)_{t_0}$ *prise par le débit souterrain au début de la crue* — qui, nous le verrons, coïncide pratiquement avec le début de la pluie (il est bien entendu que le débit souterrain croît ensuite pendant la durée de la pluie et que, par conséquent, C_r doit dépendre dans une certaine mesure de cet accroissement qui est lui-même fonction de divers autres facteurs).

On constate (fig. 5 et 5 *bis*) que pour une valeur donnée de $(Q_s)_{t_0}$, le coefficient C_r est assez peu variable, alors que tous les autres facteurs dont il pourrait *a priori* dépendre — humidité du sol dans les couches superficielles non saturées, importance et intensité des précipitations, état de la végétation, évaporation — varient, comme nous le verrons aux paragraphes *c*) et *d*) ci-dessous, dans de très larges intervalles.

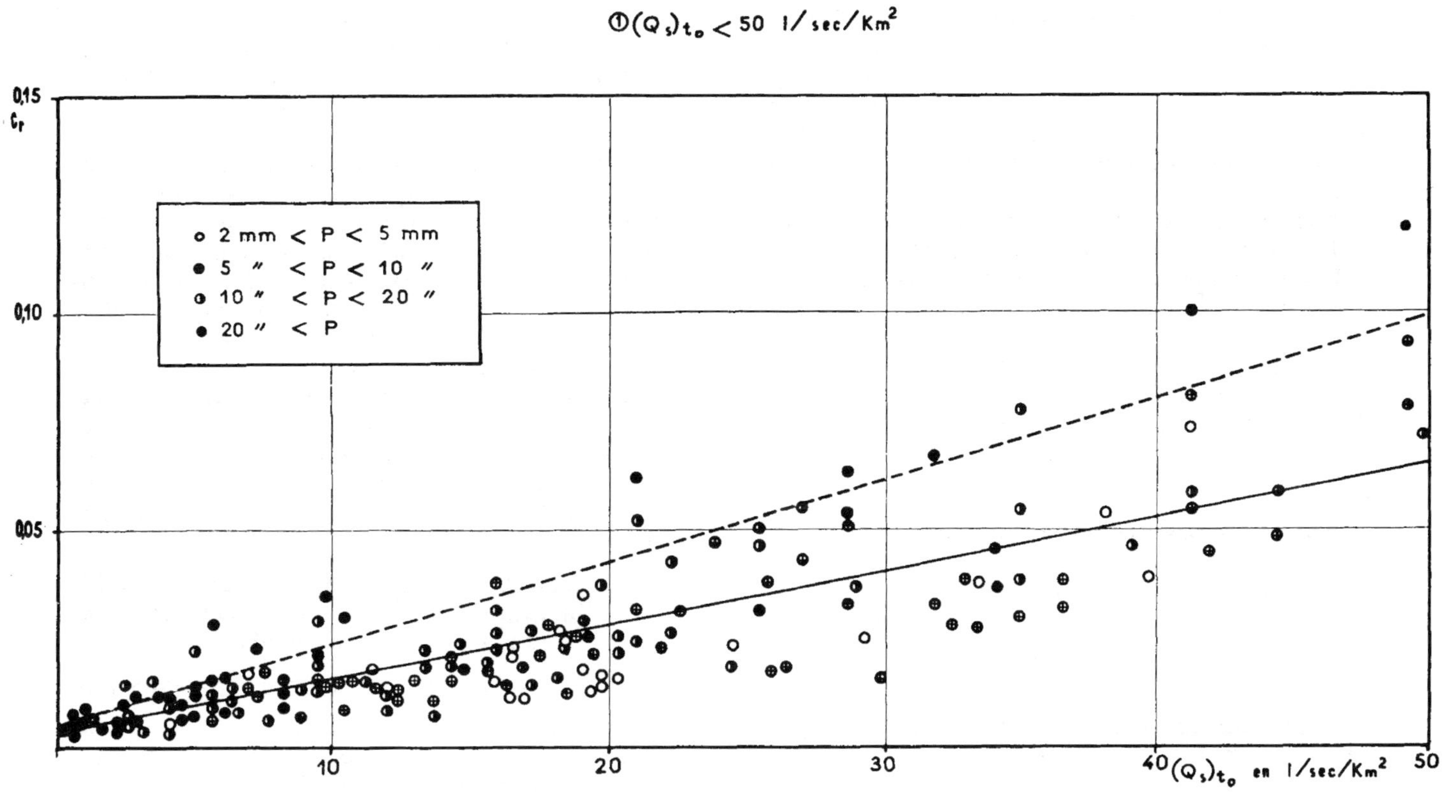

FIG. 5

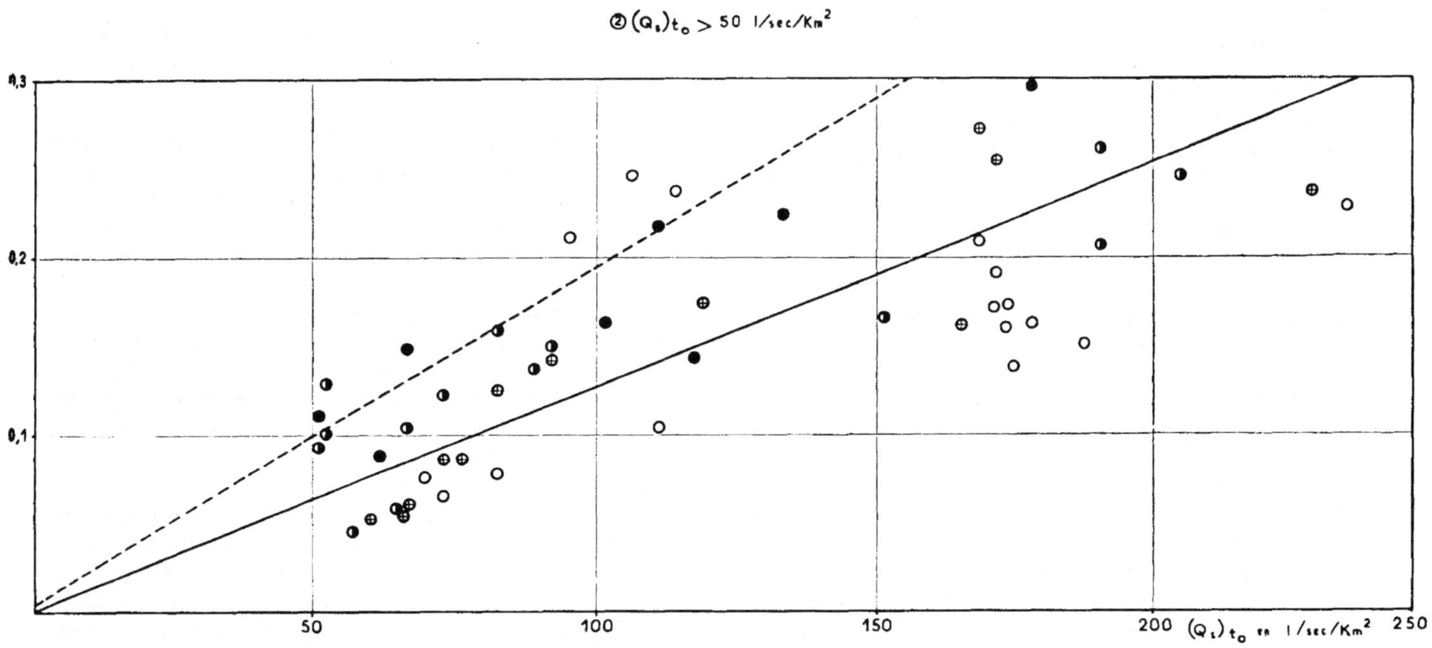

FIG. 5 b

C_r coefficient de ruissellement, en fonction de $(Q_s)_{t_0}$ débit souterrain observé immédiatement avant la crue. — Influence de l'importance des précipitations P.

Autrement dit, *le coefficient de ruissellement C_r relatif à une crue est essentiellement fonction du débit souterrain $(Q_s)_{t_0}$ mesuré à l'instant initial de la crue.*

D'autre part, comme on pouvait le prévoir, C_r *est fonction croissante de* $(Q_s)_{t_0}$.

2. Variations de C_r en fonction du niveau moyen de la nappe phréatique a l'instant initial de la crue :

Nous repérons le niveau moyen de la nappe sur le bassin versant par la moyenne H des distances entre surface du sol et surface libre de la nappe dans les cinq puits A, B, F, H et *b* (voir carte de la fig. 1).

L'examen de la figure 6 conduit à deux conclusions qui, sous des formes différentes, ne sont autres que les conclusions obtenues dans l'étude des variations de C_r en fonction de $(Q_s)_{t_0}$:

C_r *est essentiellement fonction du niveau moyen de la nappe phréatique à l'instant initial de la crue* (la dispersion est forte pour $H_0 < 60$ cm parce que le terrain est alors saturé en B et *b*).

C_r *décroît quand le niveau initial de la nappe s'abaisse.*

c) Influence sur C_r du montant P des précipitations :

A priori, C_r doit être fonction croissante de P. La valeur de $(Q_s)_{t_0}$ permet en effet de repérer le niveau de la nappe au début de la pluie; mais le niveau de la nappe s'élève à mesure qu'il pleut; il en résulte une augmentation du coefficient instantané de ruissellement et par conséquent, pour une valeur donnée de $(Q_s)_{t_0}$, une augmentation du coefficient de ruissellement C_r qui est un coefficient moyen pour l'ensemble de la pluie.

On constate (fig. 5 et 5 *bis*) que, pour une valeur donnée de $(Q_s)_{t_0}$,

— C_r apparaît comme indépendant de P tant que $P < 20$ mm;

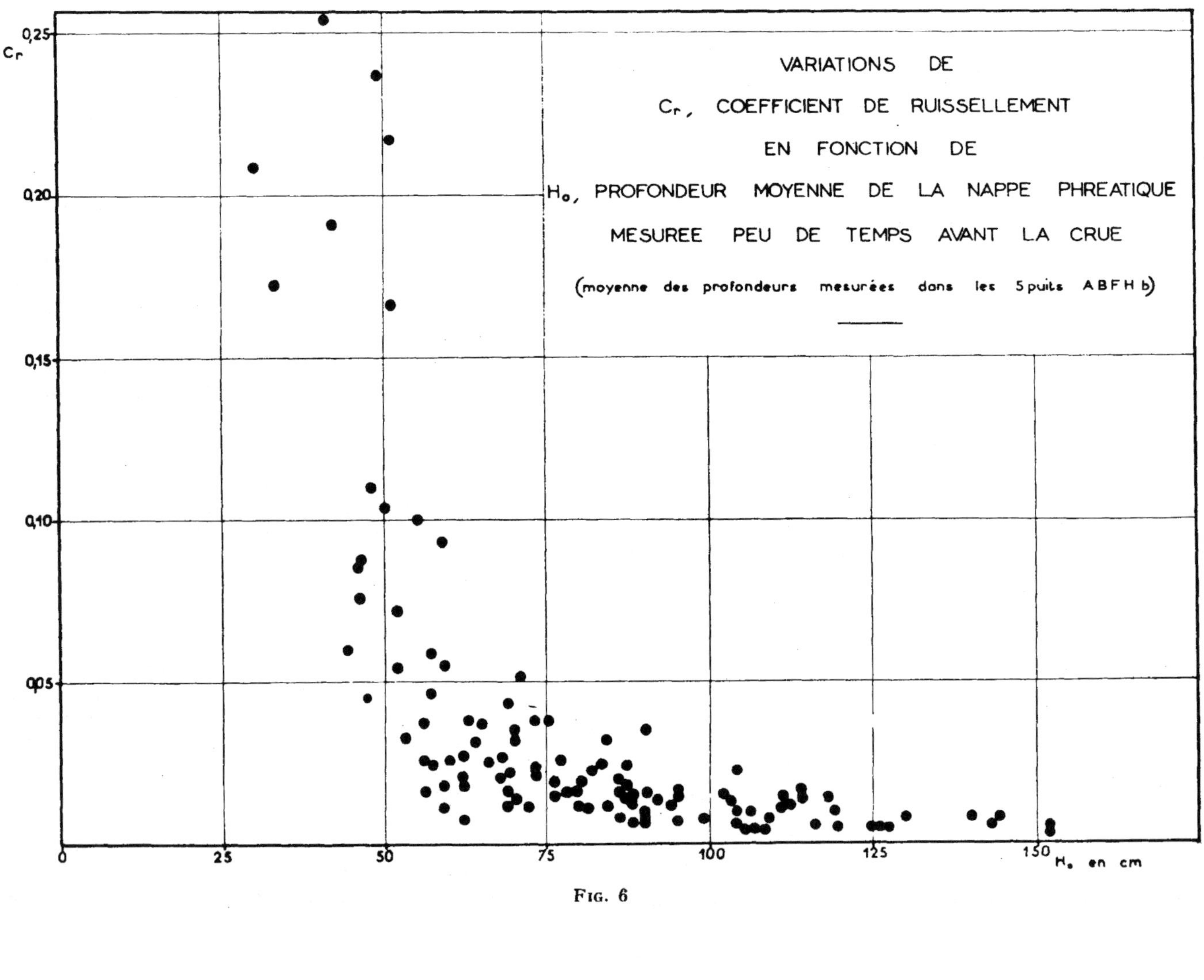

Fig. 6

— Les valeurs de C_r correspondant à $P > 20$ mm sont systématiquement parmi les plus fortes (les plus fortes valeurs de P envisagées sont de l'ordre de 50 mm).

Nous avons pu vérifier que ce deuxième résultat ne peut être attribué à l'influence d'aucun des autres facteurs dont C_r peut *a priori* dépendre.

On peut donc conclure que le coefficient de ruissellement C_r relatif à une crue est fonction croissante du montant P des précipitations qui produisent la crue. Mais l'accroissement de C_r avec P est relativement faible et n'apparaît expérimentalement que pour les pluies les plus importantes.

Les variations de C_r en fonction de P sont évidemment continues; mais les résultats expérimentaux nous permettent seulement de représenter les variations de C_r en fonction de $(Q_s)_{t_0}$ et P par deux courbes, l'une valable pour $P < 20$ mm et l'autre pour $P > 20$ mm; nous n'avions aucune raison *a priori* de choisir pour ces courbes une forme plutôt qu'une autre; les deux nuages de points expérimentaux s'accommodent assez bien de droites.

Les variations de C_r en fonction de $(Q_s)_{t_0}$ seront *donc représentées sur les deux droites D_1 et D_2* (tableau II) :

— *la droite* D_1 *pour* $P < 20$ *mm;*

— *la droite* D_2 *pour* $P > 20$ *mm* (valable au moins jusque pour $P \sim 50$ mm).

TABLEAU II

Coefficient de ruissellement C_r en fonction de $(Q_s)_{t_0}$ et P

$(Q_s)_{t_0}$			C_r	
			1 $P < 20$ mm	2 $P > 20$ mm
1	l/s/km²		0,4 %	0,6 %
2,5	»	(DC Etiage)	0,6 %	1 %
4	»	(DC 11 mois)	0,8 %	1,3 %
7	»	(DC 9 mois)	1,2 %	1,8 %
14	»	(DC 6 mois)	2,1 %	3,1 %
25	»	(DC 3 mois)	3,5 %	5,2 %
45	»	(DC 1 mois)	6 %	9 %
80	»	(DC maxi)	10 %	16 %
150	»	(DC 1 jour)	19 %	29 %
200	»		26 %	39 %
250			32 %	48 %

d) INFLUENCE SUR C_r DES AUTRES FACTEURS :

Nous avons étudié l'influence sur C_r de chacun des autres facteurs dont il pouvait *a priori* dépendre :

— *Intensité* i *des précipitations;*

— *Humidité du sol à l'instant initial de la crue dans les couches superficielles qui ne sont pas atteintes par la nappe;*

— *Etat du manteau végétal.*

— *Evaporation des eaux de ruissellement.*

Nous avons constaté que *le coefficient de ruissellement C_r est pratiquement indépendant de tous ces facteurs, bien que chacun d'eux soit très variable pour l'ensemble des crues étudiées et même pour une valeur donnée de $(Q_s)_{t_0}$.* (Noter, en particulier, que l'intensité *i* de la pluie varie, pour l'ensemble des crues étudiées, de 2 à 100 mm/h, soit dans le rapport de 1 à 50, et, pour une valeur donnée de $(Q_s)_{t_0}$, dans le rapport de 1 à 10).

Interprétation des résultats de l'étude de C_r. Nouvelle théorie du ruissellement :

a) Le Bassin expérimental peut être partagé en deux zones S_r et S_i d'étendues variables :

— *La « zone de ruissellement »* S_r, de superficie A_r, comporte, d'une part, des zones imperméables d'étendue fixe (routes, chemins empierrés, chemins de terre tassée par le passage répété des hommes ou du bétail, surfaces rocheuses, etc.) et, d'autre part, des zones d'étendue variable constituées de terrains perméables, mais à peu près complètement saturés d'eau. La pluie tombant sur la zone S_r se transforme entièrement en ruissellement superficiel ou hypodermique.

— *La « zone d'infiltration »* S_i, de superficie A_i, est constituée par les terrains perméables non saturés. Le sol de texture sableuse, qui forme les couches superficielles du Bassin expérimental est caractérisé par une « capacité d'infiltration » *f* très forte qui dépasse l'intensité de toutes les pluies pouvant tomber sur ce bassin — à l'exception seulement de pluies d'une rareté extrême —; ainsi, sauf en des cas très exceptionnels, la pluie tombant sur la zone S_i est entièrement absorbée par infiltration et ne donne lieu par conséquent à aucun ruissellement.

Le coefficient de ruissellement d'une pluie élémentaire — que nous appellerons aussi « coefficient instantané de ruissellement » — *est donc* *égal au rapport du volume d'eau tombée sur la « zone de ruissellement » S_r au volume total d'eau tombée sur l'ensemble du bassin versant.*

A une valeur donnée du niveau moyen de la nappe phréatique — ou ce qui revient au même, à une valeur donnée du débit souterrain — il correspond une valeur déterminée A_r de la superficie de la zone S_r et par conséquent une valeur à peu près déterminée du coefficient instantané de ruissellement, car la répartition de la pluie sur le bassin est toujours la même en première approximation (ce coefficient serait égal à :

$$A_r/(A_r + A_i).$$

si la répartition était uniforme).

Ainsi, le coefficient instantané du ruissellement est, en première approximation, un invariant pour un niveau donné de la nappe phréatique — ou pour une valeur donnée du débit souterrain — quand les autres facteurs de l'écoulement varient.

b) Quand le niveau moyen de la nappe phréatique s'élève — ou ce qui revient au même, quand le débit souterrain s'accroît — l'étendue de la zone S_r augmente et par conséquent aussi le coefficient instantané de ruissellement.

Mais le niveau moyen de la nappe — ou le débit souterrain — varient assez peu au cours d'une précipitation, à moins que cette précipitation ne soit vraiment très importante; l'étendue de la zone de ruissellement S_r et le cooefficient instantané de ruissellement varient donc eux aussi relativement peu au cours d'une pluie. *Aussi le coefficient de ruissellement C_r correspondant à l'ensemble d'une pluie est-il en première approximation déterminé par le niveau moyen de la nappe phréatique à l'instant initial t_0 de la pluie — ou, ce qui revient au même, par le débit souterrain $(Q_s)_{t_0}$ à cet instant initial*; il est à peu près égal à la valeur de A_r/A à l'instant initial t_0 (A étant la superficie totale du bassin; $A = A_r + A_i$).

Par contre, les grandes variations du niveau de la nappe ou du débit souterrain se traduisent par des variations importantes de l'étendue de S_r. *Le coefficient de ruissellement est donc fonction croissante du niveau initial de la nappe ou du débit souterrain initial $(Q_s)_{t_0}$.*

c) En somme, nous ne contestons pas la validité de principe de la théorie classique du ruissellement en un point déterminé du bassin versant : il est bien évident que le ruissellement instantané produit en un point est toujours la différence entre l'intensité de la pluie et la capacité f d'infiltration en ce point; il est bien certain, d'autre part, que la valeur de f en un point est déterminée uniquement par l'état d'humidité du terrain en ce point et que f est ainsi un invariant, en ce point, pour un état donné de l'humidité du terrain, quand les autres facteurs de l'écoulement varient.

Mais la théorie classique, valable en un point, n'est pas applicable à l'ensemble du bassin versant, pour la bonne raison que l'état de l'humidité du terrain n'est pas le même en tous les points du bassin versant.

D'ailleurs, les valeurs de f pour le terrain qui constitue le Bassin expérimental sont tellement fortes que le mécanisme de production du ruissellement en un point y est beaucoup plus simple que le mécanisme de principe donné par la théorie classique; il se réduit en fait à un mécanisme de « tout ou rien » : ou bien le terrain est saturé et toute la pluie ruisselle, ou bien le terrain n'est pas saturé et toute la pluie s'infiltre.

d) La nouvelle théorie du ruissellement que nous proposons permet aussi d'expliquer les divers autres résultats obtenus dans l'étude du coefficient de ruissellement C_r :

— L'indépendance de C_r vis-à-vis de *l'intensité* i *de la pluie* et de *l'humidité des couches superficielles* qui ne sont pas atteintes par la nappe: due au fait que, quelle que soit l'humidité de ces couches, f dépasse l'intensité de toutes les pluies possibles sur le Bassin expérimental, à l'exception seulement de pluies extrêmement rares; nous avons, par exemple, mesuré une absorption de 20 mm en 15′ dans un terrain moyennement humide, ce qui correspond à une capacité d'infiltration d'environ 13 dixièmes de mm par minute; une pluie dont l'intensité atteint cette valeur est extrêmement rare, à moins qu'elle ne soit de courte durée; mais les averses courtes et très intenses se produisent en général en été, et le terrain est alors plus sec et présente un taux d'absorption nettement plus élevé.

— Les variations de C_r en fonction du *montant P de la pluie :* Une pluie particulière produit une légère élévation du niveau de la nappe, donc une légère augmentation de l'étendue de la zone S_r, et par conséquent une légère augmentation de C_r; on comprend ainsi que cette augmentation ne soit sensible que dans le cas des pluies les plus importantes.

— L'indépendance de C_r vis-à-vis des *variations saisonnières vis-à-vis du développement de la végétation :* Ces variations n'ont évidemment aucune influence sur l'étendue de la zone S_r; mais si R est la lame d'eau retenue par la végétation, le coefficient de ruissellement, qui serait A_r/A en l'absence de rétention, est en réalité $A_r(P - R)/A.P$. D'après les chiffres donnés par divers auteurs pour des plantes identiques ou simplement analogues à celles que l'on rencontre avec le Bassin expérimental, on peut évaluer grossièrement la valeur moyenne de R pour l'ensemble du Bassin expérimental :

— R doit atteindre sa valeur maximum R_{max} pour les pluies supérieures à 5 mm et R_{max} doit être de l'ordre de 1 mm de mai à octobre et de 0,1 à 0,2 mm de novembre à avril.

— Pour $P < 5$ mm, R est inférieur à R_{max} et décroît avec P.

Ces ordres de grandeur de R permettent de comprendre que les variations saisonnières du manteau végétal n'aient pratiquement pas d'influence sur C_r.

— L'indépendance de C_r vis-à-vis de *l'évaporation :* L'évaporation pendant la durée du ruissellement n'a évidemment qu'une influence tout à fait négligeable sur l'étendue de la zone S_r; mais le coefficient de ruissellement, qui serait A_r/A en l'absence d'évaporation est en réalité $A_r(P - e)/A.P$, si $Ar.e$ est le volume d'eau évaporé à partir des eaux de ruissellement.

Il est facile d'évaluer *e* en tenant compte de la durée du ruissellement (pratiquement T + 6 heures), de la durée T de la pluie (pendant laquelle l'évaporation est négligeable), de la quantité de pluie retenue par la végétation (qui diminue d'autant l'évaporation de l'eau ruisselée) et de la diminution progressive, pendant la durée du ruissellement, de la surface couverte par les eaux de ruissellement (égale à A_r à la fin de la pluie, nulle à la fin du ruissellement) : *e* varie depuis des valeurs négligeables (hiver ou nuit) jusqu'à un peu plus de 1 mm (milieu de la journée en été). On comprend ainsi que les variations de l'évaporation n'ont pas d'influence appréciable sur le coefficient de ruissellement.

Conclusion à l'étude du volume du ruissellement:

Pour toutes les crues possibles sur le Bassin expérimental d'Alrance, depuis les crues pratiquement négligeables jusqu'aux crues de fréquence approximativement quinquennale, le volume V_r du ruissellement est lié au montant P des précipitations qui les produisent, par *un coefficient de ruissellement* $C_r = V_r/P$ qui présente les propriétés suivantes :

1. C_r est en général *très faible.*
2. C_r est *avant tout déterminé par le niveau de la nappe phréatique à l'instant initial de la crue* — ou, ce qui revient au même, *par la valeur du débit souterrain à l'instant initial de la crue*;

   C_r croît quand le niveau initial de la nappe s'élève — ou, ce qui revient au même, quand le débit souterrain initial augmente.
3. C_r dépend *quelque peu du montant P des précipitations,* dont il est fonction croissante.
4. C_r est *pratiquement indépendant de tous les autres facteurs* dont il pouvait *a priori* dépendre : état de la végétation, évaporation, et, ce qui est plus étonnant parce que contraire aux idées généralement admises, intensité des précipitations et humidité du sol dans les couches superficielles qui ne sont pas atteintes par la nappe.

Ces résultats sont contraires à la « théorie classique » du ruissellement, mais peuvent être expliqués par la nouvelle théorie du ruissellement que nous proposons : ruissellement produit par la totalité de la pluie tombant sur une zone imperméable ou saturée dite « zone de ruissellement », dont l'étendue varie avec le niveau moyen de la nappe phréatique.

Cette explication que nous donnons des résultats de l'étude du volume du ruissellement, permet de penser que ces mêmes résultats restent valables pour les crues supérieures à la crue quinquennale, à l'exclusion toutefois des crues provoquées par des précipitations d'intensité exceptionnelle dépassant la capacité d'infiltration dans les parties perméables et non saturées du bassin versant.

II. — Forme de l'hydrogramme de ruissellement.

En première approximation, la forme de l'hydrogramme de ruissellement produit par une pluie uniforme est définie — à une affinité près ayant pour base l'axe des temps — par les trois paramètres suivants (fig. 7) :

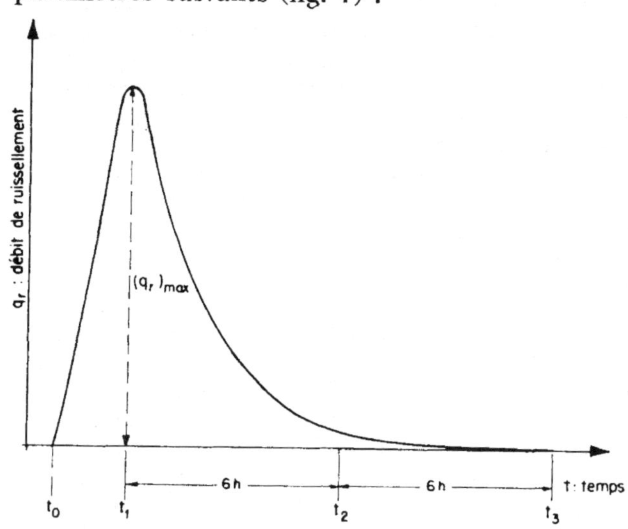

FIG. 7

Forme de l'hydrogramme de ruissellement.

1. *Le « temps de montée »* $t_1 - t_0$;
2. *Le rapport* $\rho = (q_r)_{max}/V_r$ *du débit de pointe de ruissellement au volume du ruissellement*;
3. *Le « temps de base » ou durée du ruissellement :* $t_3 - t_0$.

a) RÉSULTATS DES OBSERVATIONS (118 crues provoquées par des pluies uniformes) :

1. *Le « temps de montée »* $t_1 - t_0$:

Il est remarquable (fig. 8) que pour une valeur donnée de la durée T de la pluie :

— *Le « temps de montée » est peu variable alors que l'importance de la crue varie dans un très large intervalle* (importance repérée par la valeur V_r du volume du ruissellement);

— *Le « temps de montée » est indépendant de l'importance de la crue.*

Les variations de $t_1 - t_0$ en fonction de T peuvent ainsi être représentées par la courbe unique de la figure 10; on déduit de l'examen de cette courbe que le « temps de montée » de l'hydrogramme unitaire produit par une pluie instantanée est de l'ordre d'une vingtaine de minutes.

2. *Rapport* $\rho = (q_r)_{max}/V_r$:

Il est remarquable (fig. 9) que pour une valeur donnée de la durée T de la pluie,

— *Le rapport* ρ *est assez peu variable alors que l'importance de la crue varie dans un très large intervalle;*

— *Le rapport* ρ *est indépendant de l'importance de la crue.*

3. *« Temps de base »* $t_3 - t_0$:

Nous avons vu précédemment (sous chapitre A) que le ruissellement est complètement terminé

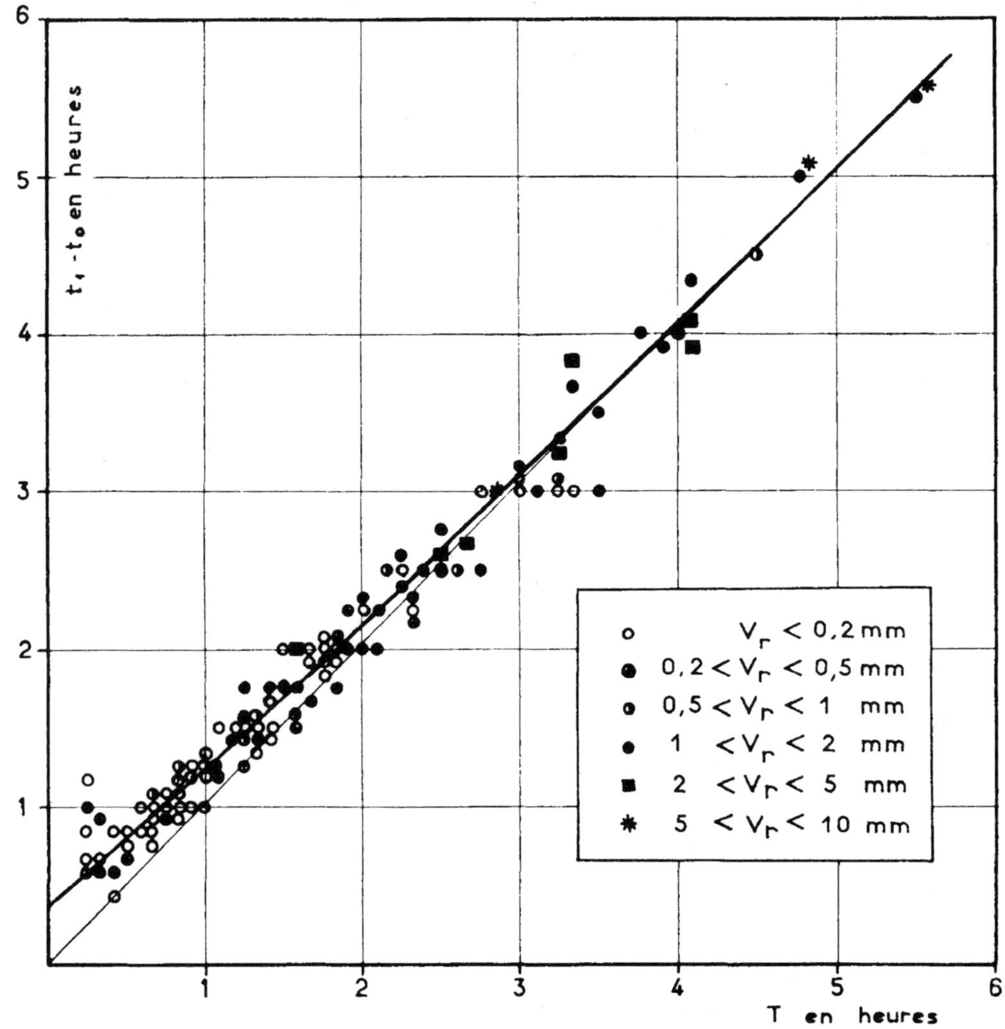

FIG. 8

Forme de l'hydrogramme de ruissellement.
Variation du temps de montée $t_1 - t_0$ en fonction de la durée T de la pluie.

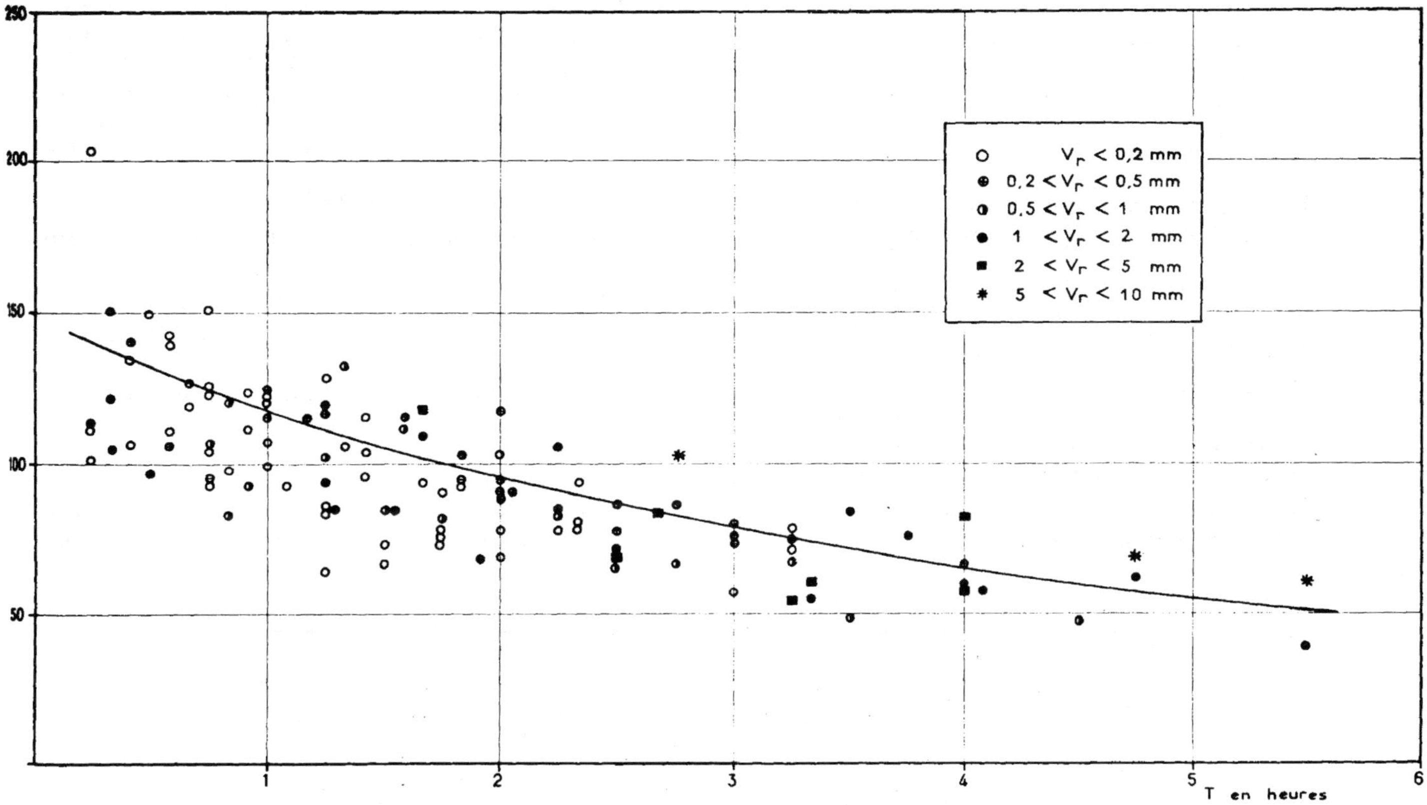

FIG. 9

Forme de l'hydrogramme de ruissellement.
Variation du rapport $\rho = (q_r)_{max}/V_r$ en fonction de la durée T de la pluie.

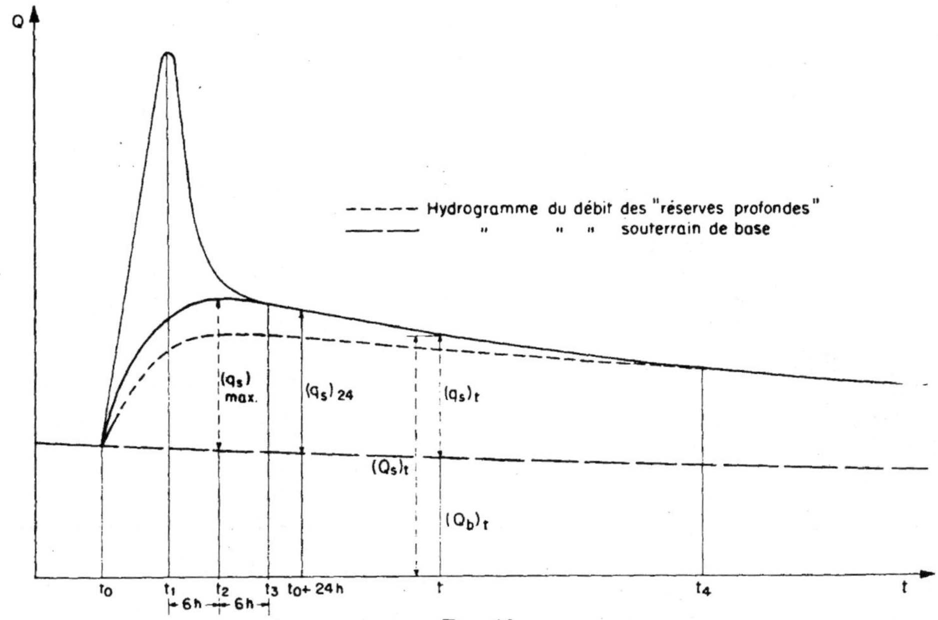

FIG. 10

Hydrogramme de l'écoulement souterrain.

douze heures après l'instant t_1 de la pointe de crue et qu'il est déjà à peu près nul six heures après la pointe de crue. *Le « temps de base » est donc égal en principe à* $t_1 - t_0$ *(« temps de montée ») + 12 heures et pratiquement à* $t_1 - t_0 + 6$ *heures.*

b) INTERPRÉTATION DES RÉSULTATS EXPÉRIMENTAUX :

On peut aisément admettre que pour une « zone de ruissellement » donnée les longueurs des parcours des eaux de ruissellement soient déterminées et qu'ainsi $t_1 - t_0$, ρ et $t_3 - t_0$ ne dépendent que de la durée de la pluie. Par contre, il peut paraître étonnant que pour une durée donnée de la pluie, $t_1 - t_0$, ρ et $t_3 - t_0$ restent les mêmes, alors que l'étendue de la zone de ruissellement S_r varie.

En réalité, sur le Bassin expérimental, *les trajets parcourus par les eaux de ruissellement dépendent asez peu de la superficie de la « zone de ruissellement »*. En effet, la nappe phréatique étant voisine de la surface du sol sur la majeure partie du bassin versant, la zone de ruissellement se présente sous la forme d'un réseau très ramifié qui s'étend pratiquement sur tout le bassin; ses ramifications suivent le fond des vallées principales, mais remontent également le moindre petit vallon; la zone S_r comporte aussi des surfaces isolées réparties sur tout le bassin. On trouve ainsi des terrains saturés jusqu'au voisinage des crêtes, même en période de sécheresse. Quand le niveau moyen de la nappe varie, les ramifications du réseau s'élargissent ou se rétrécissent; elles rejoignent des zones de ruissellement précédemment isolées ou au contraire se fractionnent en nouvelles surfaces isolées; mais les distances parcourues par les eaux de ruissellement restent à peu près toujours les mêmes.

c) COMPARAISON DES RÉSULTATS EXPÉRIMENTAUX ET DES RÉSULTATS THÉORIQUES OBTENUS PAR APPLICATION DE LA « MÉTHODE DE L'HYDROGRAMME UNITAIRE » :

1. On sait que, d'après les principes de la méthode dite « de l'hydrogramme unitaire », les hydrogrammes des crues provoquées par des pluies de même durée sont des courbes affines, la base de l'affinité étant l'axe des temps; on en déduit que le « temps de montée », le rapport ρ et le « temps de base » sont, pour une durée donnée de la pluie, indépendantes du volume de ruissellement : ce sont précisément les résultats que nous avons déduits des observations au paragraphe *a* ci-dessus.

2. D'après les principes de la méthode de l'hydrogramme unitaire, le rapport d'affinité des crues produites par des pluies uniformes de même durée est égal au rapport des intensités des pluies : c'est bien ce résultat que nous avons déduit des observations, du moins pour des crues caractérisées par la même valeur initiale $(Q_s)_{t_0}$ du débit souterrain; bien entendu, pour des crues caractérisées par des valeurs différentes de $(Q_s)_{t_0}$, les coefficients de ruissellement sont différents et le rapport d'affinité est égal au produit du rapport des intensités par le rapport des coefficients de ruissellement.

3. Toujours d'après les principes de la méthode de l'hydrogramme unitaire, des pluies successives produisent un hydrogramme obtenu par addition des ordonnées des hydrogrammes produits par chacune d'elles. Par application de ce principe, nous avons construit des hydrogrammes théoriques produits par des pluies uniformes de diverses durées; des hydrogrammes-types ainsi construits, nous avons déduit les courbes théoriques de variation de $t_1 - t_0$ et de ρ en fonction de T : *on constate que ces courbes théoriques obtenues par application de la méthode de l'hydrogramme unitaire coïncident avec les courbes expérimentales déduites des nuages de points expérimentaux des figures 8 et 9.*

d) DÉCALAGE ENTRE LA PLUIE ET LA CRUE :

Le début de la crue coïncide pratiquement avec le début de la pluie.

Ce résultat expérimental s'explique par le fait que la zone S_r qui produit le ruissellement borde le cours d'eau à partir de l'amont immédiat de la station de jaugeage : le ruissellement produit par les premiers éléments de la pluie tombant sur les parties de la surface S_r voisines du cours d'eau atteint tout de suite ce dernier et les temps de parcours dans le cours d'eau sont ensuite négligeables.

CONCLUSION A L'ÉTUDE DE LA FORME DE L'HYDROGRAMME DE RUISSELLEMENT :

Sur le Bassin expérimental d'Alrance, quelles que soient les circonstances et l'importance de la crue — depuis les crues pratiquement négligeables jusqu'aux crues à peu près quinquennales — *la forme de l'hydrogramme de ruissellement produit par une pluie uniforme est — à une affinité près ayant pour base l'axe des temps :*

— *déterminée uniquement par la durée de la pluie;*

— *indépendante de toutes les autres caractéristiques de la pluie, du temps et du terrain.*

Elle peut être obtenue — à une affinité près ayant pour base l'axe des temps — *par application de la méthode dite « de l'hydrogramme unitaire ».*

Ces divers résultats sont compatibles avec la *nouvelle théorie de ruissellement* que nous avons proposée au paragraphe I ci-dessus.

D) L'ÉCOULEMENT SOUTERRAIN DE CRUE

A partir de l'instant t_0 du début de la crue, le débit souterrain Q_s comprend deux parties (fig. 10) :

— *Le débit souterrain de base :* Q_b;

— *Le débit souterrain de crue :* q_s.

1. Il croît assez rapidement depuis l'instant t_0, où il est nul, jusqu'à l'instant $t_2 = t_1 + 6$ heures (c'est-à-dire six heures après la pointe de crue), où il atteint son maximum $(q_s)_{max}$.
2. Il décroît ensuite lentement : sa décroissance est d'abord plus rapide que celle du débit souterrain en période de tarissement; mais elle est de plus en plus lente et finit par reprendre à l'instant t_4 la lenteur caractéristique des périodes de tarissement (par définition, t_4 marque la fin de la « période de crue »).

I. — Le maximum $(q_s)_{max}$ du débit souterrain de crue (fig. 11 et 11 *bis*).

Nous étudions le coefficient K_{max} défini par :

$$K_{max} = \frac{(q_s)_{max}}{P}$$

$(q_s)_{m,x}$ exprimé en l/s/km² et P en mm.

On constate que K_{max} est assez peu variable pour une valeur donnée de $(Q_s)_{t_0}$, bien que tous les autres facteurs dont il peut *a priori* dépendre varient dans de très larges intervalles : K_{max} *peut donc, en première approximation, être considéré comme déterminé par la valeur initiale* $(Q_s)_{t_0}$ *du débit souterrain, ou, ce qui revient au même, par le niveau initial de la nappe phréatique.*

K_{max} *apparaît comme une fonction croissante de* $(Q_s)_{t_0}$ *pour les valeurs de* $(Q_s)_{t_0}$ *envisagées dans notre étude* (0 à 200 l/s/km²).

Nous avons enfin vérifié que K_{max} *est pratiquement indépendant de tous les autres facteurs dont il pouvait* a priori *dépendre.*

INTERPRÉTATION :

$(q_s)_{max}$ est en toute première approximation proportionnel à l'accroissement des réserves phréatiques produit par la pluie P; K_{max} est donc en toute première approximation proportionnel au *coefficient* C_s *d'écoulement souterrain de la pluie* P *en l'absence d'évaporation.*

Par définition, en effet, C_s est égal au rapport à P de l'écoulement souterrain V_s que produirait la pluie P en l'absence d'évaporation : C_s est donc aussi égal au rapport à P de l'accroissement des réserves phréatiques produit par la pluie P.

Dire que K_{max} ne dépend que de $(Q_s)_{t_0}$ revient à dire que le coefficient d'écoulement souterrain C_s est pratiquement déterminé — comme le coefficient de ruissellement C_r — par la valeur initiale $(Q_s)_{t_0}$ du débit souterrain ou par le niveau initial de la nappe.

Nous avons indiqué antérieurement que le Bassin expérimental se décompose en deux parties : la « zone de ruissellement » S_r, imperméable ou totalement saturée, où toute la pluie ruisselle, et la « zone d'infiltration » S_i, non saturée, où toute la pluie est absorbée par infiltration.

La zone S_i se décompose à son tour en deux parties :

— *La zone* S_a, où la nappe est assez éloignée de la surface du sol pour que la pluie infiltrée soit entièrement retenue par les couches non saturées et n'atteignent par conséquent pas la nappe (on sait en effet que l'eau infiltrée à la la surface du sol ne peut dépasser une profondeur donnée que si elle a d'abord rétabli un certain taux minimum d'humidité — le « taux de rétention » — dans toutes les couches comprises entre la surface et cette profondeur). Toute la pluie tombant sur la zone S_a est définitivement perdue pour l'écoulement.

— *La « zone d'infiltration jusqu'à la nappe phréatique »* S_p où la nappe phréatique est assez proche de la surface du sol pour que les couches non saturées ne retiennent qu'une faible partie de l'eau infiltrée. La pluie tombant sur la zone S_p rejoint pratiquement en totalité la nappe phréatique et, à moins d'être évaporée, produit un écoulement.

6*

K_{max} coefficient d'augmentation maximum du débit souterrain en fonction de $(Q_s)_{t_0}$, débit souterrain observé immédiatement avant la crue.

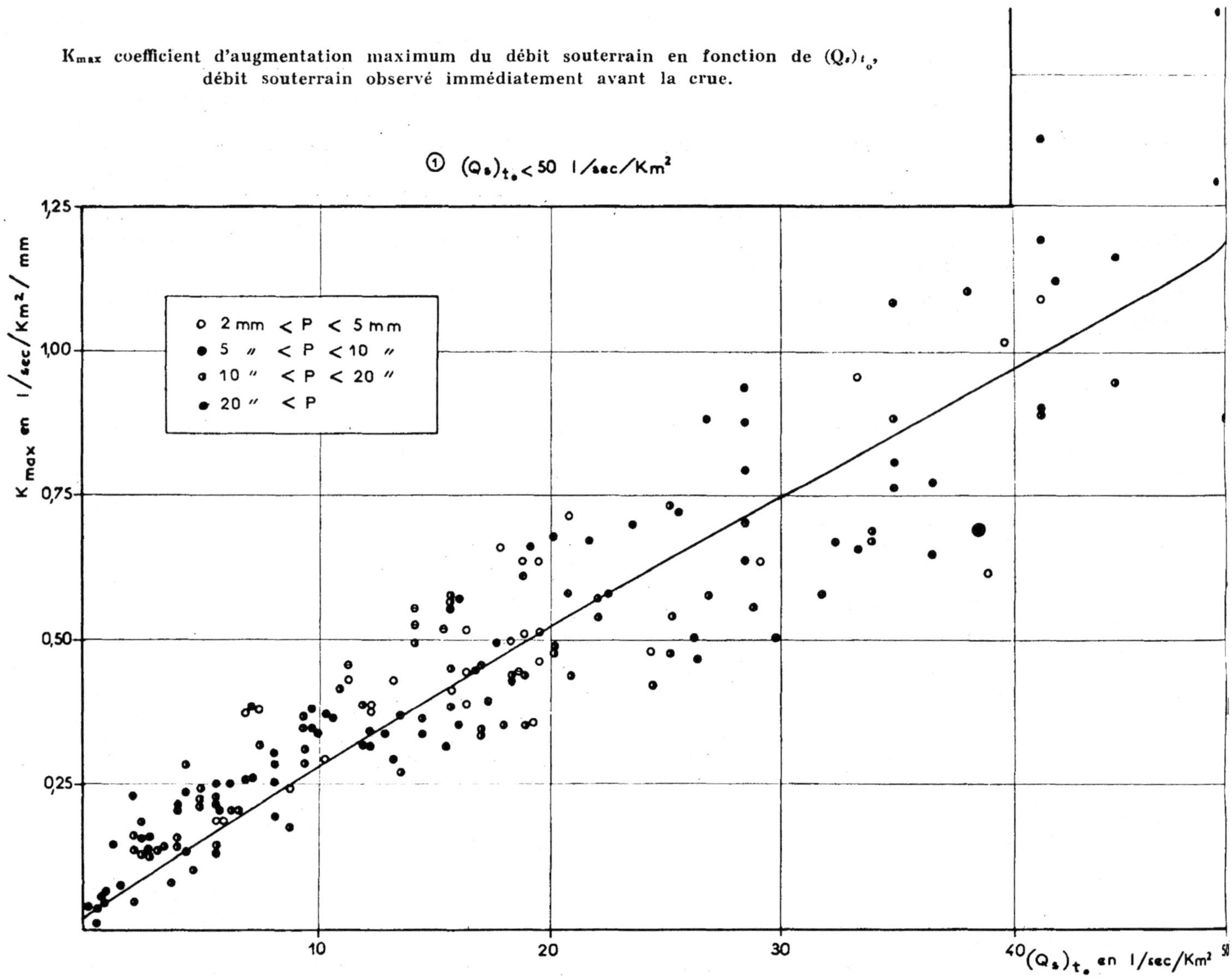

FIG. 11

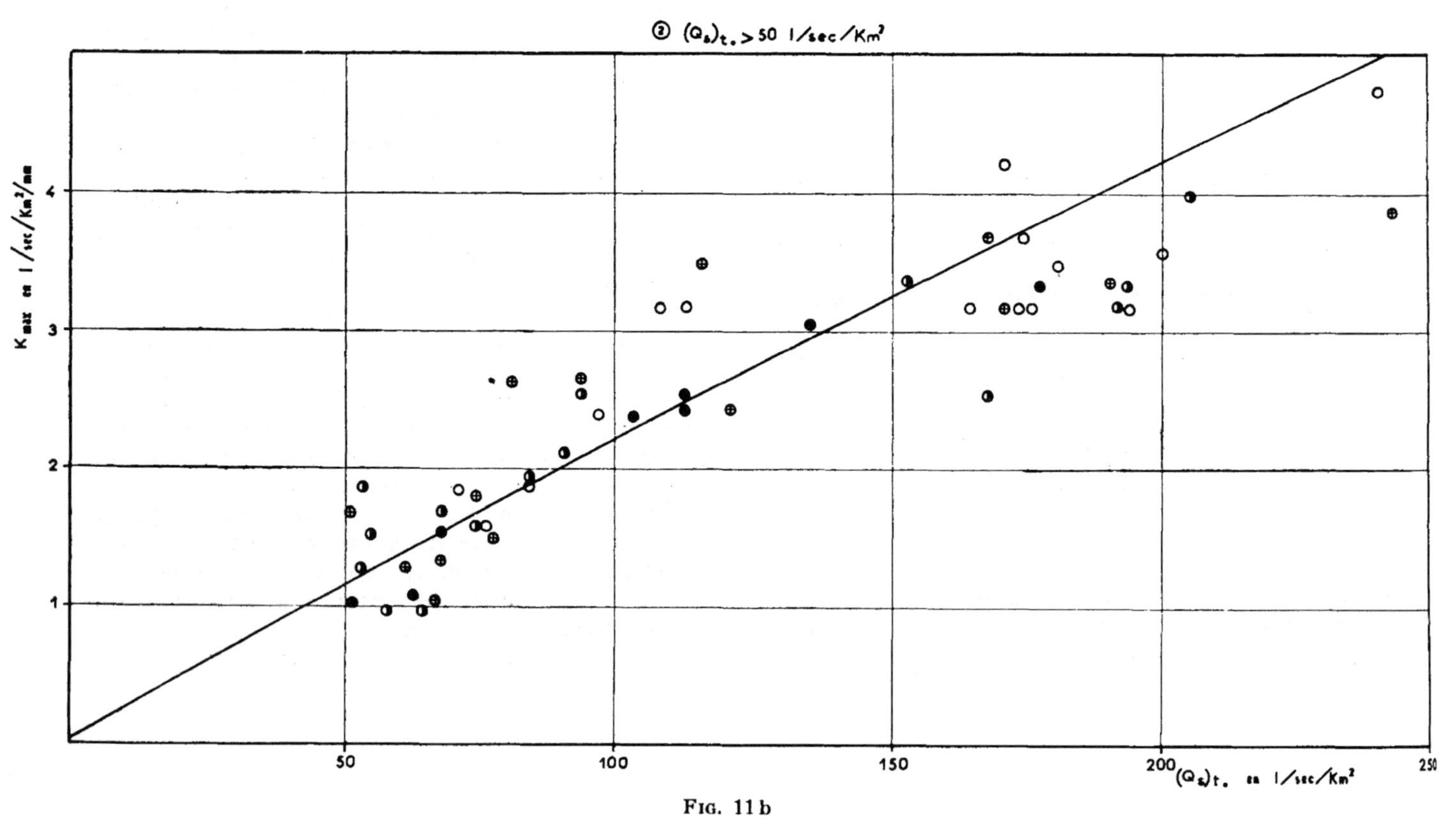

FIG. 11 b

Le niveau moyen de la nappe sur le bassin versant détermine les superficies A_a et A_p des zones S_a et S_p et par conséquent le coefficient instantané d'écoulement souterrain, parce que la répartition des précipitations sur le bassin est assez peu variable : *le coefficient instantané d'écoulement souterrain est égal au rapport de la quantité de pluie tombant sur la zone S_p à la quantité totale tombant sur le bassin.*

Quand le niveau de la nappe s'élève — ou, ce qui revient au même, quand le débit souterrain augmente — la superficie de la zone S_p et le coefficient d'écoulement instantané d'écoulement souterrain varient; mais nous avons déjà vu qu'une pluie particulière ne provoque qu'une faible élévation du niveau de la nappe : *on comprend ainsi que C_s et par suite K_{max} ne dépendent en première approximation que du niveau de la nappe à l'instant initial de la crue, ou de la valeur initiale $(Q_s)_{t_0}$ du débit souterrain.*

Pour les valeurs les plus faibles de $(Q_s)_{t_0}$, la superficie A_p de la zone S_p, et par conséquent le coefficient C_s sont fonctions croissantes de $(Q_s)_{t_0}$; mais comme la superficie de la zone de ruissellement croît quel que soit $(Q_s)_{t_0}$, il est bien évident que A_p et C_s passeront par un maximum quand les zones S_r et S_p couvriront à elles deux tout le bassin versant et qu'ensuite A_p et C_s seront fonction décroissante de $(Q_s)_{t_0}$; nous déterminerons un peu plus loin la valeur de $(Q_s)_{t_0}$ correspondant au maximum de C_s.

II. — (q_s) 24, débit souterrain de crue vingt-quatre heures après le début de la crue.

L'étude du coefficient K_{24}, défini par :

$$K_{24} = \frac{(q_s)_{24}}{P}$$

conduit à des résultats analogues à ceux obtenus pour K_{max}.

III. — Coefficient C_{s1} d'écoulement souterrain pendant le premier jour de la crue.

L'étude du coefficient C_{s1}, défini par :

$$C_{s1} = \frac{V_{s1}}{P}$$

(V_{s1} : volume de l'écoulement souterrain pendant le premier jour de la crue) conduit à des résultats analogues à ceux obtenus pour K_{max} et K_{24}.

IV. — Décroissance du débit souterrain de crue après le premier jour de la crue. Coefficients d'écoulement souterrain pendant les jours successifs de la crue. Coefficient d'écoulement souterrain.

a) CAS OU LE DÉBIT SOUTERRAIN TOTAL EST INFÉRIEUR A 50 L/S/KM² (condition réalisée en moyenne pendant plus de 350 jours par an).

1. *Décroissance du débit souterrain* q_s :

Nous avons pu étudier la décroissance de q_s, pour une trentaine de crues, pendant les deux ou trois jours qui suivent le jour de la pointe de crue (ensuite q_s devient trop faible et le débit de base trop mal déterminé pour que q_s soit connu avec une précision suffisante).

L'examen des quelques dizaines de valeurs de δ correspondantes (δ est le coefficient journalier de décroissance du débit souterrain de crue) conduit aux conclusions suivantes (tableau III) :

— δ est d'autant moins inférieur à l'unité que l'on s'éloigne du premier jour de la crue (autrement dit, la décrue est de moins en moins rapide);

— δ est d'autant plus faible que la température du jour considéré est plus forte (influence de l'évaporation);

— Pour une température et une distance à la pointe de crue déterminées, δ paraît être indépendant de q_s (résultat analogue à celui déjà obtenu pour D_1 dans le cas du débit de tarissement);

— A température égale, δ est inférieur au coefficient D_1 de décroissance journalière du débit souterrain en période de tarissement (autrement dit, la décrue est plus rapide en période de crue qu'en période de tarissement).

2. *Application : Calcul des coefficients d'écoulement souterrain pendant les jours successifs de la crue et du coefficient C_s d'écoulement souterrain.*

Nous avons vu, au paragraphe II ci-dessus, que le coefficient C_{s1} d'écoulement souterrain pendant le premier jour de la crue ne dépend que de $(Q_s)_{t_0}$; nous avons vu aussi que les coefficients δ des jours successifs de la crue ne dépendent que de la température, c'est-à-dire de l'évaporation. Par conséquent, *en l'absence d'évaporation, les coefficients d'écoulement souterrain pendant les jours successifs de la crue et leur somme, le coefficient d'écoulement souterrain C_s, ne dépendent que de $(Q_s)_{t_0}$.*

On peut admettre que l'évaporation est négligaeble pour une température de l'ordre de 5°. Connaissant les valeurs de δ correspondant à cette température pour les quatorze premiers jours de la crue, on calcule facilement les coefficients d'écoulement souterrain en l'absence d'évaporation pendant chacun des quatorze premiers jours de la crue d'abord en fonction du coefficient C_{s1} d'écoulement souterrain pendant le premier jour, puis en fonction de $(Q_s)_{t_0}$. Comme le débit souterrain de crue q_s décroît après le quatorzième jour comme un débit de tarissement, on peut calculer l'écoulement souterrain de crue après le quatorzième jour en fonction de la valeur prise par q_s à la fin du quatorzième jour en utilisant la formule mise en évidence au chapitre I ci-dessus (Etude du tarissement) : $W \sim 2{,}9\ q_s$; on en déduit le coefficient d'écoulement souterrain après le quatorzième jour de la crue en fonction de $(Q_s)_{t_0}$ (tableau III).

TABLEAU III

Coefficients d'écoulement souterrain, de ruissellement et d'écoulement total

($Q_s < 50$ l/s/km²)

$(Q_s)_{t_0}$ Débit souterrain à l'instant initial de la crue	Coefft d'écoulement souterrain pendant les 14 premiers jours de la crue	Coefft d'écoulement souterrain après le 14ᵉ jour de la crue	C_s Coefft d'écoulement souterrain	C_r Coefft de ruissellement	C Coefft d'écoulement total
1	2,5 %	2,5 %	5 %	0,5 %	5,5 %
5	6 %	6 %	12 %	1 %	13 %
10	11 %	11 %	22 %	2 %	24 %
20	19 %	20 %	39 %	3 %	42 %
30	27 %	28 %	55 %	4 %	59 %
40	35 %	36 %	71 %	5 %	76 %
50	43 %	44 %	87 %	7 %	94 %

Pour les valeurs de $(Q_s)_{t_0}$ inférieures à 50 l/s/km² (valeurs réalisées en moyenne pendant plus de 350 jours par an),

— *le coefficient C_s d'écoulement souterrain en l'absence d'évaporation, qui,* nous l'avons déjà dit, ne dépend que de $(Q_s)_{t_0}$, *est fonction croissante de $(Q_s)_{t_0}$.*

— *C_s est en gros douze fois plus fort que le coefficient de ruissellement C_r,* ce qui montre bien l'importance primordiale de l'écoulement souterrain sur le Bassin expérimental d'Alrance dans les circonstances courantes.

Noter que le coefficient d'écoulement total $C = C_r + C_s$ atteint environ 100 % pour $(Q_s)_{t_0} \sim 50$ l/s/km².

b) Cas ou le débit souterrain total est supérieur a 50 l/s/km² (condition réalisée en moyenne pendant moins de quinze jours par an).

Les données dont nous disposons au-dessus de 50 l/s/km² ne sont pas suffisantes pour nous permettre de déterminer les lois de décroissance du débit souterrain de crue au-dessus de 50 l/s/km² et les variations de C_s pour les valeurs de $(Q_s)_{t_0}$ supérieures à 50 l/s/km².

L'allure de ces lois et des variations de C_s peut cependant être déduite de divers résultats obtenus précédemment.

α. Nous venons de voir que le coefficient C d'écoulement total atteint à peu près 100 % quand $(Q_s)_{t_0} = 50$ l/s/km²; nous avons montré, d'autre part, que le coefficient de ruissellement C_r est fonction croissante de $(Q_s)_{t_0}$: *le coefficient C_s d'écoulement souterrain est donc fonction décroissante de $(Q_s)_{t_0}$ pour les valeurs de $(Q_s)_{t_0}$ supérieures à 50 l/s/km²* (fig. 12).

Autrement dit, quand $(Q_s)_{t_0}$ atteint ou dépasse 50 l/s/km², la zone S_a (nappe phréatique éloignée de la surface du sol) est réduite à zéro; la zone de ruissellement S_r (terrain imperméable ou saturé) et la zone S_p (nappe voisine de la surface du sol) couvrent à elles deux tout le bassin versant : toute la pluie tombant sur le bassin ruisselle ou rejoint la nappe; quand $(Q_s)_{t_0}$ croît au-delà de 50 l/s/km², la zone S_r s'étend et par conséquent la zone S_p se réduit : à la limite, quand le bassin versant est totalement saturé, la zone S_r couvre tout le bassin et la zone S_p est réduite à zéro, le coefficient C_r est égal à 100 % et le coefficient C_s est nul.

β. Nous avons vu aussi que le coefficient C_{s1} d'écoulement souterrain pendant le premier jour de la crue est fonction croissante de $(Q_s)_{t_0}$ pour $(Q_s)_{t_0} < 200$ l/s/km². Le coefficient C_s étant fonction décroissante de $(Q_s)_{t_0}$ pour $(Q_s)_{t_0} > 50$ l/s/km², on en déduit que *plus $(Q_s)_{t_0}$ augmente entre 50 et 200 l/s/km², et plus l'écoulement souterrain de crue se concentre sur les premiers jours de la crue.*

Ce résultat s'explique par la diminution progressive de l'étendue de la zone S_p lorsque $(Q_s)_{t_0}$ croît au-delà de 50 l/s/km².

Il est donc probable qu'il reste vrai pour les valeurs de $(Q_s)_{t_0}$ supérieures à 200 l/s/km².

A la limite, si le bassin versant était totalement saturé, tout l'écoulement se ferait sous forme de ruissellement et serait concentré sur le premier jour de la crue.

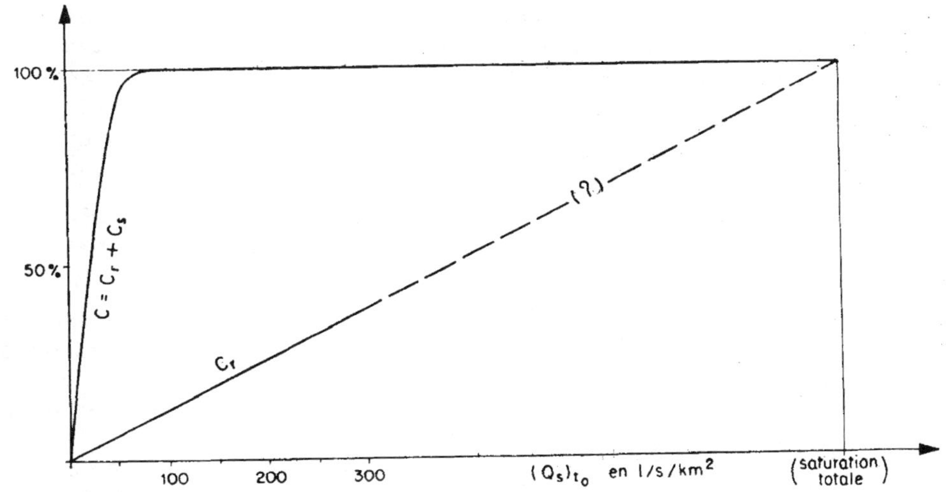

FIG. 12

Variations des coefficients de ruissellement C_r, d'écoulement souterrain C_s et d'écoulement total C (en l'absence d'évaporation), en fonction de la valeur initiale du débit souterrain $(Q_s)_{t_0}$.

Conclusion à l'étude de l'écoulement souterrain de crue.

Les résultats de l'étude de l'écoulement souterrain de crue peuvent se résumer par quelques conclusions remarquables analogues à celles de l'étude du ruissellement :

1. *L'hydrogramme de l'écoulement souterrain de crue ne dépend que d'un nombre réduit de facteurs :*

   — *2 facteurs caractérisant la pluie : le montant et la durée;*

   — *1 seul facteur caractérisant l'état d'humidité du terrain à l'instant initial de la crue: le niveau moyen de la nappe phréatique sur le bassin versant à l'instant initial de la crue,* ou bien *le débit souterrain à l'instant initial de la crue;*

   — *les valeurs de la température moyenne journalière pour les jours successifs de la crue.*

   a) Le maximum $(q_s)_{max}$ *du débit souterrain de crue* est égal au produit du montant P de la pluie par le coefficient K_{max} qui est déterminé par le niveau initial de la nappe (ou la valeur initiale du débit souterrain) : $(q_s)_{max} = K_{max} . P$.

   L'instant du maximum est déterminé par la durée T de la pluie.

   La valeur $(q_s)_{24}$ *du débit souterrain de crue 24 heures après le début de la crue* est égale au produit du montant P de la pluie par le coefficient K_{24} qui est déterminé par le niveau initial de la nappe (ou la valeur initiale du débit souterrain).

   La décroissance du débit souterrain de crue à partir du deuxième jour de la crue ne dépend que des températures moyennes des jours successifs de la crue (à moins que le maximum du débit souterrain ne dépasse 50 l/s/km², condition assez exceptionnelle : dans ce cas, la décroissance dépend aussi de la valeur de ce maximum).

   b) L'analogie entre écoulement souterrain de crue et ruissellement apparaît plus nettement si l'on remarque que le volume V_s de l'écoulement souterrain en l'absence d'évaporation est égal au produit du montant P des précipitations par *le coefficient* C_s *d'écoulement souterrain en l'absence d'évaporation,* coefficient qui ne dépend pratiquement que du niveau initial de la nappe (ou de la valeur initiale du débit souterrain $(Q_s)_{t_0}$. C_s est fonction croissante de $(Q_s)_{t_0}$ pour $(Q_s)_{t_0} < 50$ l/s/km² et fonction décroissante de $(Q_s)_{t_0}$ pour $(Q_s)_{t_0} > 50$ l/s/km².

2. *L'hydrogramme du débit souterrain de crue est pratiquement indépendant de la rétention par la couverture végétale et, ce qui est plus étonnant parce que contraire aux idées généralement admises, de l'intensité de la pluie et de l'humidité du sol dans les couches superficielles qui ne sont pas atteintes par la nappe.*

3. *Malgré leur simplicité, les lois de l'écoulement souterrain de crue mises en évidence sont valables pour toutes les crues, quelles que soient leurs circonstances et leur importance, depuis les crues pratiquement négligeables jusqu'aux crues de fréquence approximativement quinquennale* (seules les lois donnant la décroissance du débit souterrain dépendent de la va-

leur du maximum du débit souterrain quand celui-ci dépasse 50 l/s/km², condition qui n'est réalisée qu'assez rarement).

4. *Tous les résultats concernant l'écoulement souterrain de crue peuvent être expliqués par la nouvelle théorie de l'écoulement que nous proposons.*

5. *L'explication que nous donnons des résultats de l'étude de l'écoulement souterrain de crue permet de penser que ces mêmes résultats restent à peu près valables pour les crues supérieures à la crue quinquennale.*

Conclusion à l'étude des lois de l'écoulement.

Nous avons étudié successivement le tarissement (chapitre I), le ruissellement (chapitre II C) et l'écoulement souterrain de crue (chapitre II D); nous avons ainsi dégagé les diverses lois qui régissent l'écoulement du Bassin expérimental d'Alrance.

Ces lois sont remarquables à plusieurs points de vue :

a) Elles font intervenir *un nombre très réduit de facteurs : 5 au total* (en général 1 ou 2 pour une loi), les autres facteurs n'ayant pratiquement pas d'influence sur l'écoulement; *ces 5 facters sont facilement mesurables*.

1. *Le montant de la pluie :* P;
2. *La durée de la pluie : T* (on suppose les précipitations décomposées en averses partielles à peu près uniformes et de durée inférieure à six heures) (noter que l'intensité de la pluie n'intervient pas);
3. Un seul facteur permettant de caractériser l'état d'humidité du terrain : *le niveau moyen de la nappe phréatique à l'instant initial de la pluie*, ou bien *la valeur du débit souterrain à l'instant initial de la pluie :* $(Q_s)_{t_0}$;
4. *La température moyenne journalière :* θ;
5. *Le numéro d'ordre du jour considéré* par rapport à l'instant initial de la pluie : *n*.

b) Elles sont *peu nombreuses : 8 au total :*

— Pour le débit de tarissement :

Une loi donnant D_1 (coefficient journalier de décroissance du débit de tarissement) en fonction de θ.

— Pour le ruissellement :

Une loi donnant C_r (coefficient de ruissellement) en fonction de $(Q_s)_{t_0}$ et P (et qui se réduit à deux lois linéaires à une variable donnant C_r en fonction de $(Q_s)_{t_0}$, l'une valable pour $P > 20$ mm, et l'autre pour $P < 20$ mm);

Trois lois donnant la forme de l'hydrogramme de ruissellement en fonction de T (à une affinité près ayant pour base l'axe des temps);

Une loi pratiquement linéaire donnant le « temps de montée »;
Une loi pratiquement linéaire donnant le « temps de base »;
Une loi donnant le rapport ρ du débit de pointe de ruissellement ou volume de ruissellement.

— Pour l'écoulement souterrain de crue :

— Une loi donnant K_{max} (rapport du maximum du débit souterrain de crue $(q_s)_{max}$ au montant P de la pluie) en fonction de $(Q_s)_{t_0}$;

— Une loi pratiquement linéaire donnant l'instant du maximum du débit souterrain de crue en fonction de T;

— Une loi donnant δ (coefficient journalier de décroissance du débit souterrain de crue en fonction de θ, *n* et $(Q_s)_{max}$ et qui se réduit à une loi à deux variables θ et *n* quand $(Q_s)_{max} < 50$ l/s/km², condition réalisée pendant plus de 350 jours par an.

c) Elles sont de *forme très simple* (bien souvent linéaire).

d) Elles peuvent être *expliquées par la nouvelle théorie très simple de l'écoulement que nous proposons :* partage du bassin versant en trois zones dont les étendues varient avec le niveau de la nappe et dont les deux premières produisent respectivement le ruissellement et l'écoulement souterrain de crue.

e) Malgré leur petit nombre et leur simplicité, elles sont *valables à peu près en toutes circonstances* (périodes de fonte des neiges exceptées) (plus exactement, elles sont valables pour toutes les crues, depuis les crues pratiquement négligeables jusqu'aux crues de fréquence approximativement quinquennale; et l'explication que nous en donnons permet de penser qu'elles restent valables pour les crues supérieures à la crue quinquennale, à certaines exceptions près que nous avons précisées).

Commentary on Paper 5

Tsukamoto, Y. (1963) Storm discharge from an experimental watershed.

There is a long and distinguished history of hillslope hydrological and streamflow generation studies in Japan, but much of the literature has only become known relatively recently as contacts and communication has increased (though it has to be said that Japanese hydrologists have published articles in English for a very long time). This paper, from one of the pioneers of scientific hydrology in Japan (along with Isao Yamaguchi and Yooichi Noguchi), has been chosen to represent this Japanese contribution. One of the advantages of having different traditions and histories in a subject area is that different types of analyses might result, even if the scientific concepts on which they are based are similar.

Yoshinori Tsukamoto had earlier (1961, in Japanese with an English summary) produced a process study of subsurface runoff generation in an experimental catchment of Tokyo University at Aichi, arising from his PhD work. In this paper, the concepts developed earlier are extended to the complete catchment scale. Tsukamoto attempts to relate the vertical profile of soil water storage to spatial patterns of soil water storage by an analysis of incremental storage changes (cumulative rainfall inputs in excess of cumulative discharge over successive time increments, see Fig. 3 of the paper). The idea was that if the initial hillslope storage had a characteristic spatial pattern of saturation deficits (Fig. 2), then the catchment-scale storage increments could be distributed according to how much of that storage would be used to bring the profile to saturation (and which would then act as a runoff contributing area), and how much would be used to replenish the retention (or unsaturated) zone. In this way, knowing a pattern of initial deficits would allow the spatial pattern of response for successive increments of water to be determined.

These ideas were being developed at the same time as Hewlett and others at Coweeta were also emphasizing the importance of subsurface response in streamflow generation (Paper 6) and, in fact, after finishing his PhD, Tsukamoto spent some time working with John Hewlett at Coweeta. The study is essentially suggesting the concept of a variable saturated contributing area that acts as a runoff source area, later more widely known through the work of Dunne & Black and others (Paper 15). The idea of representing the tendency to saturation as a spatial function of storage deficit, also pre-dates the topographic index concept of Kirkby (1975) and wetness index of O'Loughlin (1981). Here though, there is no theory about the response of the water table, only about filling of the available storage volume on the hillslopes as the cumulative volume of rainfall increases (Fig. 3). It is not absolutely clear from the paper how these detention volumes are mapped back into space on the hillslopes in a way that properly accounts for hillslope form.

Yoshinori Tsukamoto became a Professor in Agriculture and Technology at Tokyo University where he remained throughout his career. He wrote several Japanese texts on forest hydrology and geomorphology and has been President of the Japan Society of Hydrology and Water Resources. Tsukamoto's seminal work had an important impact on later generations of Japanese hillslope hydrologists who have continued to work on small zero-order basins and hillslopes. Recent examples of this continuing influence can be seen in Kitihara (1993), Tsuboyama *et al.* (1994, 2000), and Uchida *et al.* (1999, 2002).

References

Kirkby, M. J. (1975) Hydrograph modelling strategies. In: *Processes in Physical and Human Geography* (ed. by R. Peel, M. Chisholm & P. Haggett), 69–90. Heinemann, London, UK.

Kitihara, H. (1993) Characteristics of pipe flow in forested slopes. In: *Exchange Processes at the Land Surface for a Range of Space and Time Scales* (ed. by H.-J. Bolle, R. A. Feddes & J. D. Kalma), 235–242. IAHS Publ. 212. IAHS Press, Wallingford, UK.

O'Loughlin, E. M. (1981) Saturation regions in catchments and their relation to soil and topographic properties. *J. Hydrol.* **53**, 229–246.

Tsuboyama, Y., Sidle, R. C., Noguchi, S. & Hosoda, I. (1994) Flow and solute transport through the soil matrix and macropores of a hillslope segment. *Water Resour. Res.* **30**(4), 879–890.

Tsuboyama, Y., Sidle, R. C., Noguchi, S., Murakim, S. & Shimizu, S. (2000) A zero-order basin – its contribution to catchment hydrology and internal hydrological processes. *Hydrol. Processes* **14**, 387–401.

Tsukamoto, Y. (1961) An experiment on subsurface flow. *J. Japanese Forestry Society* **43**(2), 62–67.

Uchida, T., Kosugi, K. & Mizuyama, T. (1999) Runoff characteristics of pipeflow and effects of pipeflow on rainfall–runoff phenomena in a mountainous watershed. *J. Hydrol.* **222**, 18–36.

Uchida, T., Kosugi, K. & Mizuyama, T. (2002) Effects of pipeflow and bedrock groundwater on runoff generation in a steep headwater catchment in Ashiu, central Japan. *Water Resour. Res.* **38**(7), doi: 10.1029/2001WR000261.

Storm Discharge from an Experimental Watershed.

Yoshinori TSUKAMOTO*

実験流域からの一降雨の流出

塚 本 良 則*

要旨 土壌表層がポーラスで土壌水分不足を補充する以前に地表流下や中間流出が発生することのない流域については次のことが成り立つ。地下水減水曲線を使って降雨を土壌水分不足補充分(Retention water)と降雨流出分(Detention water)とに分ける。これをいくつかの独立降雨について実行し，降雨流出分と降雨量との関係図を作る。この図の関係曲線を近似的に直線で示すと，この直線の勾配は流域を土壌水分不足量により区分する時の面積割合を示す。このことは山口[1]によりすでに指摘されているが，著者はこの方法を東京大学愛知演習林の東山試験地の量水流域に適用し，4月と8月の土壌水分不足量——面積割合を算出した。この結果に野口[2]の土壌水分の季節変動曲線を使用してこの流域の土壌水分不足量の季節変化を新しく計算してみた。この結果を実測降雨量に適用したところ，5年間の平均で計算流量は実測流量の91%となった。

Synopsis: A storm water is divided into Detention water (Ground water) and Retention water (Replenishment of soil moisture deficiency) by the normal ground water depletion curve. The figure showing the relationship between rainfall and detention water is drawn by applying the above method to isolated storms. The relationship is expressed approximately by straight lines. Then, the gradient of the lines shows the area percentage of soil moisture deficincy which is the same amount to the rainfall at the time. This is found by YAMAGUCHI[1]. The writer applied this theory to Higashiyama watershed of Tokyo Unv. Forest at Aichi and estimated the moisture deficiencies of the soil water belts and their area percentages in April and August. The seasonal variation of soil moisture obtained by NOGUCHI[2] was appleid to the above result to compute newly the seasonal variation of moisture of the soil water belts of the watershed. The amount of detention water computed by using this soil moisture was 91% of that measured actually for the five year period.

Introduction

The waters that appear in a soil profile of a watershed are illustrated in Fig. 1. The waters are ground water, capillary fringe water, intermediate water and soil water. Capillary fringe water is connected with ground water through capillary pores. Soil water is under the influnce of evaporation and transpiration. Intermediate water, which has no connection with ground water, is not influenced by evaporation and transpiration and always at field capacity. Fig. 2 is obtained by the application of Fig. 1 to a watershed profile. These waters in a watershed are tabulated in Table 1 in connection with storage and discharge.

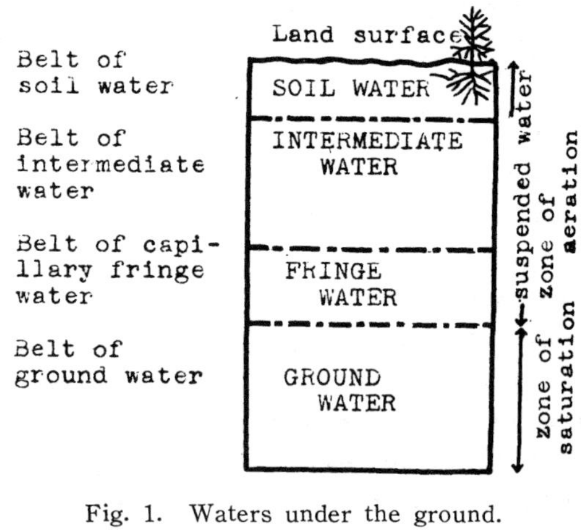

Fig. 1. Waters under the ground.

The column of discharge of Table 1 will be explained further in detail. It is assumed in this watershed that neither overland flow nor subsurface flow takes

* 東京農工大学農学部 Fac. of Agr., Tokyo Unv. of Agriculture & Technology, Tokyo

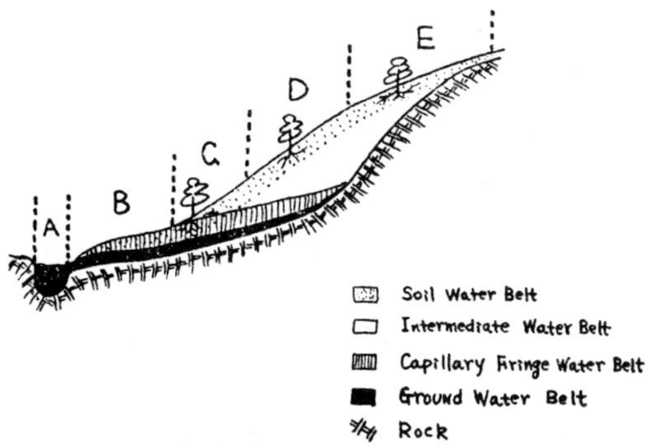

Fig. 2. A profile of a watershed.

place before soil moisture deficiency is replenished. Then, on the areas A and B in Fig. 2 storm water becomes detention water and is discharged later as ground water. The discharge of storm water from C, D and E depends on moisture deficiency of soil water belts. A soil water belt contains soils of different moisture deficiencies from place to place. Therefore, a given amount of rainfall yields detention water in some parts of the soil water belt, while the same rainfall is entirely retained for replenishment of soil moisture deficiency in other parts. Let precipitation be P and moisture deficiency of a soil water belt at this time D_s. In the part of $P<D_s$ of a soil water belt, storm water is retained in this part for replenishment of moisture deficiency. However, in the case of $P>D_s$, the amount equal to D_s of the rainfall is retained in the belts, while the remainder, $P-D_s$, goes down through the intermediate water belt into the ground water belt.

Table 1. Waters in the profiles of a watershed.

	Water a in soil profile	Locality	Storage (Reservoir)	Discharge
A	Ground water	Channel surface	Ground water (Detention) aelt	All of a rainfall
B	Fringe water Ground water	Areas near streams		
C	Soil water Fringe water Ground water	Areas close to streams next to B	Soil water (Retention) belts, and Ground water (Detention) belts which store the remainder of retention storage	Detention water (rainfall-retention storage)
D	Soil water Intermediate water Fringe water Ground water	Foot and middle parts of a mountain slope		
E	Soil water Intermediate water	Upper parts of a mountain slope		

We usually analyse the composite form of the runoffs which have passed through the soil of A, B, C, D and E. The runoffs are exerted various influences by storage functions of soil profiles. It is, therefore, interesting to investigate how the discharge of a storm is affected by the difference of moisture deficiency of a soil water belt which is composed of the areas A, B, C, D and E. And also how the moisture deficiency of a soil water belt fluctuates seasonally.

Theory

Detention water (Ground water) of a storm······ The fundamental formula expressing the relationship among rainfall, storage, loss and runoff in a watershed is written as follows,

$$P=R_o+L_g+L_s\pm\Delta S_g\pm\Delta S_s+L_l$$

in which P is rainfall, R_o actual runoff, L_g loss from the ground water belt, L_s loss from the soil water belt, ΔS_g the increment of ground water storage, ΔS_s the increment of soil water storage, and L_l leakage. All are for a given period.

When applying this formula to the practical computation of detention water of a storm, L_l may safely be assumed to be zero, and L_s to be included in ΔS_s. Thus, $P-\Delta S_s$, which is called "detention water" of a storm, is expressed as follows,

$$P-\Delta S_s=R_o+L_g\pm\Delta S_g$$

in this equation the value of each item in the

right hand side can be got easily. R_o is actual streamflow. ΔS_g is estimated by using the normal ground water depletion curve. L_g can be computed from the difference in the water depletion rates of summer and winter, and from Thornthwaite's evapotranspiration[3].

Moisture deficiency and detention water……All the storm water on the areas A and B becomes detention water. Let area percentage of these two to the watershed be α_1. The relationship between rainfall and detention water from these two areas is shown in Fig. 3, 1) by the straight line of gradient β_1 which is expressed in the equation of $\alpha_1 = \tan\beta_1$(rainfall$<r_1$). The rainfall…smaller than r_1… on the watershed is retained in the soil water belt except the areas And B for replenishment of soil moisture deficiency. This is depicted in Fig. 3, 2), in which detention water is computed by $\tan\beta_1\times$ rainfall. As rainfall becomes heavier, some parts of the areas C, D and E participate in storm water discharge. Let area percentage of these areas of C, D and E to the watershed be α_2. The relationship between rainfall and detention water from these areas is indicated in the equation of $\alpha_2 = \tan\beta_2$ as in the above case. Therefore, the total amount of detention water caused by the rainfall ($r << r_2$) is expressed by the straight line of the gradient, $\beta_1+\beta_2$, in Fig. 3, 1), and the amount of detention water in percentage is shown by the tangent of β_1 and β_2 in Fig. 3, 2). As rainfall becomes still bigger, there will be larger area of $P>D_s$ is in C, D and E. For a rainfall, $r_2<<r_3$, the relationship between rainfall and detention water from these areas is expressed in the equation of $\alpha_3=\tan\beta_3$. The total detention water is ($\tan\beta_1 + \tan\beta_2 + \tan\beta_3$) of the rainfall as in Fig. 3, 2). When P becomes equal to D_s at the largest moisture deficiency profile in the watershed, all the rainfall on the watershed begins to become detention water. At this point the gradient of rainfall-dentention water relationship becomes 1 (45°), and the following equation is concluded:

$$\beta_1+\beta_2+\beta_3+\cdots\cdots+\beta_n=\frac{\angle R}{2},$$

$$\alpha_1+\alpha_2+\alpha_3+\cdots\cdots+\alpha_n=1$$

$$(\tan\beta_1+\tan\beta_2+\cdots\cdots+\tan\beta_n=1)$$

These facts are illustrated in Fig. 3, 1) and 2) as mentioned above. In Fig. 3, 3) a watershed is divided into several sections according to the amount of moisture deficiency as $\tan\beta_1$, $\tan\beta_2$,……and $\tan\beta_3$. Moist soils (A, B and C) are placed at lower parts of the watershed and dry ones (D and E) appear at higher parts. This figure conforms well to the results of forest soil science[4)5)].

As described above, the gradient of rainfall-detention water relationship proves to be an indication of moisture deficiency of a soil water belt. This fact is found by YAMAGUCHI[1]. The writer appleid this to the estimation of seasonal fluctuation of moisture deficiency of a soil water belt.

Application of the theory to an experimental watershed, and discussion

The data of the experimental watershed (watershed A, 106.7 ha) of the Tokyo Unv. Forest in Aichi were employed. In this watershed the mother rock is granite, and the forest has been kept undisturbed over 50 years. Consequently the soil is so porus that overland flow and subsurface flow under the soilsurfce do not take place. This was proved by the experiment carries out recently[6]. Therefore, storm water moves downwards, replenishes moisture deficiency of a soil water belt, and the remainder reaches the ground water belt through the intermediate water belt. It is reported that moisture of a soil water belt in April is different from that in August in this watershed.

In this study the above mentioned theory was appleid to rainfalls in April and August. There would be a doubt about how precisely detention water can be computed. In order to make up this uncertainty, data as many as possible were employed. As this uncertainty is considered mainly due to the application of the normal ground water depletion curve, the storms were employed whose ground water runoff is entirely at the normal ground water depletion rate. The result is shown in Fig. 4. By using Fig. 4, the writer analysed the relationship between detention water and moisture deficiency of the soil water belt of the watershed. Let rainfall divide into following classes as in Fig. 3: 0∼20, 20∼50, 50∼80, 80∼110, 110∼150, 150∼180, 180∼. The gradients of these curves in April and August were computed according to these classes. Moisture deficiency of the soil water belt and its area percentage were obtained by making use of these gradients as tabulated in Table 2 indicates the moisture deficiency of the soil water belt of the whole watershed, which is 32 mm in April and 64.7 mm in August. These values in April and in August are nearly equal to those of Fig. 4. Furthermore, Table 2 indicates the following facts. In April 46% of the watershed is nearly at field capacity and 10% at the moisture deficiency of 110 mm, the largest one. In August 40% of the watershed is at field capacity, while the largest moisture deficiency is 180 mm which occupies 20% the watershed. Table 2 is obtained from Fig. 4, in which each point has a large deviation from the average curve. Fig. 5 is obtained from Table 2 by plotting the values in the Table. The average gradients of April and August were shown by straight lines as in Fig. 5. NOGUCHI[2] reported monthly variation of soil moisture of the same watershed. According to his result the writer grouped the moisture deficiency of the

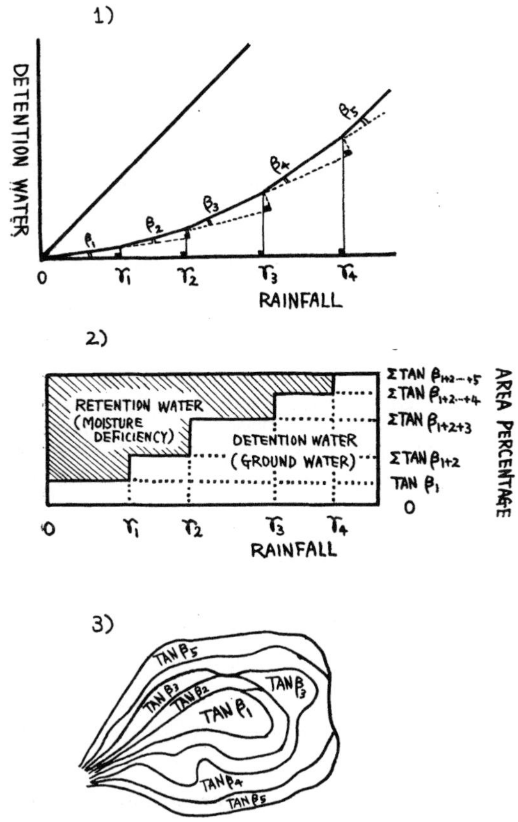

Fig. 3. Analysis of a rainfall-detention water relationthip curve.

1) Approximate expression of the relatioship curve by straight lines.

2) Explanation of the characteristic of these approximate straight liness.
The tangent of the gradients of these lines is useful to divide a storm water into retention water and detention water.

3) Division of a watershed into several sections by the amount of moisture deficiency.

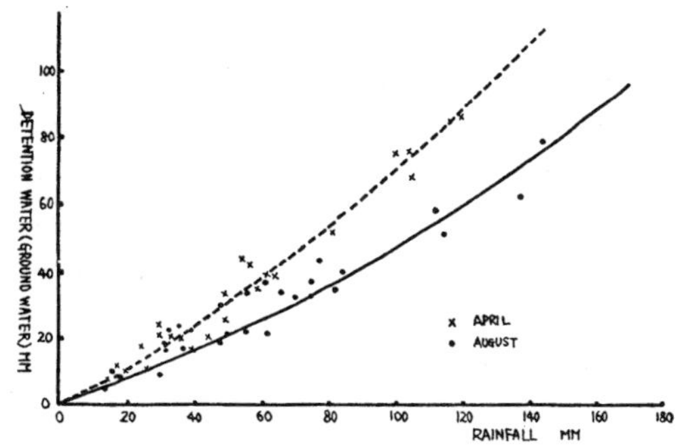

Fig. 4. Detention water of a storm in April and in August.

Table 2. Moisture deficiency and its area percentage.

April

Rainfall (mm)	Gradient	Soil moisture deficiency (mm)	Area percentage (%)
0∼ 20	0.46	0	46
20∼ 50	0.64	20	18
50∼ 80	0.75	50	11
80∼110	0.90	80	15
110∼	1.00	110	10

Total soil moisture deficiency: 32 mm

August

Rainfall (mm)	Gradient	Soil moisture deficiency (mm)	Area percentage (%)
0∼ 20	0.40	0	40
20∼ 50	0.43	20	3
50∼ 80	0.52	50	9
80∼110	0.60	80	8
110∼150	0.67	110	7
150∼180	0.80	150	13
180∼	1.00	180	20

Total soil moisture deficiency: 64.7 mm

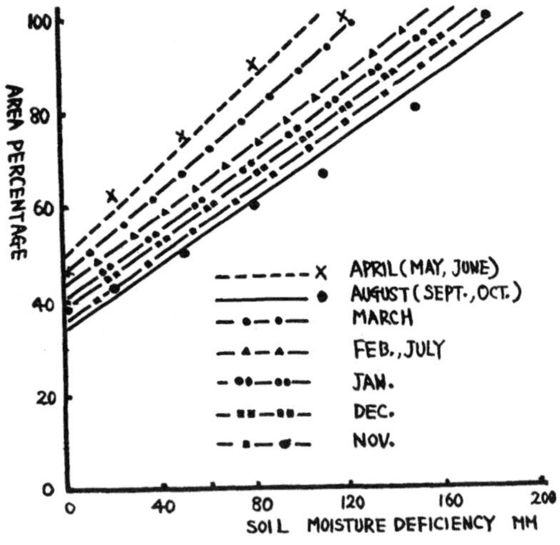

Fig. 5. The relationship between soil moisture deficiency and a area percentage.

watershed into 7 classes: April May June; March; July February; February; December; November; August September October. For these classes straight lines similar to those in April and in August were drawn in accordance with moisture differences among these 7 classes. Fig. 6 is obtained from Fig. 5. Fig. 6 shows the average moisture deficiency of the soil water belt of the watershed in each month. It is clearified that in April, at the time of maximum soil moisture, the average moisture of the soil water belt is lower than field capacity. By applying Fig. 6 to the rains from 1949 to 1952, the writer obtained the result as indicated in Table 3. According to this result, the discharge computed by Fig. 6 is 91% of the actual discharge. This proves that Fig. 6 is practically usable for estimating average discharge of a storm. The seasonal fluctuation of retention water at various rainfalls is depicted in Fig. 7 which was obtained from Fig. 6.

The writer comes to the following conclusion as already said by YAMAGUCHI[1]. The areas A and B play an important role in the maintenance of streamflow in this region of Japan. And the larger part of storm runoff seems to be composed mainly of the runoff through the areas A, B and C, and partly of that through D and E.

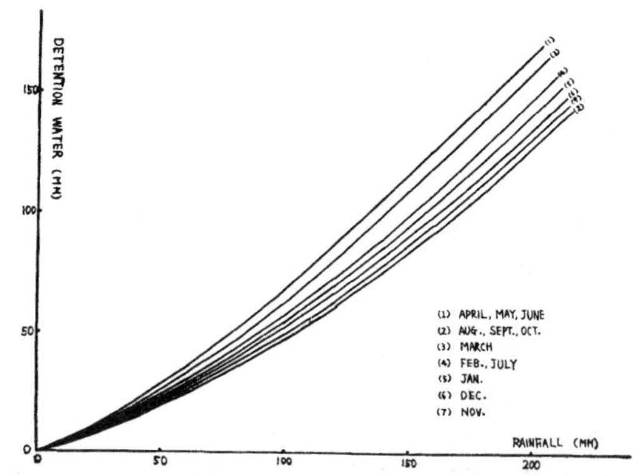

Fig. 6. Detention water (Discharge) of a storm in each month. April and August values are adjusted by the average curves.

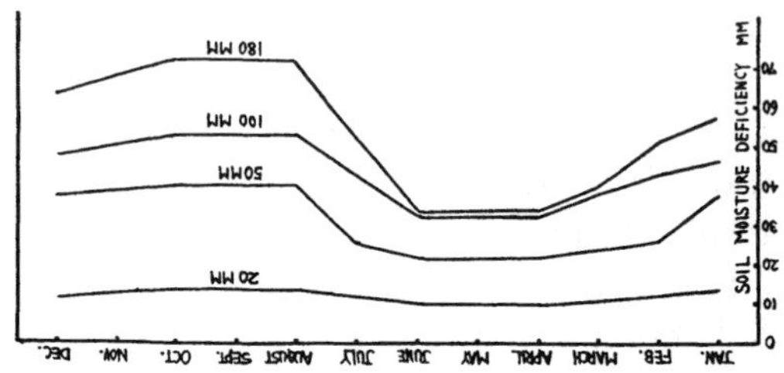

Fig. 7. Seasonal fluctuations of storm water retained in the soil water belt of the watershed at various rainfalls.

Table 3. Comparison between actual discharge and estimated discharge by Fig. 6.

Year	Estimated discharge	Actual discharge	Estimated dis. / Actual dis.
1949	963.8	977.2	0.99
1950	955.8	1081.8	0.87
1951	836.1	969.9	0.84
1952	1133.0	1269.6	0.87
	3888.7	4298.5	0.91

Acknowlegment

The writer is deeply indebted to both Dr. Y. NOGUCHI and Dr. I. YAMAGUCHI for their guidance during the study. And also to Dr. M. ASAHI who gave the writer facilities for using the basic data of the Tokyo Unv. Forest.

(Received Feb. 20. 1963)

Literature cited

1) Isao YAMAGUCHI: Suiri Kagaku, No. 27, 1962.
2) Yooichi NOGUCHI: Miscellaneous Information, Tokyo Unv. Forest, No. 12, 1957.
3) Yoshinori TSUKAMOTO: Jour. Japan. Forestry Soc. **42**-1, 1960.
4) Yoshihisa MASHIMO: Forest Soils of Japan, 11, 1960.
5) Hiroshi KAWADA: Iour. Japan. Forestry Soc. **38**-6, 1956.
6) Yoshinori TSUKAMOTO: Jour. Japan Forestry Soc. **43**-2, 1961.

Commentary on Paper 6

Hewlett, J. D. & Hibbert, A. R. (1963) Moisture and energy conditions within a sloping soil mass during drainage.

The work of Charles Hursh at Coweeta (see Paper 3), which stressed the importance of subsurface flow processes at the same time as the Hortonian model of hillslope responses, was continued and extended by a later Director at Coweeta, John Hewlett. In 1961, in a USDA Forest Service report, he summarized the results of the work at Coweeta and his perceptual model of streamflow generation that included the concept of a shrinking and expanding subsurface source area. Figure 1 of Paper 9, later reproduced in Hewlett (1974) in *Water Resources Research*, first appeared in that report.

Here I have chosen to reproduce the paper of John Hewlett and Alden Hibbert from 1963 that gives some of the detailed results from experiments on a constructed hillslope (approximately 1-m wide by 1-m deep by 15-m long; Scholl & Hibbert (1973) report on results from the larger sloping lysimeter mentioned at the end of this paper). The advantage of a constructed hillslope is that the boundaries of the flow domain are well defined. The disadvantage is that any effects of soil structure will be lost when the soil is repacked into the constructed trough. The paper analyses the drainage of the slope from saturation, suggesting that there are two regimes with an intervening transition period. The regimes are interpreted on the basis of consistent behaviour on a plot of discharge against log(time). The initial drainage period is interpreted as being the rapid drainage of saturated larger pores; the later period, the slow drainage of unsaturated soil. They note that their measured flow rates from unsaturated flow would be sufficient to maintain the baseflow if scaled up to the Coweeta catchments.

This method of analysing drainage curves on semi-log paper, and interpreting the results in process terms, had already had a long history in hydrological analysis (and would make an interesting historical study in itself). The exponential decline of discharge with time was used by Horton as a representation of the "depletion curve" (e.g. Paper 1). The concept underlying these analyses is that of an exhaustion phenomenon. Successive flow domains, or processes, continue to contribute to the hydrograph until their associated storage is exhausted and the recession becomes dominated by a new slower process. [As an aside, Horton interpreted the exponential decline of his infiltration equation in a similar way – not as a result of the profile control of infiltration rates as implied by models based on the Darcy-Richards equation, but, as he explained in his unpublished monograph on infiltration, as an exhaustion of the capacity of the soil to accept water at the soil surface due to packing, redistribution of particles, blocking of larger pores, etc. (Beven, 2004)]. Barnes (1939, 1940) went as far as to distinguish five different processes from the different slopes of an extended hydrograph recession curve (see also example in Paper 14). The method was also used by Weyman (Paper 17), Harr (Paper 22), and many others.

How should we interpret these analyses today? If these breaks in slope on this type of plot have been commonly found and reported as an empirical result (even if it might require the eye of faith to differentiate five different segments), what is their significance? A single exponential decline has a simple theoretical interpretation as the output from a linear store (though this has an infinite rather than finite duration). The sum of the outputs of several linear stores operating in parallel will give a semi-log plot with straight lines, but with some carry over from one "process" to another because of the infinite output duration. But we are now aware that neither saturated or unsaturated zones should act as simple linear stores. We are aware that outflows may be made up of a mixture of displaced pre-event and event waters. We are also aware that at the hillslope and catchment scales there may be spatial differences in the flow processes as a result of soil, geological and topographic differences that would make if difficult to say that, during drainage from a wet state, one process has finished and another started. It remains unclear as to whether the empirical straight line segments are simply a convenient approximation to a more complex characteristic drainage function that remains simple because differences from event to event have not been reported, or whether we are missing something about representing the characteristic response of complex hillslopes that becomes simple at a certain scale, and is essentially different from analytical treatments of simple hillslopes (such as, for example, Sloan & Moore, 1984; Duffy, 1996; Troch *et al.*, 2003).

John Hewlett moved from the USDA Forest Service to the University of Georgia in 1964, but maintained strong links with Coweeta. He died in 2004 (see obituaries in *EOS Trans. Amer. Geophys. Un.* 86(12), 124, 2005, and *Hydrological Processes* 19, 2093–2095, 2005). A detailed account of work at the Coweeta Laboratory is provided in Swank & Crossley (1988) and Swank *et al.* (2001).

References

Barnes, B. S. (1939) Structure of discharge recession curves. *Trans. Am. Geophys. Un.* **20**, 721–725.

Barnes, B. S. (1940) Discussion of analysis of runoff characteristics. *Trans. Am. Soc. Civ. Engrs* **105**, 104–106.

Beven, K. J. (2004) Robert Horton's perceptual model of infiltration processes. *Hydrol. Processes* **18**, 3447–3460.

Duffy, C. (1996) A two-state integral balance model for soil moisture and groundwater dynamics in complex terrain. *Water Resour. Res.* **32**, 2421–2434.

Hewlett, J. D. (1974) Comments on letters relating to 'Role of subsurface flow in generating surface runoff. 2. Upstream source areas' by R. Allen Freeze. *Water Resour. Res.* **10**(3), 605–607.

Scholl, D. G. & Hibbert, A. R. (1973) Unsaturated flow properties used to predict outflow and evaporation from a sloping lysimeter. *Water Resour. Res.* **9**(6), 1645–1655.

Sloan, P. G. & Moore, I. D. (1984) Modeling subsurface stormflow on steeply sloping forested watersheds. *Water Resour. Res.* **20**(12), 1815–1822.

Swank, W. T. & Crossley, D. A. (1988) *Forest Hydrology and Ecology at Coweeta*. Ecological Studies 66. Springer-Verlag, New York, USA.

Swank, W. T., Meyer, J. L. & Crossley, D. A. (2001) Long term ecological research: Coweeta history and perspectives. In: *Holistic Science: The Evolution of the Georgia Institute of Ecology (1940–2000)* (ed. by G. W. Barrett & T. L. Barrett), 143–163. Sheridan Books, Ann Arbor, Michigan, USA.

Troch, P. A., Paniconi, C. & van Loon, E. E. (2003) Hillslope-storage Boussinesq model for subsurface flow and variable source areas along complex hillslopes: 1. Formulation and characteristic response. *Water Resour. Res.* **39**(11), doi: 10.1029/2002WR001728.

Journal of Geophysical Research Vol. 68, No. 4 February 15, 1963

Moisture and Energy Conditions within a Sloping Soil Mass during Drainage

John D. Hewlett and Alden R. Hibbert

Southeastern Forest Experiment Station
Forest Service, U. S. Department of Agriculture, Asheville, North Carolina

Abstract. In an effort to explain the source of nonstorm streamflow in deep-soiled areas of the southern Appalachians, a 3 × 3 × 45-foot inclining concrete trough was constructed on a 40 per cent slope. The structure was packed with a natural sandy-loam soil to a bulk density of 1.3, and after thorough soaking was covered with plastic to prevent evaporation. Outflow was measured at the base, and soil moisture tension and content were recorded for 145 days. As shown by piezometers, the larger pores were substantially emptied in 1.5 days, during which outflow occurred according to the expression,

$$Q = a_1 T^{-b_1}$$

where T is time in days since day zero and a_1 and b_1 are constants. After a 5-day transition period, the logarithm of drainage for the next 80 days was again linearly related to the logarithm of time, but with new constants, a_2 and b_2, expressing unsaturated flow from the entire soil mass. An additional 60 days of flow occurred, deviating more and more from the log-log relation as the limits of drainage were approached. Soil moisture content and tension substantiated the theory that the entire unsaturated soil mass was contributing to outflow throughout the experiment. It is concluded that unsaturated flow in the earth mantle of steep watersheds cannot be ignored in hydrograph analysis, since it may well be a primary mechanism for sustaining baseflow.

Introduction. Increasing concern over water resources has focused attention on how little is known about the source of high-quality surface water from protected mountain watersheds. The generally accepted explanation of nonstorm streamflow (base flow), which accounts for 85 per cent of the yield from many watersheds of the southern Appalachians, assumes that extensive, saturated groundwater aquifers feed streams during nonstorm periods [*Wisler and Brater*, 1959; *Todd*, 1959; *Linsley et al.*, 1958], while stormflow is assumed to come from overland or subsurface flow through interconnecting channels below the soil surface [*Hursh*, 1944]. Although these concepts may apply in many areas, their application in forested mountain land is doubtful.

At the Coweeta Hydrologic Laboratory in western North Carolina, research in the dynamics of baseflow and stormflow is in its fourth year. The experimental basin has steep forested slopes, narrow incised drainage channels, and moderately deep soil over relatively tight granitic country rock. Rainfall averages 80 inches annually but frequent storms produce little or no overland flow. Groundwater wells at Coweeta have generally failed to demonstrate saturated aquifers except along streams and drainageways, where accumulating water appears as spring or seepage flow. Because of the steep topography, such areas are necessarily restricted to the immediate vicinity of the stream, a zone which appears too small to sustain streamflow through the summer period of high evapotranspiration.

In an earlier paper, *Hewlett* [1961] offered an explanation of baseflow which visualizes the entire soil mantle as a storage aquifer feeding sustained flow. In this view, narrow groundwater bodies along stream channels are not of themselves the source but rather a conduit through which slowly draining soil moisture passes to enter the stream. Explanation of the source of baseflow in mountains appears to be chiefly a problem involving the physics of unsaturated flow in soils.

The physics and theory of moisture flow in soil masses of various geometric dimensions and internal structure have progressed in recent years beyond the level available during earlier hy-

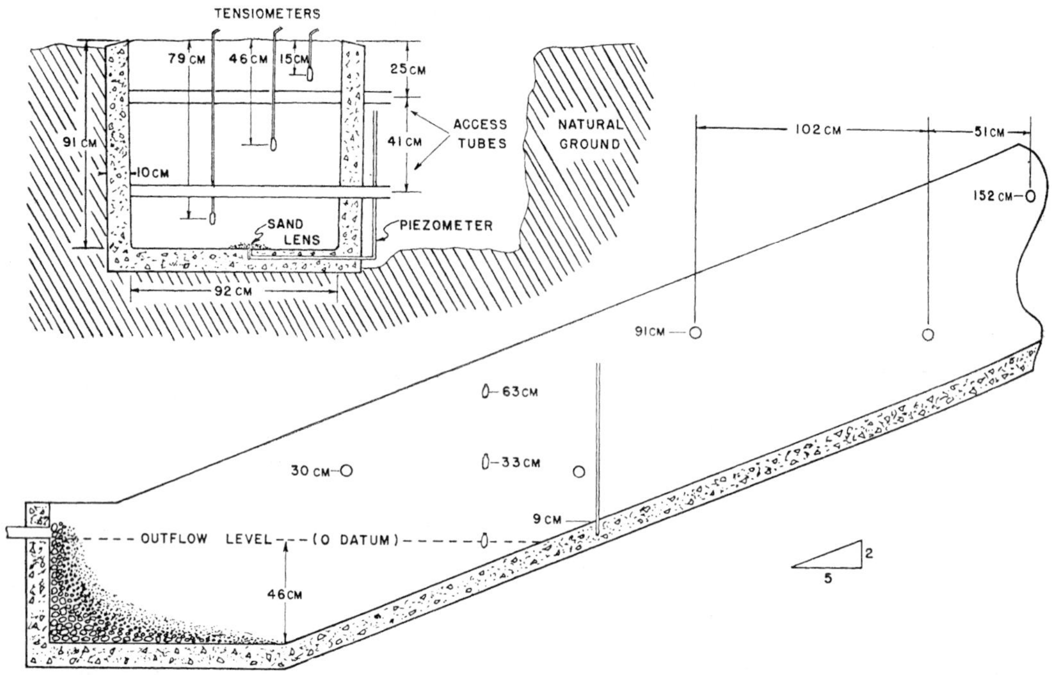

Fig. 1. Cross and longitudinal sections of the lower portion of the inclined soil column. Three additional sets of tensiometers and piezometers are located at 10- to 14-foot intervals upslope. Except where denoted by arrows, numbers indicate vertical distance above the outflow level.

drologic studies. Darcy's familiar equation for flow in saturated systems has been successfully modified to account for flow in unsaturated systems. These equations permit evaluation of moisture conductivity and diffusivity when all boundary conditions and certain physical properties of soil are known. However, as *Gardner* [1959] has pointed out, departures of real systems, particularly those involving nonsteady-state flow, from theoretical solutions indicate the importance of various natural inhomogeneities that are difficult to account for theoretically. The physics of moisture movement in soil offers a firm basis for hydrograph theories but it is doubtful that we are yet finished with field observation and model testing as a means of closing the gap between physical theory and hydrologic application.

Seeking to explain further the source of nonstorm streamflow from the slopes of the southern Appalachians, a series of inclined soil models, thought to approximate the geometry of segments of natural watersheds, have been constructed at Coweeta. *Hewlett* [1961] reported early experience with two sloping, open-topped columns, 24 and 32 feet long. These structures, when artificially soaked, produced measurable outflow for more than 2 months without further recharge. This paper deals with a larger, better-instrumented model designed to clarify the moisture and energy conditions within sloping soil profiles during drainage. Results are believed to have general application to hydrologic theory of water yield from steep land.

Methods. An inclining concrete trough measuring 3 × 3 × 45 feet was constructed on a 40 per cent slope (Figure 1) and packed with well-mixed natural soil to a bulk density of 1.3 ± 0.1. The soil was excavated nearby from the C horizon of Halewood sandy loam, a common type in the southern Appalachians. Texture analysis by the Bouyoucos hydrometer method averaged 60 per cent sand, 18 per cent silt, and 22 per cent clay. Total pore space in the packed condition was about 50 per cent by volume, very close to that in its normal undisturbed state.

After a sprinkling system had soaked the inclined surface for several days, the structure

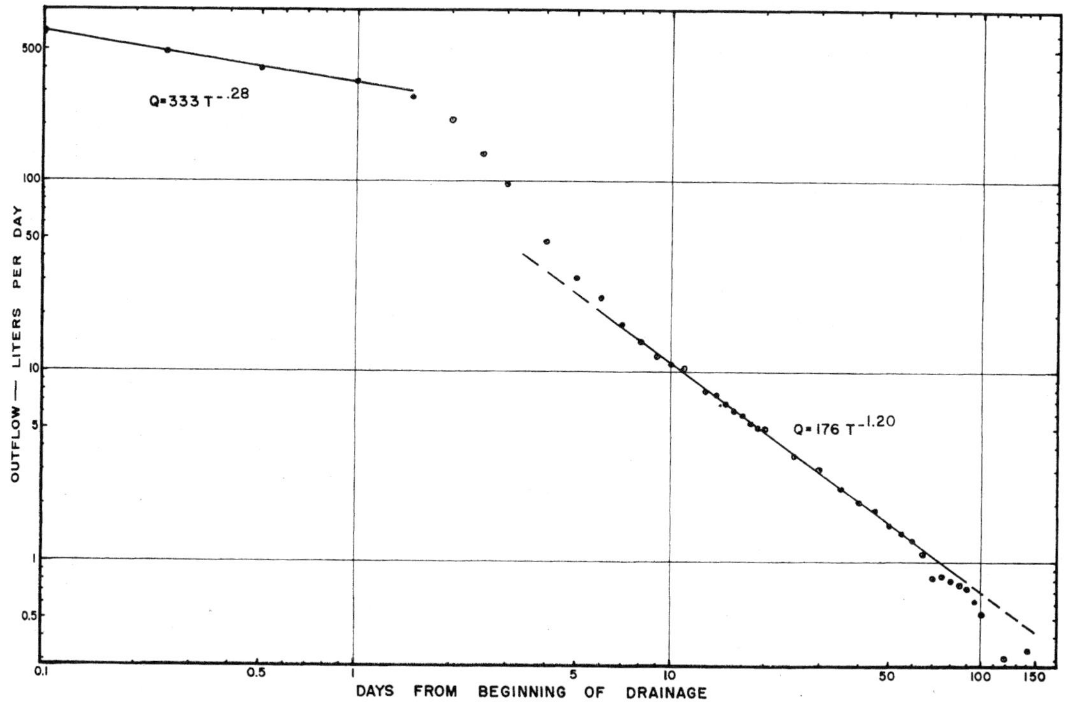

Fig. 2. Outflow since beginning of drainage, showing log-log relation to time.

was covered with plastic film and allowed to drain. Outflow was measured continuously by a water level recorder mounted on a tank at the base of the column. The outflow pipe established a free water surface which was used as the zero datum for all measurements. The soil under the free water surface was graded to sand, gravel, and rock to simulate stream bank conditions and to allow free drainage.

Seventeen access tubes to permit measurement of soil moisture by the neutron scattering method were inserted horizontally through holes in the concrete wall at various levels up the slope. In addition, there were four sets of one piezometer and three tensiometers distributed up the slope. The first tensiometer cup was located at the outflow level, or 0 cm, the first piezometer at 9 cm vertically above zero datum, the first access tubes at 30 cm, and so on up the column.

The soil was soaked on July 6, and soil moisture tension and content were recorded at first hourly, then daily, and finally weekly as the drainage rate decreased. Soil temperature and barometric pressure were recorded for the duration of the experiment.

Results. After 145 days of outflow, the experiment was terminated on November 30, 1960, and another phase of the study was begun. A total of 1.26 m^3 of water drained from 10.85 m^3 of soil above the outflow level, 76 per cent during the first 5 days, 19 per cent during the next 45 days, and 5 per cent during the last 95 days. Outflow fluctuated in a cyclic manner with temperature and barometric pressure changes, but these fluctuations were traced mainly to the influence of gases trapped within the soil below the outflow level. When averaged from week to week, outflow rate decreased smoothly with time. In Figure 2, outflow is plotted over time from beginning of drainage; both are plotted on logarithmic scales because experience has shown that the logarithm of unsaturated drainage tends to be linearly related to the logarithm of time [cf. *Ogata and Richards,* 1957; *Nixon and Lawless,* 1960]. This drainage trial, as well as replicated trials later on, tended to show two separate periods during which the log-log relation of outflow to time was linear. During the first 36 hours, while the saturated portion of the model drained rapidly, outflow could be

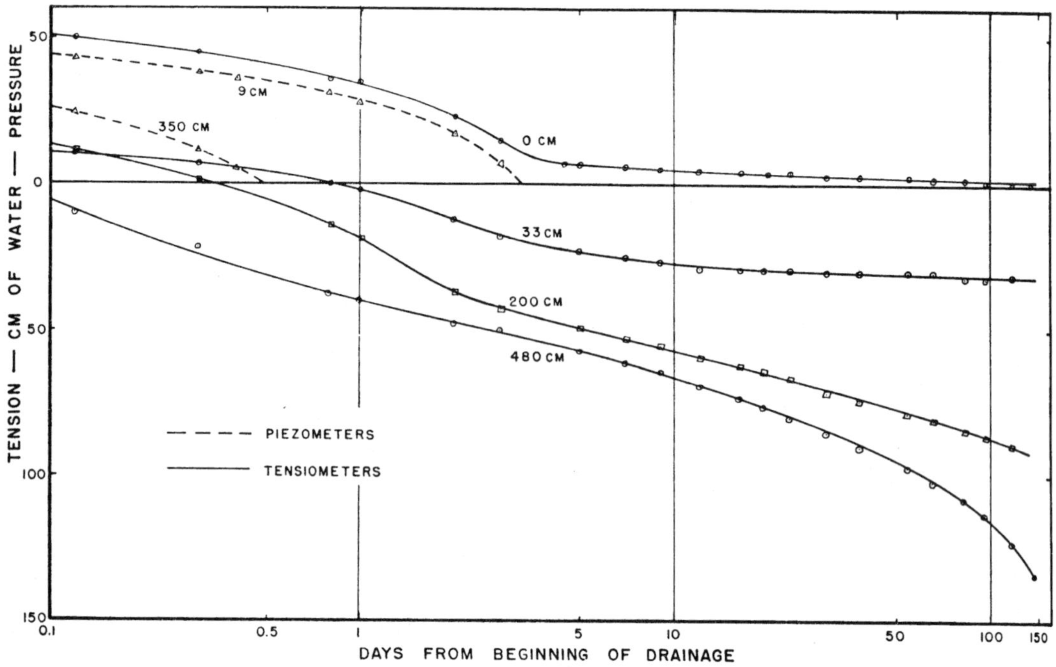

Fig. 3. Changes in pressure and tension at selected levels within the soil mass.

accurately expressed as

$$Q = a_1 T^{-b_1} \qquad (1)$$

where Q is the rate of outflow, T is time elapsed, and a_1 and b_1 are constants peculiar to the system. After a 5-day transition period, during which the larger pores of the soil mass above the outflow level were substantially emptied, the logarithm of outflow was again linearly related to the logarithm of time, but with new constants,

$$Q = a_2 T^{-b_2} \qquad (2)$$

Outflow during the last 60 days was intermittent, and it deviated more and more from equation 2 as the weather grew colder and various limitations of the system became significant.

Records from selected tensiometers and piezometers which are plotted in Figure 3 allow comparison of outflow rates to energy conditions (or matric potential) within the draining soil mass. The two major periods of drainage illustrated by equations 1 and 2 are associated respectively with saturated and unsaturated flow. During the first 36 hours, emptying of the larger soil pores masked unsaturated flow from the remainder of the soil mass, both as to amount and rate of delivery to the outflow pipe. As the free water surface retreated toward the zero datum (piezometers at 350- and 9-cm levels in Figure 3) a transitional period began and continued until about the fifth day, when all larger pores were drained. However, nearly all drainage after 36 hours was derived from the entire mass of unsaturated soil as indicated by increases in soil moisture tension at 33, 200, and 480 cm above the zero datum. There was no doubt that unsaturated conditions prevailed along the concrete bottom of the trough after the first 2 or 3 days, since piezometers were sucked dry (see Figure 1) as the free water surface passed downward.

Theoretically, at least, the soil moisture tension at any point within the model (neglecting osmotic forces) is approaching a value equivalent to the height above zero datum. Thus the tension in centimeters of water should approach the vertical distance above the outflow in centimeters. In Figure 4 the tension at 15 and 79 cm below the sloping soil surface at each of four sets of tensiometers up the slope is plotted over the vertical height above the outflow level. The data are plotted for 7 and 119 days from the beginning of drainage and the curves are shown

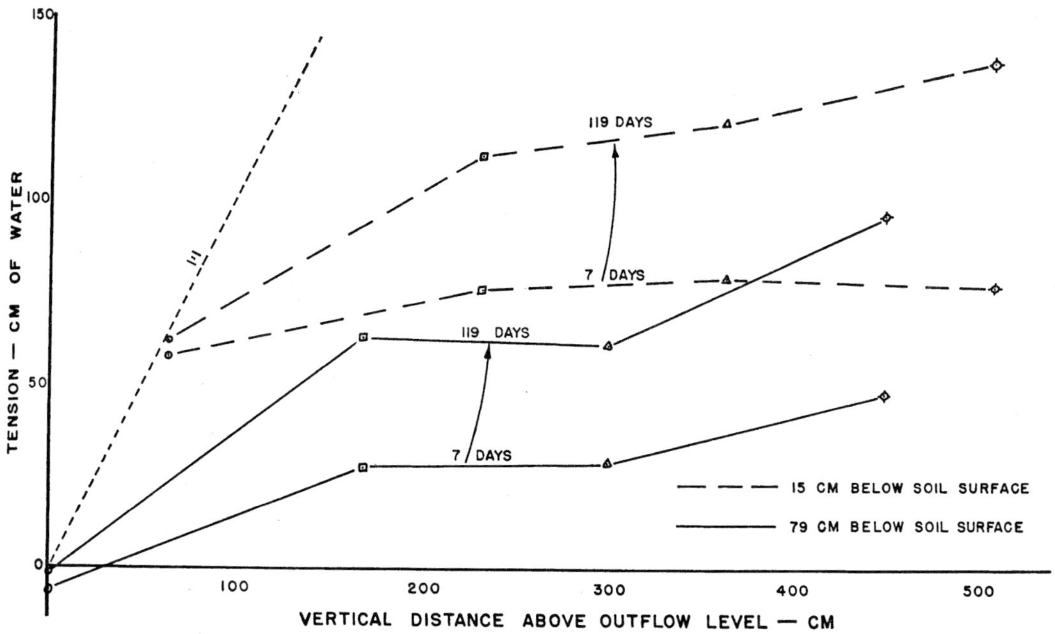

Fig. 4. Soil moisture tension increases with time and elevation above the outflow level, and decreases with depth below the soil surface.

rotating toward the 1:1 ratio of tension to height. Within each vertical profile, viewed apart from the whole structure, soil moisture tension reacts quickly to approximate a 1:1 ratio with height. Thus the difference between the 15- and 79-cm levels is roughly 64 cm all the way up the slope. However, local tension reacts more slowly to the remote datum at the outflow level. Nevertheless, all values appear to be rotating toward eventual equilibrium at the 1:1 ratio of height to tension, again neglecting possible osmotic forces. Time and elevation are shown to be factors affecting the development of tension within a draining soil mass that is protected from evaporation. Changes in the angle of slope and the physical properties of the soil may be expected to play a similar role.

Neutron meter records of soil moisture permitted quantitative determination of the source of the outflow. Moisture-content changes within the whole structure were substantially equivalent to outflow except during early drainage. Outflow exceeded soil moisture change by 31 per cent during the first 10 days. During this time, moisture concentration along the floor of the model was poorly reflected in neutron meter counts because of the location of the access tubes. After the tenth day, moisture change was well correlated with outflow. Moisture content is plotted over time in Figure 5 for the selected levels, 30, 152, and 460 cm above zero datum. The data show some scatter, but the trends are clear enough. No important change occurred in moisture content 30 cm above the outflow level. At other levels the logarithm of moisture content decreased linearly with the logarithm of time, and the sum of all changes (Q) within the draining soil mass is similarly related to time, as in equations 1 and 2. Although the data varied because of deviations in soil compaction from place to place, both the total amount of moisture drained and the rate of drainage appeared to increase with height above the outflow level.

One year later, during the summer of 1961, the whole experiment was repeated with the same structure in order to determine if weathering of the soil was affecting the results. Moisture, energy conditions, and outflow rates were virtually identical during the second experiment.

Discussion and conclusions. Soil within and below the zone of tree roots but above bedrock or a water table, when recharged regularly by

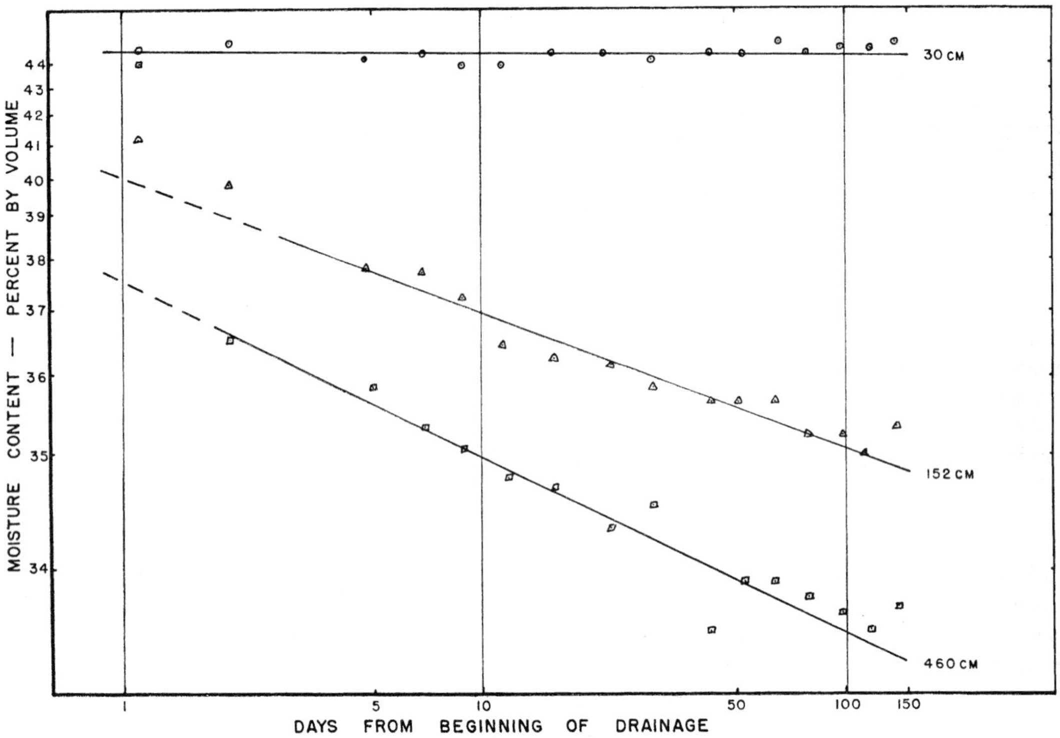

Fig. 5. Moisture content decreases with time and elevation above the outflow level.

rainfall, may constitute the chief aquifer for stored water between rains in steep terrain. *Meinzer* [1942] has referred to this zone as the 'no man's land of hydrology,' adding that water in it was to a large extent in 'dead storage.' A more realistic definition of this reservoir may be 'dynamic storage,' and, in addition to the saturated aquifers which lie below, it should be recognized as the location of important supplies of potential stream water in many areas.

Knowing where water is and how fast it is moving underlies all efforts to manage the supply. The behavior of water within the soil mass used in this study can probably be predicted by solution of the flow equation if boundary conditions are adequately defined. Definition of a natural system is possible only in general terms at present, however, because of the irregular geometry of soil masses, soil physical properties, and fluctuating rainfall. We can, of course, record outflow from a watershed, but there is a real need for improved concepts for determining the source and timing of flow by the study of the morphology of drainages. From such knowledge, watersheds can be evaluated as moderators of water flow, and future behavior under specified conditions may be predicted with greater precision. The flow equation or its modifications probably affords the best basis for hydrograph analysis and correlation of drainage morphology to outflow.

Despite differences in the size and morphology between a watershed and the soil model, the relative rates and amounts of drainage of saturated and unsaturated soil observed here, particularly when reinforced by extensive measurement of soil moisture on the Coweeta basin, lead us to conclude that unsaturated flow in mountain soils is both an important and an immediate cause of sustained baseflow in mountain streams. Even after 2 months of drainage, 10.85 m^3 of unsaturated sandy loam soil was yielding 1.42 liters of water per day. Multiplying by appropriate factors, which are, of course, rather speculative because of the lack of good estimates of total soil volumes, shows such yield to be within the order of magnitude of low flows from Coweeta watersheds, despite the fact that 2 months without recharge is an extremely rare event at Coweeta.

It is interesting to speculate whether the simplified empirical flow equations (1 and 2) are fundamental to movement of water through many soils under gravity. If so, such equations may prove to be useful tools in hydrograph analysis, since water in streams must be the sum of a series of similar components of flow. Evaluation of the parameters a and b for each soil type and drainage geometry should help to explain why watersheds which look similar often yield quite differently. Unit watershed experiments at Coweeta and comparable experiments elsewhere have brought many such differences to light [*Hewlett and Hibbert,* 1961].

Work with models and small watersheds is continuing at Coweeta in an effort to define the behavior of water in soil, particularly as its behavior affects the timing of baseflow and stormflow from mountain watersheds. It is odd to note that, whereas baseflow from the watershed model was derived from unsaturated soil, the part of flow most equivalent to stormflow came from 'groundwater' drainage. Certainly, in conventional methods of hydrograph separation this possibility is not recognized in some areas, since storm runoff is related almost by definition to overland flow and channel interception. Field observations at Coweeta suggest that stormflow may well be associated with temporary expansion of saturated aquifers along stream channels—aquifers which deplete rapidly through porous material on steep slopes. A soil model measuring 5 × 7 × 200 feet is under construction at Coweeta so that we can study the source and timing of baseflow and stormflow under natural rainfall. Mathematical models of flow will be tested on this structure and compared with adjacent gaged watersheds.

References

Gardner, W. R., Mathematics of isothermal water conduction in unsaturated soil, *Natl. Res. Council, Highway Res. Board, Spec. Rept. 40,* 78–87, 1959.

Hewlett, John D., Soil moisture as a source of base flow from steep mountain watersheds, *U. S. Forest Serv., Southeast. Forest Expt. Sta. Paper 132,* 11 pp., 1961.

Hewlett, John D., and Alden R. Hibbert, Increases in water yield after several types of forest cutting, *Quart. Bull. Intern. Assoc. Sci. Hydrol., 6,* 5–17, Louvain, Belgium, 1961.

Hursh, Charles, Appendix B—Report of sub-committee on subsurface flow, *Trans. Am. Geophys. Union, 25,* 743–746, 1944.

Linsley, Ray K., Max Kohler, and J. L. H. Paulhus, *Hydrology for Engineers,* pp. 150–160, McGraw-Hill Book Co., New York, 1958.

Meinzer, Oscar E., Ground water, in *Hydrology,* vol. 9, *Physics of the Earth,* p. 394, McGraw-Hill Book Co., New York, 1942.

Nixon, Paul R., and G. P. Lawless, Translocation of moisture with time in unsaturated soil profiles, *J. Geophys. Res., 65,* 655–661, 1960.

Ogata, Gen, and L. A. Richards, Water content changes following irrigation of bare-field soil that is protected from evaporation, *Soil Sci. Soc. Am. Proc. 21,* 355–356, 1957.

Todd, David K., *Ground Water Hydrology,* p. 152, John Wiley & Sons, New York, 1959.

Wisler, Chester O., and Ernest F. Brater, *Hydrology,* 2nd ed., pp. 181, 275, John Wiley & Sons, New York, 1959.

(Manuscript received October 11, 1962.)

Commentary on Paper 7

Betson, R. P. (1964) What is watershed runoff?

Roger Betson, who worked for the Tennessee Valley Authority (TVA) for many years, was one of the first to really question (at least in print) the widespread view of Hortonian surface runoff as the explanation for storm hydrographs. Betson starts his paper by noting "*the concept of infiltration is readily understood. Measuring and understanding the physical factors controlling the infiltration process, however, is exceedingly difficult*" (p.1541). The dominance of the Hortonian paradigm in hydrograph analysis is reflected in the structure of the paper which is based on assessment of an infiltration-capacity function based on Horton's infiltration equation. Fitting the simplest form of this function to storm data, based on infiltration-capacity alone, does not provide an adequate explanation of the observations of the volume of storm runoff (though it is not clear how the observed storm runoff volumes were calculated and in particular whether this involved a hydrograph separation). Adjustment to account for interception losses did not help much; only when it was allowed that infiltration excess might occur over only part of the basin did a good fit to the data result ($r = 0.99925$). Betson notes: "*the very low standard error, 0.02 inch* [0.05 cm], *is within an acceptable range for hydrologic data*" (p.1546)!! The good fits were repeated on a number of other catchments and suggest that, for most of these catchments, the runoff producing area is relatively constant. Betson cites a number of other studies, including the Hewlett (1961) report from Coweeta to support this contention. He also cites observations from three plots in one of the small catchments in North Carolina, to support the idea that infiltration capacities might not be exceeded over large proportions of some catchments. He shows how the concept of a constant partial contributing area can be used to derive an apparent infiltration-capacity function for a catchment, where a classical infiltration excess analysis over the whole of the catchment would fail. He also recognizes some of the limitations of the approach (see his last paragraph) but, as in so many hydrological papers, makes some final suggestions as to how the concept might be made more complex to allow better understanding of catchment responses.

Betson's approach and the partial contributing area concept appears to be very much rooted in infiltration excess concepts. There is absolutely no indication in the paper that some of the storm runoff might have been from subsurface sources, despite some of the low fractional contributing areas implied by the analysis. In this, he differs from Hewlett, even though Hewlett (1974) objects that the concept of the partial or variable contributing area originated with work at Coweeta rather than from the TVA (see discussion of Paper 6).

References

Hewlett, J. D. (1974) Comments on letters relating to 'Role of subsurface flow in generating surface runoff, 2. Upstream source areas' by R. Allen Freeze. *Water Resour. Res.* **10**(3), 605–607.

What Is Watershed Runoff?

ROGER P. BETSON

Tennessee Valley Authority, Knoxville, Tennessee

Abstract. A nonlinear mathematical model, starting with the integral of an infiltration capacity function, is developed to analytically equate the difference between rainfall and runoff to hydrologic variables. Only the three independent variables—storm rainfall, duration, and soil moisture—are used, and an equation is evolved in which the identity of the coefficients is kept intact and unusually good statistical control is maintained. The coefficients of the equation appear to be stable over a range of watershed sizes and conditions. The equation strongly indicates that runoff usually originates from a small, but relatively consistent, part of the watershed. The function can be manipulated to show a 'function of apparent watershed infiltration capacity.' This function characterizes the infiltration capacity of that portion contributing to runoff, on the average, and should prove to be a useful infiltration capacity index with which watersheds can be compared. The equation itself provides insight into why in situ measurements of infiltration capacity seldom agree with the capacity determined from rainfall-runoff data. It also indicates why storm runoff frequently is not linear with respect to causative factors.

INTRODUCTION

Infiltration, by definition, is the movement of water through the soil surface into the ground. The concept of infiltration is easily understood. Measuring and understanding the physical factors controlling the infiltration process, however, is exceedingly difficult.

The rate of infiltration of precipitation, in general, equals the rate of precipitation as long as the infiltration capacity of the soil is not exceeded. When this capacity is exceeded, the excess precipitation is either stored temporarily on the soil or begins to run off. Since an understanding of the surface runoff process is basic to many problems in engineering, agriculture, municipal planning, etc., and since surface runoff is closely related to infiltration capacities, there has been a considerable effort expended in trying to evaluate measures of infiltration capacity.

Measurements of infiltration capacity have generally followed two directions. Actual measurements of infiltration capacity are made in situ using artificial sprinkling devices or small surface runoff plots. Indirect determinations of infiltration capacity are made from measurements of storm precipitation and watershed runoff. These indirect determinations often result in more than an index of the infiltration capacity, since results frequently vary from storm to storm and seldom agree with in situ measurements. It is important to note, however, that the infiltration capacity obtained from these indirect determinations represents an average for the watershed, whereas the in situ or plot measurements approach a point, or local, measure of infiltration capacity. Point measures can vary considerably from the average.

Indirect measurements of infiltration capacity have generally been hampered by the fact that this capacity is not a constant value. It is a complex function, principally of soil moisture but also of many other factors. It may change considerably during the course of a single storm. Computations of either mathematical infiltration capacity functions or indices have generally been limited to individual storms. An analytic approach to the computation of an infiltration capacity function has been lacking because available techniques for this type of problem solution usually required linear or linear-transformable equations.

With the availability of electronic computers, the limitations of simplicity and linearity have been removed as requirements of analytic solutions. Computer techniques are now available that will evaluate almost any equation that can be devised for data adjustment. This paper shows the development of a complex function of watershed infiltration capacity using analytic techniques. This development starts with an available empirical model and continues with improvements based on the results obtained in fitting the model to actual data.

The Infiltration Process

The role of infiltration in the hydrologic cycle was probably first recognized by *Horton* [1933]. This infiltration pertained to the passage of water through the soil surface. Once the water is in the soil its further movement is defined as percolation. The maximum rate at which water can enter the soil is dependent on many factors. Horton recognized a maximum capacity and a minimum capacity. The maximum capacity for any given rain occurs at the beginning of the rain. This rate decreases rapidly at first because of changes in the structure of the surface soil and increases in surface soil moisture, and then gradually approaches a somewhat stable minimum. The decay in the capacity results from such processes as change in surface soil moisture, rain packing of the soil, closing of sun checks, swelling of crumb structure, in-washing, and breaking down of the crumb structure. The minimum infiltration capacity approaches the percolation rate of the soil profile. Under conditions of saturation the percolation rate and, in turn, the infiltration capacity can be limited by the permeability of some least pervious soil horizon.

Horton defined the infiltration capacity, $f(p)$, as the maximum rate at which a given soil can absorb precipitation as it falls. The actual infiltration during a storm is equal to the capacity only when the rainfall intensity equals or exceeds $f(p)$. Horton proposed that the relation between $f(p)$ and the rainfall duration can be expressed by

$$f(p) = f_c + (f_0 - f_c)e^{-kT} \quad (1)$$

where

$f(p)$ = instantaneous infiltration capacity in inches per hour at time T.
f_c = minimum infiltration capacity.
f_0 = maximum infiltration capacity at time zero.
T = time in hours.
e = Napierian base.
k is a positive constant.

If this infiltration capacity function can be considered constant for a given set of physical watershed conditions (i.e., cover, soil, state of tilth, etc.), the integral of this function over time would determine the maximum amount of precipitation that could be infiltrated in a given time. If the storm rainfall is intense enough to exceed the infiltration capacity during the entire storm, the difference between the storm rainfall and this volume resulting from the integration would be approximately equal to the surface runoff volume. Thus it should be possible to determine the coefficients in this equation from rainfall-runoff data. Evaluating this equation from data, however, would necessitate using nonlinear fitting techniques since the function cannot be easily linearized.

Earlier work in the Tennessee Valley Authority (TVA) has shown the feasibility of using the multivariate technique of component analysis in the solution of complex equations by the method of nonlinear least squares [*Snyder,* 1962]. The technique for solution by nonlinear least squares may be found under the name of 'method of differential corrections' [*Nielsen,* 1957]. Whenever the composite technique of nonlinear least squares and component analysis has been tried in TVA, successful evaluation of mathematical models has been achieved. All the solutions of the various forms of the developed equation were obtained from an available special-purpose IBM 704 program [*TVA,* 1963].

Development of an Infiltration Capacity Function

Basic data. The data used in the development of the watershed infiltration capacity function were collected at the Western North Carolina Cooperative Research Project located near Waynesville and Asheville, North Carolina. This project has been described in various progress reports as well as in special reports on the project [*TVA,* 1960*a*]. The purpose of this small watershed research project is to determine water-land relationships for agricultural development of some of the principal soils and crops in western North Carolina. In accomplishing the project objectives, information is being obtained on the effects of single agricultural practices on (*a*) runoff, (*b*) soil moisture, and (*c*) groundwater levels. The program involves basic research in the development of hydrologic relationships.

The four experimental watersheds are located

in the Blue Ridge subregion on the southern Appalachian soils region. The soils are all acid and low in available phosphorus and they vary in their content of potassium. Subsoils are usually permeable and well drained. Average slopes in the watersheds range between 22 and 38 per cent, some slopes being as steep as 60 per cent. The watersheds are generally bowl-shaped, varying in size from 3.7 to 5.6 acres. A metal cutoff wall is installed to a predominantly metamorphic bedrock at the lower end of each watershed. Continuous discharge measurements are made with either a 2-foot H-type flume or a 1½-foot San Dimas flume. Surface runoff measurements are also made on 0.05-acre subplots within some of the watersheds.

The development of the infiltration capacity function was based on data obtained when the watersheds were in either a moderately or heavily grazed pasture cover. These two covers were used so that changes in the structure of the soil surface resulting from either the impact of high-energy rainfall or tillage would be minimized in the initial stages of the development.

Initial assumptions. Various investigators have shown that the initial infiltration capacity is a function of soil moisture. As the soil moisture at the beginning of a storm becomes higher, the initial infiltration capacity becomes lower, approaching the percolation rate as a condition of saturation is reached. If the infiltration capacity function of a watershed can be considered relatively constant under a given watershed regime, it should be possible to use soil moisture to predict the initial infiltration capacity or, in other words, the beginning point of the function. If, in addition, storm runoff can be considered to be the flow that results when rainfall exceeds the infiltration capacity, it should be possible to evaluate a watershed infiltration capacity function from conventional hydrologic measurements. The extent to which that portion of flow known as subsurface flow, or interflow, may invalidate this assumption will be discussed later.

The soil moisture used in this study is an index computed on a special-purpose IBM 704 program [*TVA*, 1963]. The basic concept of this index is a bookkeeping system of daily values of soil moisture balance. An upper- and a lower-level soil moisture index are computed. The upper-level index, which was used in the equations that follow, has been found to agree well with near-surface measurements of soil moisture.

The initial trial to develop a watershed infiltration capacity function was based on Horton's infiltration capacity equation. The area under the infiltration capacity curve for time T is a volumetric loss. If storm runoff can be considered an excess that results when the infiltration capacity is exceeded, this loss should equal the difference between rainfall and runoff.

$$L = R - RO = \int_0^D (c + be^{-nT})\, dT \quad (2)$$

where

L = loss in inches.
R = storm rainfall in inches.
RO = storm runoff in inches.
D = storm duration in hours.
c, b, and n are coefficients to be determined. They correspond to f_c, $(f_0 - f_c)$, and k, respectively, in (1).

Integration of the equation in this form, however, starts at the maximum infiltration capacity, or initially dry condition, whereas the initial infiltration capacity for a given storm might be considerably less than the over-all maximum. In fact, when the ground is saturated, the maximum infiltration capacity at the beginning of a storm may not be much different from the minimum capacity (c). The equation, therefore, needs to be modified to incorporate soil moisture in such a manner that it indicates the time-beginning point of the equation or, in other words, the lower limit of integration. This can be done easily because the exponential function has the characteristic of being piecewise continuous in the interval $0 \leqq t \leqq T$; therefore, the exponential function can be divided into a finite number of subintervals in each of which $F(t)$ is continuous and possesses finite left- and right-hand limits at every point $0 \leqq t \leqq T$. If a time-soil moisture functional relationship can be established to determine the time equivalence of the computed soil moisture index, this soil moisture function can be added directly in order to determine the beginning time in the infiltration capacity equation, thus allowing the equation to vary from storm to storm. With this provision, the equation would be

$$L = \int_{f(sm)}^{D+f(sm)} (c + be^{-nT})\, dT \qquad (3)$$

where $f(sm)$ = function of soil moisture.

Integrating, we get

$$L = cT - (b/n)e^{-nT} \Big|_{f(sm)}^{D+f(sm)} \qquad (4)$$

The functional scaling relationship between time and the soil moisture index is unknown. The function must, however, be able to relate the index, which varies from zero for dry conditions to 1.0 for wet conditions, to an equivalent time in hours, and probably should allow for a nonlinear relation. Further, the time equivalence of the soil moisture parameter should be zero for storms that occur when the soil is dry (*sm* index equals zero). A polynomial relation can satisfy these two conditions so that

$$f(sm) = mS + gS^2 \qquad (5)$$

where S = soil moisture index and m and g are coefficients.

Substituting the functional form for $f(sm)$ given in (5) for the limits of integration in (4) yields

$$L = cD + (b/n)\{\exp[-n(mS + gS^2)] - \exp[-n(D + mS + gS^2)]\} \qquad (6)$$

Figure 1 graphically shows the relationship of the variables.

The equation in this form was fitted to watershed hydrologic data. The only storms included in the data were continuous, with relatively high rainfall intensities throughout. It was hoped, at this stage, that if any storms were included that contained some rainfall at intensities below the infiltration capacity the prediction error resulting from the fitting could be used to further modify the storm list. The results obtained in fitting this equation to watershed 1 pasture data are shown in equation 1, Table 1. Theoretically, the value of the polynomial soil moisture–time function at a soil moisture value of 1.0 should be approximately equal to the length of time it takes the exponential function to become insignificant. In other words, the exponential function ideally should drop out of the equation for soil moisture values of 1.0. Equation 1, Table 1, however, indicates that at a soil moisture value of 1.0 the time equivalent of the soil moisture is negative; this does not have a physical interpretation. In view of the low degree of adjustment achieved in fitting the equation, as indicated by a multiple correlation coefficient $r = 0.61$, it appears that the additional degree of freedom allowed by the polynomial time–soil moisture function was used to adjust for some of the large residual error. Since the polynomial relationship cannot be evaluated, a linear time–soil moisture relationship should be adequate.

Modification for interception losses. It was apparent from the results of fitting the first

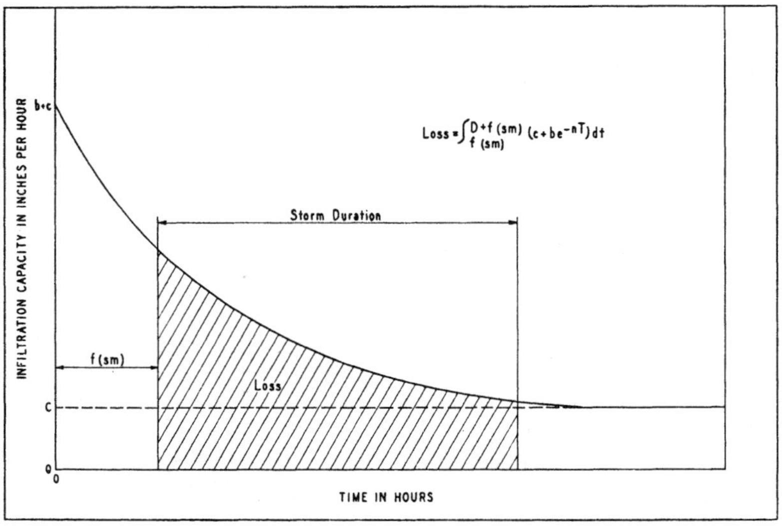

Fig. 1. Graphical representation of infiltration capacity function.

TABLE 1. Preliminary Equation Solutions

Equation 1 (watershed 1)

$$L = 0.138D + (0.0002/1.69)\,\{\exp[-1.69(7.90S - 12.18S^2)] - \exp[-1.69(D + 7.90S - 12.18S^2)]\}$$

Standard deviation $Se = 0.36$
Multiple correlation coefficient $r = 0.61$

Equation 2 (watershed 1)

$$L = 0.378 - 0.0767S + 0.0901D + (1.358/4.322)\{\exp[-(4.322 \times 1.189S)] - \exp[-4.322(D + 1.189S)]\}$$

$Se = 0.34$
$r = 0.66$

Equation 3 (watershed 5)

$$L = 0.266 - 0.068S + 0.026D + (3.87/2.82)\{\exp[-(2.82 \times 0.518S)] - \exp[-2.82(D + 0.518S)]\}$$

$Se = 0.39$
$r = 0.65$

Equation 4 (watershed 1)

$$L = -0.117 + 0.099S + 0.0016D + (0.853/5.433) \cdot \{\exp[-(5.433 \times 0.296S)] - \exp[-5.433(D + 0.296S)]\} + 0.948R$$

$Se = 0.0178$
$r = 0.99925$

where

L = loss in inches.
D = storm duration in hours.
S = soil moisture index.
R = storm rainfall in inches.
e = Napierian base.

equation to data that some major factor involved in the infiltration capacity relationship had been ignored. This factor could be interception. Interception losses represent a reduction in the effective rainfall, or throughfall. Studies by Wollny [*Baver*, 1938] in 1880 showed that rainfall interception by plants could amount to a substantial percentage of the rainfall. Since interception losses are not dependent on the duration of the storm, the equation cannot in its present form adjust for this type of loss. Adjustment could be made by the addition of a constant to the equation. It is probable, however, that this interception loss is not a constant; rather it is high when conditions are dry and less when conditions are wet. To approximate this condition it was assumed that the interception loss varied with the soil moisture as follows:

$$f(i) = a - gS \tag{7}$$

where $f(i)$ = function of interception, S = soil moisture, and a and g are coefficients. The assembled equation is now

$$L = a - gS + cD + (b/n)(e^{-nmS} - e^{-n(D+mS)}) \tag{8}$$

In this form the equation was again fitted to watershed 1 and, in addition, to watershed 5 data. The results of these two fittings are shown in Table 1, equations 2 and 3, respectively. The coefficients of both of these equations are reasonable. The indicated interception $(a - gS)$ at a soil moisture value of zero is 0.38 and 0.27 inch, respectively, on watersheds 1 and 5. For wet conditions where soil moisture equals 1.0, the interception values decrease slightly to 0.30 and 0.20 inch, respectively. The minimum infiltration capacity (c) is 0.09 in./hr on watershed 1 and 0.026 in./hr on watershed 5, and the maximum infiltration capacity for these two watersheds $(b + c)$ is 1.45 and 3.9 in./hr, respectively. These values are in substantial agreement with other measurements made on these watersheds and also agree with published data.

Despite the apparent reasonable magnitudes of the coefficients, the degree of adjustment of this equation is low for both sets of data. Figure 2, a residual error plot of watershed 1 data, shows that the degree of adjustment is low because a prediction bias exists that is related to the dependent variable loss.

Modification for partial area runoff. The

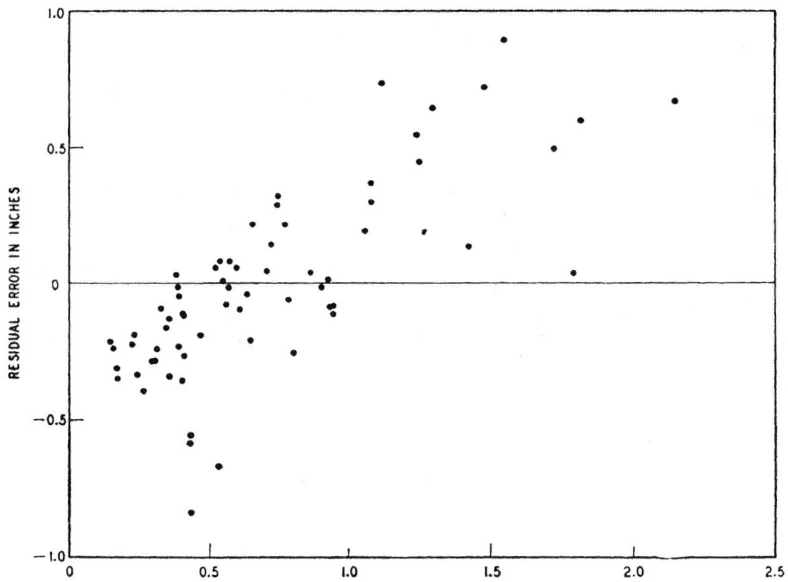

Fig. 2. Residual error of prediction versus observed loss for Waynesville watershed 1 (equation 2, Table 1).

residual error bias shown in Figure 2 in itself provides no insight as to the cause of the bias. Large residual error (R_e = observed − predicted) results when either the observed data are in error or the model does not fit the data. This could mean that either the runoff does not occur according to the infiltration capacity concept or the model itself is incorrect. Since the observed hydrologic data are considered reasonably correct, the model must be revised.

The consistency of the bias shown in Figure 2 suggests that a linear relationship can be fitted between the residual error and loss, the dependent variable of the equation. If this equation were linear, the bias could not have occurred, since the addition of such a linear loss function to the equation, in effect, amounts to a simple rescaling of each of the coefficients of the independent variables, a solution which would have been found in the fitting process. However, since the equation is nonlinear a simple rescaling of all the independent variables is not possible. This same effect can be achieved if the dependent variable, loss, is rescaled by including either storm rainfall or storm runoff as an independent variable in the equation. This fact is developed further in a following section. Actually, storm rainfall was selected as the rescaling independent variable because it is considerably larger than runoff and therefore more proportional to the loss. The equation with the linear rainfall term R added (the intercept at zero rainfall will be absorbed into the constant a of the equation) is

$$L = a - gS + cD + (b/n)(e^{-nmS} - e^{-n(D+mS)}) + hR \qquad (9)$$

The results of fitting this equation to watershed 1 pasture data are shown in Table 1, equation 4. A high degree of adjustment was achieved by the equation, as indicated by a multiple correlation coefficient of 0.99925. The very low standard error, 0.02 inch, is within an acceptable range for hydrologic data. The sign and values of all of the coefficients except the g term are within reasonable limits. The reversed sign of the coefficient g, however, does not have physical significance. It indicates that the interception loss increases with soil moisture and that the equation is probably overdetermined. This term can therefore be omitted from the equation. The physical significance of the term h will be developed in the following section.

The final form of the infiltration capacity equation is

$$L = a + cD + (b/n)(e^{-nmS} - e^{-n(D+mS)}) + hR \qquad (10)$$

TABLE 2. Equation Solutions

Watershed	Cover	Condition	Season	Coefficients						Statistics	
				a	*c*	*b*	*n*	*m*	*h*	*r*	*Se*
Watershed 1	Pasture		All	−0.032	0.0014	0.391	8.61	0.035	0.954	0.9995	0.015
Watershed 5	Pasture		All	−0.139	−0.045	1.025	1.52	0.61	0.765	0.975	0.113
Parker Branch	Complex	Calibration	Summer	−0.016	0.0045	0.394	3.31	0.21	0.782	0.995	0.040
Parker Branch	Complex	Evaluation	Summer	−0.017	0.006	0.287	2.86	0.47	0.870	0.993	0.052
Parker Branch	Complex	Calibration	Dry summer	0.001	0.011	0.351	3.55	0.53	0.772	0.996	0.048
Parker Branch	Complex	Calibration	Wet summer	0	0.002	0.258	3.91	0.37	0.862	0.996	0.018
Parker Branch	Complex	Evaluation	Dry summer	−0.006	0.011	0.133	5.59	0.10	0.921	0.9997	0.015
Parker Branch	Complex	Evaluation	Wet summer	−0.022	−0.128	0.307	.133	2.45	0.739	0.969	0.084
Parker Branch	Complex	Evaluation	Winter	−0.176	0.005	1.923	2.56	0.41	0.559	0.950	0.108
Middle Creek	Complex		Summer	−0.059	0.0006	0.785	4.58	0.12	0.856	0.985	0.104
Middle Creek	Complex		Winter	0.022	0.009	0.427	2.01	0	0.612	0.976	0.143
Watershed 2	Pasture	Moderately grazed	All	−0.0002	0.0009	0.487	4.07	0.43	0.849	0.9925	0.060
Watershed 2	Pasture	Heavily grazed	All	0.0044	0.0031	0.606	4.31	0.29	0.793	0.9905	0.066
Copper Basin 1 W	36%		All	−0.021	0.043	1.13	2.73	0.21	0.142	0.888	0.122

In this form the equation has been fitted to several sets of data from the single-practice western North Carolina watersheds as well as several other watersheds.

Interpretation of Results

Table 2 shows the results of fitting the final equation to data from various watersheds. The Parker Branch watershed, described in various progress reports and special reports on the project [*TVA*, 1960*b*], is a 1.51-mi^2 research area located in the same general region as the western North Carolina project watersheds. This complex-practice agricultural watershed was calibrated hydrologically from 1953 to 1955. Starting in 1955 the farmers were given an intensive farm development program designed to achieve optimum economic well-being of the people. Evaluation of the effects of this program was begun in 1958 and ended in 1962. The Middle Creek watershed is a 32.7-mi^2 area within Hiwassee River watershed. This area is in the Great Valley of east Tennessee. This agricultural complex-practice watershed was gaged from 1944 through 1961 to demonstrate the effect of TVA's applicable regional development integrated on a single watershed. The Copper Basin area in the southeastern corner of Tennessee was denuded primarily as a result of open-air copper ore roasting in the late 19th century. Watershed 1W, a 5-acre area, was gaged from 1943 through 1951 as part of a project to determine the effect of an erosion control program. Cover on this watershed, which was left untreated, never exceeded 36 per cent of the area. These areas are shown on the location map, Figure 3.

The coefficients of the equation for the various solutions shown in Table 2, with a few exceptions which will be explained later, show surprising consistency. This consistency generally remains despite the fact that the drainage areas for the watersheds vary from 3.7 acres to 32.7 mi^2, are located from the Blue Ridge physiographic region to the Great Valley, and vary in

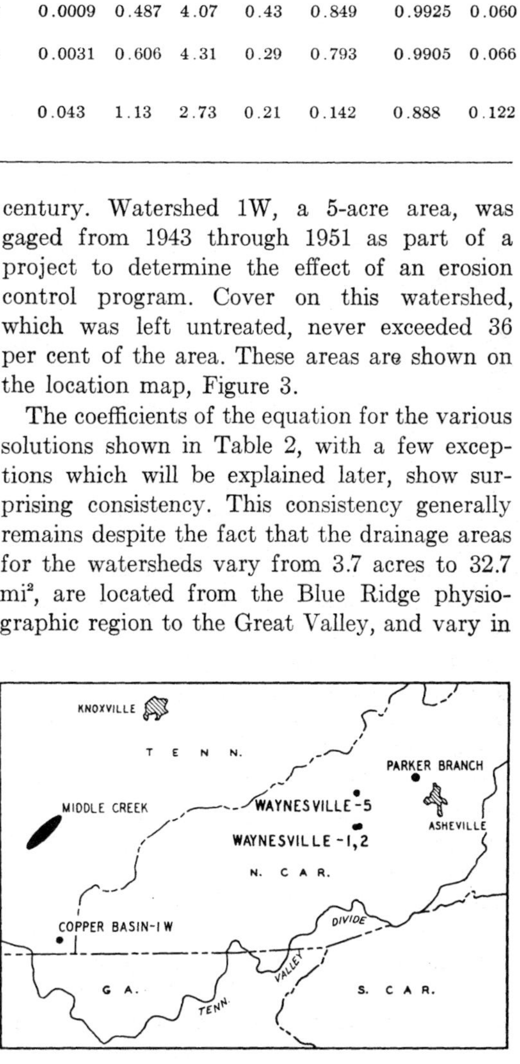

Fig. 3. Location map.

covers from single-practice pasture, to denudation, to complex practice. The storms from the Middle Creek watershed have been divided into summer and winter seasons and in addition to that division, the Parker Branch data have been subdivided into storms with significant rainfall within the three days before the storm, i.e. initial wetness and dryness.

The indicated minimum infiltration capacity for most data sets is of the order of 0.01 in./hr or less; the maximum rate is approximately ½ in./hr. This consistency also generally follows through the coefficient n which determines the shape of the exponential decay function and through the coefficient m which determines the time equivalent of soil moisture. In most cases the multiple correlation coefficient is unusually high, whereas the standard error is very low relative to the mean dependent variable, the loss, and even the mean runoff.

Despite the consistency of the coefficients and the high degree of adjustment achieved by the equation, most of the indicated infiltration capacities are low in comparison with published data. The reason for this can be better understood by studying the coefficient h. Rearranging the equation will show the rescaling nature of this term

$$\text{Loss} = R - hR - RO = a + cD + (b/n)(e^{-nmS} - e^{-n(D+mS)}) = f(ip) \quad (11)$$

$$\text{or} \quad R(1 - h) - RO = f(ip)$$

where $f(ip)$ = function of infiltration capacity.

If the reduction in rainfall indicated by the term $(1 - h)$ is real, there is an implication that not all of the storm rainfall is effective in producing runoff, if runoff is to occur according to the infiltration capacity concept. It is not obvious at this point why this is so.

The dependent variable could also have been rescaled by using runoff as an independent variable and an equally good fit of the equation would have been achieved. This can be shown if the equation is rearranged

$$R - \frac{RO}{(1 - H)} = \frac{1}{(1 - h)} f(ip) \quad (12)$$

This rearrangement indicates that if it were possible to rescale storm runoff by a relatively constant factor, the resulting rainfall-runoff relation would occur according to the infiltration capacity concept. Runoff, unlike rainfall, can be rescaled mathematically. The observed volume of watershed runoff in cubic feet per second times one day (dsf) is converted to watershed inches by dividing by the drainage area (DA) and an appropriate constant (K) as in (13)

$$\text{inches} = \frac{dsf}{DA \times K} \quad (13)$$

If both sides of (13) are divided by the term $(1 - h)$ this term becomes more meaningful.

$$\frac{\text{inches}}{(1 - h)} = \frac{dsf}{(1 - h) DA \times K} \quad (14)$$

Storm runoff can be scaled upward in converting to the watershed inch unit if the effective drainage area is reduced. The term $(1 - h)$ is actually a measure of this effective runoff-producing area of a watershed. It has been the conversion to the watershed inch runoff unit with its implication of total watershed area contribution that has been complicating these studies. If either the total storm rainfall is reduced to an amount that fell over the effective runoff-producing area or the runoff is scaled upward to the watershed inch unit over a reduced drainage area, the resulting rainfall-runoff relation does occur according to the infiltration capacity concept. The effective runoff-producing area of a watershed, then, is not the same as that delineated by the topographical divide.

That the entire watershed may not contribute to runoff during moderate-size storms is not surprising. *Amorocho and Orlob* [1961] and *Moldenhauer et al.* [1960] had indications of this possibility, and the *Southeastern Forest Experiment Station* [1961] estimated that at the Coweeta Hydrologic Laboratory in western North Carolina no more than about 40 per cent of the drainage area of their watersheds contributes to storm runoff even in a relatively heavy rainfall area. It is the low percentage of the total area that does contribute and the apparent consistency of the size of this area on some watersheds that are surprising. For the equation solutions shown in Table 2, the term $(1 - h)$ indicates an average contribution that varied from a low of only 4.6 per cent of watershed 1 to a maximum of 85.8 per cent of Copper Basin 1W.

Verification of the magnitude of the computed effective runoff-producing area of a watershed is difficult on the Parker Branch or Middle Creek watersheds because of the complex patterns of land use. Ideally, to verify that the term $(1 - h)$ is a measure of the effective runoff-producing area of a watershed, the equation should be fitted to data obtained from watersheds with known runoff-producing areas. One way to do this would be to use relatively impervious watersheds such as those in urban areas. For this type of watershed the coefficient h should have small values, so that $(1 - h)$ approaches unity. Lacking this type of data, some verification of this interpretation of the term $(1 - h)$ can be made using data from the western North Carolina project watersheds and from Copper Basin 1W watershed. These two projects represent extremes in land-use practices. Figure 4 shows a verification plot of the residual error versus the loss for watershed 1. The bias has been removed from the error, leaving a more normal pattern, and the errors themselves are reduced to acceptable values. The more normal error pattern indicates that with the dependent variable rescaled the equation does adjust to these data. The relatively small remaining error indicates that the percentage of watershed contributing to runoff is relatively constant over the range of storms included in the data set. The storm precipitation within this data set includes storm rainfall amounts varying from 0.16 in. to 2.28 in. and intensities up to 2.6 in./hr. The coefficient h indicates that, on this watershed for storms in the light-to-moderate range, runoff occurs on the average from only 4.6 per cent of the watershed, or, conversely, the storms within the range of these data did not exceed the infiltration capacity of about 95 per cent of this watershed. On this watershed, installation of a sheet-metal cutoff wall has resulted in a permanently swampy area above the flume. The swampy area covers about 1 to 2 per cent of the watershed. It is evident, by the quick response of the streamflow discharge to changes in rainfall intensity and the unusually low yield of storm rainfall, that most of the storm runoff that was measured from this watershed under pasture cover came from this swampy area. The equation indicates that most of the storm runoff results from an area about twice the size of the swampy portion of the watershed.

If the concept that, on the average, runoff usually results from a relatively fixed percentage of the watershed area is valid, further verification should be possible on watershed 2. On this watershed three subplots are maintained to measure surface runoff. These 1/20-acre subplots are located around the watershed about ⅓ to ½ of the distance up to the divide. Some subplot surface runoff was measured from about ⅓ of the storms included in this data set. This seems to contradict the result of the equation that indicates that average runoff from the watershed results from only 15 per cent of the area. The amount of storm runoff recorded, however, supports the interpretation. Only during one of the storms included in the set was a substantial volume of runoff recorded from all three plots. In general, surface runoff from these plots amounted to less than 0.01 inch, and seldom was runoff recorded from all three plots during a given storm. The only time substantial

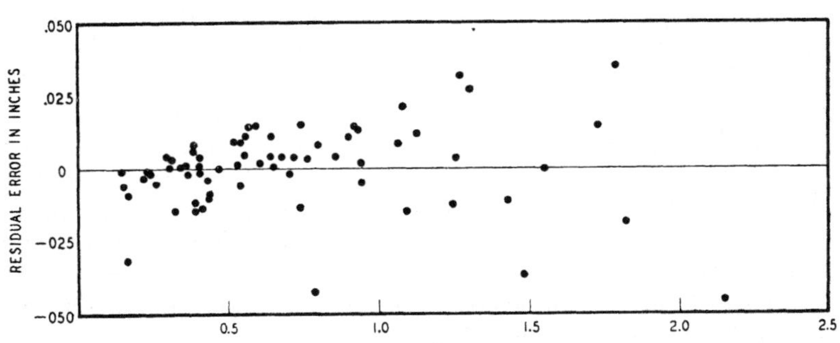

Fig. 4. Residual error of prediction versus observed loss, Waynesville Watershed 1.

runoff was measured at the plots, the storm was very intense.

Some intuitive judgment concerning the reasonableness of the term h that results from fitting the equation to Copper Basin 1W watershed data should be possible. This watershed, with 64 per cent of the area completely denuded, is subject to relatively heavy, rapid runoff through an advanced system of gullies that completely dissect the area. Streamflow is intermittent and most of the runoff can be considered surface runoff. Under these conditions, runoff should occur from a large percentage of the watershed, and the indicated infiltration capacities should be relatively low since crusting of the soil surface results from frequent, intense storms. The coefficient h for the Copper Basin 1W data (Table 2) indicates that the average watershed area contributing to runoff was 86 per cent. The indicated infiltration capacities, however, are not low. They are actually among the highest of the various sets of data used. In fact, these are the only indicated infiltration capacities, of all those determined by the equation, that are reasonable. This reasonableness suggests that the indicated infiltration capacity values are related to the amount of the watershed contributing to runoff.

The residual error distribution that results from fitting this equation to any of the data sets generally follows a seasonal pattern. High runoff (negative predictions) occurs during the winter season and low runoff occurs during the summer. This separation suggests that during the winter, when the amount of moisture stored in the soil profile is high, runoff probably occurs from a somewhat larger portion of the watershed than during the summer when this contributing area is smaller. The fitting technique sets the equation at what might be considered an average condition. This interpretation helps to explain some of the exceptions to the consistency of the coefficients found in Table 2. Inconsistencies were found in the fittings when the data contained a disproportionately high number of winter storms with 'apparently' high runoff, or when the actual infiltration capacities of the watershed were low, so that occasional high runoff resulted when more than the average watershed area contributed during a very intense storm. Both these conditions occur at watershed 5. A negative, uninterpretable coefficient for the minimum infiltration capacity resulted. The storms that were used for the Parker Branch evaluation period study, labeled 'summer data with wet antecedent conditions,' on the other hand, contained many events with high runoff that resulted because the moisture storage capacity was reduced when the upper soil horizons were near saturation. The equation resulting from these data is an almost random array of coefficients. This result must occur when large variations in the runoff occur because of conditions external to the equation. This result did not occur with the Copper Basin 1W fitting because the data were relatively 'homogeneous.' Further modifications of the equation will have to be made to account for, at least, the effect of changes in the volume of moisture stored within the soil on runoff.

If the infiltration capacity function that has been developed is to measure the watershed infiltration capacity, storm surface runoff should have been used rather than total storm runoff. Present methods of hydrograph analysis, however, do not permit reliable separation of the hydrograph into surface and subsurface flow. Because total storm runoff has been used, the losses predicted by using the function will be somewhat lower than they would have been if surface runoff only had been used. However, the degree to which this equation adjusts for a fairly wide range of hydrological events indicates that these two components of storm runoff, combined to evaluate the equation, are not incompatible as data. If this is true, the function should be a relatively stable index of the infiltration capacity of a watershed.

An Apparent Infiltration Capacity Function

It would be helpful if the coefficients that result from fitting the equation had more reasonable values. It was noted above that the magnitude of the coefficients is related to the size of the contributing area. Equation 12 shows how the coefficients might be rescaled to remove the effect of the size of the contributing area on their magnitude. By dividing through the equation by the term $(1 - h)$, the coefficients a, c, and b are scaled upward. This has the effect of scaling the coefficients of the infiltration capacity function upward to the equivalent value they would have had if runoff had occurred from the

TABLE 3. Apparent Watershed Infiltration Capacities

Watershed	Cover	Condition	Season	Coefficient *a*	*c*	*b*	*n*	*m*
Watershed 1	Pasture		All	−0.70	0.03	8.5	8.61	0.035
Parker Branch	Complex	Calibration	Summer	−0.07	0.02	1.8	3.31	0.21
Parker Branch	Complex	Evaluation	Summer	−0.13	0.05	2.2	2.86	0.47
Parker Branch	Complex	Calibration	Dry summer	0.004	0.05	1.5	3.55	0.53
Parker Branch	Complex	Calibration	Wet summer	0.0	0.014	1.9	3.91	0.37
Parker Branch	Complex	Evaluation	Dry summer	−0.08	0.14	1.7	5.59	0.10
Parker Branch	Complex	Evaluation	Winter	−0.31	0.009	3.4	2.56	0.41
Middle Creek	Complex		Summer	−0.41	0.004	5.45	4.58	0.12
Middle Creek	Complex		Winter	0.06	0.023	1.1	2.01	0
Watershed 2	Pasture	Moderately grazed	All	−0.001	0.006	3.2	4.07	0.43
Watershed 2	Pasture	Heavily grazed	All	0.021	0.015	2.9	4.31	0.29
Copper Basin 1W	36%		All	−0.025	0.050	1.32	2.73	0.21

entire watershed in a manner similar to that from the contributing area. The shape coefficient n and the time–soil moisture coefficient m remain unchanged. Table 3 shows the computed coefficients of the apparent infiltration capacity for the equation solutions included in Table 2. The solutions for watershed 5 and Parker Branch that broke down have not been included for reasons previously explained.

The value of the minimum and maximum apparent infiltration capacities (c and $c + b$, respectively) as shown in Table 3 are not too different across the set of solutions. The minimum capacity ranged from a low of 0.006 in./hr on the moderately grazed pasture of watershed 2 to 0.14 in./hr for all summer storms during the calibration period at Parker Branch. The maximum rate ranged from 8.53 in./hr on the watershed 1 pasture to 1.12 in./hr for winter storms on Middle Creek. The coefficients of the equation fitted to Copper Basin 1W data are well within the range of values obtained from other watersheds where considerably less of the watershed contributed to runoff. This consistency indicates that these coefficients should be a measure of the infiltration capacity of that area of the watershed that on the average contributes to runoff. Considering that the equation contains only the three independent variables—storm duration, rainfall, and a soil moisture index—and that no restrictions are imposed on the magnitudes of the coefficients of the equations, the range of coefficients obtained in fitting this equation is surprisingly consistent. This consistency seems to hold over watersheds ranging in size from 3.7 acres to 32.7 mi^2 and for watersheds with decidedly different topography, agricultural practices, and subclassifications of storms within a data set. It would appear, then, that the apparent watershed infiltration capacity function is a relatively stable measure with which the runoff, or, more directly, the loss, from a watershed may be characterized.

Conclusions

The stepwise development of a hydrologic equation to evaluate apparent watershed infiltration capacity has been presented. This equation can be useful both as a hydrologic tool and as a means of developing a better understanding of the storm runoff process.

The results obtained from fitting the equation to various sets of data indicate that storm runoff, at least in the geographic region of the study, frequently occurs from only a small part of the watershed area. This was found to be true on small test watersheds in pasture cover, and it appears to be true also for larger watersheds with complex land-use patterns. These results also seem to indicate that the size of the runoff-contributing area may not vary much under normal conditions. Logically, the size of the area must change as the moisture storage capacity of the soil changes, and in addition, the size of this area is also related to rainfall intensity. However, it appears that these two factors do not cause large variations in the size of the runoff-producing area of many watersheds except under unusual conditions. If this is true, the concept of partial area storm runoff helps to explain the nonlinear nature of many hydrologic relations. Mathematical models relating storm rainfall and modifying variables to runoff, peak discharge, or other dependent variables frequently underpredict large events, particularly if a wide range of storms is used. A typical storm list contains a large number of small-to-moderate events where the runoff-contributing part of the watershed is a relatively small area. Because of the number of these smaller events, the model is forced to fit this typical event. When during an extreme storm runoff does occur from a much larger area within the watershed, the small increase in the measured independent variables seldom justifies the large increase in the dependent variable. Actually, it appears that different relationships can be justified for large and small events.

The results from this study also help to explain why infiltration studies based on rainfall-runoff data seldom agree with in situ measurements. If storm runoff usually occurs from only a small part of a watershed, the infiltration capacity of a greater part of this watershed is seldom exceeded during normal storms. It appears, then, that in situ measurements of infiltration capacity will always equal or exceed measurements obtained from rainfall-runoff data. The two techniques will yield similar results only when either the in situ measurements are made on the runoff-contributing area or the function of apparent infiltration capacity is evaluated for relatively impervious or saturated areas.

The function that was developed should be useful as a means of estimating storm runoff. Because the equation was constructed logically, most of the coefficients have a physical interpretation and thus are relatively stable and reproducible. This stability is an important factor if the function is ever to be used for predicting hydrologic events on an ungaged watershed. Although only three independent variables are used, the standard errors that result from fitting are generally less than 10 per cent of the mean of the dependent variable.

The equation in its present form has a somewhat limited application. On small test areas it will work well only with data that do not contain too many storms with unusually high runoff. On larger areas it appears to fit consistently, probably because of the integrating effect of many factors. To be useful, the model requires further improvements, and these improvements are currently under development. From a practical point of view, and judging from results obtained thus far, the model should adjust for increased runoff resulting from reductions in soil moisture storage, changes in agricultural cover, and extreme rainfall intensities. If these factors can be incorporated in the model successfully, a far better understanding of watershed runoff should result.

References

Amorocho, J., and G. T. Orlob, An evaluation of inflow-runoff relations in hydrologic studies, *Water Resources Center Contrib. 41*, p. 59, University of California, Berkeley, 1961.

Baver, L. D., Ewald Wollny, A pioneer in soil and water conservation research, *Soil Sci. Soc. Am. Proc., 3*, 330–333, 1938.

Horton, R. E., The role of infiltration in the hydrologic cycle, *Trans. Am. Geophys. Union, 14*, 446–460, 1933.

Moldenhauer, W. C., W. C. Barrows, and D. Swartzendruber, Influence of rainstorm characteristics on infiltration measurements, *Trans. Intern. Congr. Soil Sci., 7th, Madison, Wis., 1*, 426–432, 1960.

Nielsen, K. J., *Methods in Numerical Analysis*, The Macmillan Company, New York, 1957.

Snyder, W. M., Some possibilities for multivariate analysis in hydrologic studies, *J. Geophys. Res., 67*(2), 721–729, 1962.

Southeastern Forest Experiment Station, *Report for 1961*, pp. 62–63, Asheville, North Carolina, 1961.

Tennessee Valley Authority, Hydrology of small watersheds in relation to various covers and soil characteristics (a pictorial brochure), in cooperation with North Carolina State College of Agriculture and Engineering, Knoxville, Tenn., 1960*a*.

Tennessee Valley Authority, An analysis of the Parker Branch watershed project 1953 through 1959 (a progress report), in cooperation with North Carolina State College of Agriculture and Engineering, Knoxville, Tenn., 1960*b*.

Tennessee Valley Authority, TVA computer programs for hydrologic analyses, *Res. Paper 3*, Knoxville, Tenn., 1963.

(Manuscript received June 15, 1963; revised December 17, 1963.)

Commentary on Paper 8

Whipkey, R. Z. (1965) Subsurface stormflow from forested slopes.

Ron Whipkey published three significant papers in the mid to late 1960s (this one, 1967, 1969) reporting one of the first widely known studies of subsurface stormflow on hillslopes (though Whipkey himself recognizes a number of earlier studies of the phenomenon, going back to Hursh). Whipkey worked for the USDA Forest Service in Columbus, Ohio, as a forest engineer. He notes here, very much from an engineering perspective, that "*so little is known about rates and quantities of subsurface stormflow that it is nearly impossible to devise mechanical or biological techniques for altering timing or volumes*" (p.74).

The experimental work was based on a trough flow collection system on a 17-m forest plot (2.44 m wide) with collectors at the surface and a number of different layers in the soil down to 1.5 m. Whipkey notes that the near-surface soil was "*well-permeated with roots, root channels, small animal and earthworm burrows and structural cracks*" (Whipkey, 1969, p.774), and in Whipkey (1967) he looks at the representation of subsurface stormflow as a non-Darcy flow, something later picked up by De Vries & Chow (1978). The slope was watered with an artificial sprinkler system, because "*natural rainfall is undependable in summer*". This allowed the reproducibility of the responses to similar rainstorms to be studied. He showed how the responses of the plot were different for dry and wet antecedent conditions. Surface runoff was very small; the bulk of the downslope seepage was collected in the layer immediately above an impeding horizon (56–90 cm). Figure 4 in this paper, showing the hydrographs from the different layers, has been widely reproduced (e.g. Paper 13; Kirkby & Chorley, 1967; Chorley, 1978).

The review of non-Darcy flow in the Whipkey (1967) paper is followed by an interesting discussion. Bob Zinke asked if Whipkey had found a way of mapping the structural openings in the soil; Whipkey replies "*no*". There is also an exchange between Alden Hibbert and Whipkey about when water will enter into structural voids. Hibbert suggests that they will not take flowing water until the soil is saturated because a positive pressure is required. Whipkey agrees, especially if there is entrapped air, but notes that water can enter and air can escape if the structural voids are open to the surface (note that Horton also presented this concept in his unpublished monograph on infiltration, see Beven, 2004). Hibbert asks if he means water ponding on the surface and running into the structural pores; Whipkey replies "*yes, or rainfall*". Hibbert notes that on deep road cuts at Coweeta, even with heavy rainfall, no seepage has been observed, but Whipkey points out that the soil there is much more homogeneous and that a shallow impeding layer may be important in the generation of saturated conditions and of subsurface stormflow. Finally, Teller asks what Whipkey hopes to do with the results. Whipkey replies in a way that reflects the predominant hydrological concerns of the time in forest engineering: "*Well it is very difficult to justify a study of basic phenomena other than to say we want to learn more about the process. But, we do ultimately want to be able to find some way, mechanically or biologically, possibly both, of shattering this impeding layer so that infiltrated water will go more deeply into the soil or groundwater zone. We have no problem with infiltration, and the surface soils are very permeable. But the problem is to reduce the rapid movement of water to the streams. We feel that it is a waste of effort when you get it infiltrated and then it runs out during the storm and becomes flood flow*" (p.260).

A graduate of Pennsylvania State University in 1950, Ron Whipkey died in 2003. The initial "Z", by the way, stands for Zebadia.

References

Beven, K. J. (2004) Robert Horton's perceptual model of infiltration. *Hydrol. Processes* **18**, 3447–3460.

Chorley, R. J. (1978) The hillslope hydrological cycle. In: *Hillslope Hydrology* (ed. by M. J. Kirkby), 1–42. Wiley, Chichester, UK.

De Vries, J. & Chow, T. L. (1978) Hydrologic behaviour of a forested mountain soil in coastal British Columbia. *Water Resour. Res.* **14**(5), 935–942.

Kirkby, M. J. & Chorley, R. J. (1967) Throughflow, overland flow and erosion. *Bull. Int. Assoc. Sci. Hydrol. Bull.* **12**(3), 5–21.

Whipkey, R. Z. (1967) Theory and mechanics of subsurface stormflow. In: *Forest Hydrology* (ed. by W. E. Sopper & H. W. Lull) (Proc. NSF Advanced Science Seminar), 255–260. Pergamon, New York, USA.

Whipkey, R. Z. (1969) Storm runoff from forested catchments by subsurface routes. In: *Floods and their Computation*, vol. 2 (Proc. Leningrad Symp., 1967), 773–779. IAHS Publ. 85. IAHS Press, Wallingford, UK.

SUBSURFACE STORMFLOW FROM FORESTED SLOPES [1]

RONALD Z. WHIPKEY,

Research Forester, U.S. Forest Service,
Central States Forest Experiment Station, Columbus, Ohio

Hydrologists recognize two main sources of runoff to streams: indirect runoff—water that infiltrates and percolates to become ground water; and direct runoff—water that reaches the straem without being temporarily stored as ground water. Although direct runoff occurs during or shortly after a storm, some of this water may travel to the nearest stream via an underground route. This underground, stormperiod flow, which we prefer to call subsurface stormflow, occurs most commonly in forested areas.

Undisturbed forest soil is generally covered by organic litter that protects the soil surface and keeps it permeable to water infiltration. In addition, the A and B horizons of forest soils are interlaced with roots, old root holes, animal burrows, structural channels, and so on, that provide a highly permeable medium for the rapid movement of water in all directions ([4,5]). So, when a relatively impermeable soil layer is reached, percolating water moves laterally to a stream, thus increasing flow during extremely wet periods. This flow is distinguished from true ground-water flow in that it flows to the stream channel before reaching the general ground-water zone.

A number of studies have been conducted to show that subsurface stormflow occurs ([2,3,7,8,9,10,11,12,13,16,17]) especially where: (*a*) the land is sloping, (*b*) surface soil is permeable, (*c*) a water-impeding layer is near the surface, and (*d*) the soil is saturated. But so little is known about rates and quantities of subsurface stormflow that it is nearly impossible to devise mechanical or biological techniques for altering timing or volumes.

Measuring flow beneath the surface is extremely difficult and many investigators have used laboratory models to test flow theories under various conditions ([15]). However, because we know so little about subsurface stormflow, a postulated model would be pure conjecture. Does the flow originate from a saturated zone or a partially saturated zone? We cannot generalize for all cases. We do know that the equation for subsurface flow is of the non-linear differential form. In solving such an equation we must make simplifying assumptions in line with field conditions. But we did not know enough about field conditions in unglaciated soil to make such assumptions, so we chose to use field plots and simulated rainstorms to gain data and insight into the process.

The site chosen for study is a slope typical of Allegheny Plateau. It is located in east-central Ohio on a sandy loam soil with a slope of 28 percent. The vegetative cover is a 60-year-old mixed oak stand and the plot surface is covered with a 5- to 10-centimeter depth of mixed-hardwood leaf litter. The soil, classified as DeKalb sandy loam, is very permeable down to the 90-centimeter depth. The upper layer is interlaced with tree roots and decayed root channels.

At depths below 90 centimeters the percentages of silt and clay particles increase and the soil itself is more compacted, containing fewer roots. Other less marked changes in soil density occur at the 56- and 120-centimeter depths (table 1).

PROCEDURE

We built a trough system to collect water seepage from these major textural layers ([18]) (fig. 1). First, a section of the slope was chosen where the depth to the textural changes was uniform. Then we dug a pit at the downhill end of the plot leaving an open soil face. This face was equipped with the troughs and made as watertight as possible by sealing it with mortar. We felt this was essential to minimize inflow gain from areas outside the plot boundaries.

Artificial plot boundaries made of inserted sheet metal or flexible plastic sheets were not used because we believe these devices lead to unnatural flow. They can provide a path along their face for concentrated flow, especially when the soil is saturated. By wetting the soil far

TABLE 1

Physical properties of forest soil used in simulated rainfall-seepage study

Soil depth (in cm)	Textural class	Sand (0.1-.05 mm)	Particle-size distribution (in percent)		Bulk density
			Silt (0.5-.002 mm)	Clay (< .002 mm)	
0-56	Sandy loam	64	28	8	1.33
56-90	Sandy loam	67	18	15	1.41
90-120	Loam	42	38	20	1.78
120-150	Clay loam	31	40	29	1.80

beyond the actual plot boundaries the possibility of much water flowing from the plot to the drier soil around was eliminated.

The plot was 17 meters long, running up and down the slope, and the trough pit was 2.44 meters wide, although we believe that the effective area drained by the troughs was slightly wider.

Artificial rain had to be used because natural rainfall is undependable in this area in the summer. So, we constructed a sprinkler system, with upright nozzle fixtures 122 centimeters apart, running the length of the plot. The width of the plot was 244 centimeters but the area wetted was 370 centimeters wide. Thus, the plot itself and an area 63 centimeters wide on either side were wetted. Rainfall depth was measured by collecting "rain" in 10 wooden troughs 2.5 by 210 centimeters positioned randomly in herringbone pattern up the slope.

Hydraulic head measurements were made at various depths during wetting and draining of the plot with multiple-unit tensiometers. This instrument, equipped with six independent, porous, ceramic measuring rings mounted coaxially on a single stem, was ideally suited for our purpose. This type of tensiometer minimizes soil disturbance and simplifies measuring vertical, hydraulic gradients at a single spot.

Our tensiometers were equipped to measure hydraulic head at 15-, 30-, 45-, 60-, 75-, and 90-centimeter depths. Tensiometers were located at up slope distances of 0.60, 3.60, 6.60, 9.60, and 12.60 meters from the face of the plot on a line equidistant from the lateral boundaries. A capped, metallic access tube for measuring soil moisture with a neutron meter was located near each tensiometer battery.

PRECIPITATION

Simulated rainstorms were made in August, September, and October 1963. The weather during late spring, summer, and early autumn of 1963 was clear, dry, warm, and windy, providing opportune conditions for high evapotranspiration. Rain intensities of our artificial storms ranged from 1.7 to 5.1 centimeters per hour, length of storms ranged from 60 minutes to 150 minutes, and total rain depths for any storm ranged from 1.7 centimeters to 10.7 centimeters (table 2). Previous studies have shown that approximately 0.33 centimeter of rainfall may be intercepted and stored by the litter covering the soil surface. This depth was subtracted from all rainfall depth readings.

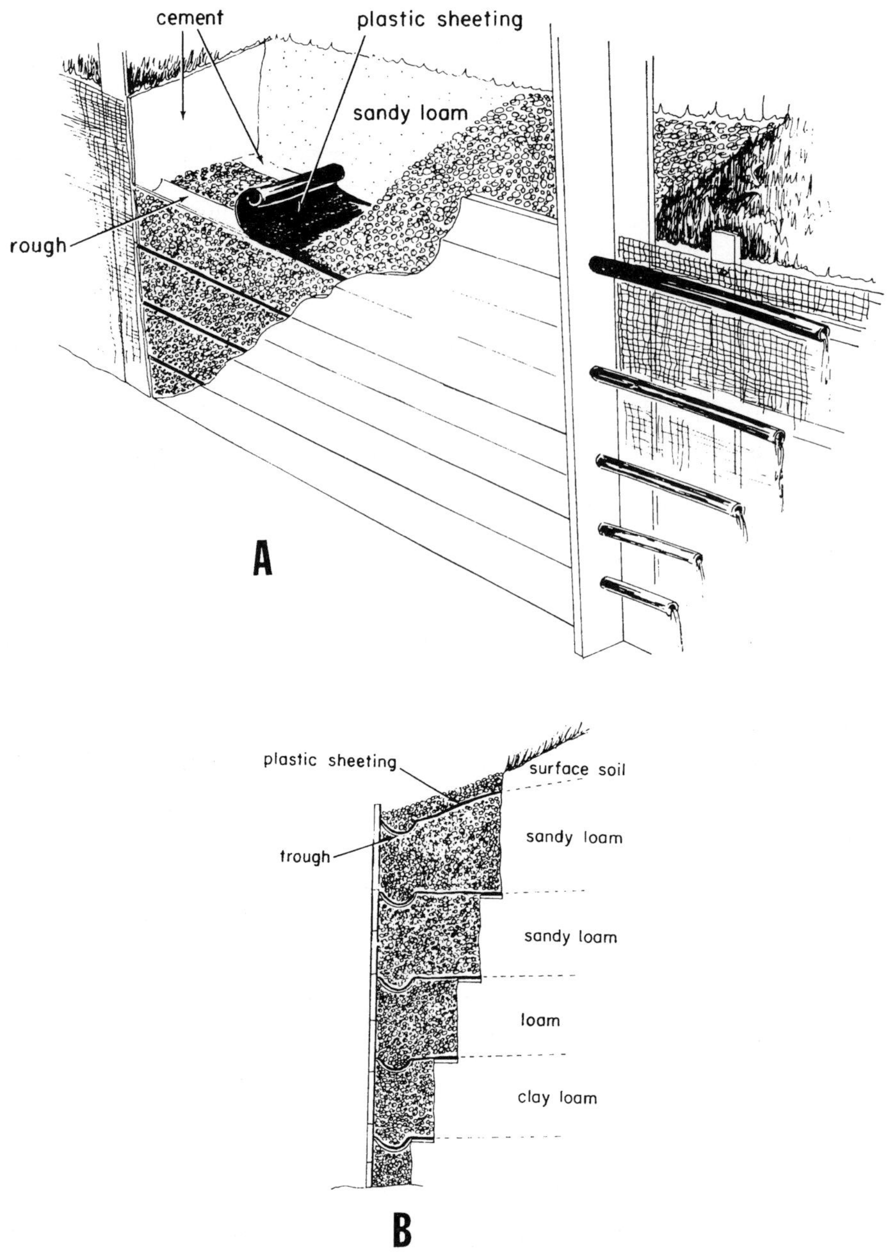

Fig. 1 — Schematic drawing of seepage collection system.

A. Open view of downslope face of soil plot showing step system and side walls plastered with cement to prevent loss of plot water or inflow of foreign water; polyethylene plastic apron cemented to step and wall and placed in trough to carry seepage from open soil face into respective troughs; and strip on either side of trough system, protected by gravel and wire mesh, to allow free outflow of water from either side of collection system.

B. Side view or profile of completed trough collection system showing wooden form containing pea gravel and plastic apron troughs at each soil depth.

TABLE 2

Summary of simulated storm and seepage results, by storms

Storm No.	Antecedent condition	Storm duration	Storm intensity	Rain depth	Seepage depth	Seepage as percent of rain depth	Percent of total seepage by soil depth (cm)					Peak seepage rate, by soil depth (cm)	
							Surface	0-56	56-90	90-120	120-150	0-56	56-90
		Minutes	cm/hr	cm	cm	Percent	Percent	Percent	Percent	Percent	Percent	cc/sec	cc/sec
1	dry (*)	100	2.6	4.3	0.2	5	—	21	79	—	—	0.6	5.5
2	dry	60	1.7	1.7	0	0	—	—	—	—	—	—	—
3	dry	100	2.3	3.8	.1	3	—	—	70	30	—	—	2.3
4	dry	100	2.1	3.5	0	0	—	—	—	—	—	—	—
5	dry	150	4.0	10.1	.4	5	—	5	75	20	—	3.7	14.2
6	wet (**)	90	3.9	5.9	.5	9	—	8	69	19	4	4.3	13.0
7	dry	120	4.4	8.8	.7	9	—	28	57	12	3	22.2	15.3
8	wet	90	3.6	5.5	.6	13	—	19	64	14	3	14.7	14.6
9	wet	90	4.0	6.0	.9	16	—	26	60	11	3	28.3	19.8
10	wet	105	2.5	4.4	.6	14	—	4	73	18	5	3.5	14.9
11	wet	75	2.3	2.9	.3	12	—	—	63	31	6	—	4.8
12	dry	120	2.5	5.0	.3	7	—	—	70	25	5	—	5.8
13	wet	120	2.0	4.0	.6	16	—	11	68	17	4	7.9	14.5
14	wet	75	2.0	2.5	.3	14	—	—	64	28	7	—	4.1
15	dry	121	1.5	3.1	.1	3	—	2	48	44	6	.1	.9
16	wet	101	2.1	3.5	.4	13	—	1	76	19	4	.5	9.9
17	wet	100	1.9	3.1	.4	16	—	1	79	15	5	.5	10.2
18	wet	100	1.8	3.1	.4	16	—	2	71	22	5	1.0	9.3
19	wet	90	2.1	3.2	.4	15	—	1	74	20	5	.3	10.1
20	dry	100	2.9	4.8	.1	2	—	—	48	38	14	—	.4
21	wet	100	3.0	5.0	.5	10	—	2	76	18	4	.3	11.0
22	wet	150	3.2	8.1	1.2	16	1	37	52	8	2	42.7	20.5
23	dry	120	5.1	10.2	1.0	10	3	31	54	10	2	42.1	20.3
24	dry	145	4.4	10.7	.9	9	4	32	53	9	1	52.4	17.3

(*) "Dry" refers to antecedent condition where plot drained more than 4 days between simulated storms.
(**) "Wet" refers to antecedent condition where plot drained less than 4 days between simulated storms.

For convenience the storms were arbitrarily grouped into intensity classes and into "wet" or "dry" condition immediately before "rain" (antecedent condition) (table 2). Soil was classified as dry if more than 4 days had passed since the last rain; wet if less than 4 days had passed since the last rain. The following tabulation shows the number of storms by this class-grouping:

Storm intensity Centimeters per hour	Intensity class	Number of storms		
		Dry antecedent condition	Wet antecedent condition	Total
< 2.2	medium-low	3	6	9
2.2-4.0	medium	4	6	10
> 4.0	medium-high	4	1	5
Total		11	13	24

RESULTS AND DISCUSSION

Seepage

Total seepage for the 24-hour period after any storm began ranged from 3 to 16 percent of rain applied. From storms of less than 5.0-centimeter precipitation on dry soil, seepage outflow was less than 5 percent of total rain applied (table 2). From such storms on a wet soil, seepage ran as high as 16 percent.

Where rainfall depths exceeded 5.0 centimeters the seepage percent in wet soils did not exceed this 16 percent. Under dry conditions, as might be expected, the highest percent seepages were found after storms that put the greatest volume of water on the plot. The seepage percent for dry soil was as high as 10 percent of total rainfall for storms over 5.0 centimeters (fig. 2).

Sixty-four percent of all seepage came from the 56- to 90-centimeter-depth zone. This zone is underlain by a compact silt layer. Seepage from all storms by depths was as follows:

Depth drained (Centimeters)	Percent of total seepage
surface	1 (2)
0-56	16
56-90	64
89-120	16
120-150	3
	100

The greatest subsurface flow came from the soil layer immediately above the flow-impeding layer. This was also true in trial runs made previously with a similar plot technique on silt loam and sandy loam soils.

78

Under all antecedent conditions, flow from the 56-centimeter depth responded quickly to simulated rainfall, peaking at or near the end of the storm and then dwindling to no flow in several hours. Seepage from the 90-centimeter depth was more stable, sometimes continuing at or near peak flow for a time and then diminishing slowly. Flow from the two lowest depths was steady and uniform, sometimes for 8 hours after seepage from the 56- and 90-centimeter depths had peaked.

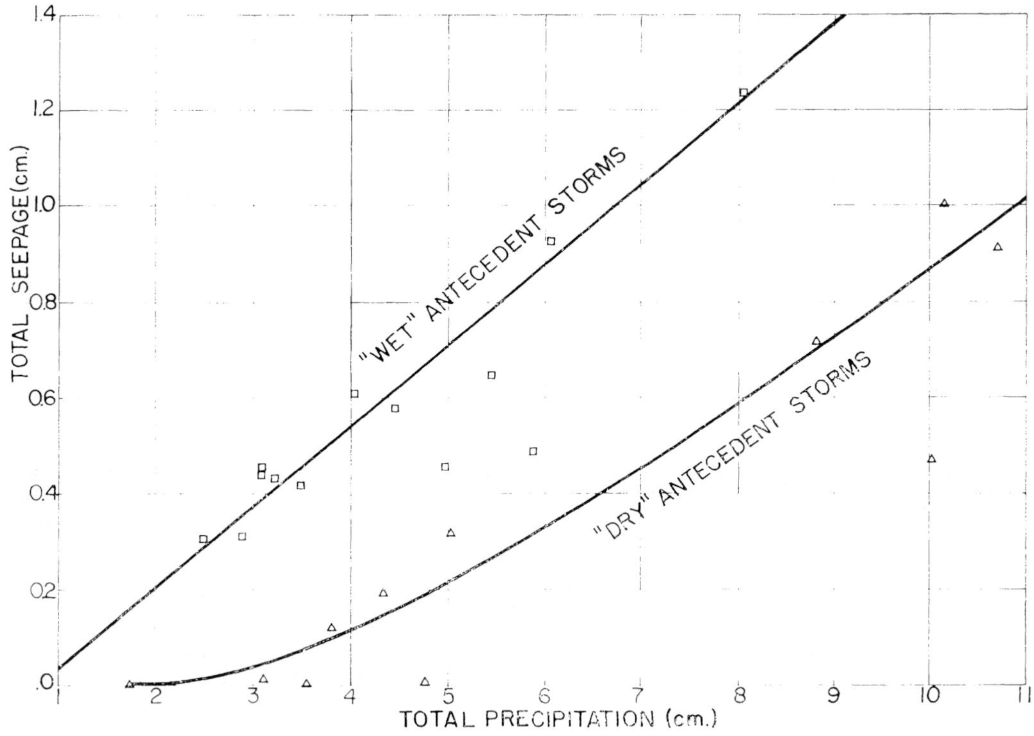

Fig. 2 — Relation between total seepage outflow and total simulated rainfall for each event. Note the difference in outflow between "dry" antecedent and "wet" antecedent storms.

In this experiment flow of infiltrated water was essentially vertical until a compacted layer was reached. At this layer there was lateral flow. We also found this to be the case in previous experiments conducted in 1961 and 1962 on soils varying from 45 centimeters to 160 centimeters in depth to the first impeding layer.

Seepage patterns indicated resistance to wetting, perhaps caused by occluded or confined air left in large pores of the coarse-textured sandy loam. Under dry conditions seepage flow started from the highest layer before it started from the layer below. This pattern occurred whenever soil-moisture-tension values (negative pressures) were greater than 300 centimeters—sometimes after only 3 days of drying. Tensiometer readings showed that a saturated zone existed above the plane of the wetting-front in these cases. Hansen ([6]) has shown that the wetting-front in irrigation of dry soils may act as an impediment to vertical water movement. Thus it was possible for lateral flow to occur above this plane even before a textural change was reached.

79

This type of subsurface flow can occur after a high-intensity rain during the growing season but on the Allegheny-Cumberland Plateau heavy rains usually occur in the dormant season during extremely wet periods. Since high-intensity storms of over 2-centimeter depth are infrequent during the dry growing season this lateral flow above the wetting-front is probably not significant here.

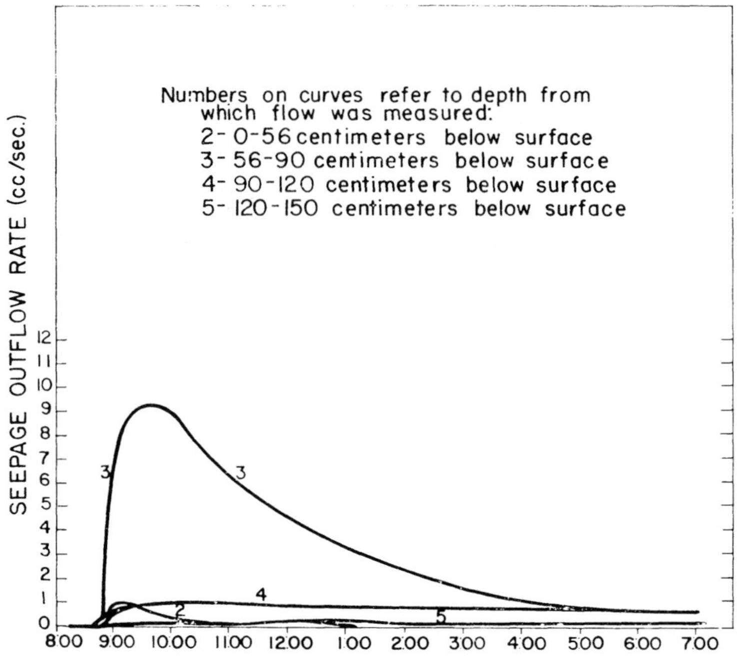

Fig. 3 — Discharge hydrograph of outflow seepage from a storm under "wet" conditions.

With higher intensity storms and/or longer durations, greatest volumes of seepage flow under wet conditions occurred from the 0- to 56-centimeter zone (table 2, fig. 4). Tensiometer readings showed that while the 56- to 90-centimeter zone was completely saturated in 17 storms, the 0- to 56-centimeter zone was never saturated to the surface. However, with increased wetting, flow occurred from this upper zone, indicating the buildup of a temporary "ground-water zone" or partially saturated depth that increased in height with increased quantities of simulated rain.

Whether prior conditions were wet or dry, the initial seepage from any zone would start from 90 to 150 minutes after the beginning of the storm. This timelag is difficult to explain, but there are three possible reasons:

1. Before saturated flow occurs some of the percolating water is withdrawn to replace soil moisture deficits caused by direct evaporation, plant use, and movement by capillary conduction to lower soil depths.

2. For outflow to occur the water must first break through the finer textured soil into the pea gravel (fig. 1). This can cause a timelag, but we do not feel this is an important consideration. The gravel in the trough-collection system was wet at all times during the simulated-rainfall season and there was little if any chance for the soil face to dry and for surface-tension forces to create a "wall" effect at the soil-gravel interface. The trough-collection systems were opened and inspected to check for erosion of the soil face, and for breaks or leaks. In all instances we found good, wetted contact between the pea gravel and the soil face.

3. The vertical distance water traveled before reaching the impeding layer helped determine the rapidity with which outflow occurred. Further, water from the principal seepage-producing zone (56 to 90 centimeters) traveled laterally through an additional 10 to 15 centimeters of soil not wetted by vertical percolation. This was a result of the step-system used in placing the collection troughs (fig. 1) to ensure a watertight collection system.

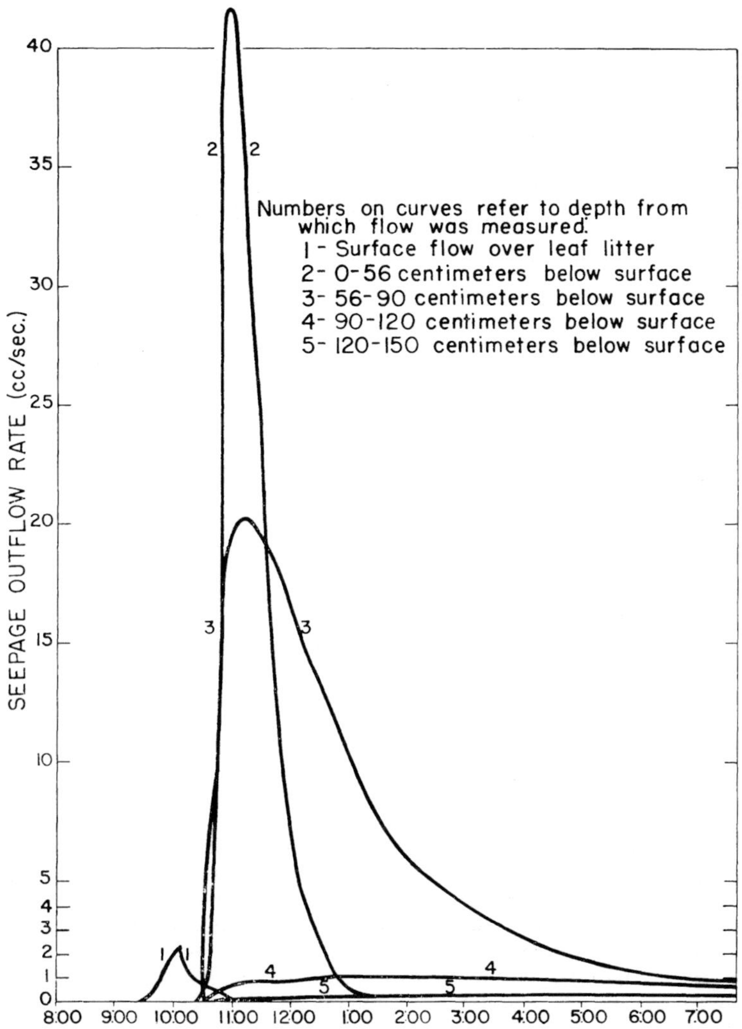

Fig. 4 — Discharge hydrograph of outflow seepage from a storm under "dry" conditions. Note similarity of pattern to that of figure 3.

Hydraulic Conductivity

We were able to compute hydraulic conductivities with the complete hydraulic head data and seepage outflow data available for those depths below 56 centimeters. No figure for the

81

0- to 56-centimeter depth was obtained because that zone was never saturated. To assure that flow from the 90-centimeter depth was coming primarily from the zone immediately above, we studied only those storms where flow had just barely started from the 0- to 56-centimeter depth. This probably indicated that the saturated zone fell within the province of the 56- to 90-centimeter zone. Peak flow rates were used for the discharge rate at saturation. Hydraulic gradients were computed from tensiometer readings plus elevation head readings, all taken at the time of peak flow.

The Darcy continuity equation was used to compute hydraulic conductivity (1). This equation, $V = K\nabla \emptyset$, is applicable to viscous fluid flow through saturated soil when K is the proportionality constant (hydraulic conductivity) between fluid flow velocity V and the gradient of the total-potential function $\nabla \emptyset$. For total potential we used the hydraulic head (pressure potential) plus elevation head (gravitational head). Head values throughout the plot were referenced to depth of the impeding layer at the first tensiometer battery located 60 centimeters up slope from the plot face.

Using the equation as $K = QL/A\Delta h$ where L = length of saturated area, A = wetted cross-sectional area, and Δh = change in hydraulic head between extreme upper and lower tensiometer batteries, the following K values were derived for the saturated zones :

Soil depth (Centimeters)	Average K (Centimeters per hour)
56 to 90	28.6
90 to 120	1.7
120 to 150	.2

These computed hydraulic conductivity (horizontal plane) values are average values. Actual values vary from point to point.

The value for the coarse-textured sandy loam (56- to 90-centimeter depth) seems reasonable considering the high sand content and the comparatively high number of root holes. The computed average K values for the subsoil layers (below 90 centimeters) compare favorably with values computed for similar-textured subsoils at the Northwestern Drainage Laboratory of the Ohio Agricultural Experiment Station (14).

Hydraulic Head Measurements

We used hydraulic head as measured by multiple-unit tensiometers as an indication of antecedent moisture conditions. When simulated storms occurred every day or every other day, antecedent moisture remained high. However, when the plots were permitted to drain for 4 or more days, tension of the surface layers exceeded 450 centimeters.

The hydraulic head readings indicate wetting pattern as well as hydraulic gradient and antecedent moisture. Hydraulic gradients by depth and time are shown for times during and after the simulated storm (figs. 5 and 6). In simulated storms where wet antecedent conditions prevailed, the tensiometer batteries at all five plot positions showed that wetting took place quickly and uniformly.

One rather interesting phenomenon indicated by tensiometer readings was a buildup or mound of soil water in the lower reaches of the plot. The depth of this zone remained somewhat uniform for 6 to 8 hours after the simulated storm with the general saturated zone gradually decreasing in length. The depth and extent of this saturated zone (as indicated by tensiometer readings) was substantiated by direct measurement of free water levels in 3.8-centimeter-diameter piezometer tubes.

Results from Similar Storms

Reproducibility is a test of the value of the method. We ran five consecutive storms of similar intensity and duration (table 2). Seepage from the first storm (dry antecedent condition) was quite low. Seepage from the next four storms (wet antecedent conditions) was higher and almost the same in total volume and discharge rate for all four storms. The hydrographs of

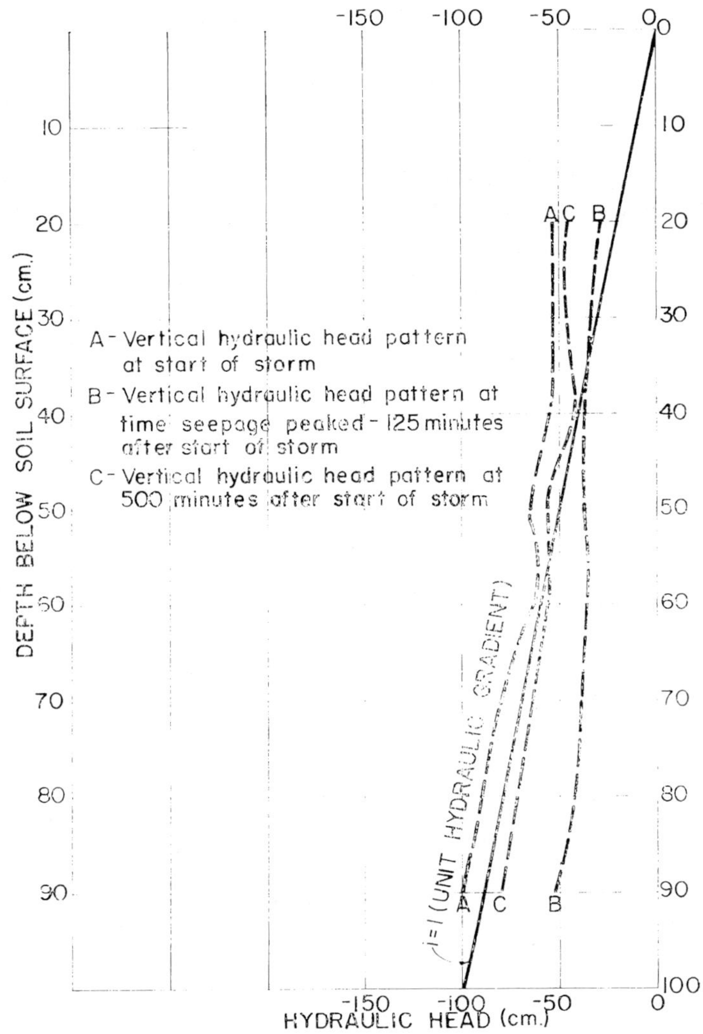

Fig. 5 — Vertical hydraulic head pattern for storm under "wet" conditions taken from readings of multiple unit tensiometer at Station about 40 cm upslope from seepage collection system (see fig. 1). Note line of unit hydraulic gradient. Head readings to left of this line are under negative pressure; readings to right of this line are under positive pressure. Unit hydraulic gradient line is line of zero pressure or zero soil moisture tension.

83

seepage from the four storms were so similar as to be almost coincident (fig. 3). This encourages us to believe that the present system is dependable enough to be used for the soil and topographic conditions we have in this area.

We do not imply that the findings of this study are typical for all regions or indicative of the subsurface stormflow process in general, but we do feel that the results do indicate, within limits, what does occur in unglaciated, hydrologically shallow, hillside soils.

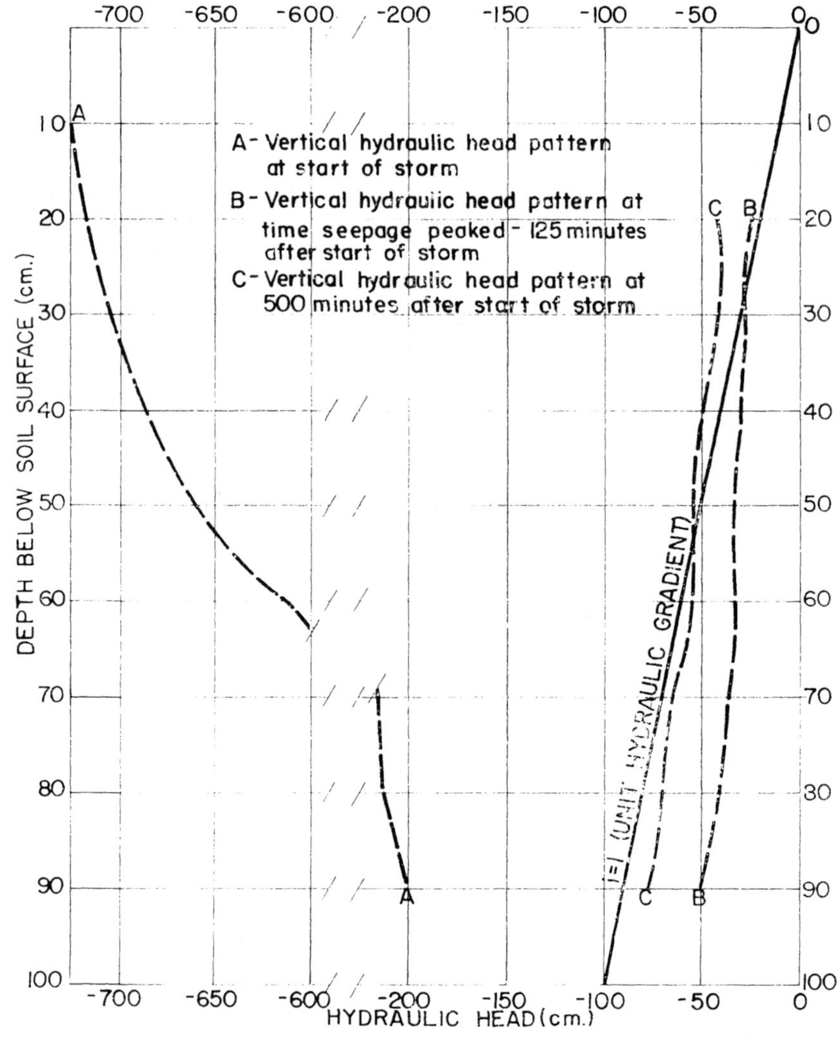

Fig. 6 — Vertical hydraulic head pattern for storm under "dry" conditions taken from readings of multiple unit tensiometer at Station about 40 cm upslope from seepage collection system. Note that soil was extremely wet within 20 cm of surface at 125 minutes after storm start. In this storm heavy seepage rates occurred from soil depth 0 to 56 cm.

84

SUMMARY

A distinct pattern of outflow occurs under different rainfall intensities and depths, hydraulic gradients, and antecedent conditions.

Seepage outflow resulting from simulated storms can be measured at soil-textural discontinuities by means of a downslope-collection-trough system.

Greatest quantities of outflow occurred from the transition point between sandy loam and silty loam—the finer textured layer served as a waterflow-impeding layer. Smallest quantities of flow, but extremely steady flow, occurred from the silt loam and clay loam layers. Erratic flow occurred from the upper layers (0- to 56-centimeter depth) of the sandy loam horizon.

Hydraulic head readings indicated the presence of a saturated zone that was thickest and hence nearer the soil surface at the downslope face. As flow peaked and then receded the depth of the "saturated" mound at the lower position of the plot remained almost constant; the saturated depth up slope, however, decreased with time until the downslope face was reached.

In coarse-textured soil the infiltrated water first travels downward, then, when a finer textured layer is reached, the water travels essentially laterally over this impeding layer. As storm depth increases, the amount of lateral flow increases.

Hydraulic gradient curves computed from hydraulic head readings confirmed the patterns of rise and recession found in physical measurements of outflow.

REFERENCES

(1) American Society of Agricultural Engineers, Drainage Research Committee (1962). Measuring saturated hydraulic conductivity of soils. Spec. pub., Sp-Sw-0262.

(2) BURYKIN, A.M., (1957). Seepage of water from soils in mountainous regions of the humid substropics. *From Pochvovedenie*, No. 12, 90-97. (English translation).

(3) COLMAN, E.A., (1953). Vegetation and watershed management. The Ronald Press, 412 pp.

(4) GAISER, R.N., (1952). Root channels and roots in forest soils. *Proc. Soil Sci. Soc. Amer.*, **16**, 62-65.

(5) GAISER, R.N. and CAMPBELL, J.R., (1951). The concentration of roots in the white oak forests of southeastern Ohio. U.S. Dept. of Agr., Forest Serv., Cent. States Forest Expt. Sta. Tech. Paper 120, 13, pp., illus.

(6) HANSEN, V.E., (1955). Infiltration and soilwater movement during irrigation. *Soil Sci.*, **79**, 93-105.

(7) HEWLETT, J.D., (1961). Soil moisture as a source of baseflow from steep mountain watersheds. U.S. Dept. of Agr., Southeast. Forest Expt. Sta. Paper 132, 11 pp.

(8) HURSH, C.R., (1944). Chairman's report to sub-committee on subsurface flow. *Trans. Amer. Geophys. Union*, Part V, 743-746.

(9) HURSH, C.R. and FLETCHER, P.W., (1942). The soil profile as a natural reservoir. *Proc. Soil Sci. Soc. Amer.*, **7**, 480-486.

(10) HURSH, C.R. and HOOVER, M.D., (1941). Soil profile characteristics pertinent to hydrologic studies in the southern Appalachians. *Proc. Soil Sci. Soc. Amer.*, **6**, 414-422.

(11) KIRKHAM, D., (1947). Studies of hillside seepage in the Iowan drift area. *Proc. Soil Sci. Soc. Amer.*, **12**, 73-80.

(12) REEVE, R.C. and KIRKHAM, D., (1951). Soil anisotropy and some field methods for measuring permeability. *Trans. Amer. Geophys. Union*, **32**, 582-590.

(13) ROESSEL, B.W.P., (1951). Hydrologic problems concerning the runoff in headwater regions. *Trans. Amer. Geophys. Union*, **31**, 431-442.

(14) TAYLOR, G.S. et al., (1961). Drainage characteristics of Toledo and Hoytville soils. *Ohio Agr. Expt. Sta. Res. Bul.* 876, 23 pp.

(15) TODD, D.K., (1953) .Investigation of unsteady flow in porous media by means of a Hele-Shaw Viscous Fluid Model. Unpublished Ph. D. diss., Univ. of Calif., 85 pp.

(16) VAN DIJK, D.C., (1958). Water seepage in relation to soil layering in the Canberra district. CSIRO, Div. of Soils, Commonwealth of Austral., Rpt. 5/58, 13 pp.

(17) VAN 'T WOUDT, B.D., (1954). On factors governing subsurface stormflow in volcanic ash soils, N.Z. *Trans. Amer. Geophys. Union*, **35**, 136-144.

(18) WHIPKEY, R.Z., (1965). Measuring subsurface stormflow from simulated rainstorms—a plot technique. U.S. Dept. of Agr., Forest Serv., Cent. States Forest Expt. Sta. Res. Note CS-29, 6 pp., illus.

(*) The study was completed with the cooperation of the Muskingum Conservancy District.

(**) In the first 21 storms only a trace of surface runoff was measured. In most instances this was the result of the "shingle" effect of litter and was not due to rainfall intensity exceeding infiltration capacity.

Commentary on Paper 9

Hewlett, J. D. & Hibbert, A. R. (1967) Factors affecting the response of small watersheds to precipitation in humid areas.

In this paper, John Hewlett and Alden Hibbert give a summary of their experience of the responses of small, forested catchments (see also Paper 6). They address the question of how so much stormflow can reach the channel by subsurface flow pathways. They note that of the storm runoff "*a fraction will be some of the actual drops that fell during the storm—that is some new rain—and the other fraction will be what we might call translatory flow or flow produced by a process of displacement*" (p.279). They present translatory flow as a type of near-piston flow process, distributed over the hillslope, and cite the tracing experiments of Horton & Hawkins (1965) to support this concept. Horton & Hawkins used tritium to trace inputs added to a 1.2 m column of sandy clay soil in the laboratory and showed that new inputs tend to displace water already stored in the profile downwards. What Hewlett & Hibbert do not record, however, is that the time scale for the displacement in the Horton & Hawkins experiments was of the order of three weeks, somewhat longer than the time scale of the storm runoff hydrograph. The concept of translatory flow did, however, get absorbed into hillslope hydrology (e.g. Chorley, 1978) and displacement processes were later confirmed by the environmental tracing experiments of Sklash & Farvolden (Paper 27).

The 1967 paper was also influential in introducing the constant slope method of separating stormflow from baseflow in hydrograph analysis. Hewlett & Hibbert proposed, from an analysis of records from 15 small Appalachian catchments, that a line of slope 0.05 $cfs/mi^2/h$ ($0.000814\ m^3\ s^{-1}\ km^{-2}\ h^{-1}$), starting at the point of rise on the hydrograph and extending to intersect the recession, would provide a suitable differentiation of a slow baseflow component. This method is obviously simple to implement and has been used widely, regardless of geology or physioclimatic region. It would not, however, appear to be meaningful in terms of hillslope flow processes. At least Hewlett & Hibbert did not make any equivalence between stormflow identified in this way and process, and noted that "*since an arbitrary separation must be made in any case, why not base the classification on a single arbitrary decision, such as a fixed universal method for separating all hydrographs on all small watersheds*" (p.280). Thus, they were only trying to quantify the fast storm response in a consistent way, given the complexity of the flow and displacement processes on real hillslopes. However, as Beven (1991) concluded in a short section on choosing a hydrograph separation method: don't!!

John Hewlett also published a number of other significant articles, including a regression analysis of storm responses from forested catchments (after separation using the constant slope method) which showed that rainfall intensity was not significantly related to peak storm runoff (Hewlett *et al.*, 1977; Bosch & Hewlett, 1982). There was also a text on forest hydrology (revised as Hewlett, 1982) and an influential comment in *Water Resources Research* (Hewlett, 1974). It was through this comment that Fig.1 became more widely distributed. The tone of the comment is that of being somewhat peeved that ideas developed at Coweeta from the time of Charles Hursh onwards, were being attributed to others (and particularly to the Tennessee Valley Authority, see Betson, Paper 8). The comment provides a useful summary of the history of his perceptual model of hillslope responses in permeable soils, particularly those with forest cover, and where surface runoff is rarely seen on the hillslopes. However, comparison with Dunne & Black (Paper 15), for example, suggests that he was not entirely correct in the representation of the ideas of others.

Hewlett also discusses the use of the terms stormflow, baseflow, and surface stormflow in this comment. He notes that stormflow and baseflow cannot be adequately defined, and reserves surface stormflow for rainwater contributing to the hydrograph that has never entered the soil, "however briefly". He therefore follows Horton's idea of the soil as a "separating surface" (see commentary on Paper 1) in arguing that the much higher velocity of flow of water over the surface, relative to that of flow through a porous medium, justifies this distinction. Hewlett was aware of the possibility of displacement of stored water by storm rainfalls but Fig. 6.1 appears to represent his assessment of the subsurface contributions to stormflow in terms of relative proportions of <u>rainfall</u> contributing to streamflow.

References

Beven, K. J. (1991) Hydrograph separation? In: *Proc. 3rd National Hydrology Symposium, British Hydrological Society*, 3.1–3.7.

Bosch, J. M. & Hewlett, J. D. (1982) A review of catchment experiments to determine the effect of vegetation changes on water yield and evaporation. *J. Hydrol.* **55**, 3–23.

Chorley, R. J. (1978) The hillslope hydrological cycle. In: *Hillslope Hydrology* (ed. by M. J. Kirkby), 1–42. Wiley, Chichester, UK.

Hewlett, J. D. (1974) Comments on letters relating to 'Role of subsurface flow in generating surface runoff. 2. Upstream source areas' by R. Allen Freeze. *Water Resour. Res.* **10**(3), 605–607.

Hewlett, J. D. (1982) *Principles of Forest Hydrology*. University of Georgia Press, Athens, USA.

Hewlett, J. D., Fortson, J. & Cunningham, G. (1977) Effect of rainfall intensity on stormflow and peak discharge from forest land. *Water Resources Res.* **13**, 259–266.

Horton, J. H. & Hawkins, R. H. (1965) Flow path of rain from the soil surface to the water table. *Soil Sci.* **100**, 377–383.

Paper 9, is reprinted from pages 275–290 of *International Symposium on Forest* Hydrology (ed. by W. E. Sopper & H. W. Lull), Pergamon, Oxford, UK, by permission of A. R. Hibbert.

FACTORS AFFECTING THE RESPONSE OF SMALL WATERSHEDS TO PRECIPITATION IN HUMID AREAS

JOHN D. HEWLETT and ALDEN R. HIBBERT

Associate Professor of Forest Hydrology, School of Forestry, University of Georgia, Athens, Georgia, and Research Forester, Coweeta Hydrologic Laboratory, Southeastern Forest Experiment Station, Forest Service, U.S. Department of Agriculture, Asheville, North Carolina

ABSTRACT

Customary separation of stream hydrographs into overland flow, interflow and base flow has little meaning when applied to most small watersheds. A revised description of runoff processes in forested headwaters relates quick rises in streamflow to variable source areas and subsurface translatory flow, or the rapid displacement of stored water by new rain. Because this makes the classification of hydrograph components difficult, if not impossible, a numerical rating system, the response factor, was developed from precipitation and streamflow records for use in classifying the hydrologic response of small watersheds in humid areas. A simple uniform hydrograph separation method was necessary to make inter-watershed and inter-regional comparison of response meaningful. Long-term hydrograph records from fifteen forested watersheds in eastern United States were separated into quick and delayed flow by computer and ranked according to mean precipitation, quick flow, and the response factors quick flow/precipitation and quick flow/total water yield. Similar data from nine agricultural watersheds allowed comparison of response among twenty-four small basins, systematically revealing important relationships not usually noted. The over-riding prevalence of subsurface flow was indicated and exceptions were pinpointed. Possibilities and advantages of mapping small watersheds or stream networks by a universal response factor were discussed.

STREAMFLOW is generated chiefly by processes operating beyond perennial stream channels. This truism alone is justification for detailed study of the means by which precipitation is transformed into streamflow. Classical hydrology has been much concerned with channel and related processes which produce flood volumes and peak stages at critical locations downstream. The main problem on the larger streams has been to relate flood magnitude to flood frequency rather than to source areas. This downstream focus has left us even today with a poor understanding of source area hydrology, and has contributed to some unnecessary controversy over the source of floods and alternate methods for controlling them.

For example, the statement or assumption that all floods are due to surface runoff has persisted in hydrology papers and even in some hydrology textbooks despite much evidence to the contrary in forestry and agricultural research. Virtually all techniques now in general use for estimating flood peaks and volumes are based on the seldom challenged assumption that a quick rise in streamflow proves that rainfall is failing to infiltrate and is running over the surface of the ground to stream channels. The term interflow was proposed, and is in fairly general use, to refer to subsurface flows which contribute flood waters in some cases, but there has been no general agreement as to what interflow is or how it works. Hursh (1944) suggested long ago that quick subsurface flow was the primary source of stormflow from forest land. But only a few hydrologists have attached any real importance to interflow as a flood phenomenon and it has remained a sort of catch-all term to cover many small watershed processes we do not understand. Despite gradual recognition of the importance of subsurface processes in streamflow, the traditional concept of stormflow as a produce of overland flow has lingered on in virtually all hydrologic classification and analysis methods. These methods have become encumbered with a confusing array of corrections and adjustments to force the natural variability among watersheds to agree with poorly defined models of upstream hydrologic processes. One of the unfortunate results is that we still appear to have no convenient method for classifying the hydrologic behavior of individual drainage basins. This paper will be an effort to re-examine some upstream hydrologic processes, particularly on forested watersheds, and to propose the use of a response factor for classifying small watersheds.

HYDROGRAPH SEPARATION

The volume of water represented by the hydrograph at any point in a perennial stream channel is usually divided into four parts—channel interception, overland flow, interflow and base flow. Definitions vary considerably among glossaries of hydrologic terms. Channel interception is usually defined as the amount of flow derived from precipitation falling from the clouds or splashing off nearby vegetation directly into the stream. Overland flow is usually defined as an amount of rainfall or snowmelt moving over the surface of the soil to the stream without infiltrating at any point. Interflow is usually thought of as the rapid movement of some subsurface water to the stream during a variably specified period, usually termed the direct runoff period. However, some hydrologists exclude interflow from direct runoff. Direct runoff is the amount of flow resulting directly from a rainfall or snowmelt event but arbitrarily limited in duration; it usually includes all three types of flow just defined. Direct runoff is separated from the fourth type, base flow, by arbitrary methods referred to as hydrograph separation. Base flow is thought of as flow from ground water aquifers, or flow that is left after direct runoff has been separated.

Hydrograph separation is one of the most desperate analysis techniques in use in hydrology. It must be so, since the definitions given above are discrete in neither space nor time. Channel interception appears easiest to deal with because it can be equated with an amount of water falling in a limited time on a definable area. However, stream channels tend to grow and shrink in an uncertain way during storms; we usually ignore shrinking and expanding channel areas in separating channel interception and also in separating overland flow from interflow and base flow. Often water will collect on an impervious area during a storm but infiltrate before it reaches streams. In other cases, water will infiltrate but come to the surface in an intermittent or ephemeral channel before it flows to the perennial channel. As we approach the perennial channel spatially, and the end of direct flow chronologically, interflow becomes hopelessly inseparable from base flow. Contrary to popular opinion these different types of flow rarely are discernible in the hydrograph trace. It is doubtful that any graphical technique applied to a hydrograph is going to put firm definitions under such concepts.

It would seem, in view of the advanced mathematical and theoretical techniques which are being applied in synthetic hydrology today, that we should be pretty well along with the observation and description phases of hydrology. But despite many records of runoff the description of processes controlling rates of flow from forests, croplands and pastures in humid areas is by no means adequate to meet the needs of modern watershed planners. How forests, mountains, croplands, pastures and urban areas contribute to floods is still a basis for vigorous argument among agencies and hydrologists whenever a flood-control program is proposed. Furthermore, the movement of agricultural chemicals and other pollutants through soils into streams is raising new questions about the source, timing and turnover rates of water from upland areas.

Experiments to control evapotranspiration by chemicals, forest cutting or forest type conversions force us to consider in detail how and when the water saved will get into streams and reservoirs. The idea of spreading or spraying extra water on forested slopes to increase low flows or purify effluents is growing; here again attention will be focused on how this water moves or is detained in the porous mantle. These are old questions which require some new answers.

The concept that streamflow from a small watershed is due to a shrinking and expanding source area—the variable source area concept —grew out of studies of the drainage of sloping soil models at the Coweeta Hydrologic Laboratory (Hewlett, 1961). In the next section, these ideas will be discussed in relation to the small, forested, upland watershed. For purposes of this paper a small watershed shall generally mean a natural basin large enough to provide flow during most of the year but not greater than 20 square miles. Larger limits might be set; the important restriction is that travel time and storage in the stream channel during storm periods should not be the major factor in determining the volume of direct runoff.

A DESCRIPTION OF FOREST RUNOFF IN RELATION TO VARIABLE SOURCE AREA

Usually a discussion of the runoff from a watershed begins with the assumption that direct runoff is the product of overland flow and that other types of flow are mere exceptions to the general rule. Perhaps the opposite approach is more logical in the case of forest land; that is, to begin with the assumption that all flow is subsurface flow until there is evidence otherwise. As anyone who has visited a well-vegetated watershed during a rainstorm knows, adequate infiltration and subsurface flows predominate during most storm events. Rainfall and runoff records from all kinds of watersheds in this region show that only the exceptional storm produces flow in amounts greater than 25 percent of gross rainfall and that the largest storm flows of record have rarely exceeded 50 percent of the rainfall that produced them. While these figures tend to set an upper limit to the percentage of rainfall which might be overland flow, they are not by any means a measure of overland flow; numerous workers have shown that much of the direct runoff is subsurface flow also. For example, on one forested Coweeta experimental watershed, direct runoff as a percentage of rainfall has reached as high as 50 percent (6 in. of direct runoff) with virtually no overland flow involved. As we shall see later these experimental watersheds are not atypical; Sopper and Lull (1965) have recently provided further evidence that similar experimental watersheds in the northeast represent flow conditions in that region quite well. If we assume that subsurface flows predominate on most wildland soils, how do direct runoff and base flow get to the channel?

Any effort to model a natural watershed must fail in some degree. However, it is apparent that the essential features of a small watershed as we have defined it are four in number: a stream channel, a slope, a water divide, and a porous mantle of some average depth to an impervious layer. Any or all of these may be indefinite on a given watershed, but despite the natural variation from place to place and region to region, most small upland watersheds may be depicted as shown in Fig. 1. The fundamental dimensions are area, channel length, slope percent and porous mantle depth. In this diagram we have assumed that the sandy-loam soil mantle to bedrock averages 6 ft, slope percent is about 30, and the flood plain is a negligible portion of the slope. Channel length will increase and slope length will decrease during prolonged rainfall. The basin is entirely forested, or at any rate has infiltration capacities which will seldom be exceeded by local rainfall intensities. These conditions are not unusual in the East,

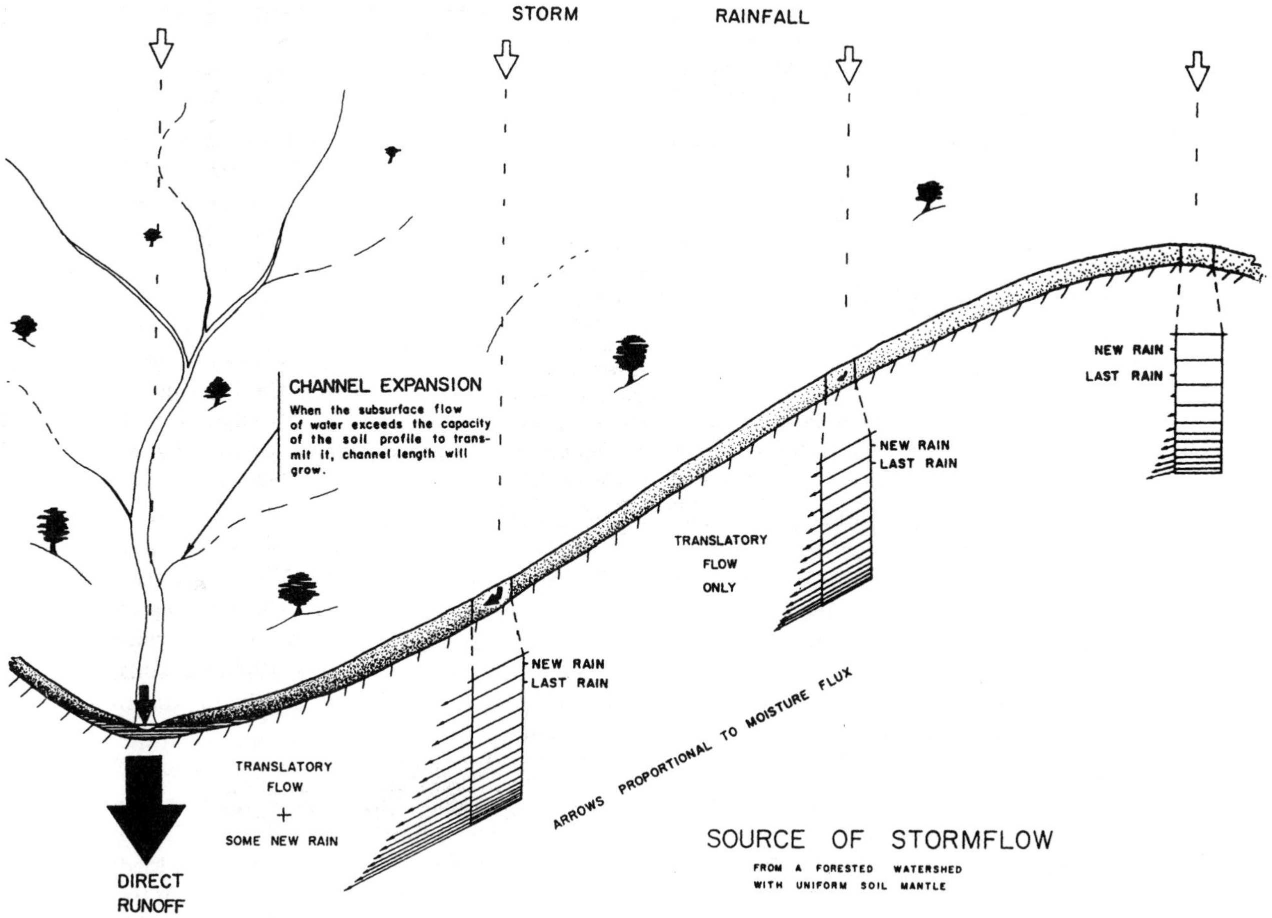

FIG. 1. Diagram showing the source of stormflow (direct runoff) from a forested watershed with a uniform soil mantle. The interaction between the factors illustrated constitutes the variable source area concept of runoffs from small watersheds.

although it would be a misstatement to call them typical.

A large rainstorm, for example, the 10-year frequency storm, is represented in amount and distribution by the blank arrows at the top, and the total direct runoff by the large black arrow at the bottom. Each blank arrow represents a unit of rainfall applied to a unit area beneath it. The smaller black arrows shown within the soil mantle represent the relative effect of a unit of rainfall upon the magnitude of direct runoff below; the black arrow over the stream represents channel interception. Upslope there is no way to calculate the relative influence of each rainfall unit on direct flow, but it is apparent that, unless the entire slope is nearly saturated, this influence must decrease fairly rapidly. If it did not then the volume of water appearing as direct runoff would approach the volume of net rainfall delivered to the forest floor, an extremely rare occurrence in this region as records show. Assuming average antecedent wetness a 4-in. storm would release about 4 in. of water in and immediately adjacent to the stream channel, probably 1 in. one-third up the way upslope, and very little from there on up to the water divide, adding up to a total of about 1 in. of direct runoff or 25 percent of the rainfall. In effect, therefore, the yielding portion of the watershed shrinks and expands, depending on rainfall amount and antecedent wetness of the soil. Areas near the water divide would detain rainfall long enough to prevent it from materially influencing direct runoff, but would supply moisture to sustain base flow or evapotranspiration during the subsequent weeks.

As rainfall continues on these fairly deep soils perhaps half of the area would be influencing direct flow. Part of the expansion in the area supplying flow would occur as an extension of the perennial channel into intermittent and ephemeral channels. It might be argued at this point that the extension of channels is evidence of overland flow, but most or all of the water may still be entering these channels as subsurface flow. The distinction is important because small delays caused by brief infiltration offer large opportunities for transmission into deeper layers. A better description of this phase is to say that when the subsurface flow of water from upslope exceeds the capacity of the soil profile to transmit it, the water will come to the surface and channel length will grow. This in essence is the variable source area concept. Because direct runoff is produced by a watershed whose channel and effective area are varying rapidly in size, it is little wonder that techniques such as the unit hydrograph work only on selected storms and tend to yield anomalous results when applied to other storms on the same or another small watershed.

Usually we think of subsurface flow as moving too slowly to contribute much volume to direct runoff. Certainly it moves more slowly than overland flow if we consider only the movement of a drop of water from some point on the slope to the channel. To understand how infiltrated water can produce such rapid rises in streamflow from forest land, it is helpful to discriminate between two types of flow which, although not entirely separable, are more descriptive than the usual interflow, overland flow and base flow. As we proceed upslope in Fig. 1, each unit of rainfall contributes more to temporary storage and less to direct runoff. However, of the part contributed to direct runoff, a fraction will be some of the actual drops that fell during the storm—that is, some new rain—and the other fraction will be what we might call *translatory flow* or flow produced by a process of displacement. This is a contribution to direct flow of water already stored in the soil mantle before rainfall began. It will be released in large quantities only when the soil is within the field capacity range or wetter. Above the zone of saturation, we may regard such movement as due to thickening of the water films surrounding soil particles and a resulting pulse in water flux as the saturated zone is approached. The process under rainfall is varying everywhere in a most complex way but such movement can be verified in an elementary manner by allowing a soil column to drain to field capacity in the laboratory and slowly adding a unit of water at the top. Some water will flow from the bottom almost immediately, but it will be apparent that it is not the same water added at the top. Translatory flow as a factor in direct flow will prevail chiefly in the lower and mid-slope portions of our model watershed. Upslope translatory flow may be regarded as a pulse in soil moisture which will migrate slowly downhill.

Fortunately some experimental evidence for the translatory process is available. Horton and Hawkins (1965) have recently performed tritium experiments on 4-ft soil columns which suggest the prevalence of translatory movement in natural soil profiles. Referring to the profile diagrams in Fig. 1, Horton's work showed how a 1-in. layer of tritiated water moved through the soil after it had been drained to equilibrium (field capacity). The first water to flow out contained no tritium. Thereafter 1 in. of pure water was added each day to the top of the column and the effluent was analyzed daily for tritium. The tagged inch of water "bumped" its way through the soil pushing ahead 87 percent of all the original water contained by the column before the tritium was added. Horton and Hawkins concluded that infiltered rainwater moves primarily through capillary pores, avoiding the larger ones, and that each rain tends to displace most of the water ahead of it. In Fig. 1 the profile diagrams indicate how this principle might apply to flux rates during storm flow.

Up to this point we have not dealt with rainfall variables, overland flow, evapotranspiration, land use effects, semi-permeable layers in the soil, or other such factors controlling runoff, but have confined attention to the basic morphological watershed features which will operate to control water yield regardless of other factors. By sidestepping the usual arbitrary classifications of the hydrograph, a fresh look at some upstream hydrologic processes becomes possible. For example, recognition that translatory flow is involved reduces the mystery of how forest land can produce large amounts of storm flow without appreciable overland flow. It would not be impossible to treat overland flow as an exceptionally rapid extension of the channel system into areas where the soil cannot transmit the water as subsurface flow. This seems more appropriate than treating all direct runoff as overland flow. A better idea of how streamflow is actually produced should allow improvement in synthetic models of watershed processes and in the selection of variables for their solution. Some recent work by Snyder (1962) is already a step in this direction. Finally, the view of a watershed as a topographic pattern of soil water storage and availability is already well established in soil survey, site and vegetation studies, and there will be some advantage in bringing hydrologic theory in line with these views.

RESPONSE FACTORS FOR SMALL WATERSHEDS

Although it is virtually impossible to separate direct runoff from base flow on a physical basis, we must for practical reasons classify flows which run quickly from watersheds versus flows which are delayed or well controlled. The main trouble with elaborate hydrograph separation methods is that an arbitrary classification of the rate of flow is usually added to another arbitrary classification of the source of flow. That is, we usually decide subjectively which rates of flow are storm flows and then arbitrarily divide these rates into direct runoff and base flow. Since an arbitrary separation must be made in any case, why not base the classification on a single arbitrary decision, such as a fixed, universal method for separating all hydrographs on all small watersheds? A method capable of being applied by a computer to all hydrographs would meet the requirements.

The simple terms quick and delayed flow, rather than the usual direct runoff and base flow, will be used to avoid confusion with the variable definitions and techniques associated with the latter terms. Many intricate methods for separating quick and delayed flow might be programmed for computers but all would have the same basic limitation; they are completely arbitrary. We wanted a method which would be simple yet would apply the same mathematical rule to all hydrograph events. After examining many hydrograph events from about 200 water-years of record collected on fifteen small forested watersheds in the Appalachian–Piedmont region, we decided that a line projected from the beginning of any stream rise at a slope of 0.05 cubic feet per second per square mile (csm) per hour until it intersected the falling side of the hydrograph would be a simple and satisfactory method of separating streamflow into quick and delayed flow. The following requirements are met: (1) The computer ignores stream rises caused by normal diurnal

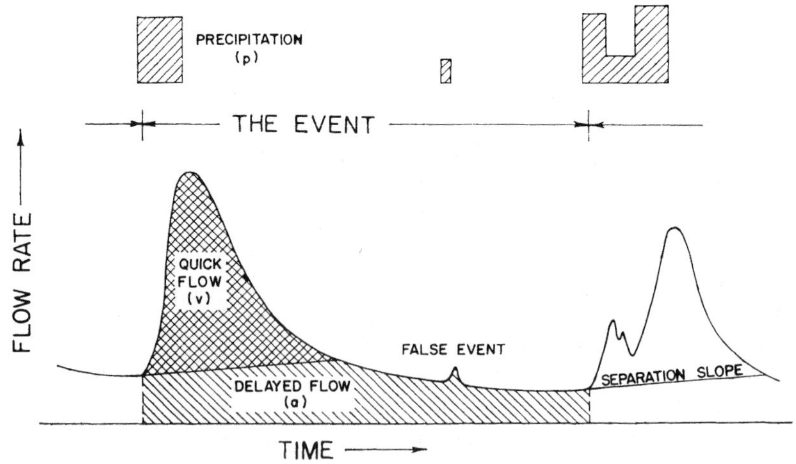

FIG. 2. Diagram of a typical small watershed hydrograph, showing the method used for computer separation of hydrograph events into quick and delayed flow. The constant separation slope is 0.05 cfs/mi²/hr.

fluctuation in flow. (2) On the largest single-peaked storm hydrographs of record, quick flow was shut off within a time lapse of 4 or 5 days. For watersheds less than 20 square miles, experience shows that 5 days will include all unusual, damage-producing flows. (3) Large rainfalls separated by a period of about 3 days would be calculated as separate events. (4) All separations examined appeared reasonable and in line with normal practice.

Figure 2 shows the relation of terms to the hydrograph. A Fortran IV program was written* for the IBM 7094 computer to separate and list various features of each stream rise and depletion period, including v, the quick-flow volume, and a, the delayed-flow volume, both in inches. The event is defined as the time-rate of flow from the beginning of one stream rise to the beginning of the next, limited to events where v is at least 0.001 in. of flow from the basin. If fv is less than 0.001 in., the computer deletes the "false" event and adds the flow to the concurrent "true" event. The minimum acceptable volume of v could be made any size desired. Within the region under discussion, there may be anywhere from about 10 to 100 events per year, varying from 0.001 to over 7.5 in. of quick flow.

The total amount of quick flow (V) yielded by periods (normally one year) is a characteristic of natural watersheds, controlled chiefly by precipitation and the four basic features discussed in the previous section. Although informative for rating watersheds according to the volumes of quick flow they produce, average V alone tells little about a watershed's ability to control the precipitation that falls upon it or its relation to the total yield of water after evaporative losses are deducted. Using V as the numerator, two response factors may be calculated, one expressing the fraction of the total precipitation yielded as quick flow, and the other the fraction of total water yield which is quick flow. We designate these Rp and Ry, respectively, and define them:

$$Rp = \frac{V}{P} \tag{1}$$

* The authors are indebted to Dr. Jerome Clutter, University of Georgia, for programming this technique, which proved more intricate than its simple conception indicated. Details on the flow separation program may be obtained from the authors on request.

and

$$Ry = \frac{V}{P - E} \qquad (2)$$

where V is the sum of all quick flow in inches from a basin during a specified period of time (usually a year), P is gross precipitation in inches and E is the actual evapotranspiration in inches. The denominator used to calculate Ry is the total amount of water yielded to streams and groundwater by the watershed, including any that may escape measurement at a stream gage. In both cases, a concrete runway will show a response close to 1 since it quickly sheds all the water that falls upon it, whereas a deep, permeable forest soil might show a response close to 0. These two methods for classifying the response of a watershed will be compared and evaluated.

It is hypothesized at the beginning that all small watersheds, if undisturbed by man, have a long-term mean response factor which may be determined from records and which reflects its hydrologic behavior in relation to other basins. Thus the performance of basins may be compared through a response factor even though widely separated in distance and rainfall characteristics. Furthermore, the effect of man's complex activities on watersheds may be evaluated and classified partly in terms of changes produced in "natural" response factors, assuming these may be estimated in some manner.

The simplest response factor is Rp, expressing the fraction of rainfall and snowmelt which flows off as quick flow. Such comparisons are not new; direct runoff, separated from base flow by various methods, has often been reported as a fraction of individual storm rainfalls. But because of the lack of a universal hydrograph separation method, there appears to have been no effort to extend this simple concept to a watershed rating system.

Ry is more difficult to calculate because it requires either knowledge of actual evapotranspiration or streamflow records which are known to include the total yield of the watershed. If the stream gage catches and measures all water represented by the term $(P-E)$, then we may substitute $(V+A)$, where A is the sum of the delayed flow over the same period. Usually, however, gaging is not that successful and some delayed flow will pass under the gaging station or leak into or out of the basin elsewhere. Fortunately small errors in the denominator of equation (2) will not affect Ry seriously. If the errors are large, P may be estimated from a nearby rain gage and E calculated by one of several theoretical methods available.

Either or both of these response factors might be used for rating and mapping the hydrologic behavior of small watersheds. In this preliminary report, we have chosen to devote most attention to the meaning, variability and predictability of Ry since it may be calculated entirely from the hydrograph if good quality records are available. Whenever possible, Rp has also been calculated and compared with Ry.

Table 1 identifies fifteen forested watersheds for which at least 3 years of high-quality, continuous streamflow and precipitation records could be secured. These watersheds were expected to show low response factors. We have been unable so far to process suitable records from agricultural watersheds. Since Ry is supposed to reflect not so much the land use and cover conditions as the morphology of basins and their ability to control the water that falls on them, there is some advantage in limiting the first computations to basins with little overland flow. Analysis should provide the original or natural Ry for these watersheds and provide a test whether Ry is sensitive to basin morphology or is merely a reflection of land use, overland flow and precipitation differences. Some preliminary estimates of response factors on agricultural watersheds will be introduced further on.

Ry is obviously a variable and will fluctuate from season to season and year to year. How

TABLE 1

Some Physiographic and Other Data from Fifteen Forested Watersheds of Eastern United States

Watershed and location	No. yr	Source	Area (acres)	Mean elev. (ft)	Mean* slope %	Soil material	Forest type†	Forest cover
Coweeta 2, Franklin, N.C.	27	USFS	31	2800	30	Granitic sandy-loam	O-H	100
Coweeta 21, Franklin, N.C.	24	USFS	60	3250	34	Granitic sandy-loam	O-H	100
Coweeta 18, Franklin, N.C.	27	USFS	31	2700	32	Granitic sandy-loam	O-H	100
Coweeta 14, Franklin, N.C.	28	USFS	152	2890	21	Granitic sandy-loam	O-H	100
Coweeta 8, Franklin, N.C.	24	USFS	1877	3130	22	Granitic sandy-loam	M-H	100
Coweeta 28, Franklin, N.C.	13	USFS	360	3940	33	Granitic sandy-loam	M-H	100
Coweeta 36, Franklin, N.C.	21	USFS	114	4250	47	Granitic sandy-loam	M-H	100
Bent Creek 7, Asheville, N.C.	5	USFS	735	3100	22	Granitic sandy-loam	O-H	100
Copper Basin 2, Tenn.	5	USFS	88	1900	27	Granitic sandy-loam	O-H	100
Union 3, S.C.	10	USFS	22	570	7	Granitic sandy-clay	Pine	95
Fernow 4, Parsons, W. Va.	11	USFS	96	2700	18	Sandstone silt-loam	M-H	100
Leading Ridge 1, Pa.	3	Penn.	303	1200	19	Shale silt-loam	M-H	100
Dilldown Creek, Pa.	4	USGS	1530	1900	4	Sandstone sandy-loam	Sc-O	100
Burlington Brook, Conn.	3	USGS	2637	900	3	Glacial stony-sand	N-H	85
Hubbard Brook 4, Laconia, N.H.	3	USFS	89	1965	26	Glacial stony-loam	N-H	100

* Calculated by dividing the change in elevation from the weir to the highest point by the horizontal length of the watershed.

† Abbreviation key: Oak-hickory (O-H); mixed hardwoods (M-H); scrub oak (Sc-O); northern hardwoods (N-H).

much does it vary and how many years of record are necessary to estimate the long-term mean? Several watersheds from the Southeast with from 11 to 27 years of record were used to estimate the time required to reduce *Ry* to an acceptable estimate of the true mean. It appears that 6 years of record are required to estimate mean annual *Ry* within 15 percent of its true value about 95 percent of the time. Although not as good as was hoped, this variation will not prevent an orderly ranking of watersheds. Calculated from 207 years of record on 15 watersheds, the weighted average standard deviation of *Ry* was 0.04. As might be expected, *Rp* varied similarly. The same number of years are needed to estimate *Rp* to similar limits. However, the numerical value of *Rp* is usually about $\frac{1}{2}Ry$. Also, the standard deviation of *Rp* averages only about 0.02.

To determine how much deviation in annual precipitation from local mean precipitation affects *Ry* and *Rp*, regressions were calculated and tested for significance. The pooled regression coefficient in the following equation was significantly different from 0 at the 0.01 level.

$$(Ry - \bar{R}y) = 0.0015(P - \bar{P}). \qquad (3)$$

The coefficient is so small that precipitation must deviate from the mean by 35 in. to change annual *Ry* by 0.05 points. Equation (3) may be used to adjust *Ry* determined on short records, but fortunately we may conclude that *Ry* is fairly stable with respect to local annual precipitation variation. Similar regressions of *Rp* on annual precipitation produced a higher (significant) regression coefficient.

$$(Rp - \bar{R}p) = 0.0020(P - \bar{P}). \qquad (4)$$

Because the numerical value of *Rp* is smaller than that of *Ry*, equation (4) represents a substantial influence on the value of mean *Rp* determined on short records. If the local long-term mean precipitation is known, adjustment of *Rp* by equation (4) is recommended. Mean *Rp* then becomes a response factor for the watershed at mean precipitation.

TABLE 2

Annual and 6-month Values of Ry *on Three Watersheds*

Season	Year																				
	42	43	44	45	46	47	48	49	50	51	52	53	54	55	56	57	58	59	60	61	62
	COWEETA 21																				
May–Oct.	0.09	0.07	0.05	0.05	0.06	0.06	0.09	0.09	0.11	0.06	0.04	0.04	0.05	0.08	0.04	0.09	0.05	0.08	0.08	0.08	0.08
Nov.–Apr.	0.09	0.10	0.09	0.08	0.12	0.08	0.09	0.11	0.07	0.06	0.13	0.14	0.12	0.11	0.08	0.11	0.08	0.09	0.06	0.09	0.09
Nov.–Oct.	0.09	0.09	0.08	0.07	0.10	0.07	0.09	0.10	0.08	0.06	0.11	0.11	0.10	0.10	0.07	0.11	0.07	0.09	0.07	0.09	0.09
	COWEETA 36																				
May–Oct.	—	0.16	0.10	0.13	0.18	0.19	0.23	0.28	0.33	0.13	0.08	0.10	0.12	0.21	0.08	0.22	0.08	0.25	0.15	0.14	0.14
Nov.–Apr.	—	—	0.25	0.17	0.26	0.18	0.21	0.26	0.20	0.19	0.27	0.27	0.27	0.27	0.20	0.28	0.22	0.20	0.14	0.19	0.25
Nov.–Oct.	—	—	0.22	0.15	0.24	0.18	0.21	0.27	0.25	0.17	0.24	0.23	0.23	0.25	0.17	0.26	0.19	0.22	0.14	0.17	0.23
	FERNOW 4																				
May–Oct.	—	—	—	—	—	—	—	—	—	0.44	0.24	0.43	0.43	0.39	0.45	0.46	0.44	0.24	0.24	0.19	0.45
Nov.–Apr.	—	—	—	—	—	—	—	—	—	—	0.32	0.22	0.20	0.30	0.35	0.36	0.29	0.25	0.37	0.26	0.35
Nov.–Oct.	—	—	—	—	—	—	—	—	—	—	0.31	0.28	0.30	0.32	0.39	0.37	0.35	0.25	0.35	0.25	0.37

Table 2 contains representative values of annual and 6-month *Ry* from three watersheds with fairly long records. Season has some affect on the numerical value of mean *Ry* but not as much as might be expected. Winter season *Ry* correlates most closely with the annual mean. However, means and variances between seasonal and annual *Ry* were tested and found non-significantly different in four out of five watersheds; Fernow 4 was the exception. An incomplete analysis showed that *Rp* tends to be significantly higher in winter than in summer, a fact to be expected because of the higher contributions of summer rainfall to soil-moisture recharge and evapotranspiration.

The major factor causing variation in both response factors from period to period and year to year appears to be the number and size of the larger rain storms occurring within the period. Storm rainfalls in excess of 6 in. tend to release about half their volume to quick flow. Two or more such storms in any period will tend to elevate the response factor for that period beyond two standard deviations from the mean response. Fortunately this seldom happens, and because of the moderating influence of watershed morphology on all storms, both response factors have a strong central tendency.

There is no question that the slope separation coefficient used to compute quick flow will influence the numerical value of response factors. In order to determine whether a range in slopes would change the ranking of watersheds and thus defeat the whole purpose of the response factor concept, slopes of 0.03, 0.07 and 0.09 were added in the separation program and response factors were computed again. Table 3 shows that

TABLE 3

Changes in the Uncorrected Annual Response Factor Ry *as the Separation Slope is Increased from* 0.03 *to* 0.09, *Averaged from One Wet and One Dry Year of Record*

Watershed	Slope coefficient			
	0.03	0.05*	0.07	0.09
	Ry			
Coweeta 2	0.13	0.11	0.10	0.09
Coweeta 14	0.13	0.12	0.10	0.10
Coweeta 8	0.16	0.15	0.13	0.12
Coweeta 36	0.29	0.25	0.23	0.21
Fernow 4	0.41	0.35	0.32	0.29
Hubbard Brook 4	0.43	0.37	0.34	0.32

* Slope 0.05 used as the standard in this paper

increasing the slope sharply reduces the numerical value of *Ry*. The effect on *Rp* was similar. However, the ranking of watersheds remains virtually unchanged throughout the range of slope tested. An increase of 0.01 unit in slope produces about 5 percent decrease in *Ry*. Some justification for selecting a standard slope of 0.05 was obtained from calculations of the standard deviation of the individual event *Ry* for several years of data from six watersheds. This analysis showed a tendency toward minimum variation at slope 0.05, indicating that higher or lower slopes might cause additional variation in mean *Ry* by periods. While more study of the influence of the separation slope is needed, particularly in relation to variation in response factors with watershed size, we feel that a slope of 0.05 will prove to be a satisfactory standard for small watersheds.

WATERSHEDS COMPARED BY RESPONSE FACTOR

The second column in Table 4 shows uncorrected *Ry* for fifteen forested watersheds of the Appalachian and Piedmont region. Watersheds are ranked and grouped according to the corrected $\bar{Ry}$, calculated by equation (2) from precipitation and evapotranspiration estimates listed as $\bar{P}$ and $\bar{E}$. The standard deviation of uncorrected $\bar{Ry}$ and the mean annual quick flow in inches are also shown. The estimate of $\bar{E}$ is admittedly crude; we merely used maps of the eastern United States prepared by Thornthwaite, Mather and Carter (1958), showing estimated average potential evapotranspiration ($\bar{Ep}$) and average annual water deficit ($\bar{D}$) in inches, and assumed that $\bar{Ep}$ minus $\bar{D}$ is equal to actual evapotranspiration $\bar{E}$. The only justification for such a crude method is the limited effect of rather large errors in ($\bar{P}-\bar{E}$) on *Ry*. Only three response factors were significantly changed. In the case of the Union 3 and Fernow 4 watersheds, *Ry* calculated directly from the hydrograph was probably too high because some of the total water yield escapes measurement by the gaging station. Bent Creek 7 was corrected to a higher *Ry*, indicating perhaps that some water may be

TABLE 4

Uncorrected and Corrected Mean Ry *for Fifteen Forested Watersheds, Ranked and Grouped According to the Corrected* Ry

Watershed	Uncorrected *Ry*	S.D.†	$\bar{V}$	$\bar{P}$	$\bar{Ep}$	$\bar{D}$	$\bar{E}$	Corrected *Ry*
				inches				
Coweeta 2	0.09	0.03	2.8	72	32	0	32	0.07
Coweeta 21	0.09	0.02	4.5	78	32	0	32	0.09
Coweeta 18	0.09	0.02	3.4	72	32	0	32	0.09
Coweeta 14	0.10	0.03	3.7	72	32	0	32	0.09
Coweeta 8	0.12	0.03	5.4	77	31	0	31	0.12
Bent Creek, N.C.	0.09	0.05	2.7	48	31	1	30	0.15*
Coweeta 28	0.15	0.03	8.5	88	29	0	29	0.15
Coweeta 36	0.21	0.04	13.9	90	28	0	28	0.22
Copper Basin, Tenn.	0.22	0.07	6.2	60	33	1	32	0.22
Dilldown, Penn.	0.20	0.09	6.0	50	25	1	24	0.23
Fernow, W. Va.	0.32	0.05	7.9	58	25	1	24	0.23*
Union, S.C.	0.44	0.10	3.7	47	34	2	32	0.25*
Leading Ridge, Pa.	0.28	0.04	4.4	40	25	1	24	0.28
Burlington Brook, Conn.	0.27	0.12	7.5	47	23	1	22	0.30
Hubbard Brook, N.H.	0.33	0.10	8.4	46	22	1	21	0.33

* Correction in *Ry* significant at the 0.05 level.   † S.D. stands for standard deviation.

added to the basin from adjacent areas. These are fairly small basins; on watersheds of several square miles, water exchange by underflow, deep seepage or lateral movement should not usually be large enough to influence the numerical value of Ry. Turning to Rp, it is apparent that failure to measure evapotranspiration plays no part in this response factor; only gross error in estimating precipitation and quick flow are involved in Rp.

Both Coweeta 2 and 36 are contained within the larger watershed Coweeta 8, demonstrating the fact that subwatersheds may have response factors both larger and smaller than the main basin. The moderating effect of size on response makes possible both internal and external comparisons of hydrologic behavior, area by area or basin by basin.

An outstanding feature of the response factors listed in Table 4 is that none of these forested catchments yield more than one-third of their total yield as quick flow. Hubbard Brook 4, Burlington Brook and Leading Ridge 1 group together as quickest yielders of flow, probably due chiefly to shallow soil mantles. Coweeta 2, a steep but deep-soiled catchment, empties only 7 percent of its total yield as quick flow, while Coweeta 36, only a mile away, dumps 22 percent. At the same time, Coweeta 36, the steepest and highest watershed in the set, yields an average of 13.9 in. of quick flow per year, five times as much as Coweeta 2, which yielded only 2.8 in. In Coweeta 36 we have an example of a completely forested catchment which, although demonstrating no appreciable overland flow, produces much larger volumes of potential flood waters than most pastures and cultivated areas.

Although the records of most agricultural watersheds are not yet in shape to run through the separation program, we have selected nine experimental watersheds from those listed and described by the Agricultural Research Service (1965) and made a rough estimate of Ry and Rp for comparison with the fifteen forested catchments (Table 5). Records were selected which showed only negligible amounts of delayed flow—that is, runoff ceased within a day or two after rainfall ceased—and thus all measured flow was assumed to be quick flow. If anything, the response factor so calculated should be biased upward slightly because the separation slope was not applied to flow during the runoff period. Most of these small basins are under some form of improved management. The startling exceptions are the Beaver Creek,

TABLE 5

Estimated Response Factors for several Agricultural Watersheds, derived chiefly from "Hydrologic Data for Experimental Agricultural Watersheds in the United States", ARS, U.S. Department of Agriculture, May 1965

Watershed	Area (acres)	Crop	$\bar{V}$*	$\bar{P}$	$\bar{E}p$ (in.)	$\bar{D}$	$\bar{E}$	Estimated $\bar{R}y$	Estimated $\bar{R}p$
Coshocton 135, Ohio	73	improved past.	0.6	33	25	1	24	0.06	0.02
College Pk. W-6, Md.	4	fertilized past.	0.9	41	28	1	27	0.06	0.02
Blacksburg W-III, Va.	19	90% cultivated	0.5	36	29	1	28	0.06	0.02
Watkinsville W-1, Ga.	19	good grass past.	1.6	49	35	2	33	0.10	0.03
Coshocton 185, Ohio	7	grain and meadow	1.2	33	25	1	24	0.13	0.06
Coshocton 127, Ohio	2	cultivated, corn	3.5	36	25	1	24	0.29	0.15
Oxford W-10, Miss.	5530	62% idle pasture	7.5	48	34	1	33	0.50	0.16
Beaver Crk., Mo.†	8960	60% oak, 40% crop	7.1	38	—	—	25	0.55	0.19
Oxford WC-1, Miss.	4	cultivated, poor	18.1	54	34	1	33	0.86	0.34

* Watersheds selected because no delayed flow was involved; all flow was assumed to be quick flow (V).

† From Whipkey and Fletcher (1959); quick flow was determined by hydrograph separation, and the total average measured flow (13 in.) was assumed to be the total water yield.

TABLE 6

Twenty-four Small Watersheds from Eastern United States ranked according to Mean Annual Precipitation, Quick Flow and the Response Factors Ry *and* Rp

Watershed	$\bar{P}$ (in.)	Watershed	$\bar{V}$ (in.)	Watershed	$\bar{R}y$	Watershed	$\bar{R}p$
Coshocton 135	33	Blacksburg W-III	0.5	Coshocton 135	0.06	Coshocton 135	0.02
Coshocton 185	33	Coshocton 135	0.6	College Prk. W-6	0.06	College Prk. W-6	0.02
Coshocton 127	36	College Park W-6	0.9	Blacksburg W-III	0.06	Blacksburg W-III	0.02
Blacksburg W-III	36	Coshocton 185	1.2	Coweeta 2	0.07	Watkinsville W-1	0.03
Beaver Creek	38	Watkinsville W-1	1.6	Coweeta 18	0.09	Coweeta 2	0.04
Leading Ridge 1	40	Bent Creek 7	2.7	Coweeta 14	0.09	Coshocton 185	0.04
College Park W-6	41	Coweeta 2	2.8	Coweeta 21	0.09	Coweeta 18	0.05
Oxford W-10	43	Coweeta 18	3.4	Watkinsville W-1	0.10	Coweeta 14	0.05
Hubbard Brk. 4	46	Coshocton 127	3.5	Coweeta 8	0.12	Coweeta 21	0.06
Burlington Brk.	47	Coweeta 14	3.7	Coshocton 185	0.13	Bent Creek 7	0.06
Union 3	47	Union 3	3.7	Bent Creek 7	0.15	Coweeta 8	0.07
Bent Creek 7	48	Leading Ridge 1	4.4	Coweeta 28	0.15	*Union 3*	0.08
Watkinsville W-1	49	Coweeta 21	4.5	*Coweeta 36*	0.22	Coweeta 28	0.10
Dilldown	50	Coweeta 8	5.4	Copper Basin 2	0.22	Coshocton 127	0.10
Oxford WC-1	54	Dilldown	6.0	Dilldown	0.23	Copper Basin 2	0.10
Fernow 4	58	Copper Basin 2	6.2	Fernow 4	0.23	Leading Ridge 1	0.11
Copper Basin 2	60	Beaver Creek	7.1	*Union 3*	0.25	Dilldown	0.12
Coweeta 2	72	Burlington Brk.	7.5	Leading Ridge 1	0.28	Fernow 4	0.14
Coweeta 18	72	Oxford W-10	7.5	Coshocton 127	0.29	*Coweeta 36*	0.15
Coweeta 14	72	Fernow 4	7.9	Burlington Brk.	0.30	Oxford W-10	0.16
Coweeta 8	77	Hubbard Brk. 4	8.4	Hubbard Brk. 4	0.33	Burlington Brk.	0.17
Coweeta 21	78	Coweeta 28	8.5	Oxford W-10	0.50	Hubbard Brk. 4	0.18
Coweeta 28	88	Coweeta 36	13.9	Beaver Creek	0.55	Beaver Creek	0.19
Coweeta 36	90	Oxford WC-1	18.1	Oxford WC-1	0.86	Oxford WC-1	0.34
Average	54		5.4		0.23		0.10

Missouri, and the Oxford, Mississippi, watersheds; 18.1 in. of quick flow per year was yielded by the Oxford WC-1 cornfield, the highest response we have noted. However, perhaps equally surprising is the relatively low response on other agricultural watersheds when compared to the forested watersheds. Table 6 ranks all twenty-four experimental watersheds first according to local mean precipitation $\bar{P}$, then according to each of the parameters $\bar{V}$, $\bar{R}y$ and $\bar{R}p$. No particular runoff relation is revealed in ranking by $\bar{P}$ except that some areas receive more rainfall than others. Ranking by quick flow amount shows which of these watersheds are the greatest source of potential flood waters. Whether these waters do indeed produce or add to floods depends on many other factors, not the least of which is the troublesome subjectivity underlying the usual definition of a flood. There is some relation between average precipitation and the amount of quick flow produced but not as much as might be expected. This indicates the natural variability among watersheds which we seek to index by a response factor. *Ry* indicates how the watershed controls its total water yield, *Rp* indicates how it controls the precipitation received. In general, Table 6 shows that a similar ranking is achieved by both factors. Watersheds which shifted ranking position significantly between *Ry* and *Rp* are in italics; an inadequate period of record was available to determine whether or not other shifts were important. Union 3 moved up in the ranking by *Rp*, which may indicate nothing more than that we have over-estimated actual evapotranspiration in calculating *Ry*. In the case of Coweeta 36, however, the similarity between uncorrected and corrected *Ry* indicates that the total water yield has been fairly well estimated and that the shift in ranking is probably due to greater influence of precipitation on Coweeta 36. The shift downward into the group of northeastern forest watersheds indicates that this steep catchment, while it moderates most of its yield quite well, is unable to handle all of the excessive rainfall that occurs upon it. Each inch of rainfall above its average of 90 in. produces about an additional $\frac{1}{2}$ in. of quick flow. The wettest water year (November 1 to October 30) of record on Coweeta 36 delivered 130 in. of precipitation, producing an *Rp* of 0.22, equivalent to the average at Beaver Creek, Missouri. Coweeta 36 may be fairly representative of watersheds above 4000 ft elevation in the southern Appalachian Mountains, an area characterized by high rainfall.

Which watershed characteristics affect these response factors most strongly? Variation in the flow separation criteria does not appear to change relative ranking, and, once standardized, appears to play little part in variation among watersheds. While there is some variation among seasonal response factors, the difference is not consistent. Such evidence as available indicates that watershed size, within the 20-square-mile limit we set, plays little part in response factors. Larger watersheds will have to be examined to determine the applicability of response factor concept to them. The following factor complexes, in the order of their natural primacy, largely determine the response factor of small watersheds within humid regions.

1. The average soil mantle depth, or depth to a relatively impervious layer, is probably foremost among watershed factors affecting response. Soil physical properties—stoniness, permeability, retention and so forth—may be included here as a set of associated factors.

2. Average land slope and its effect on the average length of slope from the channel to the water divide, influences response by controlling hydraulic potentials for flow and by sustaining depletion rates.

3. The average size and number of the larger storms—or the average annual storm—influence response by the obvious effect on high flow and rates of infiltration. Average annual precipitation and its distribution may be included here, as it influences response through control of antecedent wetness conditions and variable source areas contributing subsurface flow.

4. Land use is listed last, not because it is minor, but because its effects on the time distribution of flow are superimposed on the effects of the other factors. Land-use effects are obvious but are perhaps best evaluated in relation to original or natural response factors for the area, rather than as a prime hydrologic characteristic. The effects of land use on flow rates are not always as drastic as they appear and it is perhaps best to deal with these effects in relation to response in the absence of land use.

Any one of these factor complexes may dominate the response of a particular watershed but that fact need not destroy the essential logic of this list of causes. Unfortunately, we do not yet have sufficient data to calculate regressions of the response factor on soil depth, slopes and storm sizes. In the meantime, this list will serve as a guide for the selection of variables.

The calculation of *Ry* is difficult because of the lack of precise methods for estimating total water yield, as well as the scarcity of high-quality streamflow records. We can get around the first difficulty by using theoretical methods and it is to be hoped that there may be similar short cuts for determining the average amount of quick flow. Since *Ry* seems to be less dependent on precipitation than *Rp*, it may be more responsive to morphological features of watersheds. However, the simpler factor *Rp* is easier to visualize and, if corrected to mean precipitation, may tell about the same story. Either factor would meet an obvious need of researchers and planners for a convenient numerical handle for describing the general hydrologic behavior of a region, basin by basin.

There are about 50,000 small watersheds of 20 square miles in the humid eastern parts of the United States; each of these has a characteristic response factor which, if even approximately known, would aid planners in the allocation of funds and in design criteria for water resource development. Experimental results and prediction methods might be grouped and extended partly by response classes, as well as on the basis of geological formations, soil types, vegetal cover and climate. For example, the application of predictive techniques such as runoff curve numbers, the unit hydrograph, hydrologic condition classes and multiple regression methods might be improved in accuracy by routinely including or adjusting for local numerical response factors. Mapping hydrologic response by regions would be no small job, but it is to be hoped that every stream reach will not have to be gaged to provide such a map. Regressions of response on morphological factors will help extend experimental data. On smaller management blocks, skeleton diagrams of stream networks might be color coded according to response factor classes, revealing at a glance the response characteristics of working compartments. Such hydrologic response maps would be analogous to soil type maps, with which they may prove to have some useful correlations. A scattering of small gaged watersheds within each physiographic province may be sufficient to rough in preliminary maps; many such records exist but not always in a form amenable to flow separation by computer.

As the preliminary data in Table 6 suggest, some unusual relations may be expected to show up when hydrologic behavior is compared by numerical ranking. If further data bear us out, it may surprise many that some improved croplands appear to control the water that falls on them more efficiently than some forest lands. Also, if the experimental units in Table 6 are representative of the eastern United States, then in the long run only 10 percent of the precipitation (23 percent of the total water yielded) is released as quick flow. Only a fraction of this 10 percent can be classed as channel interception and overland flow, since much direct runoff from forested and other well-vegetated land must be subsurface flow. While it is risky to place an average figure on channel interception and overland flow in this region, it must be greater than 1 percent, the approximate percentage of the land occupied by channels, but is probably less than half the average quick flow, or roughly 2 in. out of 54 in. of precipitation on these twenty-four watersheds. This conceals many important exceptions, but nevertheless the crude average tends to put overland flow in some perspective as a component of the total hydrologic cycle in the East.

Often it is not averages but the largest probable event that most concerns the hydrologist. Mean response factors alone will not define flood peaks or flood frequency in a quantitative sense. But on the average we may expect that the largest probable event in any area will produce higher peaks, larger flood volumes, and do more damage and be harder to control in basins with high response factors. The response factor is a simple concept, basically a mapping technique, which should aid the flow of ideas and experience between watershed managers and researchers.

LITERATURE CITED

AGRICULTURAL RESEARCH SERVICE (1965) *Hydrologic data for experimental agricultural watersheds in the United States*. U.S. Dept. Agr., Misc. Pub. 994. 496 pp.

HEWLETT, JOHN D. (1961) *Soil moisture as a source of base flow from steep mountain watersheds*. Southeast. Forest Expt. Sta., Paper 132. 11 pp.

HORTON, J. H. and R. H. HAWKINS (1964) *The importance of capillary pores in rainwater percolation to the ground water table*. E. I. du Pont de

Nemours and Co., Savannah River Plant, DPSPU 64-30-23. 13 pp.

HURSH, CHARLES R. (1944) Report of the subcommittee on subsurface flow. *Amer. Geophys. Union Trans.*, Part V, 743-6.

SNYDER, W. M. (1962) Some possibilities for multivariate analysis in hydrologic studies. *Jour. Geophys. Res.* **67**, 721-9.

SOPPER, WILLIAM E. and HOWARD W. LULL (1965) Streamflow characteristics of physiographic units in the Northeast. *Water Resources Res.* **1**, 115-24.

THORNTHWAITE, C. W., JOHN R. MATHER and D. B. CARTER (1958) *Three water balance maps of eastern North America.* Resources for the Future, Inc. 47 pp.

WHIPKEY, R. Z. and P. W. FLETCHER (1959) *Precipitation and runoff from three small watersheds in the Missouri Ozarks.* University of Missouri, Res. Pub. 692. 26 pp.

Commentary on Paper 10

Gregory, K. J. & Walling, D. E. (1968) The variation of drainage density within a catchment.

During the "quantitative revolution" in physical geography in the 1960s, there was an ongoing debate about measures for characterizing the variability in topographic characteristics of the landscape and their relationship to different dynamic variables such as peak discharges. Drainage density, used by Horton in his classic geomorphological paper of 1945, was an interesting example in that the measure depends on the definition of stream heads. There were a variety of studies trying to assess the sensitivity of drainage density to different ways of defining the stream head and it was recognized that just taking the blue lines on a map was not an adequate way to define drainage density. However, this study by Gregory & Walling was perhaps the first to look at the hydrological variability of drainage density. They demonstrated quite clearly that, for a number of small catchments in Devon, England, there were significant variations in the length of flowing channels with discharge that could be described by a power law relationship.

This work was later extended by Blyth & Rodda (1973) in the 1856-ha River Ray catchment at Grendon Underwood, near Oxford in England. This catchment, largely underlain by heavy clay soil, showed an order of magnitude change in channel lengths with discharge but with a more asymptotic behaviour at high flows. Under drought conditions the network was highly constrained.

There is a resonance in these studies with the work of John Hewlett in the USA. His concepts of the role of the drainage network in the response of permeable basins is most clearly articulated in the comment in *Water Resources Research* (1974, see commentary on Paper 6).

Ken Gregory later became Professor of Physical Geography at Southampton University and Warden of Goldsmith's College, London. After his PhD, Des Walling stayed in the Department of Geography at Exeter as a lecturer and, later, Professor.

Reference

Blyth, K. & Rodda, J. (1973) A stream length study. *Water Resour. Res.* **9**, 1454–1461.

THE VARIATION OF DRAINAGE DENSITY WITHIN A CATCHMENT

K. J. GREGORY and D. E. WALLING
Department of Geography, Exeter University

ABSTRACT

Actual discharges from two experimental catchments are related to the corresponding measurements of the length of channel flow; within a single catchment this is equivalent to drainage density. Drainage densities ranging from less than 1.0 to nearly 10.0 are associated with discharges ranging from 0.15 to nearly 20.0 c.f.s. per square mile. In each catchment all discharge values (Q) and drainage density values (D_d) are related by a function approximately of the form $Q \propto D_d^2$, which is then discussed in relation to studies made of the variation of streamflow and drainage density between catchments. It is suggested that both peak flow and base flow are related to drainage density in the same way. Studies which include drainage density as a basin characteristic and which relate it to water yield should acknowledge that its value varies within any one catchment, and therefore that the drainage density values derived by a particular method can be related only to one type of flow. In Great Britain the watercourses shown on the 1:10,560 maps relate to specific low flow discharges.

INTRODUCTION

Drainage density, since first used by Horton (1945) has been extensively utilised in many hydrological studies. As Gray (1965) noted "The pattern and arrangement of the natural stream channels determine the efficiency of the drainage system. Other factors being constant, the time required for water to flow a given distance is directly proportional to length". Although a better geometrical description of the stream net is desirable (Wooding, 1966), drainage density, defined as the total length of stream channels per unit area, may be used as the best available index to describe a particular stream network. As a parameter in catchment studies drainage density can be used in three main ways. Firstly, it is related to watershed or physiographic characteristics such as relief ratio, rock type and basin shape. Secondly, it is related to input and output of the drainage basin system. Chorley and Morgan (1962) concluded that maximum runoff, which may reflect high intensity rainfall, is a function of drainage density. Melton (1957) has shown that for particular drainage basins, relationships exist between drainage density and both input (precipitation-effectiveness index) and output (runoff-intensity frequency). Thirdly, drainage density may be useful in relation to studies of past (e.g. Schumm, 1965) and future (e.g. Strahler, 1964) conditions. Increased flood discharges in Wales have been partly attributed to recent changes of drainage density, (Howe, Slaymaker, and Harding, 1966). If the relationships between present drainage density and its controlling variables are understood then deductions may be made about past drainage densities and perhaps about the development of this particular basin characteristic in the future.

The usefulness of drainage density as a parameter is necessarily limited by the method used to delimit the drainage net (Gregory, 1966) and by the maps upon which the net is based. A considerable range of values of drainage density has now been recorded from different areas (e.g. Cotton, 1964) but such variations must be viewed to some extent against differences in map scale, map accuracy, methods of construction of the drainage net, and operator variation. Calculation of values of drainage density is rather time-consuming and so various alternative measures have been used and these can provide data which is similar, or related, to the density of streams. A grid method of determining mean-flow distance in a drainage basin was suggested by Busby and Benson (1960) and they showed that results obtained by this method were within $\pm 2.7\%$ of measurements made for the total basin. The length of the longest watercourse has been used as a parameter in many studies of the variation in discharge from drainage basins (e.g. Taylor and Schwarz, 1952). However, this measure is not independent of area and of basin shape, and similarly total stream length in the basin which has been used by Morisawa (1959) and by Brush (1961) does not separate the density of stream channels from their total length.

Many characteristics of water and sediment yield from drainage basins have now been related to drainage density, usually represented as one factor in multivariate analyses. Peak discharge has been related to drainage density in several studies and Benson (1960) related floods of various recurrence intervals to 9 factors, one of which was drainage density. Mean annual runoff, adjusted to mean annual precipitation, was related to drainage density by Hadley and Schumm (1961). In a valuable discussion of drainage density and streamflow, Carlston (1963) demonstrated a relationship between drainage density and average minimum monthly flow. Hydrograph characteristics were analysed by Hickok, Keppel, and Rafferty (1959) who demonstrated that watershed lag time could be expressed as a multiple correlation function of watershed area, average landslope and drainage density. Variations in sediment yield also have been related to values of drainage density, and Hadley and Schumm (1961) demonstrated that sediment yield increased as drainage density increased. Although drainage density has, therefore, been related to peak, mean and low flow rates and also to sediment yield, in some areas it has been shown that it is not easily or significantly related to variations in either water or sediment yield. Bigwood and Thomas (1955) showed that the relationship between mean annual flood and drainage area for Connecticut was not improved by using stream density, and Hidore (1965) demonstrated that spatial variations in runoff in 16 basins were not well-accounted for by variations in drainage density. In the light of these differing views it is perhaps surprising that few studies have been made of the variations in drainage density which occur within a particular catchment. It is the aim of this paper to describe variations in stream length which occurred in two very small instrumented catchments and it is hoped that this will provide an illustration of how drainage density varies while other basin characteristics are maintained constant.

Experimental results

Several small catchments have been instrumented in south-east Devon, and two of these were selected for a study of the variation of drainage density within a particular catchment, especially in relation to streamflow. Discharge from each catchment was measured by sharp-crested weirs with water levels continuously recorded by vertical water-stage recorders, and stream length was measured, in feet, by reference to pegs set at predetermined positions. On each occasion a record of the discharge at the weir was obtained and then the contributing stream length was measured. Stream length was defined as the total length of concentrated detectable channel flow contributing to the discharge measured at the weir. Each of the two catchments studied includes channels which are perennial, intermittent and ephemeral in nature. Stream length was found to vary considerably with different rates of discharge and so an attempt was made to obtain length measurements associated with a wide range of flows in each catchment.

The initial results of this study show (fig. 1) that there is a definite relationship between stream length and discharge in each of the two catchments. Since area remains constant within the individual catchments, these graphs in fact represent relationships between drainage density and discharge. The range of drainage density values for catchment *A* is from 1.2 to 5.7 and for catchment *B* from 1.4 to 10.5. The difference in the range of drainage density shown by the two catchments may be explained by contrasts in land use and slight differences of rock type. Catchment *A* is almost completely unenclosed moorland and occurs on the outcrop of Upper Greensand (sandstone) covered by clay-with-flints. The same rock type occurs under the highest parts of catchment *B* but the basin also includes outcrops of the Keuper marl underlying the Greensand, and 75% of the area of the catchment (*B*) is occupied by enclosed land, which includes arable land and grassland used for grazing. Infiltration is initially greater in catchment *B* than in *A*, but peak runoff tends to be greater in *B* where there is less vegetation to intercept precipitation, and this is further emphasised by the presence of a thin peat layer in *A*. The same contrasts in catchment characteristics, especially differences in infiltration rates, probably explain why drainage density increases with discharge at a greater rate in *B* than in *A*. In *A*, discharge varies according to $L^{2.9}$ (total length of channel flow) but in *B* according to L^2. The summer values (fig. 1) do not cover as great a range as the winter ones because evaporation is increased, precipitation intensity is generally less, and vegetation growth, particularly in catchment *A* where bracken (*Pteridium aquilinum*) is extensive, helps to intercept precipitation.

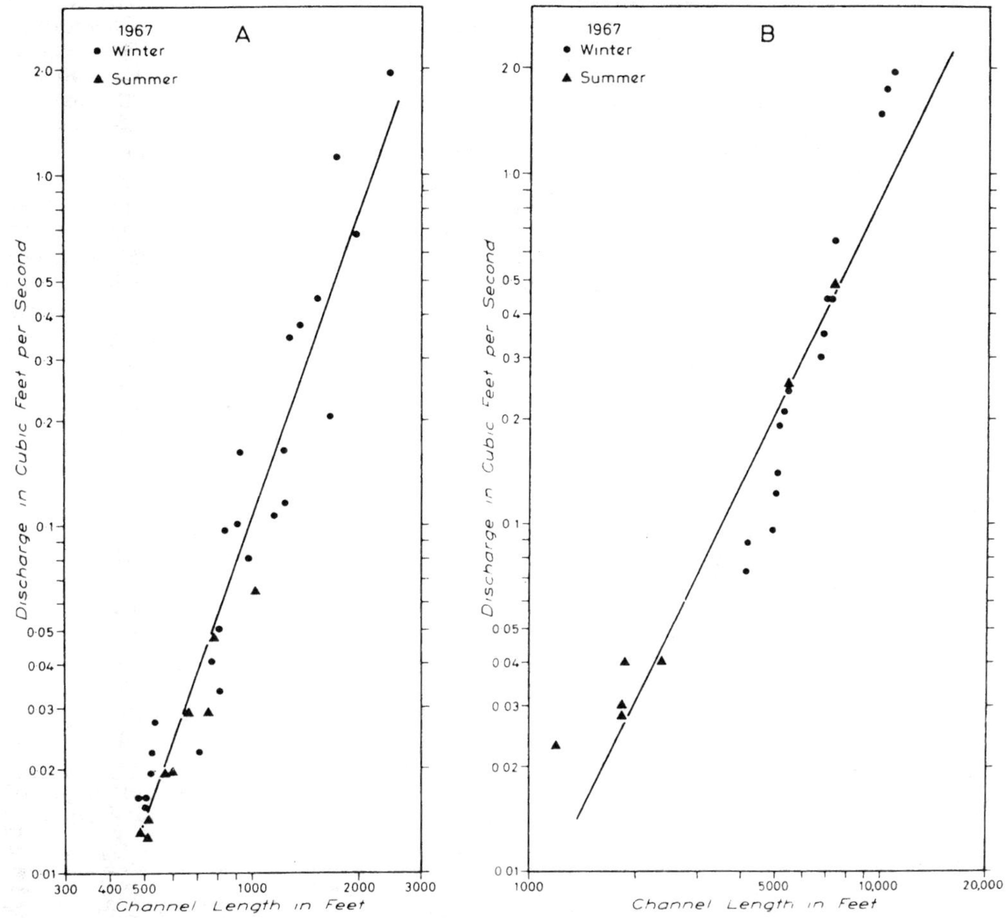

Fig. 1 — The relationship of channel length contributing to actual discharges from two experimental catchments in south-east Devon. The summer values are April-September inclusive. The best fit lines are:

$$A{:}Q = 9.7D_d^{2.9}$$
$$B{:}Q = 8.3D_d^{2.1}$$

The discharge values plotted (fig. 1) range from very low summer flows to high flows resulting from heavy winter rainfall, and they should therefore include the various elements of flow, namely surface runoff, throughflow, interflow, and base-flow. It it probable, therefore, that all three types of flow, individually or combined, are related to wetted stream length and thus to drainage density by the same type of function. This conclusion is illustrated when the discharge values, which occurred during low flow conditions on August 23rd, 1967, from these two catchments are combined with the flows from three other adjacent instrumented catchments, and the five values are plotted against the corresponding drainage densities operative at that time (fig. 2*A*). These minimum flows, demonstrated by the hydrographs to be base flow, show that in this case (fig. 2*A*), base flow discharge varies according to $D_a^{2.2}$. This relationship is based upon values from five catchments which vary in area from 0.08 to 2.5 square miles, have slightly different land use and watershed characteristics, but are on broadly similar rock types. Despite the variations which might be expected, drainage density is seen (fig. 2*A*) to be related to this particular base flow in a way very similar to the relationship obtained for all types of flow in catchments *A* and *B* (fig. 1).

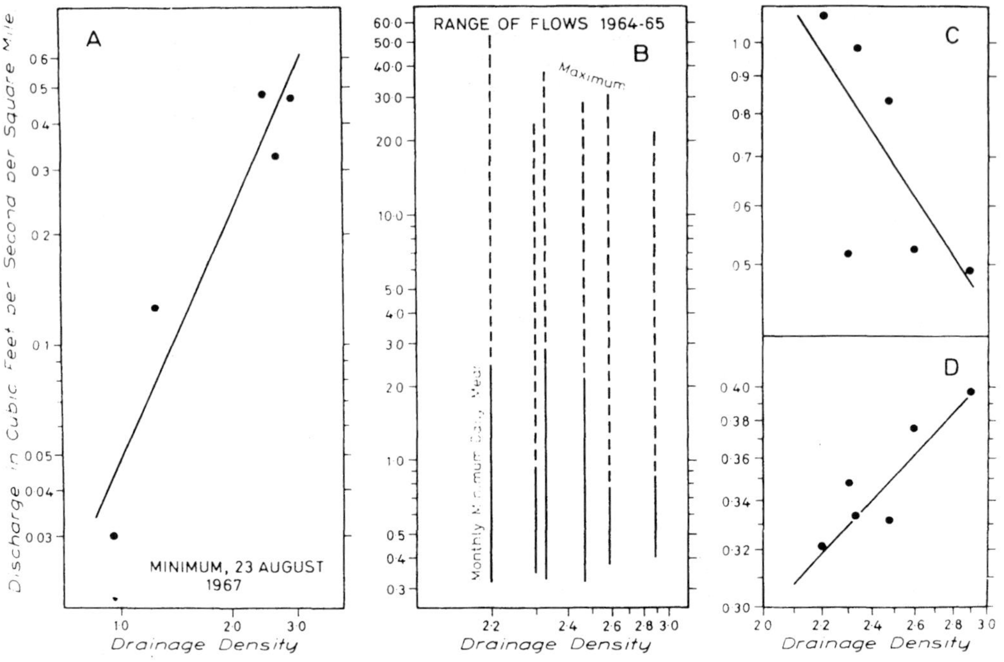

Fig. 2*A* — The relationship of minimum flow to corresponding drainage densities for 5 experimental catchments in south-east Devon.
Fig. 2*B*-*C*-*D* — The relationship of drainage density, derived from the water-courses shown on 1:10,560 maps, to range of flows 1964-65 water year (*B*), average monthly minimum flow (*C*), lowest daily mean for October, 1964 (*D*). Relationships are:

$$A: Q = 0.05\ D_a^{2.2}$$
$$C: Q = 6.57\ D_a^{-2.5}$$
$$D: Q = 0.17\ D_a^{0.8}$$

IMPLICATIONS

There are several general implications which arise from the results of the preceding study and these will be significant particularly when drainage density is to be used as a parameter to com-

64

pare different catchments. Firstly, drainage density within a particular catchment is not a static value. When drainage density is to be related to different runoff values (peak, mean or base flow) these flows may often be associated with a range of values of drainage density for each catchment rather than with a single value which has been used in many previous studies. Secondly, the rate of change of drainage density within a catchment may be related to precipitation intensity and to the characteristics of the catchment. Thirdly, it appears that within these catchments streamflow of any origin is related to drainage density by the same power function which is approximately $Q \propto D_d^2$. Therefore when attempts are made to relate drainage density and streamflow in several catchments it is necessary to consider carefully the discharge value which is to be related to particular values of drainage density, since this field study has shown that drainage density values from 1 to 10 are associated with discharge variation from 0.15 to nearly 20 c.f.s. per square mile.

C.W. Carlston (1963, 1965) studied the relationship between drainage density and streamflow for 15 basins in the United States and concluded that peak flow, designated by the 2.33 year flood, was related to D_d^2 whereas baseflow (Q_b) was inversely proportional to D_d^2(or $Q_b \propto D_d^{-2}$). That two distinct relationships should exist, one for peak runoff and one for base flow, initially seems surprising and would mean that any relationships between mean flow and drainage density might be very difficult to detect. In his analysis, Carlston treated drainage density as a static parameter and related the same density values to both peak flow and base flow. The drainage density values used were derived (for up to 25% of each basin area) by measuring all the marked drainage lines shown on the 1:24,000 maps, supplemented by the extensions into small rills indicated by contour crenulations. These values of drainage density probably represent the upper part of the range of drainage density values in each basin. When water is flowing in all the rills the corresponding discharges will be high. However some workers (e.g. Giusti, 1962; Carlston, 1963) have noted that rivers with the highest peak discharges are frequently those with the lowest baseflows. Therefore by using the same drainage density values an apparently inverse relationship of drainage density and baseflow may be indicated. The highest flood runoff values are associated with the highest drainage densities but these catchments will also have the lowest baseflows.

A further problem concerns the index of base flow which is plotted against drainage density. Although mean annual flood affords a very satisfactory index of peak discharge, Carlston (1965, p. 63) and others have noted that no completely satisfactory method has been devised to accurately designate the base flow or ground-water portion of a stream's discharge. Carlston used the average minimum monthly discharge as one of a number of possibilities for expressing base flow. Although drainage density is difficult to relate to discharge in catchments greater in area than 100 square miles, figure 2*C* attempts to show the relationship between drainage density (based upon the watercourses shown on 1:10,560 maps) and low flow-discharge for six basins in Devon, England, ranging in size from 7.8 to 232 square miles. The greatest range of flows, in the 1964-65 water year for example, occurs in the catchments with the lowest drainage density (fig. 2*B*). Thus when mean monthly minimum flow for the year is plotted against drainage density (fig. 2*C*) the function is of the form $Q \propto D_d^{-2.5}$, but when the actual minimum daily mean discharges are plotted for particular months (e.g. October 1964 which was the month in the 1964-65 water year with the lowest daily mean discharges) the relationship is of a different form (fig. 2*D*), namely $Q \propto D_d^{0.8}$. Thus in Great Britain results obtained using average minimum monthly discharge may be of a different form from those obtained by using the actual low flow discharges for particular months. Thus when relating drainage density to discharge it is important to ensure that both the discharge values and the drainage density values from different catchments are directly comparable.

Carlston interpreted his results in the light of the Jacob water table model expressed as:

$$T = \frac{a^2 W}{2 h_0}$$

where
h_0 height of water table above draining stream as measured at the ground water divide;
a horizontal distance from water table divide to stream;
W rate of accretion to water table (recharge);
T Transmissibility.
He then made a proportional to length of overland flow (L_0) and as L_0 is approximately equal to $\frac{1}{2}D_d$ then

$$T = \frac{W}{8\,D_d^2\,h_0}$$

He then deduced that as ground water discharge into streams would vary directly with transmissibility, ground water discharge (Q_b) should vary according to:

$$Q_b = \frac{W D_d^{-2}}{8\,h_0}$$

Then, assuming that W and h_0 remain constant, base flow should be related to drainage density in the form $Q_b \propto D_d^{-2}$ and as flood runoff varies in magnitude inversely with the magnitude of base flow then, on the same assumptions, $Q_{2.33} \propto D_d^2$. Sir Charles Cotton (1964) showed that drainage density does vary substantially with precipitation and therefore in a subsequent paper Carlston (1965) modified the original equation to be $Q_b \propto W\,D_d^{-2}$, as recharge (W) is roughly proportional to precipitation. But in any one catchment drainage density varies, as shown by the results described above, and therefore the value of L_0 must vary and furthermore L_0 is a function of both a and h_0. Therefore from the original Jacob water table model it is difficult to deduce a model for different types of stream flow, because the drainage density values which apply to peak flows will not apply to base flows (fig. 2*B*).

These problems are particularly relevant because it is important to know to which discharge values the drainage density values, as shown on topographic maps, really relate. From British maps it is very difficult to obtain the precise equivalent of the drainage densities measured in the United States where rills indicated by contour crenulations can be used with a reasonable degree of accuracy. In Great Britain the landscape has been altered by a long period of human occupance, and the associated pattern of lanes, hedges, and ditches all affect the expanded drainage density obtained during a heavy storm. In general terms Johnson (1966) has suggested that there may be an average of five miles of ditch per square mile of England and Wales, and in some cases the expanded drainage net cannot be detected where it includes field drains. Furthermore dry valleys, well-developed in the areas in which this study was made (figs. 1, 2), and glacial drainage channels are represented unevenly over England and Wales. These elements may be completely independent of the present drainage net and therefore pose a major problem in constructing drainage nets according to the contour crenulations shown on topographic maps (Gregory, 1966). In the light of these problems drainage density for at least some areas in England and Wales is probably best derived from the water courses shown on topographic maps, although man-made diversions which do not affect discharge values can be ignored.

Map scale introduces a further problem as there are variations in the density of watercourses indicated on maps of various scales. Catchment *B* (fig. 1) has a drainage density of 2.3 on the 1:25,000 map (Prov. Edn. 1963) but a drainage density of 3.6 on the 1:10,560 map (Regular Edition, 1963). Catchment *A* (fig. 1) has a drainage density of 1.0 on both the 1:25,000 and 1:10,560 maps but the highest discharges are associated with a drainage density of 5.7. The 1:10,560 map provides the best picture of actual surface drainage, and according to field experience in Devon this 1:10,560 net would seem to be that operating during periods of low flow. There is still some variation between basins, however, for whereas the 1:10,560 drainage density for catchment *A* is that pertaining to very low flows, the 1:10,560 density for catchment *B* (fig. 1) corresponds to a discharge ten times as great as those of very low flows. However in five small catchments in Devon (fig. 2*A*) and in five larger ones (fig. 2*D*) the values of drainage density, which are broadly those of the 1:10,560 maps, seem to be related most effectively to base flow expressed by specific low flow values. Thus in figure 2*D* specific low flow values are related to drainage densities derived from the watercourses shown on the 1:10,560 maps. This gives a relationship similar to that indicated by the small catchments (fig. 1). However, if these same drainage density values are plotted against other runoff indices such as average monthly minimum flow (fig. 2*C*) or maximum recorded discharge (fig 2*B*) a fundamentally different relationship is obtained. The drainage density values which are associated with high discharges in this part of Devon cannot be satisfactorily indicated by contour crenulations, however, because these give values still appreciably less than those of the completely expanded drainage net.

Conclusions

Drainage density is a very useful index of basin characteristics and, as Kirkby and Chorley (1967) have recently noted, it affects the magnitude of streamflow from a drainage basin. The relationship of drainage density to actual discharges, representing all types of flow, in two small catchments is shown to be of the form $Q \propto D_d^2$. The fact that discharge varies with D_d in a particular catchment means that when drainage density and streamflow are related for different catchments, consideration must be given to the index of streamflow used and to the method used to derive the drainage density values. These values vary according to the map scale used and also according to whether the net is based upon the valley network, indicated by the contour crenulations, or upon the stream net shown on maps. Although the former method is perhaps the more practical for use in the case of catchments in North America, the analyses in Devon indicate that the stream densities shown on British 1:10,560 topographical maps correspond approximately to those occurring with low flows. Drainage density is the dynamic manifestation of varying inputs into the drainage basin system and its relation to particular inputs reflects the basin characteristics; it is through drainage density that these characteristics influence the output of water and sediment from the system. The relationship between drainage density and streamflow, both within one particular catchment and between catchments, would appear from the above to be of same general nature regardless of the type of flow, and can be expressed in the form $Q \propto D_d^n$.

Acknowledgement

One of the authors (D. E. W.) is grateful to the Natural Environment Research Council who provided a research studentship.

REFERENCES

Benson, M. A., 1960, Areal flood frequency analyses in a humid region: *Internat. Assoc. Sci. Hydrology Bull.*, No. 19, pp. 5-15.

Bigwood, B. L. and Thomas, M. P., 1955, A flood-flow formula for Connecticut: *U. S. Geol. Survey Circ.*, No. 365, 16 pp.

Brush, L. M., 1961, Drainage basins, channels, and flow characteristics of selected streams in Central Pennsylvania: *U. S. Geol. Survey Prof. Paper*, 282-F.

Busby, M. W. and Benson, M. A., 1960, Grid Method of determining mean flow-distance in a drainage basin: *Internat. Assoc. Sci. Hydrol. Bull.*, No. 20, pp. 32-36.

Carlston, C. W., 1963, Drainage density and streamflow: *U. S. Geol. Survey Prof. Paper*, 422-C, 8 pp.

Carlston, C. W., 1965, The effect of climate on drainage density and streamflow: *Internat. Assoc. Sci. Hydrol. Bull.*, v. 11, No. 3, pp. 62-69.

Chorley, R. J. and Morgan, M. A., 1962, Comparison of morphometric features, Unaka Mountains, Tennessee and North Carolina, and Dartmoor, England: *Geol. Soc. America Bull.*, v. 73, pp. 17-34.

Cotton, C. A., 1964, The control of drainage density: *New Zealand Journ. of Geol. and Geophys.*, v. 7, pp. 348-352.

Giusti, E. V., 1962, A relation between floods and droughts in the Piedmont Province in Virginia: *U.S. Geol. Survey Prof. Paper*, 450-C, pp. C.128-C.129.

Gray, D. M., 1965, Physiographic characteristics and the runoff pattern: *Proc. of Hydrol. Symp. No. 4, Research Watersheds, National Research Council of Canada*, pp. 147-164.

Gregory, K. J., 1966, Dry valleys and the composition of the drainage net: *Journ. Hydrology*, v. 4, pp. 327-340.

Hadley, R. F. and Schumm, S. A., 1961, Sediment sources and drainage basin characteristics in Upper Cheyenne River Basin: *U.S. Geol. Survey Water Supply Paper* 1531, part B, pp. 137-196.

Hickok, R. B., Keppel, R. V. and Rafferty, B. R., 1959, Hydrograph synthesis for small arid-land watersheds: *Agricultural Engineering*, Oct. 1959, pp. 608-611 and 615.

Hidore, J. J. 1965, Landform characteristics affecting watershed yields on the Mississippi-Missouri interfluve: *Oklahoma Acad. Sci. Proc.*, v. 45, pp. 201-203.

Horton, R. E., 1945, Erosional development of streams and their drainage basins: *Geol. Soc. America Bull.*, v. 56, pp. 275-370.

Howe, G. M., Slaymaker, H. O., and Harding, D. M., 1966. Flood hazard in mid-Wales: *Nature*, v. 212, (5062), 584-585.

Johnson, E. A. G., 1966, Land drainage in England and Wales: *River Engineering and Water Conservation works*, Ed. R. B. Thorn, Butterworths, pp. 29-46.

Kirkby, M. J. and Chorley, R. J., 1967, Throughflow, overland flow and erosion: *Internat. Assoc. Sci. Hydrol. Bull.*, v. 12, No. 3, pp. 5-21.

Melton, M. A., 1957, An analysis of the relations among elements of climate, surface properties, and geomorphology: *Technical Report 11, Department of Geology, Columbia University.*

Morisawa, M. E., 1959, Relation of morphometric properties to runoff in the Little Mill Creek, Ohio, drainage basin: *Technical Report*, No. *17*, *Department of Geology, Columbia University.*

Schumm, S. A. 1965, Quaternary Palaeohydrology: *The Quaternary of the United States*, Eds. H. E. Wright and D. G. Frey. pp. 783-794.

Strahler, A. N., 1964, Quantitative geomorphology of drainage basins and channel networks: *Handbook of Applied Hydrology*, Ed. Ven Te Chow, pp. 4:39-4:76.

Taylor, A. B. and Schwarz, H. E., 1952, Unit hydrograph lag related to basin characteristics: *Am. Geophys. Union Trans.*, v. 33, (1952), pp. 235-246.

Wooding, R. A. 1966, A hydraulic model for the catchment stream problem. III Comparison with runoff observations: *Journ. Hydrol.*, v. 4, pp. 21-37.

Commentary on Paper 11

Ragan, R. M. (1968) An experimental investigation of partial area contributions.

Bob Ragan's studies of near-stream hillslope responses and their interactions with the stream itself were influential in the following decade (1970s), in particular helping to frame explanations of how the subsurface system could provide runoff into the stream sufficiently quickly to contribute to the stream hydrograph. One of the most important aspects of the Ragan experiments was the careful way in which he attempted to determine all the boundary fluxes to the system. As noted in the introduction, this is difficult in any study of streamflow generation and such attempts are rare, even today.

The study was cited by both Dunne (1978) and Chorley (1978), and consequently gained a wider impact. A number of later papers, looking at piezometer measurements in valley bottoms or wetlands, showed that rapid subsurface responses were possible (e.g. Palkovics *et al.*, 1975; O'Brian 1980). The results gave support to the "groundwater ridging" hypothesis of fast groundwater responses that was later used by Sklash & Farvolden (Paper 27) as a possible explanation of their isotopic hydrograph separation results.

Bob Ragan later became a Professor in the Department of Civil Engineering at the University of Maryland, USA.

References

Chorley, R. J. (1978) The hillslope hydrological cycle. In: *Hillslope Hydrology* (ed. by M. J. Kirkby), 1–42. Wiley, Chichester, UK.

Dunne, T. (1978) Field studies of hillslope flow processes. In: *Hillslope Hydrology* (ed. by M. J. Kirkby), 227–294. Wiley, Chichester, UK.

O'Brian, A. L. (1980) The role of groundwater in stream discharge from two small wetland-controlled basins in Massachusetts. *Ground Water* **18**(4), 359–365.

Palkovics, W. E., Petersen, G. W. & Matelski, R. O. (1975) Perched water table fluctuations compared to streamflow. *Soil Sci. Soc. Am. Proc.* **39**, 343–348.

AN EXPERIMENTAL INVESTIGATION OF PARTIAL AREA CONTRIBUTIONS

Robert M. RAGAN
Associate Professor, University of Maryland
College Park, Maryland
U.S.A.

Abstract

The paper describes a detailed experimental-numerical investigation of the concept of partial area contributions to storm hydrographs. A computer solution of a numerical flood routing technique was used to isolate the time-spatial distribution of local runoff entering the main channel of a small watershed. Extensive experimental information collected in the vicinity of the 619 foot length of second order stream provided a means of interpreting these land-phase hydrographs.

Analysis of a series of storms showed that only a small portion of the watershed ever contributed flow to the storm hydrograph. The contributing area was found to be a function of the storm duration and intensity and, rather than being uniformly distributed along the length of the channel, it existed in the form of localized zones of intense contribution. In a given storm the contributing area was found to fluctuate with changes in the rainfall intensity. During periods of low intensity, most of the flow came from channel precipitation and rain falling an the wet areas surrounding a series of seeps. If a period of high intensity occurred, flow developed through the forest litter on the hillsides and thereby created a larger contributing area. No interflow in the soil mass above the water table was encountered. A rapid response of the ground water at some points along the channel, however, might have been interpreted as interflow if extensive measurements had not been taken.

The behavior of the watershed was quite logical when the fundamentals of the individual processes were considered. The results of the study illustrate that there is a need for a re-eveluation of some of the traditional methods used for runoff computations. Further, any parametric model developed for the synthesis of hydrologic events should be able to reflect partial area contributions.

Résumé

L'objet de ce travail est une étude numérique et expérimentale détaillée de la conception selon laquelle une partie seulement du bassin contribue aux crues causées par les averses. Les solutions que donne une ordinateur quant à la prévision des inondations ont été utilisées pour isoler les distributions temporelle et spatiale de l'écoulement local pénétrant dans le lit principal d'un bassin hydrographique. Les renseignements expérimentaux considérables ramassés au voisinage des 650 pieds d'un cours d'eau secondaire ont permis d'interpréter ces phénomènes d'infiltration.

L'analyse d'une série d'averses a montré que seule une petite partie du bassin hydrographique ne contribuait jamais aux crues d'averses. On a trouvé que l'étendue de la zone en question était fonction de la durée et de l'intensité de l'averse et plutôt que d'être distribuée uniformément le long du lit, elle existait sous la forme de zones localisées à contribution intense. Pendant une averse donnée, on a trouvé que la zone de contribution fluctuait selon le changement d'intensité de la chute de pluie. Pendant les périodes de faible intensité, la majeure partie de la crue venait des chutes de pluie, sur le lit même et sur la zone détrempée qui entoure un réseau de sources. Si une période de forte intensité se produisait, l'écoulement se faisait à travers les taillis des versants des collines créant ainsi une zone de contribution plus importante. Aucune intercommunication des eaux n'a été trouvée dans le sol au-dessus du niveau supérieur des eaux souterraines. Une réaction rapide des eaux souterraines en quelques endroits le long du lit, aurait pu, néanmoins, être interprétée comme une intercommunication si des mesures excessivement précises n'avaient été prises.

Le comportement du bassin hydrographique était assez logique si l'on considère les principes des phénomènes individuels. Les résultats de l'étude illustrent le besoin qu'il y a de re-évaluer quelques-unes des méthodes traditionnelles employées pour l'estimation des crues. De plus, n'importe quel type de paramètre utilisé menant à une synthèse des phénomènes hydrologiques devrait pouvoir refléter ce facteur de la contribution partielle.

Introduction

Many hydrologists believe that only a portion of the watershed area contributes runoff to the storm hydrograph. The findings of a number of recent watershed studies have provided increased justification for this belief. Several of these studies have been directed by staff members of the Tennessee Valley Authority. Betson (1) examined the concept of watershed runoff through the use of a series of mathematical models based on the integral of an infiltration capacity function. Using data from two small watersheds and testing his hypotheses through multiple correlation analysis, he concluded that contributions to storm hydrographs originated from a small, but relatively consistent, part of the watershed. An examination of the behavior of two watersheds in the Elk River Basin of Tennessee (2) led to the conclusion that the watershed was a dynamic one and could vary in size as a storm progressed. It was believed that most of the flow came from the valley floors where the soil moisture was high. Hewlett and Hibbert (3), considering a forested watershed, proposed a dynamic model that derived its storm flow from valley areas near the stream and a variable source area located on the lower slopes of the hillsides. Amerman (4) was led to the conclusion that partial area contributions were significant in a study conducted on an agricultural research watershed in Ohio. He found that the runoff producing areas could be distributed in a seemingly random fashion on ridge tops, in valley floors, and on valley slopes. Zavodchikov (5) considered the role of partial area contributions in his definition of an "effective area". He considered runoff production to be controlled by an effective area which was a function of the general topography of the watershed and the soil moisture.

The present study was undertaken to gain additional insight into the details of partial area contributions. An experimental-numerical procedure was concerned with the identification of the major components of flow entering a natural channel during a period of runoff. The object was to isolate the components of flow and to determine their magnitude, time-sequence of arrival, and their origin. The results clearly show the significance of partial area contributions in the production of runoff in one watershed. Further, the information obtained provides an opportunity to assign a physical interpretation to flows coming from several sources that were located by either experimental measurement or computation.

Approach

Site selection and mathematical model

When undertaking any field investigation, the research hydrologist must have a mathematical model that has sufficient logic to properly define the quantities with which he is concerned and it must have sufficient flexibility to be applied in nature. Further, if a meaningful physical interpretation is to be derived from his results, the site chosen for investigation must have sufficient uniformity to allow the limited experimental data that it is possible to collet to be meaningful and representative of the dominant processes.

Flow in the channel system of a watershed during a period of runoff is unsteady and spatially varied as a result of local runoff entering along the length of the stream. Stoker (6) has shown that a mathematical description of this flow system requires the solution of the differential form of the continuity and momentum equations given as

$$\frac{\partial A}{\partial t} + \frac{\partial (AV)}{\partial x} = q \tag{1}$$

and

$$V\frac{\partial V}{\partial x}+g\frac{\partial y}{\partial x}+\frac{Vq}{A}+\frac{\partial V}{\partial t}=g(S_0-S_e) \quad (2)$$

where the terms have their usual meaning. Many investigators have used numerical solutions of these equations for flood routing and other problems of unsteady flow. The author (7) has shown that a numerical solution can be used to synthesize the time and spatial distribution of lateral inflows entering a channel. To be used to synthesize inflows to the channel, however, the numerical solution of equations (1) and (2) requires the precise definition of such quanities as the local channel slope, roughness, cross sections, and the initial distribution of inflows. Because of these difficulties, it was decided not to attempt to apply the complete equations to the problem of a stream draining an upland watershed. Rather, it was decided to use a model based only on equation (1) where most of the components would be determined experimentally.

After an extensive survey of watersheds in the region, the area selected for this study was a 619 ft. length of second order stream draining a 114 acre portion of a forested watershed near Essex Center, Vermont. The stream was selected because of the relative uniformity of its channel geometry and the lack of major complications in the geomorphology of the surrounding watershed. The watershed had formed as a series of deltas deposited during the step-wise recession of a post-glacial lake. As a result, the topography of the watershed could be broken into two relatively uniform soils consisting of approximately 80 ft. of uniform sand deposited on a horizontal layer of gray silt of unknown depth. This minimized the difficulty of non-homogeneous soil groups and geologic structures, and the sandy nature of the soil allowed easy intallation of all experimental equipment. The drainage network had eroded through the sand to the more dense gray silt. This led to the formation of an extensive and well defined ground water table.

A detailed survey of the channel led to the model shown schematically by figure 1. Q_U represents the flow entering the channel over an upstream control structure, and Q_D represents the flow leaving the channel over a downstream control structure. The other quantities represent the local inflows that could be identified. Q_P is the channel precipitation determined from the length and width of the reach and Q_B is the baseflow entering the reach before the storm. Q_W represents the flow coming from a series of small wet areas and entering the channel at eight points along its length-hereafter refered to as seeps. The remaining term, Q_L, represents the other components of runoff This latter term could not be measured directly and it was recognized that it could include overland flow, interflow and ground water flow. The concepts represented by figure 1 allowed equation 1 to be written as

$$[Q_B+Q_U+Q_P+Q_W+Q_L]-Q_D = \Delta S/\Delta t \quad (3)$$

where $\Delta S/\Delta t$ represents the change in channel storage during the time period Δt. Since Q_L was the only component of inflow that could not otherwise be identified, it was determined on an IBM 1130 digital computer be rearranging equation 3 as

$$Q_L = Q_D-Q_U-Q_P-Q_W-Q_B+\frac{\Delta S}{\Delta t} \quad (4)$$

In this context, Q_L is the average ungaged lateral inflow entering the channel during the time period Δt. The origin and significance of Q_L was to be determined through extensive land-phase intrumentation. Other quantities in equation 4 were to be determined from channel-phase instrumentation.

243

Channel-phase instrumentation

Extensive instrumentation in the channel was necessary to determine the quantities necessary for the solution of equation (4). The upstream and downstream discharges were measured by 90° V-notch weirs equipped with continuous water level recorders.

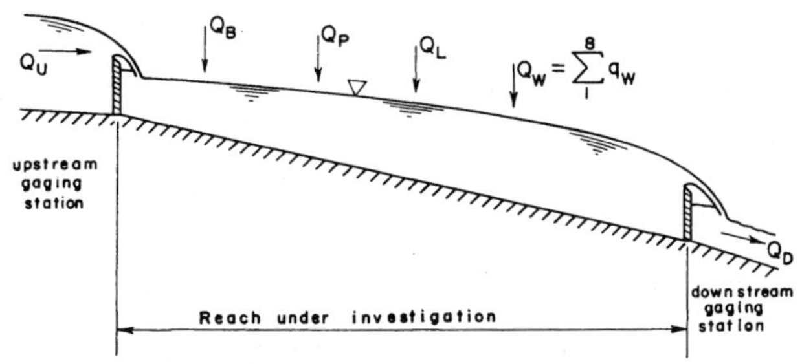

Fig. 1 — Schematic of inflow elements.

Inflows from the eight seeps along one side of the stream were obtained by installing galvanized iron V-notch weirs at their points of discharge. Three of these weirs were equipped with continuous water stage recorders. The remaining seeps were equipped with crest gages. A relationship between the recording weirs and the individual non-recording weirs was obtained by making continuous measurements during a series of storms. In the use of any routing technique based on the continuity equation, one of the most difficult parameters to estimate is the term $\Delta S/\Delta t$, or the rate of change of storage. Often, this accomplished through the use of coefficients determined from past records which relate the storage of the channel to either the downstream depth or a weighted average of the upstream and downstream depths. Such a step leads to considerable uncertainty. An examination of the channel profile revealed that it could be broken into four reaches having relatively uniform slopes. Eighteen stations were established at which cross sectional areas were obtained. These sections were not distributed uniformly along the length of the channel, rather, they were chosen to reflect significant changes in the cross section and to provide reasonable coverage of the four segments of the channel. These stations were mapped during the steady state conditions and, through the use of a portable tape recorder, during the actual passage of hydrographs. With this information the changes caused by bed movement could be incorporated into the program. Depth determinations made during storms showed that the behavior of the stations near the lower end of the channel could be monitored by the recorder on the downstream weir. Similarly, the four stations near the upper end of the channel were related to the depths measured at the upper weir. The depths at stations in the two intermediate reaches were reflected by two water level recorders placed at the section mid-points. As a result, it was possible to determine the rate of change in storage in the channel during the passage of the flood through the use of the four water level recorders and the related cross-sections at eighteen points.

Land phase instrumentation

Gross rainfall was measured through the use of one recording gage and one USWB 8-inch standard gage located in open areas. A relationship to estimate the quantity of rainfall intercepted by the trees was developed through the use of 20 ground level gages under the canopy.

If a meaningful physical interpretation was to be given the computed Q_L, measurements reflecting changes in the soil moisture and ground water had to be obtained. In addition, some technique to evaluate the significance of overland flow and interflow had to be devised. Four nuclear access tubes were used to estimate changes in the soil moisture storage. This was accomplished through the use of a nuclear probe by taking readings before storms, during, and for several days following a period of rainfall. These stations were located in the immediate vicinity of the stream, further out in the floodplain, and on the hillsides.

A portable siesmograph was used to determine the location and general configuration of the ground water table. This was quite successful since the soil was relatively homogeneous and no geologic obstructions existed. Ground water variations in the vicinity of the stream were monitored through the use of fifty-four piezometers and eighteen observation wells arranged in two lines perpendicular to the channel. Twenty-four additional wells were arranged in interlocking triangles to detect changes in direction. During a storm, water levels in those observation wells not equipped with recorders were measured as often as possible through the use of an electric probe. Piezometric pressures were measured before, one time during the storm, and after the storm to obtain general changes in the character of the equipotential lines. Lag times in the piezometers showed that the hydraulic conductivity of the sand was approximately of 80 feet per month while the silt below the channel was approximately 11 feet per month.

To determine if overland flow or interflow was occurring on the hillsides, a three-foot long interception structure was constructed. The device is shown schematically in figure 2 and evolved from a similar unit reported by Whipkey (8). Sheet metal

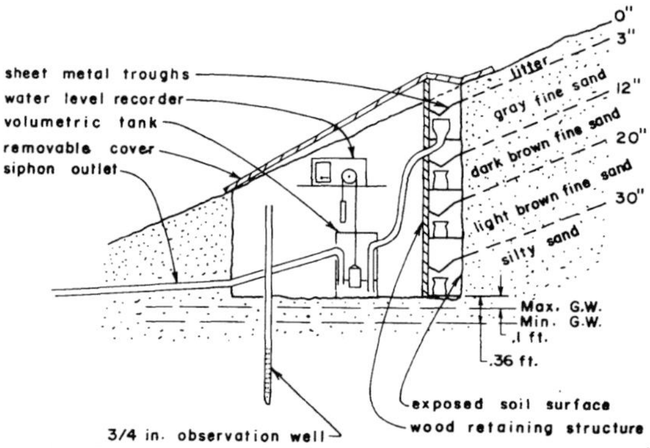

Fig. 2 — Schematic of surface and Subsurface Flow collection system.

troughs were inserted into the exposed soil surface at each apparent change in texture. It was recognized that such a device would detect only a saturated interflow, but since the unsaturated hydraulic conductivity of a sand drops so sharply with decreasing moisture content, it was assumed that any lateral movement at less than saturation would be insignificant. The trough placed immediately below the layer of litter was to collect any surface flow and deliver it to a volumetric tank equipped with a water level recorder.

Discussion

During the period of investigation, June through October, 1966, eighteen storms having a precipitation in excess of .2 inches occurred. The maximum storm had a precipitation of 1.32 inches and the maximum rainfall intensity was six inches per hour. After subtracting the rainfall interception and measuring the changes in soil moisture storage, it was found that the contributing area ranged between 1.2 and approximately 3 percent of the total watershed area. This is consistent with the literature previously cited. All of the storms have been analyzed to determine the runoff volumes involved. Seven storms have been analyzed by the proposed model of equation 4.

Figure 3 is indicative of the response of the various inflow components. This figure shows: the total inflow entering the channel; the sum of the discharges for the eight seeps along the length of the channel; the computed values of Q_L; the average flows for

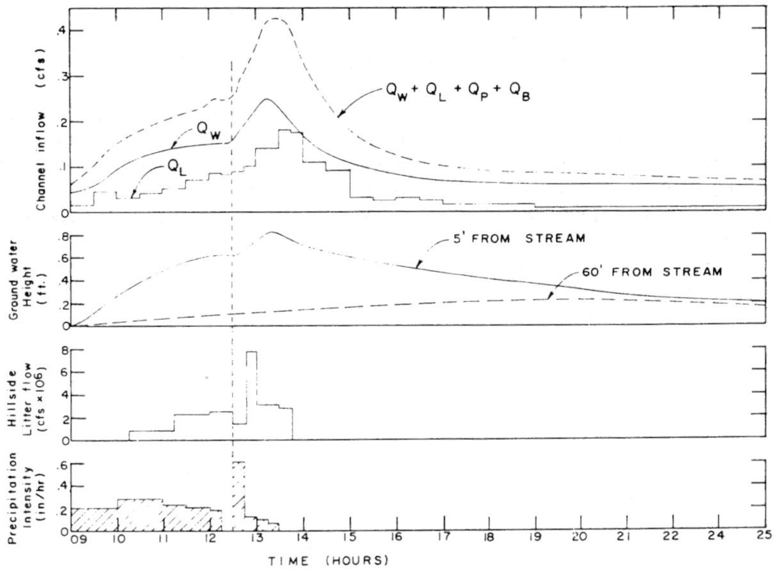

Fig. 3 — Response to storm of sept. 4, 1966, Rainfall 1.10 in. Duration 4,5 Hrs

individual time periods measured in the forest litter by the interception structure shown in figure 2; the change in elevation of the ground water table measured by recording wells at points 5 feet and 60 feet from the stream center line; and the rainfall intensity for the different time periods. Analysis of the storms by equation 4 shows that the channel precipitation varies between two and five percent of the total volume of runoff. The flow from the seeps varies from 55 to 62 percent of the runoff and the ungaged lateral inflow Q_L varies between 36 and 43 percent. These quantities are dependent on the antecedent moisture conditions and the rainfall volume, intensity, and duration. A multiple correlation analysis will be conducted to determine the details of this relationship.

The seeps were characterized by small wet areas that formed near the toe of those slopes that terminated near the stream. One seep draining into the stream, however ,was formed at the toe of a slope that terminated approximately 150 feet from the channel. All seeps showed an immediate response to precipitation and, for light to moderate rainfalls, the contributing area was equal to the wet area surveyed around the seep. During an extended period of precipitation the rate of change of rainfall input was approximately equal to the rate of change of outflow from the seep. This indicated a relatively stable drainage area. It was observed, however, that at those times when flow developed in the forest litter, as measured by the interception structure of figure 2, the instantaneous rate of change in the seep discharge increased significantly and in some cases exceeded the rate of change of rainfall input. This indicated a growing watershed area resulting from flow coming down the hillside. At those times when the rainfall intensity decreased, and flow ceased on the hillside, the rate of change in discharge of the seep would immediately drop—even though it was still on the rising side of the hydrograph. This finding indicated that the drainage area contracted back to the wet area after hillside flow ceased. During these periods ground water observation wells arranged in interlocking triangles were monitored in the area of the seeps. No shifts in ground water flow toward the seeps during any period of the rainfall were detected.

The flow measured by the interception structure cannot be considered as an overland flow. The water collected was clear and did not indicate any erosional processes. The rate of flow was also very small and did not exceed 7.8×10^{-6} cfs. An investigation was undertaken to determine more of the nature of this flow. A profile was cut during one of the storms to determine if the flow was actually occurring through the needles or through a mixture of litter and decomposed organic soil immediately above the fine sand. The cross-section showed approximately 2 inches of loose, damp, pine needles followed by a 1.5 inch mixture of dark organic soil and needles. This lower layer was saturated and when squeezed water ran out quite rapidly. The uniform sand below was moist, but it was less than saturation. It was concluded that the flow was occurring as a saturated porous media flow within the lower zone of the litter. By assuming a hydraulic conductivity equal to 10 feet per month, using the slopes of the local hillside, and cross sectional areas of flow, a discharge of 2.3×10^{-6} cfs. was computed. Considerable assumptions are involved, but the magnitude of the flow is of the same order as that measured. In addition, the general behavior is consistent with the concepts of porous media flow. The flow seemed to occur after long periods of rainfall of after periods of intense precipitation which, perhaps temporarily, was in excess of the rate at which it could enter the soil. After the storage was satisfied and the media became saturated a flow developed. Since the lower boundary was on an unsaturated soil mass under negative pressures, any flow developing in this media would be subject to considerable loss due to infiltration. When the rainfall intensity decreased the losses through the lower boundary into the unsaturated media caused the flow to cease quite rapidly. When the hillside flow was occurring, however, it was computed that the entire length of the slopes would constitute approximately 10 percent of the flow entering the channel from the seeps.

As shown in figure 3 the response of the ungaged lateral inflow was quite rapid. In most storms this inflow lagged the hydrograph peak at the lower end of the channel. Under classical interpretations this would be disignated as either interflow or perhaps part interflow and part overland flow. Except for the hillsides, the micro-relief of the basin precluded any extensive overland flow. There was also an absence of soil lenses which might aid in the formation of a saturated interflow. It was observed, however, that the recording observation well located five feet from the channel responded very quickly to rainfalls. In fact, a ridge in the ground water table, similar to that observed under infiltration basins, formed along the length of the south side of the stream. This is indicated in figure 4. The formation of the ridge is explained as follows. As seen from figure 4 the ground surface breaks fairly sharply approximately 15 feet from the stream. Near the stream the ground water table is within two feet of the ground surface while at sixty feet from the stream the ground water is 8 feet below the ground surface. In the area of the high ground water table a small quantity of rainfall infiltrated the soil and converted the capillary fringe into a saturated zone and produced a very rapid, but localized, response to ground water table pictured by figure 4. As the rainfall decreased this ridge drained and produced a more typical ground water

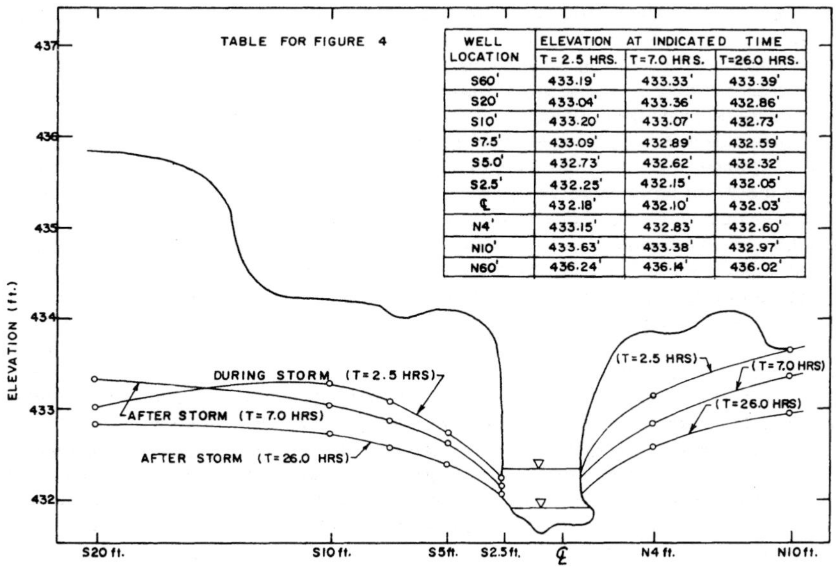

WELL LOCATION	ELEVATION AT INDICATED TIME		
	T = 2.5 HRS.	T=7.0 HRS.	T=26.0 HRS.
S60'	433.19'	433.33'	433.39'
S20'	433.04'	433.36'	432.86'
S10'	433.20'	433.07'	432.73'
S7.5'	433.09'	432.89'	432.59'
S5.0'	432.73'	432.62'	432.32'
S2.5'	432.25'	432.15'	432.05'
℄	432.18'	432.10'	432.03'
N4'	433.15'	432.83'	432.60'
N10'	433.63'	433.38'	432.97'
N60'	436.24'	436.14'	436.02'

Fig. 4 — Ground water table Fluctuations Piezometer cross section one Storm of sept. 4, 1966, Rainfall: 1.10 inches, Duration: 4,5 hours, Scale: 1'' = 4' (Horizontal) 1'' = 1' (Vertical),

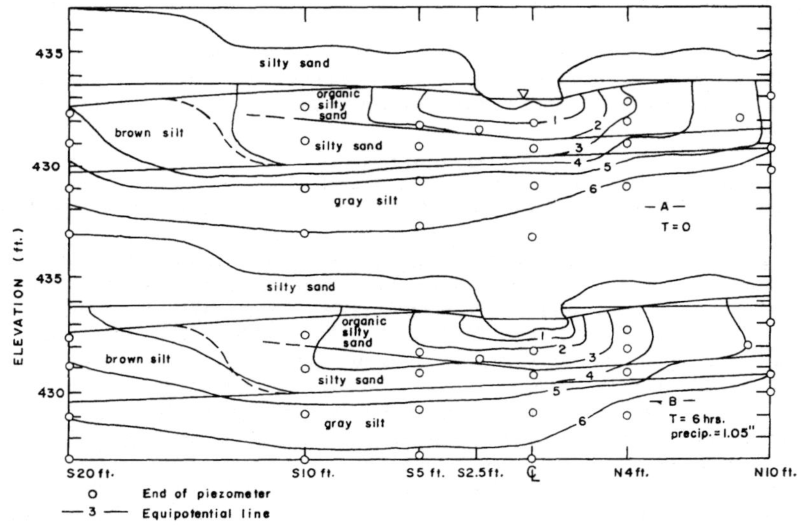

Fig. 5 — Equipotential lines before and during storm of Sept. 15, 1966,

248

profile. The ground water table away from the stream did not respond as rapidly because of the large depth of overlying soil. Repeated tests so check well lag-times showed all to be working properly. It was found that the latral inflow as computed from equation 4 was closely related to the depth variations recorded by the five-foot well. Using one set of data, the average drainage from this ridge and the more uniform grond water on the north side was computed as 2.6×10^{-2} cfs. During the same period the value of Q_L varied between $1.1 \times 10_{+}^{2}$ and 6.2×10^{-2} cfs.

In the storm represented by figure 3, measurements of soil moisture in the vicinity of the stream indicated that very little of the incoming rain water was stored in the mantle. Measurements of soil moisture away from the stream, at points where the ground water table was eight feet and fifteen below the ground surface, indicated that 70 percent of the through-fall was stored in the soil mantle. The experimental data also showed that the volume of the ungaged lateral inflow could be accounted for by considering the rain falling a strip paralleling the length of the channel. This strip was found to vary between 31.2 and 42 feet wide and is consistent with the experimental data. Further indications that a large flow increase occurred in the ground water table is reflected by the equipotential lines shown in figure 5. Figure 5-A shows the equipotential lines immediately preceding a storm and figure 5-B shows the equipotential lines six hours after the beginning of the storm. Very little change in pressure took place in the silty layer because this flow is controlled by conditions some distance from the stream. The potential lines became closer together in the vicinity of the stream in the upper layer of soil thus indicating a larger flow rate.

From this experiment it was concluded that no overland flow existed and all lateral inflow not measured in the seeps or determined as channel precipitation came from the rapid response of the ground water table adjacent to the stream.

Conclusion

The paper shows a feasible technique for isolating inflows and shows the results on one watershed. The paper shows that partial area contributions do exist and that the phenomenon is quite logical when the fundamentals of the individual runoff processes are considered. The results of this study and those cited illustrate that there is a serious need for a re-evaluation of some of the traditional methods used for runoff computations. Further, any parametric model developed for the synthesis of hydrologic events should be able to reflect partial area contributions.

REFERENCES

(1) Betson, R.P., What is Watershed Runoff? *Jour. of Geophysical Research*, Vol. 69, No. 8, AGU, April, 1964.
(2) *Bradshaw Creek - Elk River*; *A Pilot Study in Area-Stream Factor Correlation*, Research Paper No. 4, TVA, Knoxville, Tenn., April, 1964.
(3) Hewlett, J. D. and Hibbert, A.R., Factors Affecting the Response of Small Watersheds to Precipitation in Humid Areas, Paper presented International Symposium on Forest Hydrology, Pa. State Univ., State College, Pa., August 29 - Sept. 10, 1965.
(4) Amerman, C.R., The Use of Unit-Source Watershed Date for Runoff Prediction. *Water Resources Research*, AGU, Vol.1, No. 4, First Quarter, 1965.
(5) Zavodchikov, A.B., Computation of Spring High Water Hydrographs Using Genetic Formula of Runoff. *Soviet Hydrology-Selected Papers*, AGU, No. 5, 1965.
(6) Stoker, J.J., Numerical Solution of Flood Prediction and River Regulation Problems. *Rept. 1*, IMM-NYU-200, New York Univ., October, 1953.
(7) Ragan, R.M., The Determination of Local Inflows Entering a Channel. *Proc. International Hydrology Symposium*, Colorado State Univ., Ft. Collins, Colo., September, 1967.
(8) Whipkey, R.Z., Subsurface Streamflow from Forested Slopes. *Bulletin of the International Assoc. of Scientific Hydrology*, June, 1965.

DISCUSSION

Intervention of Dr. P. MEYBOOM

Question:

Would the rapid response of the water table be similar to what has been described as the LISSE EFFECT?

Intervention of J. BULEK

Question:

I wonder, if the all observed area was alluvial, or if some part of that area was nonalluvial?

Answer

It was a water deposited sand resting on silt. In the largest flows there was some bed movement. There was no seepage loss to the ground from the channel as we generally assume with the classical alluvial channel since the water table at all points and all times was higher than the free surface of the channel.

Intervention of J.H. FLEMING

Question:

What were the characteristics of the vegetation over the catchement area?

Answer

The watershed was completely forested and was operated for two purposes—municipal water supply and logging. In the easily accessible, flat areas outside the stream valley, there was controlled planting and harvesting of pine. In the valley, which was not suitable for logging, there was a mixture of oak, maple, small white birch and pine.

Intervention of Henry W. ANDERSON

Question:

Did you do any tracing of water which would verify your conclusion that "no interflow in the soil mass above the water table was encountered?" Then you found no areas of relatively impermeable soil ,either at the surface or within the soil, that would cause lateral flow?

Answer

We did not add any tracers at anytime. Our flow collection system that cut through the soil mass on the hillside would have detected any saturated movement—as it did at the surface. We could not with this device, however, detect any lateral movement occurring as an unsaturated porous media flow. There was not sufficient soil stratification to create any "perched" saturated zones. We were able to detect vertical movement of water through the unsaturated soil mass through the use of single tube neutron meters. It is very doubtful that any significant unsaturated flow moved laterally because this was a fine sand and thus had the characteristic rapid drop in hydraulic conductivity as the percent saturation decreases.

Intervention of L.L. HARROLD

Question:

On the steep forested areas, there was surface flow during periods of very high rainfall. Were there places where this flow failed to reach the stream—that is, did it seep into the sandy soil?

Answer

The major portion of this flow reached seep areas at the tops of the slope and then moved on the stream. Where the slope terminated on the flat flood plain, we had no indication that the flow continued on to the stream. I think the nature of this flow precluded the continuation across the flat plain. As stated in the paper, this was a porous media flow taking place in a thin layer at the base of the forest litter. As a result it was found only in those areas where there was a substantial gradient. Thus, in the flat areas, it simply infiltrated into the soil.

Intervention of William C. ACKERMANN

Question:

You had many instrument in your watershed, how was the time of these synchronized?

Answer

Time agreement within one minute was achieved by synchronizing all times before a storm. It was only possible because the investigators lived on the watershed.

Commentary on Paper 12

Patric, J. H. & Swanston, D. N. (1968) Hydrology of a slide-prone glacial till soil in Southeast Alaska.

Studies of streamflow generation in arctic and sub-arctic conditions have been relatively rare. Conditions in the North American arctic and sub-arctic vary very widely, from the steep mountain slopes of the coast range and islands in the west, to the low slope, glacially influenced areas of central Alaska and the Canadian shield. This study was carried out on a steep, clear-cut slope in the Maybeso watershed on Prince of Wales Island, southeastern Alaska.

An instrumented slope was studied under artificial rainfalls. Very large amounts of water were added, yet no overland flow was seen, and no sign of significant slope failure was observed despite the steepness of the slope and the continual wetting of the soil. The authors conclude that the roots of the cut trees were effective in holding the soil together, while root channels were effective in transport of water through the soil. They also observed small streams of water and small seeps emerging from bedrock cracks. Their analysis inferred that some 1/3 of the water added in one of the experiments was lost to deep seepage or evapotranspiration.

This short paper is a precursor of a range of later studies on preferential flows and high hydraulic conductivities through forested soils (e.g. De Vries & Chow, 1978), on the role of bedrock fractures on downslope flows and slope stability (e.g. Anderson *et al.*, 1997; Uchida *et al.*, 2001, 2002), and on the saturation of the soil profile by rapid flows from upslope. For an example study of the response of catchments in the low arctic see Roulet & Woo (1988).

The USDA Forest Service site at Maybeso has since been re-instrumented and used for studies of flow and sediment transport and hillslope-channel linkages (Gomi *et al.*, 2001, 2004). Tragically, Jim Patric and his wife were killed in a car crash in 2005.

References

Anderson, S. P., Dietrich, W. E., Montgomery, D. R., Torres, R., Conrad, M. E. & Loague, K. (1997) Subsurface flow paths in a steep, unchannelled catchment. *Water Resour. Res.* **33**(12), 2637–2653.

De Vries, J. & Chow, T. L. (1978) Hydrologic behaviour of a forested mountain soil in coastal British Columbia. *Water Resour. Res.* **14**(5), 935–942.

Gomi, T., Sidle, R. C., Bryant, M. D. & Woodsmith, R. D. (2001) Characteristics of woody debris and sediment in headwater streams, southeast Alaska. *Can. J. Forest Res.* **31**, 1386–1399.

Gomi, T., Sidle, R. C. & Swanston, D. N. (2004) Hydrogeomorphic linkages of sediment transport in headwater streams, Maybeso Experimental Forest, southeast Alaska. *Hydrol. Processes* **18**(4), 667–684.

Roulet, N. T. & Woo, M.-K. (1988) Runoff generation in a low arctic drainage basin. *J. Hydrol.* **101**, 213–226.

Uchida, T., Kosugi, K. & Mizuyama, T. (2001) Effects of pipeflow on hydrological processes and its relation to landslides: a review of pipeflow studies in forested headwater catchments. *Hydrol. Processes* **15**(11), 2151–2174.

Uchida, T., Kosugi, K. & Mizuyama, T. (2002) Effects of pipeflow and bedrock groundwater on runoff generation in a steep headwater catchment in Ashiu, central Japan. *Water Resour. Res.* **38**(7), doi: 10.1029/2001WR000261.

Hydrology of A Slide-Prone Glacial Till Soil In Southeast Alaska

J. H. Patric
and
D. N. Swanston

ABSTRACT.—Heavy irrigation caused no surface runoff, erosion, or debris avalanches on well-drained Karta soil, a tentative series producing much of the commercial timber in southeast Alaska. Interpreting measured rainfall, streamflow, and piezometric head in terms of Darcy's equation showed how this slide-prone soil accommodates large amounts of water. About ⅔ of the water applied drained laterally through permeable surface layers to a stream adjacent to the study area. The remaining ⅓ presumably drained deeply into highly fractured bedrock. Less permeable soil, less fractured bedrock, or longer irrigated slopes probably would have caused saturated soil under heavy watering. It appears that Karta soil must be saturated to cause debris avalanches, a condition which may occur naturally when much larger areas are wetted by much smaller rainfall.

DEBRIS AVALANCHES (*3* pp. 61-63) are a common form of soil mass movement in southeast Alaska. Generally they are caused by reduced internal friction in soils on steep slopes. Clearcutting the forest cover, when it is followed by heavy rain, apparently accelerates slide occurrence (*1*). The influence of rain is clearly expressed by Sharpe (*3*): "The typical debris avalanche . . . is almost invariably preceded by heavy rains which increase the weight of unadjusted material and aid in its lubrication."

Topographic, geologic, and soil conditions usually determine where debris avalanches occur, but rainfall is probably the principal triggering force determining when they occur. The study of this form of soil instability is thus at the mercy of local weather. Study during dry weather is unrewarding because internal friction forces within the soil are near maximum and even the steepest slopes are practically stable. During wet weather, soils may be studied under conditions of imminent sliding but at more cost and certainly with greater danger. The following experiment was conducted to initiate debris avalanching using irrigation equipment to simulate heavy rainfall. One outcome was new information on the hydrology of steep, logged slopes and an indication of the role of rainfall in triggering debris avalanches.

THE AUTHORS are, respectively, research forester, Timber and Watershed Laboratory, U. S. Forest Service, Parsons, W. Va., and research geologist, Pacific Northwest Forest and Range Expt. Sta., Institute of Northern Forestry, Forest Service, U. S. Dept. Agric., Juneau, Alaska. Patric was at the Institute of Northern Forestry at the time this work was carried on.

The Study Sites

The study was conducted on a large clearcut area logged in 1955. The slope chosen was an unstable area described by Bishop and Stevens (*1*), near Hollis on Prince of Wales Island and about 45 miles west of Ketchikan, Alaska. A moist, temperate climate prevails with average monthly rainfall ranging from 3 inches in July to 17 in October (*5*). A graywacke bedrock is overlain with compact glacial till which ranges in thickness from 2 to 30 or more feet. Western hemlock and Sitka spruce dominated the prelogging forest cover. Seedlings of these species, with low shrubs, now form a dense cover from 3 to 8 feet high on the clearcutting.

Stevens (*4*) classified this study site soil as a member of the tentative Karta series: a well-drained, permeable soil derived from glacial till. Textures ranged from silt loam to gravelly loam. Bulk density of B horizon soils averaged about 1.10, surface organic matter was 3-12 inches deep, and slopes ranged from 16° to nearly vertical. The Karta series is described (*4*) as "thick organic surface layers underlain by thin, grayish, leached horizons and thicker, splotchy, dark brown and dark yellowish brown, iron-stained horizons." The weathered soils at the study site varied in depth from 21 to 48 inches. Trees were rooted densely in the organic layers, and roots occasionally were concentrated along the bottom of the B horizon. A steep, deeply incised stream originated above the study area and flowed in a ravine along one edge (Fig. 1).

The Karta series is the most extensive soil supporting commercial timber in southeast Alaska. A third of the commercial timber on Prince of Wales Island and a much higher proportion of commercial-accessible timber grows on Karta soil. The Karta series also was

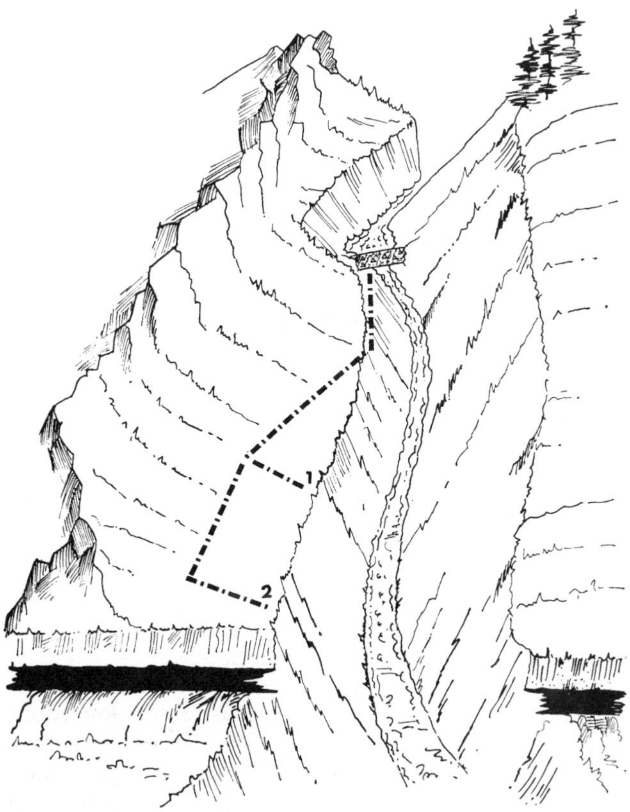

Fig. 1.—Water was piped by gravity from impoundment (upper center) to study sites 1 and 2. Streamflow was measured below study site 1 and below the washed-out logging road (foreground).

the most widespread timbered soil in a recent survey on Chichagof Island, more than 150 miles northwest of the study area.[1]

Methods

A portable irrigation system was installed as shown in Figure 1. Water from an impoundment above the study area flowed by gravity through a 3-inch pipeline to the study sites. Rotating sprinklers mounted at 20-foot intervals on a 2-inch lateral line wetted overlapping circles of 45-foot radius. The designed rate of water delivery was 0.62 inch per hour; randomly placed rain gauges indicated actual delivery closer to 0.70 inch per hour.

A strain gauge probe (*6*) and a piezometer were installed at site 2. The strain gauge probe, firmly fixed in the compact till, furnished information on creep of overlying soil. The piezometer provided information on depth of saturated soil above compact glacial till. It was also possible to measure streamflow above and below the irrigated area. At points where the stream flowed over solid rock, the entire flow could be caught in a barrel and the filling time measured by stopwatch. The road ditch (Fig. 1) cut off seepage moving along the compact till surface and carried it to the stream above the lower measuring point.

Results

The irrigation system operated from August 28 until October 20, although frequently shut down for study site examination or repair. From August 28 to August

[1] Personal communication, Freeman Stephens, regional leader for soils, Region 10, U. S. Forest Service, Juneau, Alaska.

Fig. 2.—Study site 1. The two-inch lateral irrigation line closely followed the large white log (arrow). Note number and distribution of stumps and accumulation of logging debris. An excavated stump (x) is in the center of the photograph.

Fig. 3.—A hydraulically excavated cedar stump 30 inches top diameter. Some compact glacial till has been removed to show root penetration into this dense, unweathered material (left).

Fig. 4.—Study site 2. The wetted area was about 200 feet long and 90 feet wide. Note dense brush cover, lack of logging debris and stumps. The lower sprinkler almost overhangs the ravine. The upper sprinkler is close to the ridge dividing the study area from slopes in background. White circles identify some points common to this Figure and to Figure 6 (below).

30, an 80-foot, 5-sprinkler line delivered 11 inches of water to the 38° slope at study site 1 (Fig. 2). Examination of this study area showed that heavy logging debris had protected the soil from wetting in some places while, elsewhere, a stick could easily be inserted 12 or 15 inches into soil softened almost to mud. No surface runoff had occurred but examination of the ravine wall below the sprinkled area showed eight well-defined seeps of clear water. A stream of clear water about 2 inches in diameter emerged from the rocky ravine wall about 60 feet downstream from the nearest sprinkler and perhaps 30 feet below the soil surface. This stream reduced to a trickle 12 hours after irrigation ceased, and resumed full flow 2 hours after irrigation recommenced. When another 19 inches of irrigation water caused no apparent change in soil stability on this study site, we concluded that further watering was futile.

The presence of several large stumps on study site 1 (Fig. 2) suggested that tree roots bound soil so firmly that no amount of watering would render it unstable. To explore this idea, a fire hose was fitted to the main water line for root excavation. The hydraulically excavated cedar stump shown in Figure 3 rested directly on glacial till. Most of the roots were laterals in the upper two feet of soil but deeper roots were vertical sinkers. Lateral roots spread uniformly in all directions. The tree had rooted freely into compact till but these roots were less than 1-inch in diameter. Roots in the solum tended to be straight and cylindrical while roots in the dense till were flattened and very crooked. Occasional root grafts were found. Root remains and root channels were common in both the solum and compact till. These observations led us to conclude that tree roots served a dual hydrologic function; (1) they added greatly to soil stability on steep forest land and (2) their channels facilitated water movement within the soil.

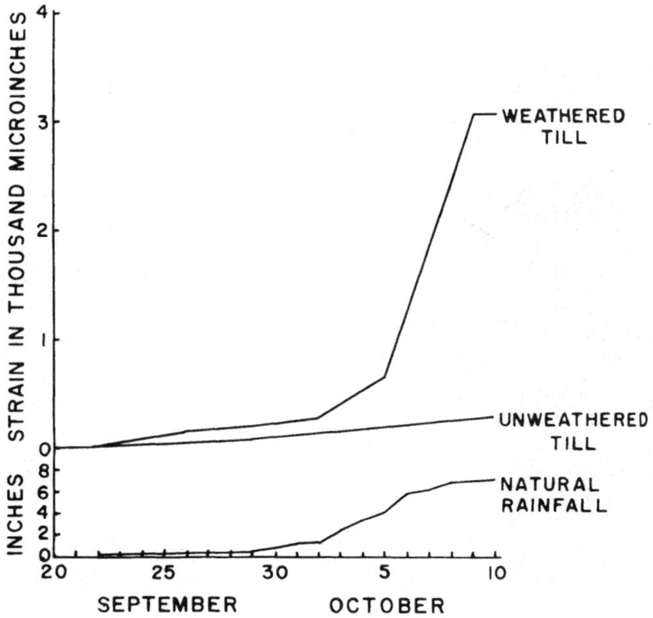

Fig. 5—Coincidence of natural rainfall with increased strain on a probe in unstable soil (study site 2). Since irrigation water was applied at a constant rate throughout this period, strain increases probably were caused by natural rainfall. Minor strain changes were within the expected range of error and are insignificant.

On August 31 the main line was extended to study site 2 and the lateral line lengthened to 120 feet with 7 sprinklers on a 26° slope. This system operated almost continuously from September 1 to September 8, delivering 107 inches of water at 60 pounds pressure. Study site 2 contained fewer stumps and little logging debris (Fig. 4), and soils wet uniformly. Four hours after irrigation started, five small streams of clear water and numerous smaller seeps were emerging from bedrock cracks below the study area. Flow also increased greatly in the ditch above the washed-out logging road (Fig. 1). At the beginning of watering, the piezometric level rose sharply after the short lag period required to saturate the soil profile. Thereafter, the piezometer indicated a fairly constant saturated zone 11 inches above the compact till.

The strain gauge probe evidenced soil movement during the first natural heavy rainfall after irrigation had started (Fig. 5).[2] Strain gauges at the base of the probe, firmly seated in compact till, showed little strain buildup. Gauges located halfway up the probe, however, showed a sharp increase in strain about one foot above the compact till, probably caused by downhill soil movement. Despite a continuing application of 16 inches of irrigation water per day on the study area, a much smaller natural rainfall over the entire watershed seemed a more effective cause of soil movement. Unfortunately, an adequate calibration had not been developed for this probe to relate strain increases directly to soil movement.

[2]Unpublished file report by David Barr, at the Institute of Northern Forestry, Juneau, Alaska.

Fig. 6.—Downstream view above study site 2. Darkened portion of ravine wall was wetted directly by sprinklers or by seepage. A ditch above the washed-out logging road (background, center) effectively cut off downslope drainage and carried seepage water to the stream. Measuring points (x) in the stream were in the center foreground and just below the road.

Streamflow measurement above and below study site 2 (Fig. 6) established a downstream flow of 4,339 gallons per hour more than passed the upstream measurement point. Since 6,720 gallons per hour were being applied as irrigation water, about 2/3 of this amount reached the stream channel as subsurface flow. The remaining 1/3 presumably was lost to deep seepage and to evapotranspiration.

Several pits in and near study site 2 exposed a continuous saturated zone 6 or more inches above compact till. Irrigation water was absorbed immediately into surface layers, then sank rapidly to the level of saturation. Within the saturated zone, water was observed to follow root holes, rock crevices, or whatever channels were open. Surface runoff never was observed on this study site.

Continuous sprinkling was resumed on September 21, delivering over 400 inches of water to study site 2 until shut down on October 20. An additional 20 inches of natural rain also fell during this final period of irrigation. On October 19, a sprinkler broke from the 2-inch lateral line. A funnel-shaped column of water rose about 20 feet, wetting a circular area approximately 15 feet in diameter. This application approached a rate of 6 inches per hour, far exceeding the 100-year, one-hour rainfall of 1.4 inches for southeast Alaska (*2*). A small, clear stream about 1 inch deep and 12 inches wide began near the broken sprinkler and ran perhaps 100 feet downhill before sinking into the soil. There was no indication of erosion. This interesting accident showed that a tremendous amount of water is needed to produce surface runoff on litter-covered, uncompacted, well-drained forest soil.

Discussion

Several internal features of Karta soils are believed to influence their ability to dispose of large amounts of water. These features include 3 to 10 inches of litter, a deep and porous solum, frequently fractured bedrock, steep and broken topography permitting rapid outflow of absorbed water.

Clearcutting probably caused a transistory reduction in the ability of Karta soil to absorb and transmit large amounts of water. Trampling, falling timber, and log skidding redistributed the litter and must have compacted the solum in some places. In the meantime, needlefall from slash quickly recovered some bared soil while frost action probably broke up compaction during the first winter after logging. A dense brush cover was established two or three years after logging. As demonstrated in this experiment, absorbed and transmitted quantities of water far exceeded natural rainfall.

Water transmission in the soil is also influenced by geology. Locally, transmission is aided by the discontinuous and heterogeneous till deposits and the highly fractured and jointed bedrock surface. This discontinuity results from variation in primary deposition of the glacial materials, tree root growth, and subsequent disturbance by mass wastage. Bedrock jointing is caused by load release, fracture during metamorphism, and frost wedging during and after glaciation. Thus,

Fig. 7.—Diagram showing relation of flow rate (Q) to cross-sectional area (A) in an idealized channel. Length and weight of lines indicate velocity and magnitude of flow in successively larger channels.

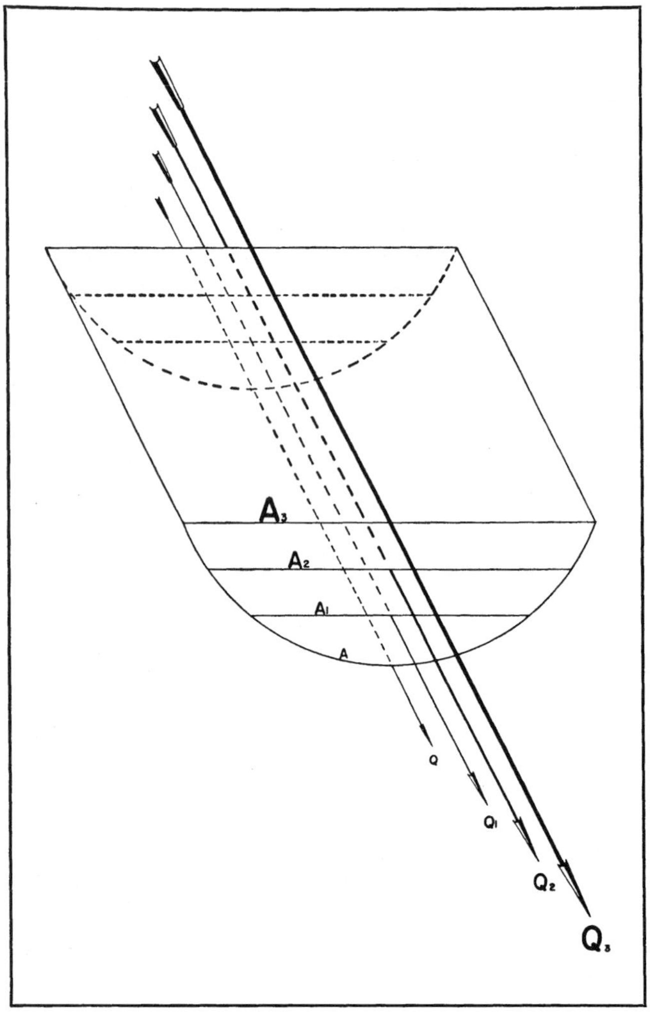

where till thins or coarsens in texture, and bedrock is near the surface, the rate of flow is increased and water entering these soils passes quickly into shallow rock passages. Streams of water flowing from bedrock cracks below the study area demonstrated such deep seepage. Surface seeps also are common low on the surrounding till slopes and usually mark gravel lenses in the till or zones of shattered bedrock near the surface.

The character of the soil water movement and probable reasons for failure to produce a debris avalanche in the study area may be deduced from Darcy's equation:

$$Q = KiA \quad (1)$$

Q = flow rate of soil water
K = permeability coefficient of soil
i = flow surface gradient or head loss
A = cross-sectional area of flow zone

In the study site K is assumed constant for a saturated soil and i equals slope of the study site surface. A is the saturated area perpendicular to flow direction. Flow rate (Q) is directly proportional to cross-sectional area (A) assuming that K and i are constant. When water is applied to partly saturated soil, the piezometric (saturated) surface rises as long as infiltration exceeds flow rate (Q) within the soil. As piezometric head increased, Q increased about in proportion to A (Fig. 7). With a constant supply of water to the soil, the piezometric surface stabilized and outflow (Q) balanced infiltration. An equilibrium condition was thus achieved, manifest on the study site by a constant piezometric head and an apparently constant outflow into the stream.

Stabilization of the piezometric surface at a low level (6 to 11 inches above unweathered till) with a high simulated rainfall (0.70 inch/hr. or 16.8 inches/day), indicates a high flow rate (Q) for the study area. Root channels, rock crevices, and a discontinuous till cover also provided the necessary channels for about 1/3 of the applied water to pass into bedrock passages. Lacking this drainage into bedrock, irrigation water applied to the study area would have passed into the stream entirely by flow within the soil, presumably resulting in a considerably higher piezometric head. Only small areas on short slopes were wetted on these study sites. On larger study sites, subsurface flows from soils wetted upslope would augment irrigation water absorbed into downslope soils. Under these conditions, piezometric heads would increase downslope, perhaps even saturating entire profiles, certainly causing maximum buoying effects and minimizing internal friction within soils. Figure 5 shows increasing strain with increasing natural rainfall suggesting the effect of a larger accumulation area on soil water level and accompanying decreases in soil strength. If this reasoning is sound, instability and slope failure should increase on long slopes.

It appears that landslide production in future irrigation experiments will depend on locating longer slopes in an area underlain by continuous impermeable till where piezometric level will rise most effectively. In areas of recent natural slide activity, the compact till cover is relatively uniform over long, continuous slopes. Piezometers on these slopes show heads in excess of 30 inches during natural rainfall with increasing piezometric levels downslope.[3]

[3]Swanston, D. N. Geology and slope failure in Maybeso Creek valley, Prince of Wales Island, Alaska: unpublished Ph.D. dissertation on file at Department of Geology, Michigan State University, East Lansing.

Literature Cited

1. Bishop, Daniel M., and Mervin E. Stevens. 1964. Landslides on logged areas in southeast Alaska. Res. Paper NOR-1, U. S. Forest Serv., Juneau, Alaska. 18 pp., illus.
2. Miller, John F. 1963. Probable maximum precipitation and rainfall-frequency data for Alaska. Weather Bur., U. S. Dept. of Com., Washington, D. C. 69 pp.
3. Sharpe, C. F. Stewart. 1960. Landslides and related phenomena. Pageant Books, Inc., N. J. 137 pp.
4. Stevens, M. E. 1965. Interim soil management report, Hollis area, South Tongass National Forest. Mimeogr. Rep., Region 10, U. S. Forest Serv. 47 pp.
5. U. S. Weather Bureau. 1965. Climatic summary of the United States, Alaska. Climatography of the U. S. No. 86-43. Suppl. for 1951 through 1960., U. S. Govt. Printing Office, Washington, D. C. 68 pp.
6. Williams, P. J. 1962. An apparatus for investigation of the distribution of movement with depth in shallow soil layers. Building Note No. 39, Nat. Res. Council, Ottawa, Canada. 10 pp.

Commentary on Paper 13

Kirkby, M. J. (1969) Infiltration, throughflow and overland flow.

Mike Kirkby was taught as a student at Cambridge University by Dick Chorley, who also supervised his PhD on modelling the form of hillslopes. He also spent time as a post-doc with Luna Leopold at the USGS and with Reds Wolman at Johns Hopkins University. He made several contributions to the book edited by Chorley in 1969, but the one reproduced here is significant because of the way in which it provided evidence for the possibility of different types of hydrological response dominating in different environments, and also for different types of events in the same catchment, depending on antecedent conditions, land surface conditions and event magnitude (see also Kirkby & Chorley, 1967). This is summarized succinctly in Fig. 5.5.1 of the paper that compares cumulative potential infiltration rates with cumulative rainfall rates of different return periods. This demonstrates quite clearly that in some cases we should expect Hortonian overland flow, in some cases we should not. A similar analysis, but based on different data, was provided by Freeze (1972) in support of his simulation study of subsurface responses, and by Dunne & Dietrich (1980) in a tropical catchment (see also Beven, 2004, for an analysis of data from Horton's own Hydrological Laboratory). This was important because, despite the best efforts of Hursh, Hewlett, Whipkey and others, the dominant paradigm of streamflow generation of the time was still that of Horton.

Kirkby cites the work of Hewlett & Hibbert (Paper 6) and Betson (Paper 8) and uses the data from Whipkey (Paper 7) to illustrate the possibility of fast subsurface responses and notes that, due to the flow velocities of subsurface water, this implies that the water appearing in the hydrograph from subsurface sources must be water displaced from the soil by the rainfall inputs during the time scale of the hydrograph. It follows that this water "*has been stored in the soil for a matter of weeks or months and so has been able to come to chemical equilibrium with the soil*" (p.96).

Mike Kirkby went on to become a Professor of Physical Geography at the University of Leeds, being appointed Emeritus Professor on his retirement in 2002. He was the originator of the topographic index concept (Kirkby, 1975) that was later used as the basis for the TOPMODEL rainfall–runoff model (Beven & Kirkby, 1979). He was (and is) also the founding editor of the journal *Earth Surface Processes and Landforms*.

References

Beven, K. J. (2004) Infiltration excess runoff at the Horton Hydrological Laboratory (or not?). *J. Hydrol.* **293**, 219–234.

Beven, K. J. & Kirkby, M. J. (1979) A physically-based variable contributing area model of basin hydrology. *Hydrol. Sci. Bull.* **24**(1), 43–69.

Chorley, R. J. (ed.) (1969) *Water, Earth and Man*. Methuen, London, UK

Dunne, T. & Dietrich, W. E. (1980) Experimental study of Horton overland flow on tropical hillslopes. 1. Soil conditions, infiltration and frequency of runoff. *Z. Geomorph. Suppl. Bd* **35**, 40–59.

Freeze, R. A. (1972) Role of subsurface flow in generating surface runoff. 2. Upstream source areas. *Water Resour. Res.* **7**(4), 929–941.

Kirkby, M. J. (1975) Hydrograph modelling strategies. In: *Processes in Physical and Human Geography* (ed. by R. F. Peel, M. D. Chisholm & P. Haggett), 69–90. Heinemann, London, UK.

Kirkby, M. J. & Chorley, R. J. (1967) Throughflow, overland flow and erosion. *Int. Assoc. Sci. Hydrol. Bull.* **12**(3), 5–21.

I. Infiltration, Throughflow, and Overland Flow

M. J. KIRKBY
Department of Geography, Bristol University

When the rainfall that has not been intercepted by vegetation reaches the ground surface part of it fills small surface depressions (depression storage), part percolates into the soil, and the remainder, if any, flows over the surface as overland flow. Each component of this equation is highly variable, and depends not only on the intensity of the rainfall but also on soil, vegetation, and surface gradient. The amount of water intercepted by vegetation depends on the type of plants and their stage of growth, but it is usually close to a value given by all of the first 1·0 mm of rainfall and 20% of the subsequent rainfall in any one storm. Rainfall reaching the soil surface has to fill the small depressions on the surface before any overland flow can occur, even on a totally impermeable surface. This depression storage does not vary with the amount of rainfall but with the nature of the surface, especially with slope gradient, vegetation cover, and land-use practices. Under natural conditions depression storage absorbs about 2–5 mm of rainfall in any one storm. Contour ploughing is particularly effective in increasing depression storage by as much as ten times.

1. Infiltration

Infiltration rate is defined as the maximum rate at which water can penetrate into the soil. For initially moister soils the infiltration rate is lower throughout storms, and for all soils it decreases during the course of a storm. The rate at which water can travel through the soil depends on the number and size of pore spaces in the soil and the distribution of water within them. In effect, the infiltrating water has two components, a transmission component, which is constant and represents a steady flow through the soil; and a diffusion component, which is an initially rapid, and then an increasingly slow, filling-up of air-filled pore spaces, from the surface downwards. These components can be expressed in the infiltration equation (Philip, 1957):

$$f = A + B \, . \, t^{-\frac{1}{2}} \tag{1}$$

where f is the instantaneous rate of infiltration;
t is time elapsed since the beginning of rainfall;
A is the 'transmission constant' of the soil; and
B is the 'diffusion constant' of the soil.

In this equation the transmission and diffusion terms can be identified with the two components of an idealized model. The transmission term represents unimpeded laminar flow through a continuous network of large pores. The diffusion term represents flow in very small discrete steps from one small pore space

TABLE 5.1.1 Variation of minimum infiltration rates with soil grain size, initial moisture content, and vegetation cover

(*a*) The effect of grain size in initially wet soils without vegetation cover

Grain size class	*Infiltration rates* (mm/hr)
Clays	0–4
Silts	2–8
Sands	3–12

(*b*) The influence of moisture content for Illinois clay-pan soils (after Musgrave and Holtan, 1964)

Initial moisture content (%)	*Infiltration rates* (mm/hr)		
	Good grass cover		*Poor weed cover*
	Topsoil > 13 in. thick	*Topsoil < 13 in. thick*	*Topsoil < 13 in. thick*
0–14	17	19	6
14–24	7	7	4
24 +	4	4	3

(*c*) The influence of ground cover for Cecil, Madison, and Durham soils (after Musgrave and Holtan, 1964)

Ground cover	*Infiltration rate* (mm/hr)
Old permanent pasture	57
Permanent pasture; moderately grazed	19
Permanent pasture; heavily grazed	13
Strip-cropped	10
Weeds or grain	9
Clean tilled	7
Bare ground crusted	6

to the next, in a random fashion. The only reason that a net diffusion flow results is that more pores are dry lower down, so that there is greater opportunity for downward movement than for upward. In an actual soil the two phases of this model cannot be separated, as all pores show a combination of the two types of behaviour, but equation (1) remains a good approximation for measured infiltration rates.

Table 5.1.1 shows some of the range of variation which can be expected in infiltration rates under a range of vegetation and moisture conditions, and in fig. 5.1.1 a comparison is made between typical infiltration rates and expected storm rainfall rates.

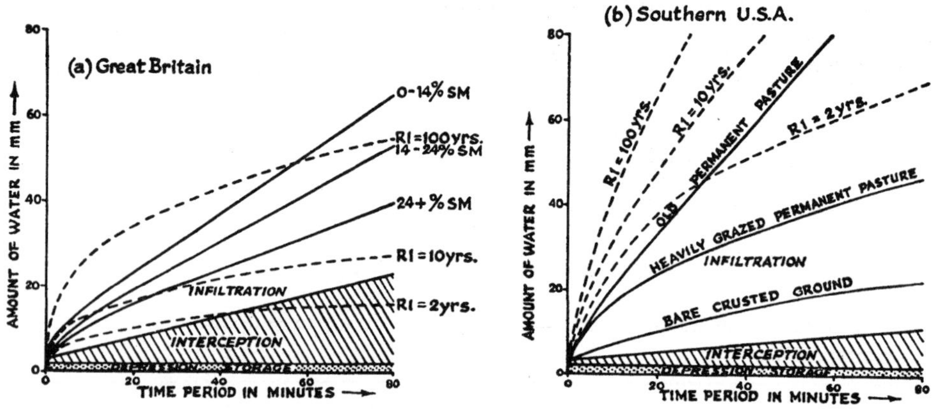

Fig. 5.1.1 Comparison of expected rainfall intensities with infiltration rate, interception, and depression storage, with representative values to demonstrate the relative frequency of overland flow in Great Britain and southern United States.

(*a*) Great Britain: rainfall (Data from Bilham, 1936), interception values for 100% vegetation cover, infiltration rates for Illinois clay-pan soils with good grass cover (Data from Musgrave and Holtan, 1964).

(*b*) Southern United States: rainfall (Data from Yarnell, 1935), interception values for 50% vegetation cover, infiltration rates for Cecil, Madison, and Durham soils (Data from Musgrave and Holtan, 1964).

2. Overland flow and throughflow

'Horton overland flow' is defined as overland flow which occurs when rainfall intensity is so great that not all the water can infiltrate, and is described by Horton [1945]. This type of overland flow is a fairly common phenomenon in semi-arid climatic conditions, but is relatively rare in humid and humid-temperate conditions. The role of vegetation is thought to be a critical cause of this distinction. Vegetation increases the infiltration rate by promoting a thicker soil cover, a better soil texture, and by breaking the impact of raindrops on the surface. Its effect on soil structure is mainly to build up an organic-rich A horizon with a relatively open pore structure and high permeability. If raindrops strike the surface without being impeded by vegetation fine material is thrown into suspension by the impact and is redeposited as an almost impermeable surface skin which can lower infiltration by as much as ten times. Vegetation therefore has a controlling influence on Horton overland flow by increasing both the initial depression storage and the infiltration rate, so that where a dense vegetation cover is established Horton overland flow is very unusual. Soil compaction by animals and vehicles reduces the infiltration rate while increasing depression storage, so that its net influence is problematic.

Within a small drainage basin, where soils are more or less homogeneous, it may be expected that interception, depression storage, and infiltration rate will not vary greatly, so that the conditions for Horton overland flow will be satisfied by comparable intensities and durations of rainfall, and overland flow will occur simultaneously all over the basin. Typical velocities for overland flow are about 200–300 m/hr, so that in a rainfall of 1 hour water from all points of a basin (200–300 m is a typical distance from divide to stream in Britain) will be reaching

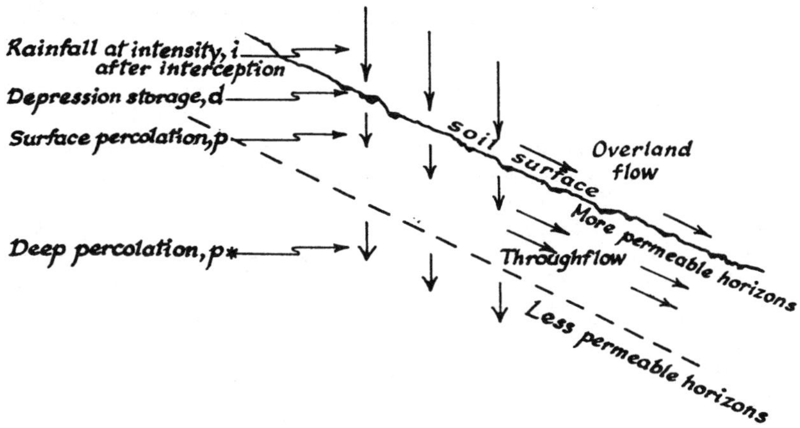

Fig. 5.1.2 Flow components over and within the soil.

a stream channel, and flow over the slope surface will have reached a steady state, represented by the equation:

$$q_0 = (i - f) \, . \, a \qquad (2)$$

where q_0 is the overland flow discharge per unit contour length;
i is the rainfall intensity after interception;
f is the infiltration rate; and
a is the area drained per unit contour length (equal to the distance from the divide if all the contours are straight lines).

Thus, provided that the rainfall intensity is high enough (or the infiltration rate low enough) for this type of overland flow the actual magnitude of the flow will be strongly dependent on the area or distance of overland flow, and will be almost independent of the storm duration, provided that it exceeds a reasonable minimum value. This is Horton's [1945] classic overland flow model.

Some of the water which infiltrates into the soil passes downward to recharge the water-table, and some, usually the greater part, flows down the hillside within the soil layers as 'throughflow' and ultimately contributes to streamflow (fig. 5.1.2). Within the soil, permeability varies and is generally highest in the open-textured organic A_0 horizon and the eluviated A_1 horizon. B horizons tend to be less permeable because clays are washed down into them, and in some cases because of the development of a hardpan. Conditions in the parent bedrock vary widely from limestones, with open solution fissures, to totally impermeable consolidated clays and shales. Wherever permeability is decreasing downward within the soil, and this occurs most commonly at the base of the A horizon, part of the water which is percolating downwards cannot penetrate into the lower layers fast enough, and is deflected laterally within the upper layer, as throughflow. This is similar to the production of Horton overland flow at the surface, except that within the soil the reduction of permeability is usually gradual, leading to a progressive deflection of throughflow. Table 5.1.2 shows an example of the progressive decline of permeability through the soil profile for a hillslope in Ohio, U.S.A.

TABLE 5.1.2 Variations of permeability and soil type with depth in an Ohio forest soil at a gradient of 15° (after Whipkey, 1965)

Soil depth (cm)	*Textural class*	*Bulk density* (gm/cc)	*Saturated permeability* (mm/hr)
0–56	Sandy loam	1·33	—
56–90	Sandy loam	1·41	286
90–120	Loam	1·78	17
120–150	Clay loam	1·80	2

If rainfall continues for a long time soil layers become saturated and throughflow is deflected closer and closer to the surface, so that the upper, more permeable soil layers are filling up from their bases because the throughflow is unable to carry away the water fast enough. In time the soil will become saturated right up to the surface and 'saturation overland flow' will occur. Under steady rainfall this condition will ultimately be attained under rainfall intensities much lower than are required to produce Horton overland flow. Since soil thickness and the velocity of throughflow vary much more over a small area than does permeability, and since the base of a slope tends to become saturated sooner than the divide, certain parts of a hillside are likely to produce saturation overland flow preferentially, in contrast to the rather widespread production of Horton overland flow, when it occurs at all.

Throughflow, travelling through soil pore spaces rather than over the ground surface, moves at very much lower velocities than overland flow. Rates of 20–30 cm/hr for throughflow are of the order of a thousand times lower than overland flow rates, so that periods of about 1,000 hours rainfall are needed for a steady state of flow to be achieved throughout an average basin. In practice, such a steady state is never attained for throughflow, and equation (2) for overland flow must be replaced with equation (3) for throughflow:

$$q_T = (p - f_*) \cdot v \cdot t \qquad (3)$$

where q_T is the throughflow discharge per unit contour length;

p is the rate of surface percolation, equal to i or f, whichever is smaller;

f_* is the rate of infiltration at the base of the more permeable soil;

v is the velocity of throughflow; and

t is the time elapsed (strictly $v \,.\, t$ should be replaced by the area/unit contour length within a distance $v \,.\, t$ upslope).

In this equation the *time elapsed* is the most important control over the flow, in place of the *distance from the divide* for Horton overland flow. In reality, the flows are delayed by the time of transmission from the surface to a zone of decreasing permeability, a period which may be a matter of minutes or a few hours.

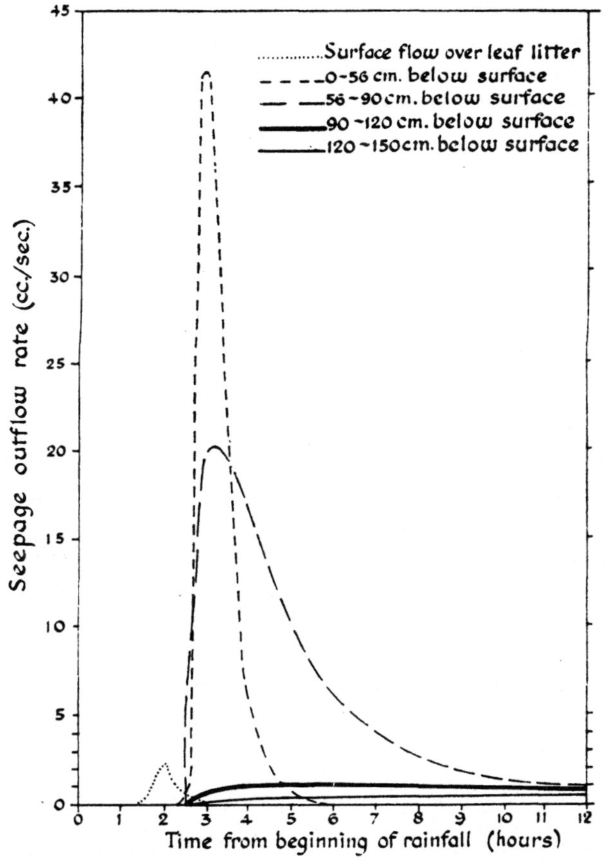

Fig. 5.1.3 Discharge hydrographs of flow within the soil resulting from a simulated storm of 5·1 cm/hr, lasting 2 hr, on a 16° slope which had previously drained for more than four days. The rapid, although small, Horton overland flow results from the initially low permeability of the dry surface soil, which rapidly increases with wetting. The lag before throughflow begins is the time taken for rain to infiltrate vertically to the 90-cm-deep, less-permeable interface (After Whipkey, 1965, p. 81).

If outflow from various depths in the soil is measured throughout a storm of uniform intensity there will be an initial transmission period of little or no flow, followed by a period in which the flow is increasing rapidly (though not linearly, because v in equation (3) is itself varying) until the moment when rainfall ceases, after which the flow will decrease more slowly than it increased. In other words, the flow should broadly resemble a flood hydrograph of a stream, despite the absence of surface run-off. In fig. 5.1.3 (Whipkey, 1965) actual values are shown for a rainfall intensity of 5·1 cm/hr falling for 2 hours on the soil whose properties are described in Table 5.1.2. It is apparent from fig. 5.1.3 that even at this high intensity Horton overland flow is negligible, and that throughflow from shallower soil layers is later than from deeper layers due to the time lag for transmission of water down through the soil, followed by saturation from the base up.

Equation (3) contains an unknown quantity, namely the velocity of throughflow, v. In order to evaluate the throughflow more exactly, Darcy's law, which states that flow through a permeable medium is proportional to the pressure gradient, must be combined with the continuity equation, which states that differences between inflow and outflow must be accommodated by changes in moisture content. For a soil layer of uniform permeability and moisture content (in depth) Darcy's law for soil on a slope is

$$Q = z \,.\, \cos \alpha \left\{ K \,.\, m \,.\, \sin \alpha - D \,.\, \frac{\partial m}{\partial x} \right\} \tag{4}$$

and the continuity equation is

$$\frac{\partial m}{\partial t} + \frac{\partial Q}{\partial x} = i \tag{5}$$

where Q is the downslope discharge measured in a horizontal direction;
x is the distance downslope measured in a horizontal direction;
K is the soil permeability (a function of soil moisture);
D is the soil diffusivity (a function of soil moisture);
α is the surface slope angle;
m is the soil moisture content;
z is the thickness of the soil layer;
i is the rainfall intensity after interception; and
t is the time elapsed.

Solution of these equations is necessarily numerical, but in a simple actual example of a uniform soil on a uniform gradient (Hewlett and Hibbert, 1966) estimates of permeability and diffusivity were as shown in Table 5.1.3; and calculated soil-moisture patterns during (*a*) uniform rainfall and (*b*) uniform drainage were as shown in fig. 5.1.4. It can be deduced from the data of Table 5.1.3 that the permeability (K) term in equation (4) is the more important for intermediate values of moisture, while the diffusivity (D) term becomes dominant at extreme low and high values of moisture. The low-moisture case is of little interest, but the high-moisture case, when the moisture content is approaching

TABLE 5.1.3 Values of permeability (K) and diffusivity (D) for varying soil moisture, computed from the results of an experiment carried out during soil drainage of a trough of soil inclined at 22° (Hewlett and Hibbert, 1963)

Moisture content (expressed as % of saturated moisture content)	*Permeability* (K) (cm/hr)	*Diffusivity* (D) (cm/hr)
68	0·000	0·0
73	0·004	3·2
78	0·081	12·5
83	0·46	19·3
88	1·58	40·8
93	4·00	100·4
98	8·67	304·0
99	10·64	517·0
100 (saturated)	12·08	Infinity

saturation, is of great hydrologic significance. In this important case soil moisture is constant at saturation, so that equation (5) becomes extremely simple, namely

$$\frac{\partial Q}{\partial x} = i \qquad (6)$$

What this means in practice is that the near-saturated soil zones respond very rapidly to changes in rainfall intensity, even before overland flow begins. Most

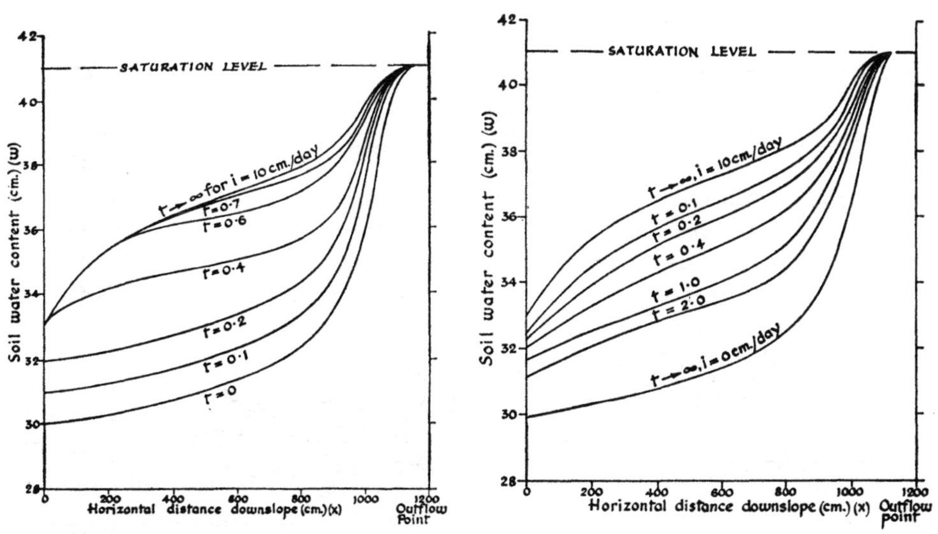

Fig. 5.1.4 Calculated moisture distributions in a 12-m-long soil trough, inclined at 22° during: (*left*) rainfall at a constant rate of 100 mm./day, and (*right*) drainage after indefinite rainfall of this intensity.

The data are calculated from the results of an experiment by Hewlett and Hibbert (1963), important is the saturated zone which is often at the side of flowing streams, for the outflow from this zone is given by:

$$q_T = i \cdot x_s + q_T^* \tag{7}$$

where x_S is the width of the saturated zone;
q_T is the throughflow discharge per unit contour length;
q_T^* is the throughflow contribution from farther upslope; and
i is the rainfall after interception.

Equations (4) and (5) also provide a basis for assessing the influence of changing gradient or (with slight modification) of contour curvature on the throughflow discharge within the soil. The following four regions of a slope are the most likely to become saturated, and hence provide more rapid response to rainfall and more frequent saturation overland flow:

1. areas adjacent to perennial streams;
2. areas of concave upwards slope profile;
3. hollows (areas of concave outward contours);
4. areas with thin or impermeable soils.

3. Areas contributing to stream flow

The relative infrequency of overland flow in humid regions, together with the very low velocities of throughflow, suggest that most of the rainfall which falls on hillslopes is unable to reach a channel until long after the rainfall has stopped and the stream flood peak has passed. In other words, only water from a relatively small 'contributing area' is able to reach a channel in time to contribute to the flood hydrograph of the stream. In its simplest form, for a rainstorm of constant intensity, the contributing area, A_c, is defined as

$$A_c = \frac{\text{Stream discharge}}{\text{Rainfall intensity}} \tag{8}$$

Where the rainfall intensity varies during the storm its value in the equation is necessarily an average one, weighted towards the most recent intensities. The contributing area is continuously changing during the storm (fig. 5.1.5), and is generally at its greatest at about the same time as the peak discharge in the stream. For basins measured in North Carolina the maximum contributing area varied relatively little from storm to storm, but from basin to basin it ranged from 5 to 85% of the total drainage area. These measurements of contributing area can be compared with models derived from the two types of hillside flow: 1. 100% overland flow, and 2. 100% throughflow. During 100% overland flow water from all parts of the basin commonly reaches a channel within about an hour, so that for a storm lasting longer than an hour the contributing area, expressed as a percentage of the total drainage area, is

$$\frac{\text{Rainfall intensity} - (\text{Infiltration and surface losses})}{\text{Rainfall intensity}} \times 100\%$$

Under conditions where vegetation and soil are thin or absent, this contributing area may be large, and the value of 85% contributing area refers to an abandoned copper strip-mining area with less than 36% vegetation cover.

During 100% throughflow the contributing area consists of:

1. the area of the stream channels themselves, which is usually 1–5% of the drainage area;
2. the areas of saturated or near-saturated soil, mainly adjoining channels, which respond rapidly to changes in rainfall intensity as is shown in equations (6) and (7);
3. a narrow strip of hillside around the saturated areas, the width of which is determined by the slow rates of throughflow in unsaturated soils.

In the North Carolina basin with a 5% contributing area, the actual channel area and a swampy area backed up behind the stream-gauging installation

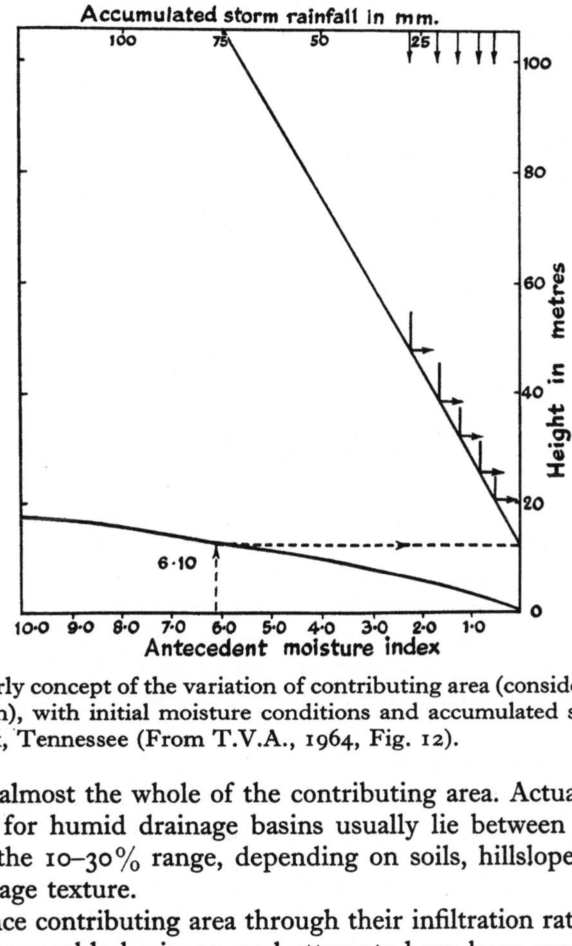

Fig. 5.1.5 An early concept of the variation of contributing area (considered as area below elevations shown), with initial moisture conditions and accumulated storm rainfall, for Bradshaw Creek, Tennessee (From T.V.A., 1964, Fig. 12).

accounted for almost the whole of the contributing area. Actual values of contributing area for humid drainage basins usually lie between these extremes, commonly in the 10–30% range, depending on soils, hillslope gradients, land use, and drainage texture.

Soils influence contributing area through their infiltration rate and the thickness of their permeable horizons, and attempts have been made to prove that contributing areas coincide with thin soil areas in a small drainage basin. Slope gradients influence the rate of throughflow (equation (4)), and hence the distribution of saturated soil and saturation overland flow. Vegetation cover and cultivation practices strongly affect the permeability of surface soil layers, mainly through the effect of vegetation in reducing rainsplash impact and through the effect of cultivation on depression storage and soil structure. In a basin with a high drainage density a relatively large area of hillslope is close to a channel, so that under throughflow the contributing area will also be relatively large; and under Horton overland flow there will be a relatively short time lag between the start of rainfall and a condition of maximum contributing area. Clearly these factors are not independent of one another, but each has a separate influence on contributing area.

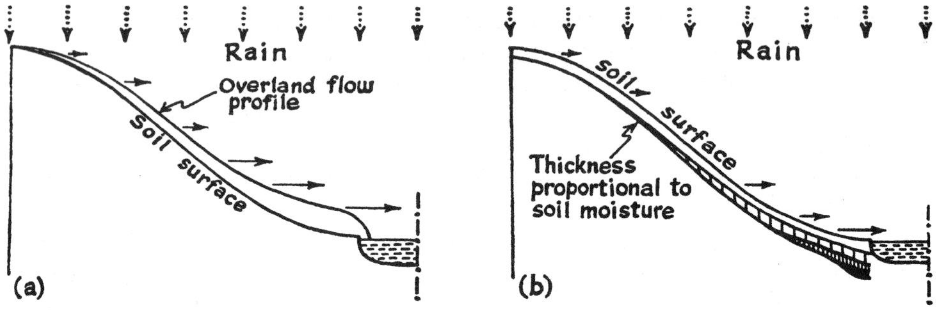

Fig. 5.1.6 Patterns of hillslope flow during Horton overland flow and throughflow. Arrow lengths show relative discharges over or through the soil.

(*a*). Horton overland flow (After Horton, 1945, p. 316). Thickness of water layer on surface is drawn proportional to actual thickness.

(*b*). Throughflow. Thickness of water layer below surface is drawn proportional to soil moisture content. Soil moisture from progressively earlier rainfalls is shown by progressively darker shading. The subsurface layer does not indicate the depth of infiltration into the soil.

Part of stream baseflow is derived from the water-table, which is itself supplied by deep percolation of water from the soil, but this contribution to baseflow is probably large only where well-defined aquifers are present. A large part of baseflow also comes from throughflow in the soil, which will take months to reach a channel from interfluve areas, and produces sufficient water to supply the measured baseflows in many areas. Since the same rain-water, much of it via throughflow, is responsible for both high and low flows in streams, all of the factors described above, which tend to produce high contributing areas during rainstorms, and hence a large proportion of total runoff during storms, also lead to reduced storage of rain-water after the flood flows have subsided, and hence to lower baseflows.

A final important contrast between the overland flow and throughflow models is that, whereas in overland flow it is the rain-water which is actually falling that flows into the stream during a rainstorm (fig. 5.1.6(*a*)), in throughflow, much of the water flowing into the channels is not physically the same as the rain-water which is currently falling (due to the time lag involved). It has been shown in infiltration experiments that almost all water flowing through the soil flows out in the order in which it flows in. This means that infiltrating water has to displace all of the soil water downslope before it can itself flow into the stream (fig. 5.1.6(*b*)), so that most water flowing into the stream, even at high flows, has been stored in the soil for a matter of weeks or months, and so has been able to come to chemical equilibrium with the soil. This soil water storage has obvious implications for interpreting the dissolved load of streams.

4. Summary

There are two extreme models of hillside water flow; the Horton overland flow model and the throughflow model. Horton overland flow occurs when rainfall intensity exceeds infiltration rate, and when it occurs at all in a basin, it is widespread. It is most common in semi-arid climates, and only occurs at progressively higher rainfall intensities under progressively thicker soil and vegetation covers. Throughflow occurs whenever the soil permeability decreases with increasing depth in the soil within some portion of the soil profile, most commonly at the base of the A horizon. Throughflow is probably the predominant mode of hillside flow in humid and humid–temperate areas, but it is of lesser importance under more arid or less-vegetated conditions with thin soils. When throughflow saturates the soil profile up to the surface, then saturation overland flow occurs. It occurs at much lower rainfall intensities than Horton overland flow, and is usually much more localized in its distribution, being commonest near streams. Under suitable conditions, both overland flow and throughflow may occur at any point although their relative frequencies will vary greatly from point to point. The separation of hillside flow into its two components, overland flow and throughflow, and a recognition of the distinct properties of each, allows a clearer understanding of the mechanisms of both streamflow and hillside erosion.

REFERENCES

BETSON, R. P. [1964], What is watershed runoff?; *Journal of Geophysical Research*, **69**, 1541–52.

BILHAM, E. G. [1936], Classification of heavy falls in short periods; *British Rainfall*, **75**, 262.

HEWLETT, J. D. and HIBBERT, A. R. [1963], Moisture and energy conditions within a sloping mass during drainage; *Journal of Geophysical Research*, **68**, 1081–7.

HORTON, R. E. [1945], Erosional development of streams and their drainage basins: hydrological approach to quantitative morphology; *Bulletin of the Geological Society of America*, **56**, 275–370.

HUDSON, N. W. and JACKSON, D. C. [1959], Results achieved in the measurement of erosion and runoff in Southern Rhodesia; *3rd Inter-African Soil Conference, Dalaba*, Paper No. 63.

JENS, S. W. and MCPHERSON, M. B. [1964], Hydrology of Urban Areas; In Chow, V. T., Editor, *Handbook of Applied Hydrology* (New York), Section 20, 45 p.

KIRKBY, M. J. and CHORLEY, R. J. [1967], Throughflow, overland flow and erosion; *Bulletin of the International Association of Scientific Hydrology*, **12**, 5–21.

LINSLEY, R. K., KOHLER, M. A., and PAULHUS, J. L. H. [1949], *Applied Hydrology* (New York), 689 p.

MUSGRAVE, G. W. and HOLTAN, H. N. [1964], Infiltration; In Chow, V. T., Editor, *Handbook of Applied Hydrology* (New York), Section 12, 30 p.

PHILIP, J. R. [1957–8], The theory of Infiltration; *Soil Science*, **83**, 345–57 and 435–48; **84**, 163–77, 257–64, and 329–39; **85**, 278–86 and 333–7.

TENNESSEE VALLEY AUTHORITY [1964], Bradshaw Creek – Elk River: A pilot study in area-stream factor correlation; *Office of Tributary Area Development, Knoxville, Tennessee*, Research Paper No. 4, 64 p. and 6 appendices.

TENNESSEE VALLEY AUTHORITY [1966], Cooperative Research Project in North Carolina: Annual Report for Water Year 1964–1965; *Division of Water Control Planning, Hydraulic Data Branch*, Project Authorisation No. 445.1, 31 p.

WHIPKEY, R. Z. [1965], Subsurface stormflow from forested slopes; *Bulletin of the International Association of Scientific Hydrology*, **10**, 74–85.

YARNELL, D. L. [1935], Rainfall intensity-frequency data; *U.S. Department of Agriculture, Misc. Publ. No. 204.*

Commentary on Paper 14

Pinder, G. F. & Jones, J. F. (1969) Determination of the ground-water component of peak discharge from the chemistry of total runoff.

In this paper, George Pinder and John Jones note the arbitrary nature of previous attempts to separate the groundwater contribution to stream discharge, and explore the use of geochemical information in distinguishing different sources of the storm hydrograph using a simple two component mixing model. They were not the first to do so; the technique appears to have been pioneered in the Soviet Union (Sklash, 1990, p.406), but this paper, appearing in *Water Resources Research*, led to a more widespread appreciation of mixing models as a technique for separating hydrographs. Note that in this paper, Pinder & Jones are trying to distinguish groundwater from direct runoff, not event water from stored or old water as in some later studies (see Paper 27). This requires that the chemical characteristics of groundwater and direct runoff be known throughout an event. Pinder & Jones argue that the chemical composition of groundwater will not vary much over time and therefore can be represented by measured values prior to an event, while the direct runoff characteristics can be represented by peak flow samples at headwater sites. Their samples suggested that concentrations were less and changed less with discharge at the upstream sites (Fig. 4), suggesting that at peak flows the concentrations should be representative of direct runoff. Given the total discharge and values of concentration for both groundwater and direct runoff, the mass balance equations can be solved to determine the groundwater discharge through the event. Doing this for a number of anions and cations resulted in different estimates for the groundwater that were then averaged to give a composite estimate.

The study was one of the first (see also Kunkle, 1965) to demonstrate that the groundwater contribution to the hydrograph may be significant (though notice the log scales used in Fig. 5a,b,c in the paper), and that responses of the identified groundwater contribution could be seen on the same time scale as the hydrograph peak, even in this case for events at the end of the summer. A similar technique was used by Anderson & Burt (1982) at Bicknoller Combe (see also Paper 23) with similar results. At the time such a conclusion was highly controversial in that the Hortonian model of runoff generation by an infiltration excess mechanism, with an implication that storm runoff is event water, was still generally accepted (see commentaries on Papers 1 to 3).

The results raised an issue about how the groundwater response could be so fast. Pinder & Jones suggest that this may be due to the importance of flow in fracture systems (incidentally, they note on p.438 that they also calculated groundwater discharges by correlation with observation well levels, but no results are given to compare with the geochemical separations). Another reason could be that the groundwater contribution is simply overestimated because any direct runoff water rapidly acquires the chemical characteristics of stored water. This could be because it picks up solutes available at mineral surfaces in contact with the flowing water, or because the event water locally mixes with and displaces stored water, with similar chemical concentrations to groundwater, into direct runoff. Kennedy (1971), for example, suggests that reaction times for rainwater to interact with soil and increase concentrations in silica and other minerals can be of the order of minutes. Nakamura (1971) and Pilgrim *et al.* (1979), for example, tried to take reaction times into account in such separations. Sklash & Farvolden (Paper 27) and others (see Sklash, 1990; Kendall & McDonnell, 1998), later provided a more secure foundation for significant contributions of stored water using environmental isotope data. A number of papers have compared the results of isotope and geochemical separations (e.g. Kennedy *et al.*, 1986; Wels *et al.*, 1991).

George Pinder went on to publish extensively on the modelling of groundwater systems. He was Professor and Chairman of the Department of Civil Engineering at Princeton University, before moving to be Director of the Research Center for Groundwater Remediation Design and Dean of the College of Engineering and Mathematics at the University of Vermont. He was awarded the Horton Award of the American Geophysical Union in 1969, and the Meizner Award of the Geological Society of America in 1975. He is a past President of the Hydrology Section of the AGU.

References

Anderson, M. G. & Burt, T. P. (1982) The contribution of throughflow to storm runoff: an evaluation of a chemical mixing model. *Earth Surf. Process. Landf.* **7**, 565–574.

Kendall, C. & McDonnell, J. J. (eds) (1998) *Isotope Tracers in Catchment Hydrology*. Elsevier, Amsterdam, The Netherlands.

Kennedy, V. C. (1971) Silica variation in stream water with time and discharge. In: *Nonequilibrium Systems in Natural Water Chemistry. Advances in Chemistry Series V* **106**, 94–130.

Kennedy, V. C., Kendall, C., Zellweger, G. W., Yerman, T. A. & Avanzino, R. J. (1986) Determination of the components of stormflow using water chemistry and environmental isotopes, Mattole River basin, California. *J. Hydrol.* **84**, 107–140.

Kunkle, G. R. (1965) Computation of groundwater discharge to streams during floods, or to individual reaches during baseflow, by use of specific conductance. *USGS Prof. Pap. 525D, D207-10.*

Nakamura, R. (1971) Runoff analysis by electrical conductance of water. *J. Hydrol.* **14**, 197–212.

Pilgrim, D. H., Huff, D. D. & Steele, T. D. (1979) Use of specific conductance and contact time relationships for separating flow components in storm runoff. *Water Resour. Res.* **15**, 329–339.

Sklash, M. G. (1990) Environmental isotope studies of storm and snowmelt runoff generation. In: *Process Studies in Hillslope Hydrology* (ed. by M. G. Anderson & T. P. Burt), 401–436. Wiley, Chichester, UK.

Wels, C., Cornett, R. J., Lazirte, D. B. (1991) Hydrograph separation: a comparison of geochemical and isotope tracers. *J. Hydrol.* **122**, 253–274.

Determination of the Ground-Water Component of Peak Discharge from the Chemistry of Total Runoff

GEORGE F. PINDER AND JOHN F. JONES

Nova Scotia Department of Mines, Halifax, Nova Scotia

Abstract. The ground-water component of stream discharge may be determined from the chemical characteristics of the stream water. A chemical mass-balance is used to relate total, direct, and ground-water runoff. To solve the mass-balance equation, it is necessary to estimate the chemical composition of the ground-water and direct-runoff components. The solute concentration of ground water is determined from total runoff during baseflow; the chemical characteristics of direct-runoff are estimated from samples of total runoff collected from selected locations in a basin during peak discharge periods. In three small watersheds in Nova Scotia ground-water runoff constituted from 32 to 42% of peak discharge for the period of analysis.

INTRODUCTION

The major problem in the separation of total streamflow into direct-runoff and ground-water components has been the inability to measure directly ground-water discharge during high water and floods. To estimate the ground-water component of total runoff, numerous empirical techniques have been developed. The usual approach has been a graphical separation of the discharge hydrograph by extrapolation of the ground-water recession curves beneath the flood peak. Graphical methods of analysis vary from the very simple, as in Figure 1 [*Linsley and Franzini,* 1964] to the more complex, as in Figure 2 [*Barnes,* 1940].

Chernaya [1964] made a comparative evaluation of the various methods of determining base flow during floods; he considered ground-water discharging as springs above the stream surface as well as seepage into the channel bottom. A quantitative evaluation of these techniques is difficult, because the methods of separation are arbitrary, unless the exact amount of the base flow can be determined.

The relationship between the quantity of base flow and stream chemistry was investigated by *LaSala* [1967] and *Archer et al.* [1967]. Their objective was to estimate the quality of streamflow on the basis of quantity and quality of both groundwater discharge and direct runoff. They calculated the total dissolved solids concentration of total runoff from the relationship

$$C_{tr} = (Q_{dr}C_{dr} + Q_{gw}C_{gw})/Q_{tr} \quad (1)$$

where

C_{tr} is the total dissolved solids concentration in total runoff (ppm);

C_{dr} is the total dissolved solids concentration in direct runoff (ppm);

C_{gw} is the total dissolved solids concentration in ground-water runoff (ppm);

Q_{tr} is the total runoff (cfs);

Q_{dr} is the direct runoff (cfs);

Q_{gw} is the ground-water runoff (cfs).

Ground-water discharge was computed from a ground-water rating curve based on correlations of streamflow data with water levels in observation wells. The chemical composition of ground-water flow was determined from samples collected during periods of baseflow; the composition of direct runoff was estimated from samples from small streams, rills, and very high streamflow.

A comparison of their calculated values with determinations of total dissolved solids from samples of total runoff indicates that this technique provides satisfactory results.

Purpose and scope of investigation. The objectives of the present study were: (1) to investigate the variation in the chemical composition of total runoff in three small drainage basins in Nova Scotia; and (2) to determine the ground-water component of discharge during periods of high flow.

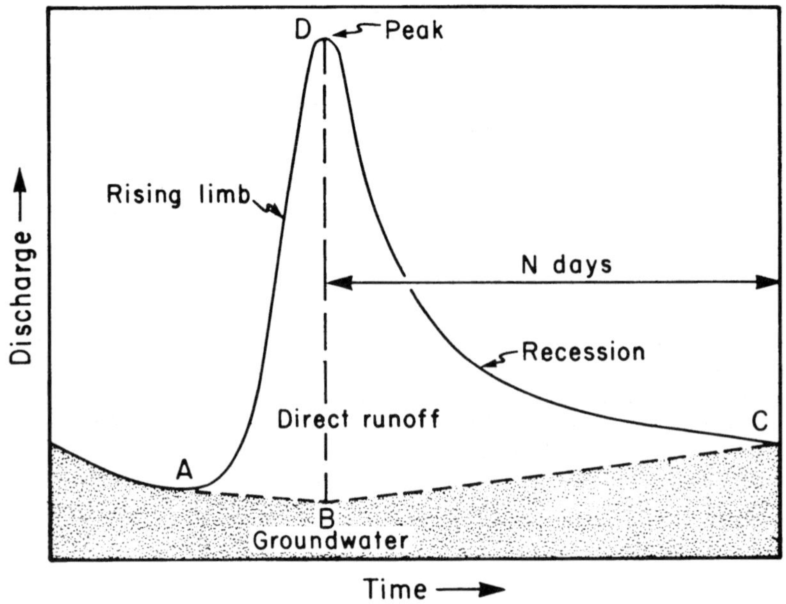

Fig. 1. Typical flood hydrograph showing method of separating direct and ground-water runoff [after *Linsley and Franzini*, 1964].

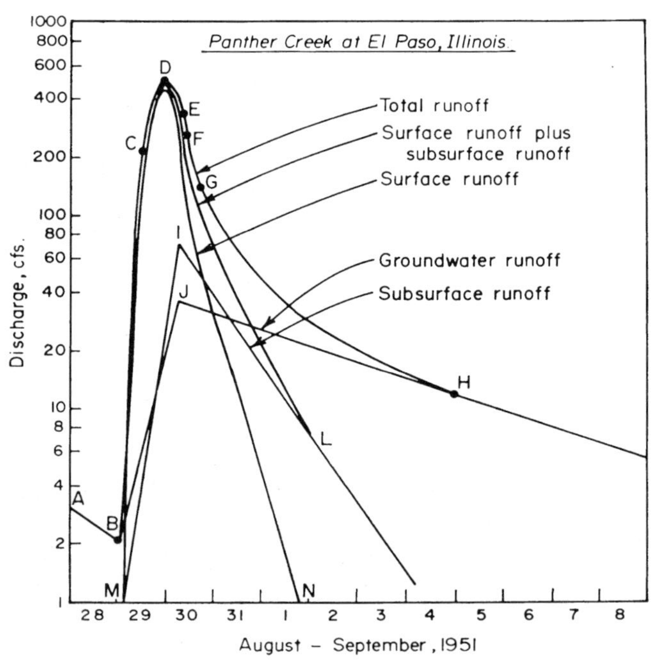

Fig. 2. Semilogarithmic plotting of a hydrograph, showing separation of runoff components [after *Chow*, 1964].

To study in detail the relationship between stream chemistry and discharge, it was necessary to select basins where precise and continuous discharge hydrographs were available; the three International Hydrologic Decade watersheds in Nova Scotia fulfilled this requirement. The sampling program was initiated September 24, 1967, after a period of low flow lasting approximately one month. Water samples were taken from three locations along each stream (upper, middle, and lower segments) every two days over a six-week period; a complete chemical analysis of the 189 samples was provided by the Water Quality Division, Inland Waters Branch, Canada Department of Energy, Mines, and Resources. The statistical analysis of the data and the calculation of the ground-water components were performed on the IBM 360 Model 50 digital computer at Dalhousie University.

The water-sampling program mentioned above was designed as a 3 × 3 × 21 factorial experiment. The three factors considered were: (1) the over-all influence of the geology and hydrology of each basin; (2) the position of the sampling location along the reach of the stream; and (3) the time at which the samples were selected (effect of the quantity of total runoff).

Figure 3 illustrates the type of information available for the basins over the period of analysis.

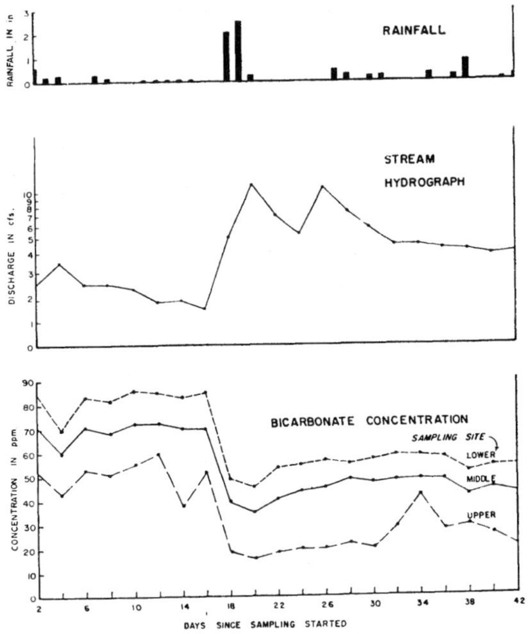

Fig. 3. Precipitation, total runoff, and the concentration of the bicarbonate ion over the period of analysis for April Brook basin; the discharge and chemical hydrographs are developed from instantaneous data.

HYDROGEOLOGY OF THE REPRESENTATIVE WATERSHEDS

Since the quantity and quality of ground-water discharge is related to the mineralogy and hydraulic characteristics of the hydrologic system through which it flows, a brief description of each representative watershed follows.

April Brook Basin. April Brook drainage basin is mainly forested with an area of 2.5 square miles and a relief of approximately 900 feet. The entire basin is underlain by rocks of Mississippian Age; bedrock in the upper segment of the basin is gently folded interbedded shale, siltstone, and sandstone of the Horton Group. The lower 15% of the basin is underlain by gypsum and limestone of the Windsor Group; sinkholes and karst topography are common in this area. Sandy, glacial till covers the highland areas to a depth of a few feet and reaches a maximum of 30 feet in the central part of the basin. Bedrock is exposed along the lower segment of the stream. There are little or no flood-plain deposits.

Ground-water discharge probably occurs in large part as seepage above the stream surface; springs are common throughout the basin, occurring in both bedrock and surficial deposits [*Pinder*, 1966].

Fraser Brook Basin. Fraser Brook drainage basin is mostly forested with an area of 3.5 square miles and a relief of approximately 310 feet. The area is underlain entirely by gently folded nonmarine shale, siltstone, and sandstone of the Canso Group of Mississippian Age. Sandy glacial till covers about 80% of the watershed and in some areas overlies a clay till exposed over approximately 15% of the drainage basin; the drift sequence varies from 5 to 45 feet in thickness. Bedrock is exposed along most of the lower segment of the brook. There are little or no flood-plain deposits.

Ground-water discharge occurs as seepage above the stream surface, but upward moving water is indicated in several locations by closed basins where swamp deposits and phreatophytes are found [*Hennigar,* 1967].

Sharpe Brook Basin. Sharpe Brook drainage basin has an area of 5.2 square miles and a relief of 700 feet. The basin is covered mainly by second-growth forest, but the lower part of the basin includes some areas that are farmed. The lower segment of the basin is underlain by gently dipping Triassic sandstones and shales mantled by glaciofluvial deposits often over 50 feet thick. The middle third of the basin is underlain by slates and quartzites of the Lower Ordovician Meguma Group. The upper part of the basin is underlain mainly by Devonian porphyritic granite. Sandy till, commonly less than 10 feet thick, overlies the granite, clayey till, sometimes over 30 feet thick, mantles the slate. Flood-plain deposits are present only along the lower segment of the basin. Bedrock outcrops are common only where Sharpe Brook cuts the granitic and metamorphic rocks.

Much of the ground-water discharge along the middle and upper parts of the basin is from springs and seeps above the stream level. In the lower segment of the basin, groundwater discharges mostly into the flood plain near the brook (Trescott, in preparation).

THEORETICAL DEVELOPMENT

The concentration of an ion in solution in total runoff consists of contributions from ground water and direct runoff. Assuming that there is no chemical interaction when solutes from ground water and direct runoff combine and that each component of runoff has a density of 1 gm/cc, the relationship between runoff components may be expressed in the form of a mass-balance equation as

$$C_{tr}Q_{tr} = C_{gw}Q_{gw} + C_{dr}Q_{dr}. \qquad (1)$$

Simplifying and solving for ground-water discharge, we obtain

$$Q_{gw} = (C_{tr}Q_{tr} - C_{dr}Q_{dr})/C_{gw}$$

or

$$Q_{gw} = [(C_{tr} - C_{dr})/(C_{gw} - C_{dr})]Q_{tr} \qquad (2)$$

where

$$Q_{tr} = Q_{gw} + Q_{dr}$$

and C_{tr}, C_{dr}, and C_{gw} are now concentrations of the individual ions in ppm.

To solve (2) the quantity of total runoff (Q_{tr}) was obtained from the discharge hydrograph; at the same time a sample of total runoff was collected for the determination of the chemistry of total runoff (C_{tr}). The chemistry of ground water (C_{gw}) was taken as the value of C_{tr} at the lowest recorded flow over the period of analysis and was assumed to be a constant over the period of the storm hydrograph.

During the period of analysis there was little variation in the discharge or the solute concentration of the ground-water component over the baseflow interval (see Figure 3). If sampling over an extended period of baseflow indicates that ground-water chemistry is a function of ground-water discharge, a suitable equation relating these two parameters must be employed in conjunction with (2) to determine the ground-water component.

ESTIMATION OF THE CHEMICAL COMPOSITION OF DIRECT RUNOFF

The assumption that ground water constitutes total runoff at baseflow implies that direct runoff includes subsurface flow. The inclusion of the subsurface component in direct runoff may be assumed in hydrograph analysis [*Chow,* 1964].

One method of estimating the chemical composition of direct runoff is to analyze samples collected from ephemeral streams; this technique, however, neglects the subsurface-flow component and may result in erroneously high values for ground-water discharge.

Another approach is to obtain samples from the main channel during periods of peak flow. In basins with extensive flood-plain deposits and ground-water discharging principally as vertical seepage into the main channel, a reversal of hydraulic gradient occurs during flood stage. During this period water moves from the river into the adjacent fluvial deposits (bank storage), and ground water cannot discharge vertically into the channel bottom. In basins indicative of this type of hydrologic system, samples obtained during peak discharge should provide reasonable estimates of the chemical characteristics of the direct-runoff component. As indicated above, however, the three representative watersheds are not characterized by

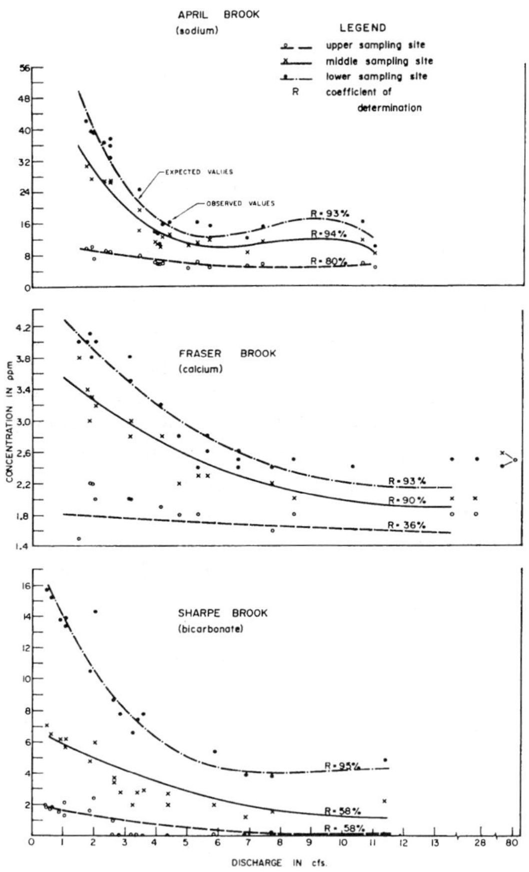

Fig. 4. Concentration in ppm of a selected ion from each brook under consideration and discharge in cfs. Expected-value curves are determined from best-fitting polynomials.

this type of basin hydrology. Ground-water discharge commonly occurs as springs above the stream surface, and bank storage is probably effective only in the lower segment of the basins. There is no reason to assume, therefore, that samples collected from the main channel of these basins during peak flow accurately represent the direct-runoff component.

The concentration of most of the ions studied increased significantly downstream. This relationship is illustrated in Figure 4 and was verified in the analysis of variance of the factorial experiment conducted at a 95% confidence level.

A multiple regression technique was used to determine the best-fitting polynomials relating total discharge and the solute concentration of total runoff (see Figure 4). The error term for testing the improvement of fit of the n + first-order polynomial was calculated from the sum of squares of deviations from regression for the nth order equation.

The observed increase in solute concentration downstream may be attributed to: (1) ground water containing a higher concentration of ions in the lower segment of the basin, due to a longer flow path in the natural hydrologic system; and (2) a decrease in ground-water discharge on the upper segment of the basin, due to a lower hydraulic gradient toward the main channel.

The consistency of the above relationship over the entire range of the stream hydrograph suggests that ground-water discharge continues during periods of peak flow. The relatively flat

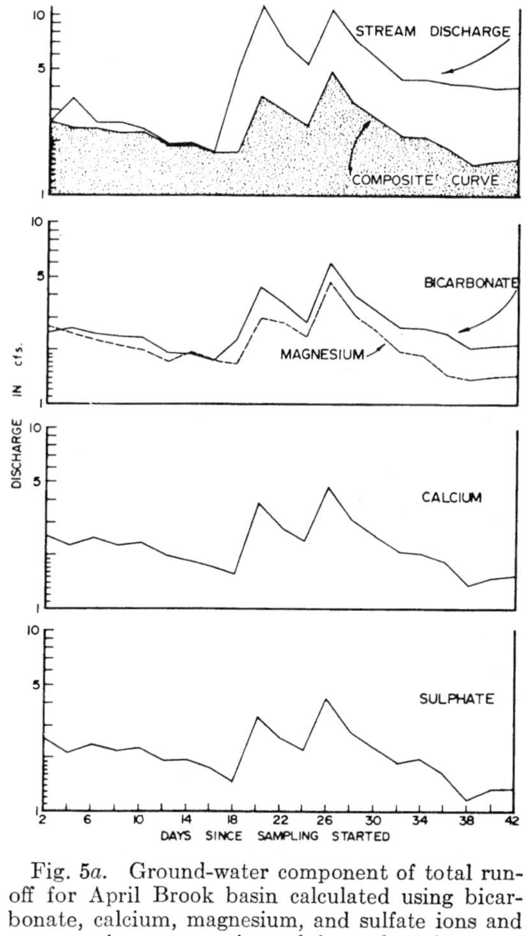

Fig. 5*a*. Ground-water component of total runoff for April Brook basin calculated using bicarbonate, calcium, magnesium, and sulfate ions and a composite curve estimated from these ions.

curves for the upper sampling sites in Figure 4 indicate that total discharge has a minor influence on the solute concentrations of total runoff along the upper segment of the stream. This implies that the chemical characteristics of ground water and direct runoff are similar in the upstream areas and that samples collected from the uppermost station during periods of peak discharge would provide a good estimate of the chemical characteristics of direct runoff. The use of this value for C_{dr} in equation 2 will yield an underestimate of ground-water discharge for ions more concentrated in the ground-water component; for ions more concentrated in direct runoff this value for C_{dr} will provide an overestimate. The calculation of a composite curve (see Figure 5) using the mean

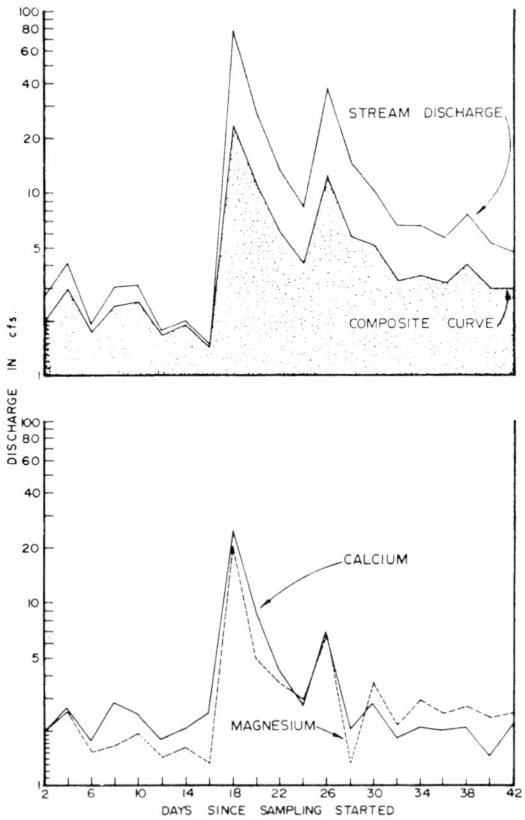

Fig. 5*b*. Ground-water component of total runoff for Fraser Brook basin calculated using calcium and magnesium. Composite curve is estimated from bicarbonate, calcium, magnesium, and sodium.

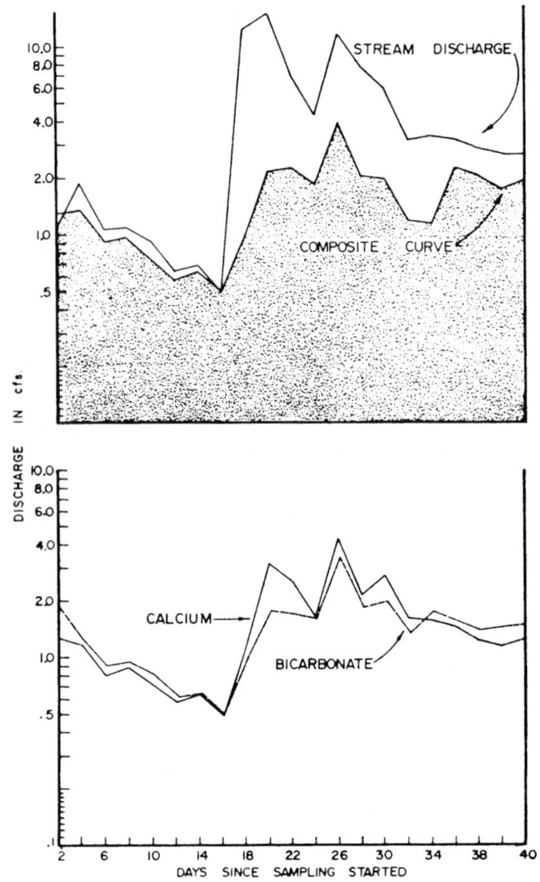

Fig. 5*c*. Ground-water component of total runoff for Sharpe Brook basin calculated using bicarbonate and calcium. Composite curve is estimated from bicarbonate, calcium, magnesium, and sodium.

of discharge values from several ions should minimize this source of error.

DETERMINATION OF THE GROUND-WATER COMPONENT

To obtain the ground-water discharge component of total runoff beneath the storm peak, equation 2 was solved using silica and bicarbonate, chloride, nitrate, sulfate, calcium, iron, magnesium, potassium, and sodium ions. Figure 5 shows the ground-water component calculated for several representative ions from each basin; the upper graph shows total runoff and the composite ground-water discharge hydrograph. In general, ions that provided consistent and reasonable values for ground-water discharge displayed a strong correlation between chemistry and total discharge as suggested by a high value for the coefficient of determination (see Figure 4).

CHEMICAL CHARACTERISTICS OF GROUND WATER AND DIRECT RUNOFF

Table 1 shows the low-flow solute concentrations at the lower sampling site and the peak discharge concentrations for the upper sampling station. These values were used in the calculation of the ground-water component of the hydrograph. Although the values are generally low for Fraser and Sharpe brooks, April Brook shows higher concentrations for the precipitate-forming ions due to the limestone and gypsum units underlying the basin. Ions that were more concentrated in direct runoff than ground water include nitrate and iron; values for sulfate are also higher in direct runoff in Fraser and Sharpe brooks.

Decaying organic matter is probably the primary source of nitrate in the basins, since fertilizer is not widely used. The solution of oxidized iron sulfide minerals occurring in the glacial till may account for the high iron and sulfate concentrations in runoff. Cation exchange and, to a lesser extent, anion exchange may remove some of these ions from the downward percolating solution before it reappears as ground-water discharge.

DISCUSSION OF RESULTS

The ground-water component of stream discharge estimated from the chemical characteristics of total runoff indicates that in the three representative watersheds ground water forms a large part of peak flow. The ground-water component at the peak stage investigated was calculated to be 42% of total runoff for April Brook, 40% for Fraser Brook, and 32% for Sharpe Brook. It is apparent that hydrograph separation assuming negligible ground-water flow at peak stream discharge (see Figure 1) cannot be applied to basins where bank storage is not an important hydrologic parameter.

The calculated ground-water discharge hydrographs indicate that in the basins under consideration the ground-water reservoir responds very quickly to the effect of rainfall. This rapid response suggests that fracture permeability may play an important role in the three basins under consideration.

Ions that occurred in the highest concentrations in solutions of total runoff were found to provide better hydrograph separations. The best ground-water discharge hydrographs, therefore, were obtained using calcium, magnesium, sodium, chloride, sulfate, and bicarbonate ion concentrations from April Brook basin.

The hydrograph separations and the solute concentration-discharge relationships presented are for a 6-week period of analysis at the end of the summer; the hydrogeology of the watersheds is typical for small basins in eastern Canada. Long-term sampling programs in areas of varied hydrogeology are necessary to determine the general applicability of the techniques and relationships presented here.

TABLE 1. Low Flow Solute Concentrations for the Lower Sampling Site and Peak-Discharge Concentrations for the Upper Sampling Station Determined for a Selected Number of Ions Over the Period of Analysis (concentrations in ppm)

	Ground Water			Direct Runoff		
Ions	April	Fraser	Sharpe	April	Fraser	Sharpe
Bicarbonate	84.8	18.0	15.8	18.8	0.7	0.0
Chloride	43.0	4.8	6.7	6.7	2.9	7.8
Nitrate	0.05	0.05	0.05	1.2	2.1	2.1
Sulphate	451.0	3.7	4.1	4.0	5.8	5.6
Silicate	6.2	3.8	11.0	4.1	2.1	6.6
Calcium	193.0	4.0	4.5	5.5	1.8	1.3
Iron	0.1	0.2	0.3	0.8	0.4	0.7
Magnesium	9.5	0.4	1.2	1.2	0.9	0.8
Potassium	1.2	2.9	0.3	0.5	0.4	0.6
Sodium	42.0		3.9	4.8	1.5	3.2

Acknowledgments. We would like to thank the Water Quality Division, Inland Waters Branch, Canada Department of Energy, Mines, and Resources, the Canada Department of Forestry and Rural Development (ARDA), and the Department of Mines, Government of the Province of Nova Scotia, whose support made this study possible. This investigation also forms part of the Canadian International Hydrological Decade Program.

REFERENCES

Archer, R. J., A. M. LaSala, Jr., and J. C. Kammerer, Chemical quality of streams in the Erie-Niagara basin, New York, *New York State Water Resources Comm. Basin Planning Rept. ENB-4,* in press, 104 pp., 1967.

Barnes, B. S., Discussion on analysis of run-off characteristics by O. H. Meyer, *Trans. Am. Soc. Civil Engrs., 105,* 104–106, 1940.

Chernaya, T. M., Comparative evaluation of graphical methods of separation of ground-water components of streamflow hydrographs, *Soviet Hydrol., 5,* 454–465, 1964.

Chow, V. T. (ed.), *Handbook of Applied Hydrology,* 14-1 to 14-13, McGraw-Hill Book Company, New York, 1964.

Hennigar, T. W., Report on the geology and hydrology of the Fraser Brook watershed, Colchester Co., Nova Scotia, *Nova Scotia Dept. Mines Open File Rept.,* 30 pp., 1967.

LaSala, A. M., Jr., New approaches to water-resources investigations in upstate New York, *Ground Water, 5*(4), 6–11, 1967.

Linsley, R. K., and J. B. Franzini, *Water-Resources Engineering,* p. 40, McGraw-Hill Book Company, New York, 1964.

Pinder, G. F., Geology of the April Brook basin, *Nova Scotia Dept. Mines Open-File Rept.,* 9 pp., 1966.

Trescott, P. C., Geology of the Sharpe Brook basin, *Nova Scotia Dept. Mines Open-File Rept.,* in preparation.

(Manuscript received October 9, 1968; revised November 12, 1968.)

Commentary on Paper 15

Dunne, T. & Black, R. D. (1970) Partial area contributions to storm runoff in a small New England watershed.

In 1970, the results of a seminal study were published in two papers in *Water Resources Research*, one of which is reproduced here. The work was drawn from the thesis of Tom Dunne, who had been registered for a PhD under the supervision of Reds Wolman at John Hopkins University after studying as an undergraduate with Dick Chorley at Cambridge University. Much of Tom's time as a PhD student, however, was spent in Vermont working with Richard Black at the USDA Agricultural Research Service Centre in Danville and in the field on the Sleepers River experimental basins, in particular in the basin above gauge WC-4. In the upper part of the basin, Tom and Richard dug a throughflow trench to drain three adjacent sections of hillslope, one divergent, one convergent and one straight in plan. The trench was in total some 70 m long and 2.5 m deep with collectors at the surface, at the base of the root zone and "*in the zone of perennial groundwater seepage*" (p.1299). In addition the basin was equipped with several V-notch weirs, including one to monitor a seep on the hillslope opposite to the trench, transects of piezometers, and neutron probe access tubes to measure soil moisture. Thus, the instrumentation was similar to that used earlier by Whipkey (Paper 7) but bigger and better!! What is even more remarkable is that the trench was dug, by hand, in the December of a Vermont winter. Apparently each morning commenced by thawing the surface soil for the next section of trench with the aid of some gasoline and a match! More detail of the trench experiment is given in the earlier paper of Dunne & Black (1970).

Dunne & Black studied the response of the slope segments and the seep under both natural and artificial rainfalls. They were really the first to demonstrate that fast surface runoff might occur on variable saturated areas of soil, maintained by flows from upslope. They mapped the patterns of saturation in different conditions in relation to the associated discharge hydrographs. They note the history of partial area or variable source area concepts in the work of Betson (Paper 8) at the Tennessee Valley Authority and Hewlett & Hibbert (Paper 9) at Coweeta, though "*these writers did not emphasize the importance of the threshold between the surface and subsurface flow systems*" (p.1309). Thus, Dunne & Black were the first (at least in the USA, see also Paper 4) to identify what is now known as "saturation overland flow", "saturation excess overland flow" or "Dunne overland flow" on a dynamic but partial area of the catchment (and therefore different from the Betson concept of a partial area of infiltration excess overland flow, and Hewlett's concept of an expanding stream network fed by variable subsurface source areas). They (and others) went on to demonstrate that this was not just a local phenomenon (see for example, Dunne *et al.*, 1975).

It is interesting that their last sentence notes that this concept provides a point of departure for new models "*not based on infiltration theory*" and for studies of nutrient losses from watersheds as well as sources of agricultural pollution. The legacy of this study is seen in the stimulus to the modelling studies of Al Freeze (1972), the topographic index concepts of Mike Kirkby (1975) and Emmett O'Loughlin (1981), the development of TOPMODEL from Beven & Kirkby (1979) onwards, and the critical source area concepts of nonpoint source pollution of Gburek & Sharpley (1998).

Tom Dunne went on to teach at the University of Nairobi (Kenya), McGill University (Canada), the University of Washington (USA) and the University of California at Santa Barbara (USA). Many prominent researchers started their research careers with him and the intellectual stimulation of his teaching, field trips, and evening paper discussion sessions. His contribution to the *Hillslope Hydrology* book edited by Mike Kirkby still remains one of the very best summaries of field studies of hillslope flow processes ever written (Dunne, 1978). He received the Horton Award of the Hydrology Section of the American Geophysical Society in 1987, and has recently led a NASA Interdisciplinary Science Team of the EOS Amazon Project. Richard Black went on to teach at Cornell University.

Unfortunately, the Dunne & Black hillslope no longer exists. It was destroyed when Route 2 from Danville to St Johnsburg was driven through the basin in 1975.

References

Beven, K. J. & Kirkby, M. J. (1979) A physically-based variable contributing area model of basin hydrology. *Hydrol. Sci. Bull.* **24**(1), 43–69.

Dunne, T. (1978) Field studies of hillslope flow processes. *Hillslope Hydrology* (ed. by M. J. Kirkby), 227–294. Wiley, Chichester, UK.

Dunne, T. & Black, R. D. (1970) An experimental investigation of runoff production in permeable soils. *Water Resour. Res.* **6**(2), 478–490.

Dunne, T., Moore, T. R. & Taylor, C. H. (1975) Recognition and prediction of runoff producing zones in humid regions. *Hydrol. Sci. Bull.* **20**, 305–327.

Freeze, R. A. (1972) Role of subsurface flow in generating surface runoff. 2. Upstream source areas. *Water Resour. Res.* **7**(4), 929–941.

Gburek, W. J. & Sharpley, A. N. (1998) Hydrologic controls of phosphorus loss from upland agricultural watershed. *J. Environ. Qual.* **27**(2), 267–277.

Kirkby, M. J. (1975) Hydrograph modelling strategies. In: *Processes in Physical and Human Geography* (ed. by R. F. Peel, M. D. Chisholm & P. Haggett), 69–90. Heinemann, London, UK.

O'Loughlin, E. M. (1981) Saturation regions in catchments and their relation to soil and topographic properties. *J. Hydrol.* **53**, 229–246.

Partial Area Contributions to Storm Runoff in a Small New England Watershed

THOMAS DUNNE[1] AND RICHARD D. BLACK[2]

Agricultural Research Service, U. S. Department of Agriculture, Danville, Vermont 05828

Abstract. During an experimental study of runoff producing mechanisms in a small drainage basin, the major portion of storm runoff was produced as overland flow on a small proportion of the watershed. Where the water table intersected the ground surface before and during a storm, water escaped from the soil surface and ran quickly to the stream at velocities 100 to 500 times those of the subsurface system. Direct precipitation onto the saturated area was another major contributor of storm flow. Storage of water within these source areas was small and travel times out of them were short. Runoff from them was largely controlled by rainfall intensity. These partial areas contributing quick runoff could expand or contract seasonally or during a storm. Their position and expansion can be related to geology, topography, soils, and rainfall characteristics. In the study basin, water that remained below the ground surface on its way to the main stream channel was a relatively minor contributor to the storm hydrograph. The response of subsurface flow to rainfall was heavily damped by storage and transmission within the soil. Measurements of water table elevation and runoff from experimental plots in another small watershed with different geologic conditions confirmed the results outlined above. Runoff records from a large number of small basins of the Sleepers River experimental watershed indicate the more general applicability of these findings.

The applicability of traditional concepts of runoff production in humid areas [*Horton*, 1933, 1945] has been questioned in recent years. The absence of empirical evidence of sheet flow on hillsides, and the measurement of high rates of infiltration into the permeable, vegetated soils of such regions have cast doubt on the usefulness of the Hortonian infiltration theory in analyzing runoff during the vast majority of storms. Some hydrologists have postulated the mechanism of subsurface storm flow as the main source of storm runoff. *Whipkey* [1965, p. 74] defined subsurface storm flow as underground, storm period flow that reaches the stream channel without entering the general groundwater zone. *Hewlett and Hibbert* [1965] described their concept of 'translatory flow' in the following way: 'Above the zone of saturation, we may regard such movement as due to thickening of the water films surrounding soil particles and a resulting pulse in water flux as the saturated zone is approached.' *Kirkby and Chorley* [1967, p. 7] used the term 'throughflow' to denote 'flow which takes place physically within the soil profile.' Studying data from *Whipkey* [1965] and a paper on soil drainage by *Hewlett and Hibbert* [1963], *Kirkby and Chorley* [1967, p. 8] claimed that '. . . throughflow is capable of producing runoff peaks in river hydrographs.'

Recent experimental work by *Dunne and Black* [1970] has thrown doubt on the role of subsurface storm flow from hillsides as a major contributor to storm hydrographs in upland watersheds of northern Vermont. They found that the importance of a hillslope as a producer of storm runoff seems to depend largely on its ability to generate overland flow. These findings were in general agreement with those of various recent investigations of the partial area concept of runoff production, which states that storm runoff is generated on only a small portion of a watershed. *Betson* [1964, p. 1541], using a mathematical model based on Hortonian infiltration theory, concluded that 'runoff usually originates from a small, but relatively consistent, part of the watershed.' Development of this idea by the *Tennessee Valley Authority* [1965] using another mathematical model suggested

[1] Now at Department of Geography, McGill University, Montreal, Canada.

[2] Now at Department of Agricultural Engineering, Cornell University, Ithaca, New York.

that the contributing area was a dynamic one and could vary in size during the course of a storm. *Amerman* [1965, p. 504] analyzing runoff data from small watersheds in Ohio found that 'Runoff-producing areas were located in seemingly random fashion on ridge-tops, valley-slopes, and valley bottoms.' The runoff producing areas were not necessarily connected to the perennial stream by continuous surface flow. Surface runoff was often absorbed before reaching a main channel. *Hewlett and Hibbert* [1965] proposed a conceptual model of runoff production in a forested watershed, whose major feature was the contribution of storm flow by development and expansion of saturated zones along valley floors and the lower portions of hillsides. Experimental studies by *Ragan* [1967] also showed that only a small but variable proportion of the watershed ever contributed flow to the storm hydrograph.

The study reported here was conducted at the Sleepers River experimental watershed, near Danville, Vermont, in the headwaters of the Connecticut River system. The watershed was chosen by the U. S. Department of Agriculture as being representative of the glaciated upland of New England. Infiltration capacities of the soils of the experimental watershed exceed rainfall intensity in the vast majority of measured storm events. Overland flow generated as Hortonian precipitation excess on large areas of hillside is not observed. Total yields of storm runoff are small. The maximum recorded rate of runoff on the Sleepers River watershed is 0.10 inch per hour from a rainfall of 3.37 inches in 38 hours in a 16-square-mile catchment. Runoff hydrographs, however, have the same general features as those from areas where overland flow is observed. They have the same general shape as hydrographs from these areas and respond quickly to rainfall. They show many of the same kinds of relationships between rainfall, runoff, and watershed condition as are recognized in areas known to experience overland flow.

OBJECTIVES

In spite of the assumptions made to facilitate hydrograph separation, the mechanisms of storm runoff production and their relative importance in humid areas are generally unknown. The primary aim of this study was to define the mechanisms of storm runoff production operating in a small watershed, and to measure directly the amounts of flow contributed by each of the forms of runoff encountered. The timing of these contributions and their relationship to the channel hydrograph were also studied. Measurements of the status of water in the soil were made to determine the physical controls of the various processes. Studies of runoff production were made during natural and artificial rainstorms.

EXPERIMENTAL AREA

The watershed in which the study was conducted has an area of approximately 60 acres, but a railroad divides the lower 10 acres from the rest of the basin. Attention will be confined to this lower part of the basin, which is carved into a glaciolacustrine terrace lying against a steep till covered bedrock hillside of calcareous schist. Figure 1 is a north–south geologic cross section of the northern part of the basin, where detailed studies were carried out. Sand 22 feet deep overlies 38 feet of varved silt and sand that rest on dense silty clay till. Dissection of the terrace has led to the exposure of till and varved lake deposits in the valley floor and on

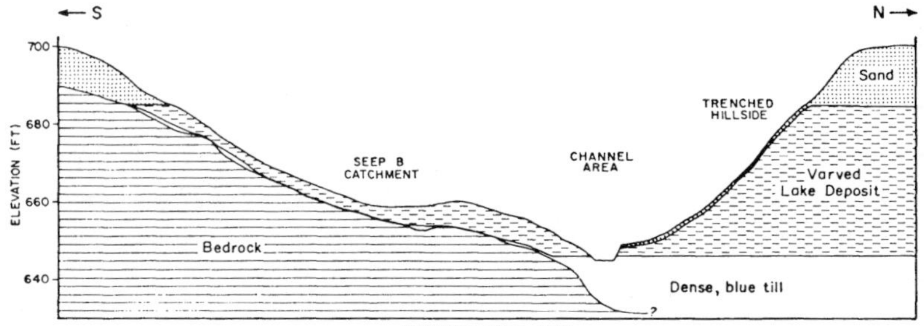

Fig. 1. Geologic cross section through the northern part of the watershed.

the lower portions of some hillsides. On slopes where a cover of sand has survived, mass movement has resulted in the deposition of a 1- to 3-foot-thick layer of highly permeable sand over the denser varved deposits.

The topography is typical of a dissected terrace (Figure 2). A small proportion of the upper terrace remains within the watershed, and is separated from the valley bottom by slopes that range in gradient from 10% to 100%. The valley floor varies from 7 to 70 feet in width and is marshy all year except for its upper portion above WC-1 (Figure 2) where flow is ephemeral. The stream flows sluggishly in a marshy depression. It is fed by several seeps that emerge from bedrock joints. Some of these seeps originate close to the stream but seep *B* emerges from the ground approximately 100 feet upslope from the point at which it is gaged. The channel draining the study basin is not typical of the major channels of the Sleepers River basin, but is representative of the channels draining small tributary valleys that occur at the head of and along the main streams.

The distribution of soils reflects the pattern of dissection and exposure of the various parent materials. The steep slopes are covered with well drained Brown Podzolic and Brown Forest sandy loams. The major difference between these soils is the presence of an impeding layer at a depth of 12 to 30 inches in the Brown Podzolic soil [*Dunne*, 1969]. Both have highly permeable A-horizons with abundant root holes and worm holes. Silt loams cover the valley floor and the lower side slopes, wherever till or varved deposits are exposed. The silt loams vary from

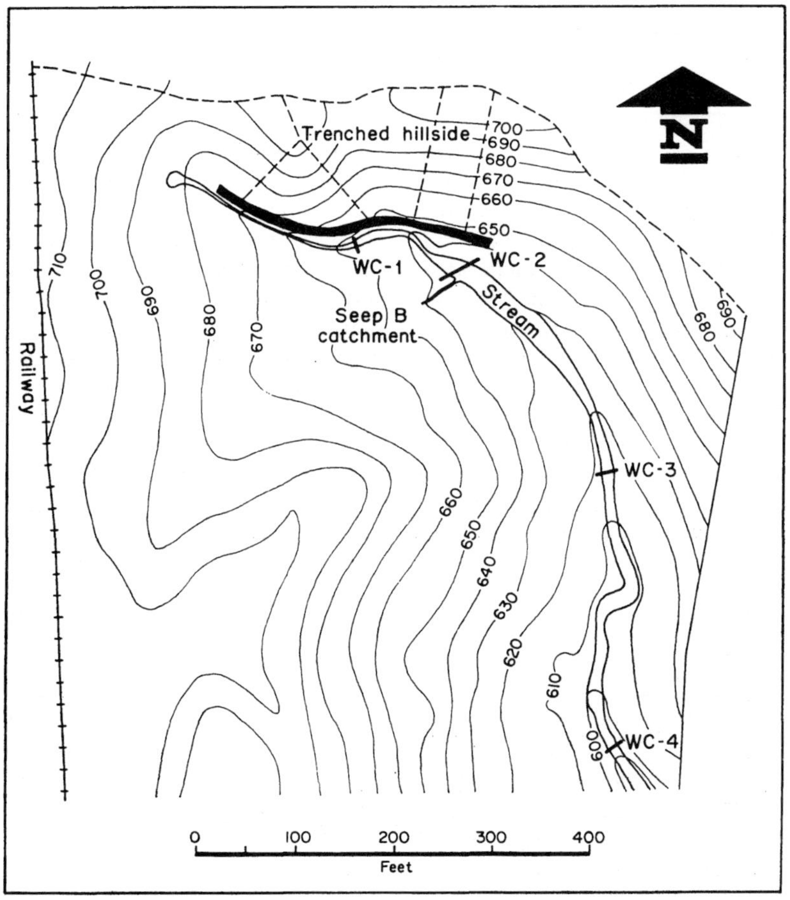

Fig. 2. Map of the lower 10 acres of the watershed studied.

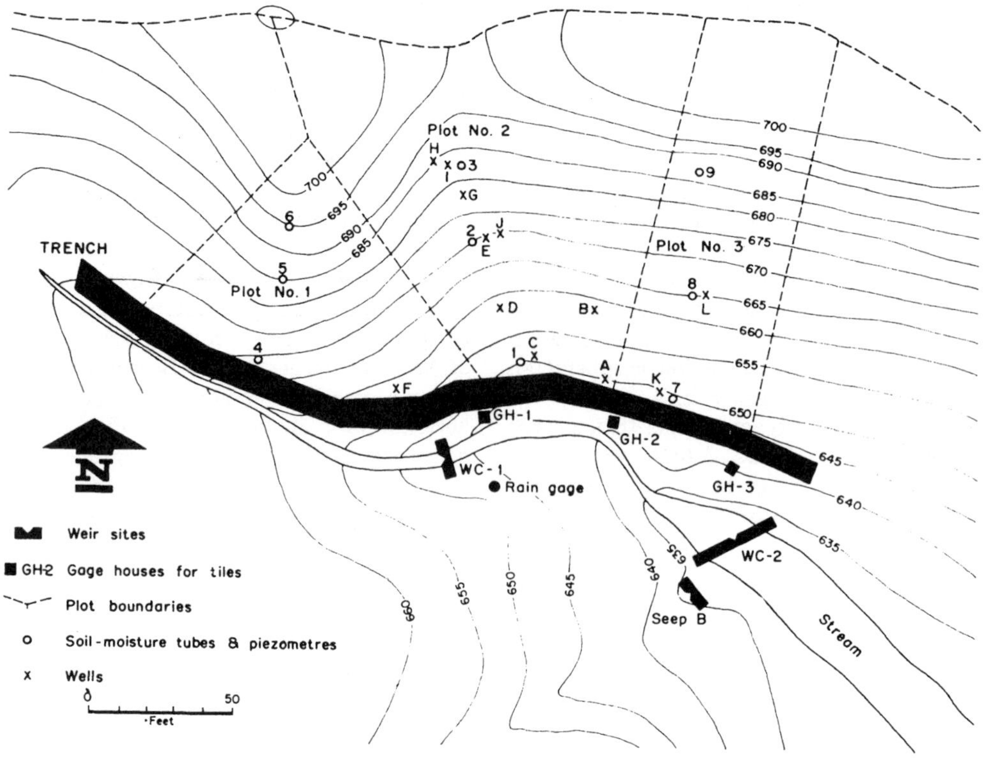

Fig. 3. Map of the trenched hillside.

poorly drained on the valley floor to moderately well drained on slopes of up to 15%. They are characterized by a shallow (3- to 12-inch) friable A-horizon containing abundant grass roots, root holes, worm holes, and structural channels. Below this is a sharp change to a dense grey mottled silt loam of the undisturbed till and silty varved deposits. Although the watershed is now in pasture–pine woodland covered the area until it was destroyed by fire approximately 30 years ago.

Within the watershed, three small areas were instrumented for intensive study. Their locations are shown in Figures 1 and 2. The first of these was a 0.6-acre hillside, chosen to represent the steep permeable well drained slopes of the basin (Figure 3). This area will be referred to as the trenched hillside. The second site chosen for study was a 0.6-acre area of low slope and moderate to poorly drained soils in the catchment of seep *B* (Figure 4). The upper 600 feet of the poorly drained valley floor above weir WC-3 (Figure 2) was the third study area.

INSTRUMENTATION

Details of the instrumentation and the data collected have been given by *Dunne* [1969]. Runoff from the various soil horizons of the trenched slope was measured by means of an interceptor trench in which drains were installed [*Dunne and Black*, 1970]. Runoff was measured separately at the ground surface, the base of the root zone, and in the zone of perennial groundwater seepage (at a depth of 5 to 8 feet). Three sections of hillside, with planforms that are respectively convex, concave, and straight were gaged separately (Figure 3). The discharge from seep *B* was measured continuously at a 45° sharp-crested V-notch weir. Other seeps in the watershed were gaged volumetrically throughout the study. In the main channel, flow was measured continuously at four H-flumes, whose positions are indicated in Figure 2 as WC-1, WC-2, WC-3, and WC-4.

Precipitation was measured with a 6-inch 24-hour weighing rain gage. During artificial storms, a random network of thirty 4-inch-

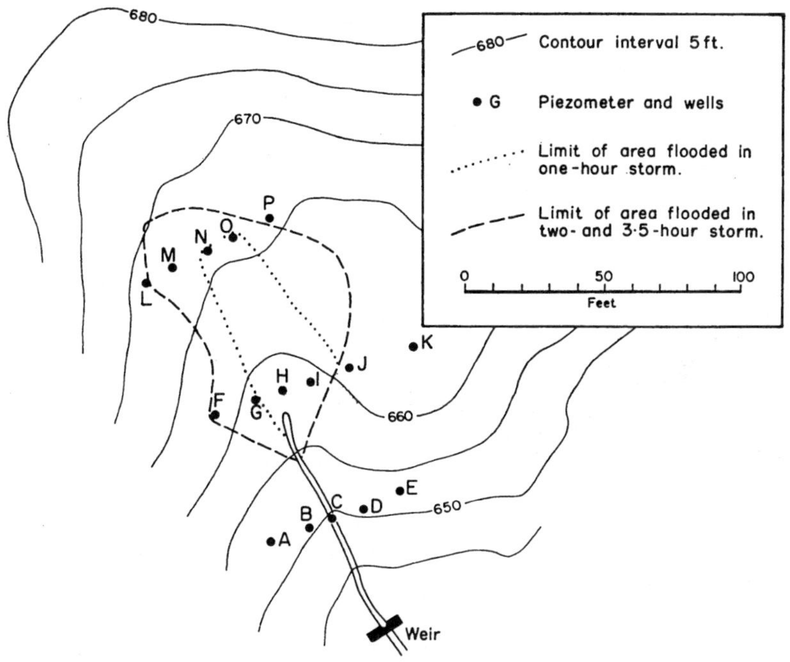

Fig. 4. Map of the seep *B* catchment.

diameter cans was used to measure rainfall. Soil moisture was measured to a depth of seven feet at nine locations on the trenched hillside (Figure 3). Measurements of water table elevation and piezometric head were made in ¾-inch-diameter pipes at locations shown in Figures 3 and 4.

EVENTS STUDIED

Data were collected in the following types of events:

1. Natural rainstorms.
2. Artificial storms of high return period on the steep, permeable trenched hillside (including the natural rainstorms that occurred so soon after the simulated storms that they could be considered as part of the same event).
3. Artificial storms on the channel area.
4. Artificial storms on the shallow soils of the seep *B* catchment.

RESULTS

Natural storms. Storms ranging in amount up to 1.83 inches were recorded during the course of the study. The maximum 20-minute intensity of these storms was 3.15 inches per hour. No significant amount of storm flow was produced on the deep permeable hillsides represented by the trenched slope during natural storms, even when the water table over the lower part of the slope was within 1.3 feet of the surface before a storm. The only exceptions to this statement were two storms that occurred soon after artificial ones [*Dunne and Black,* 1970]. Throughout the summer months, rainfall was stored in the upper three feet of soil during storms, and drained slowly to the water table during the next few days. In the spring and autumn when some rainfall did cause a reaction of the water table, the resultant subsurface discharge was too small, too slow, and too insensitive to the fluctuations of rainfall intensity to supply storm flow to the channel hydrograph (see *Dunne and Black* [1970] for details).

Areas of poorly drained soils on low gradients, represented by the seep *B* catchment, supplied small amounts of baseflow to the stream throughout most of the year, but were usually dry between late July and mid-October. Storm discharge from the seep was extremely sensitive to the fluctuations of rainfall intensity (Figure 5).

(It is a central thesis of this paper that runoff production in humid areas is spatially nonuniform. Therefore, expressing runoff volumes and rates in inches and inches per hour is misleading for small watersheds. In this paper runoff volumes are expressed in cubic feet and runoff rates in cubic feet per minute (cfm).) Hydrographs often had recession limbs that were steeper than their rising limbs and peaked within a minute or so of the end of a burst of rainfall. Storm flow volumes and peak discharges were small compared with the total amount and rate of rainfall onto the catchment of the seep. This area varied from 100 to 1000 square feet according to antecedent conditions, and to rainfall intensity and duration. In this part of the seep, which remained wet throughout most of the year, the storage capacity of the shallow topsoil was quickly exhausted. The water table intersected the land surface and a small partial area reacted to rainfall essentially as an impervious area. The contribution to peak runoff rate and total volume of storm flow from this area did not usually exceed 25% of the storm flow at gaging station WC-2.

Channel runoff at each of the gaging stations on the main stream of the basin was extremely sensitive to rainfall intensity fluctuations. Discharge began to increase within a minute of the onset of rainfall. Hydrograph peaks were sharp and occurred within several minutes of the end of a burst of rainfall. Any decrease of rainfall intensity produced a sharp decline of discharge. Many hydrographs had recession limbs that were steeper than their rising limbs. Dry weather flow was remarkably constant over periods of several days. In the following discussion, data from stations WC-1 and WC-2 will be used as an illustration. Analysis of data from stations WC-3 and WC-4 yielded similar results.

The major portion of the storm flow measured at WC-1 and WC-2 in many natural storms was found to be precipitation falling directly on the stream channel. The proportion of storm flow supplied by this channel precipitation varied with the intensity and the duration of rainfall. The area of the stream surface between these two weirs was 1000 square feet when the stream was flowing above WC-1, and 760 square feet when the source of the stream was at its late summer position, immediately below WC-1. The saturated floor of the stream depression in this reach had an area of 1310 square feet (including the area of the stream surface itself). Contributions from the floor and the steep wet banks of the stream depression, from the impervious roof over the trench (Figure 3), and from subsurface inflow along the stream banks could not be measured separately. Evidence from the channel gaging stations and from the subsurface collectors on the trenched hillside showed that these contributions were small in most short summer storms. If the saturated floor of the stream depression was considered to have yielded to the channel all the rain falling onto it, the computed runoff was greater than the observed runoff for many storms during the summer months.

Many storms were analyzed, but data on only two are presented here as an illustration. Data from others are included in the report by *Dunne* [1969]. The storms studied represented a range of types from short to long duration, high to low intensity, on wet to dry antecedent conditions, and producing both simple and multipeaked hydrographs.

A 0.75-inch storm on July 12, 1967, included three separate bursts of rain, each of which produced a separate and virtually simultaneous hydrograph peak at each of the four weirs in the watershed. Data from WC-1 and WC-2 are presented in Figure 6. Each rise of the hydrograph began within one to four minutes of the beginning of an intense burst of rain. Each recession limb was less steep than the one

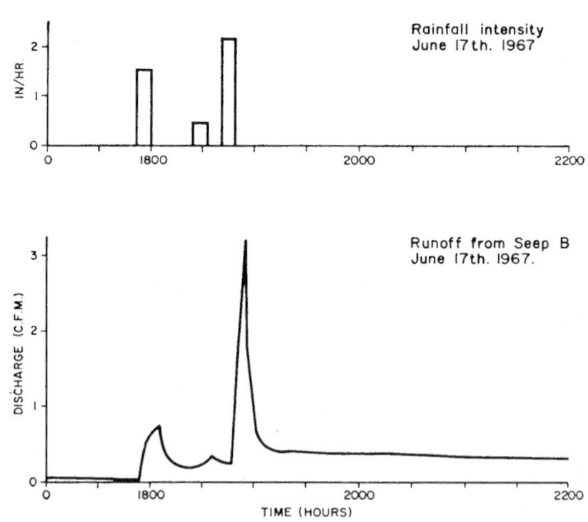

Fig. 5. Response of flow from seep *B* to a 0.49-inch rainfall.

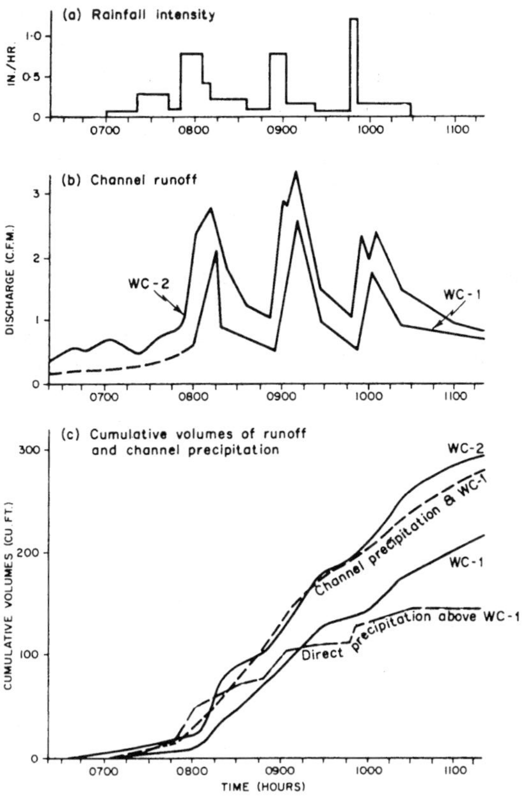

Fig. 6. Rainfall and runoff data from WC-1 and WC-2 during the 0.75-inch storm of July 12, 1967. In the lowest graph, solid lines represent cumulative volumes of storm flow at the two weirs. Dashed lines represent the contribution of rainfall onto the previously saturated areas along and at the head of the channel.

preceding it, suggesting that in the later peaks some runoff was being supplied by mechanisms involving more storage than those supplying the first peak. This conjecture was confirmed by plotting cumulative amounts of storm flow (base flow subtracted) and channel precipitation for each reach. Travel time through the reach (measured to be approximately four minutes) was ignored in plotting this graph. Figure 6 shows that the major source of storm runoff between the two weirs was channel precipitation. By the time the second rise of the hydrograph began, storm flow out of the reach amounted to only 88% of channel precipitation. By the time the third rise began, channel precipitation accounted for 93% of storm flow. This figure declined to 73% by the end of rainfall. The volume of rain falling onto the saturated floor of the channel depression during the rainstorm was 81.5 cubic feet, but only 65.5 cubic feet of storm flow were generated in this reach during the time interval covered by Figure 6. Above WC-1, 140 cubic feet of precipitation fell onto the previously measured saturated area. This figure represents 65% of the total storm runoff measured at WC-1. The remainder was presumably supplied by subsurface storm flow in the vicinity of the stream and by precipitation onto the wet banks of the marshy depression occupied by the stream. Field observations and the records of runoff collectors on the trenched slope showed that no runoff was supplied by the hillslopes flanking the stream. Soil moisture measurements before and after rainfall showed that all the precipitation was stored within the upper three feet of soil on the trenched hillside.

On the night of October 10–11, 1967, several bursts of rainfall (with intensities up to 0.63 inches per hour) occurred during a 1.27-inch storm of generally low intensity. Each of these bursts produced a small peak on a general rise of the stream hydrograph (Figure 7). Antecedent conditions were dry, and the stream did not extend above WC-1 at the beginning of the storm. Small amounts of runoff were produced on the trenched slope, where soil moisture had been increased by previous artificial storms (runoff from channel 2-1 in Figure 7 consisted of overland flow, and runoff from tile 2-2 was flow from the root zone in the central hollow of the trenched slope). Rainfall intensity was low and the storm was of relatively long duration. Time was therefore available for larger contributions from channel banks, the saturated floor of the stream depression, and subsurface lateral inflow. Observations made during the storm indicated that the saturated channel banks and stream depression were the most important of these sources. Hydrographs of subsurface storm flow from the trenched slope when the water table was within a few inches of the ground surface indicate that subsurface lateral inflow was a late and minor contributor to the channel hydrograph in this kind of watershed. At the end of rainfall, true channel precipitation (as defined previously) accounted for only 56% of the storm flow generated in this reach. Precipitation that fell between the high wet banks of the stream depression, however, exceeded the amount of storm flow emerging from the reach until one hour after the end of rainfall. Only 19% of the precipitation falling into the channel depression above WC-1 flowed out before the rain stopped.

Data from these and other storms in this and other small watersheds reinforce the conclusion that under antecedent conditions, which prevail throughout most of the year, the magnitude and general shapes of hydrographs from first order drainage areas are controlled by precipitation that falls onto the stream and onto wet areas along and at the head of the stream. As these areas allow virtually no infiltration and little storage, the response of runoff from them has two important characteristics. First, the ratio of runoff to rainfall is close to one; second, the runoff from these small areas is extremely sensitive to fluctuations of rainfall intensity. Subsurface flow from both the permeable A-horizon of the soil and from deep seepage proved to be too small, too late, and too insensitive to fluctuations of rainfall intensity to account for the rapid rise and fall of stream discharge.

Artificial storms of high return period on the trenched hillside. Data collected during natural and artificial rainstorms with return periods of two to several hundred years are presented in a separate report [*Dunne and Black*, 1970]. Only a brief outline of the conclusions of the study is given here. Subsurface storm flow from the trenched hillside was measured during these storms. Most of it originated on the central concave plot (plot 2 in Figure 3) in the permeable topsoil over the dense varved silt. Where there was no impeding layer, as in the eastern plot (plot 3), storm runoff was negligible, even in storms with a return period of several hundred years. Even the subsurface storm flow from the concave plot, however, was of minor importance compared with rates of channel runoff during smaller, natural rains.

In some of the storms, large amounts of storm runoff were produced on the concave plot. They were generated on small areas of the central hollow where topography and soil profile characteristics favored a rise of the water table to the ground surface. Intersection of the ground surface by the water table allowed the return of water to the soil surface where its velocity increased by a factor of from 100 to 500. Without the impedance of many feet of subsurface flow, water from larger areas of hillside could contribute runoff to the stream channel. In addition to this return flow, direct precipitation onto the saturated area of the hillside contributed significant amounts of runoff, varying from 24% to 58% of the peak rate of surface runoff from the plot. Although the soil profile became completely saturated in most artificial storms and the hillside was steep (30%–100%), subsurface storm flow was a minor contributor to the storm hydrograph. Only overland flow on saturated soils could contribute large amounts of runoff to the stream channel during the period of a storm. Furthermore, only runoff from this source showed the sensitivity to fluctuations of rainfall intensity that is characteristic of streamflow from small watersheds in the area studied.

Artificial storms on the channel area. Data from natural storms showed that the major part

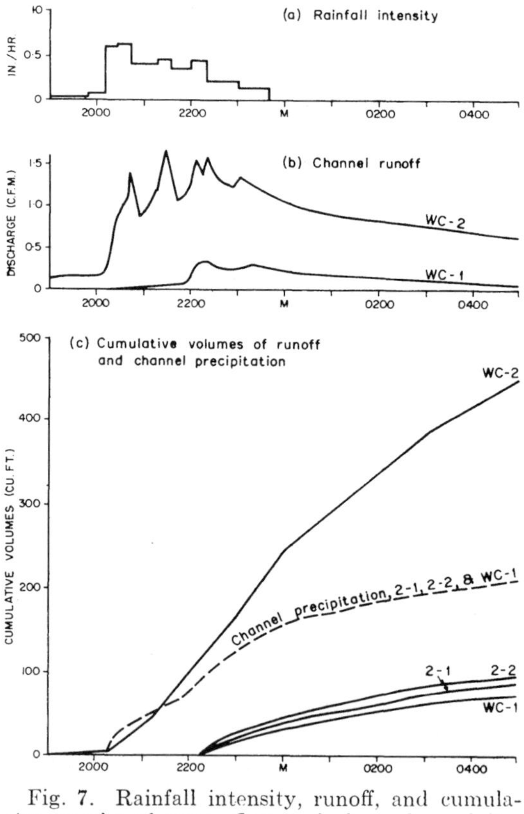

Fig. 7. Rainfall intensity, runoff, and cumulative graphs of storm flow and channel precipitation between weirs WC-1 and WC-2 for the 1.27-inch storm of October 10–11, 1967.

of storm runoff from the basin under study was generated in the area of the stream channel. Four artificial storms were therefore produced on a strip of land along the channel. The storms varied in duration from one half to six hours, and in intensity from 0.46 to 0.67 inch per hour. The specific aim of this study was to measure storm flow from this area, when the remainder of the watershed was not producing any storm runoff.

The irrigated reach of channel extended from the head of the stream to station WC-3 (Figure 2), a distance of 600 feet. Sprinkler heads were placed at 30-foot intervals along a pipe laid down the center of the channel. Because of the varying width of the stream, the strip of irrigated land on either side of the channel ranged from 15 to 40 feet. The total irrigated area was approximately 35,000 square feet. Rainfall was measured in 44 cans scattered over the valley floor and its amount was estimated from isohyetal maps. Distribution of rainfall over the stream depression was quite uniform, but toward the edges of the irrigated area there was a steep rainfall gradient. The intensities calculated for each storm were measured in the stream depression only. The other data collected during these storms were the gage records at stations WC-1, WC-2, and WC-3. The whole floor of the stream depression was covered with water and its area was measured.

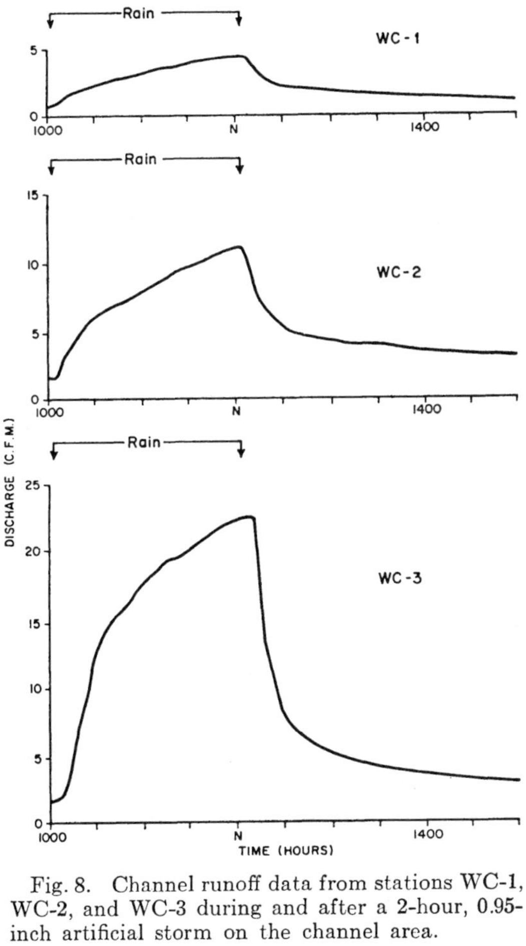

Fig. 8. Channel runoff data from stations WC-1, WC-2, and WC-3 during and after a 2-hour, 0.95-inch artificial storm on the channel area.

The hydrographs produced by irrigations of the channel area had many features similar to those from natural storms. As shown in Figure 8, the stream began to rise immediately after the onset of rainfall and reached a peak within a few minutes of the end of the storm. Recession limbs immediately after the peak were steeper than the rising limbs. This generalization was also true of hydrographs from natural storms that ended abruptly. On graphs of peak storm flow rate versus maximum 30-minute rainfall intensity, constructed from the few suitable natural storms available, points representing the two shorter artificial storms fell within the envelope for natural storms. The same was true for graphs of total rainfall versus total storm runoff. Points from the two longer storms fell outside the envelopes on these graphs because they had relatively high yields for a given rainstorm.

Some other features of the artificial storms differed from those of natural storms, mainly because of the duration of some of the former. Most of the hydrographs recorded in natural storms were produced by intense bursts of rain for periods of 30 minutes or less. The artificial storms on the channel area were less intense than many of these bursts, but most of them were longer. The increased duration of rainfall of uniform intensity allowed the runoff to approach a steady state in which the rate of runoff equaled the rate of rainfall over the whole area being irrigated. The rate of rainfall over the area above each weir and the peak rate of storm flow are presented in Table 1. They show the tendency for the ratio of rates of storm flow to rainfall to approach unity as the duration of the storm increases. In the longer storms, the rainfall rates below WC-1 seem to have been underestimated by a few percent.

TABLE 1. Peak Rates of Storm Flow* at Weirs WC-1, WC-2, and WC-3 and Rainfall Rates over the Irrigated Areas above Each Weir in the Artificial Storms on the Channel Area

	Above WC-1		Above WC-2		Above WC-3		
Duration of Storm, hrs	Rainfall Rate, cfm	Peak Storm Flow Rate, cfm	Rainfall Rate, cfm	Peak Storm Flow Rate, cfm	Rainfall Rate, cfm	Peak Storm Flow Rate, cfm	Approximate Return Period of the Storm
0.5	7.87	1.84	13.07	5.00	21.71	12.89	less than 1 year
1.0	9.47	2.89	15.60	6.91	27.87	16.28	less than 2 years
2.0	5.20	3.40	9.07	9.26	19.47	20.67	2 years
6.0	6.93	5.56	12.40	13.04	23.87	24.80	greater than 100 years

* Peak storm flow rate is peak runoff rate minus base flow rate.

Flow from the weirs WC-2 and WC-3 would be expected to approach a steady state earlier than that from WC-1 because the proportion of the irrigated area covered by the wet channel depression increased from 10% above WC-1 to 19% above WC-3. Also, much of the water falling onto the wet banks of the stream depression between WC-1 and WC-3 was observed running directly into the channel. In the reach above WC-1, a larger proportion of the rain fell onto shallow permeable soils that were not saturated, and displaced subsurface water which then entered the channel as slower seepage. Toward the end of the longer storms, however, even these became fully saturated and much of the rainfall was reaching the channel as overland flow. Comparison of the shapes of the hydrographs, and particularly of the recession curves from various gaging stations, reflects the differences in the character of the valley floor.

The role of channel precipitation in the formation of storm hydrographs was dominant in the shorter irrigations and became less important (though still significant) in the longer events. This was in agreement with observations in natural storms. Field measurements showed that channel velocities (of the order of 1000 to 2500 feet per hour) were greater than those of overland flow from the bank areas (of the order of 20 to 200 feet per hour) and of subsurface flow (of the order of 1 foot or less per hour). Runoff from rain falling onto the saturated floor of the stream depression, therefore, increased rapidly and came to a steady state more quickly than contribution from the banks. As the storm duration increased, a greater proportion of the runoff originated from overland and subsurface flow from the banks [*Dunne*, 1969]. Although it was not possible to separate these two contributions, field observation, data from the trenched hillside, and rough calculations of subsurface contributions showed that the subsurface contribution was small.

A simple routing technique was used to compare the temporal relationships of the various inflows to each gage during the artificial storms. Cumulative volumes of storm flow at each gage were plotted on a relative time scale (Figure 9). The position of the curves for WC-1 and WC-2 on this time scale were adjusted to show the approximate time at which flow from these upper gages contributed to the hydrograph at WC-3. Tracing of salt solutions (Calkins and Dunne, unpublished data, 1969) had established the fact that the velocity of flow through the steep, marshy depression was approximately constant over the range of discharges studied. The travel time through each reach was used to lag the cumulative volumes of flow from WC-1 and WC-2 behind the curve for WC-3. The cumulative curve of channel precipitation in each stretch of channel was then lagged by half the travel time for that reach (to represent an average time for the water to flow out of the reach during each time increment used in the computation) and added to the inflow from the upper gage.

In the early part of each storm, drainage of channel precipitation accounted for almost all of the storm flow produced. In the half-hour-long

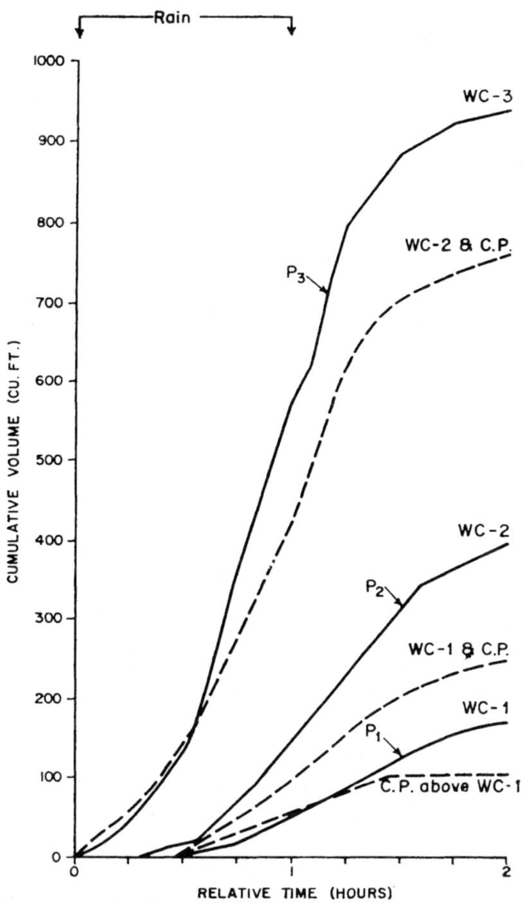

Fig. 9. Cumulative volumes of storm flow and channel precipitation (CP) during and after a 1-hour, 0.66-inch storm on the channel area. The graphs are plotted on a relative time scale, indicating the approximate time at which the upstream weirs and channel precipitation contributed flow to WC-3. P_1, P_2, and P_3 indicate the times of peak flow at weirs WC-1, WC-2, and WC-3, respectively.

storm, it dominated the hydrograph throughout. As the duration of the storms increased, lateral inflows (mainly overland flow from the banks) contributed larger amounts of storm runoff (see *Dunne* [1969] for other examples).

Artificial storms on the shallow soils of the seep B catchment. Evidence from natural rainstorms suggested that the hydrologic characteristics of the shallow silt loams covering the seep *B* catchment were intermediate between those of the trenched hillside and the valley floor. Storm runoff from the seep was produced by virtually all rains, but in spite of the rapid response of this runoff, its total yield was generally low. Three artificial rainstorms were produced on the lower half acre of the catchment draining to the seep. Their intensities (from 0.58 to 0.60 inch per hour) were less than many of the natural bursts of rainfall that produce hydrographs and their durations were greater (1, 2, and 3.5 hours). The storms had return periods varying from less than 2 to 50 years. Field observations and measurement indicated that the processes operating in these artificial storms were the same as those occurring in natural events.

At each of 16 stations (*A* to *P* in Figure 4), two piezometers were driven to depths of 2 and 4 feet. Two wells, one perforated to within a foot of the surface and one to within 0.25 foot of the surface, were also installed at each location. Piezometric and water table elevations were measured every few minutes throughout and after each storm. These measurements showed that significant subsurface flow was confined to the upper foot of soil. Only at the head of the channel (locations *G* and *H*) did sharp changes of piezometric head occur at depths of 2 feet. Here, weathering of the thin varved deposit and the underlying till was associated with the point of emergence of ground water. At these locations strong vertical potential gradients developed in each storm, and water issued from the soil surface. Elsewhere, the till and the varved silt below a depth of 1 foot were effectively impermeable during the course of a storm.

In each storm, a perched water table developed in the A-horizon of the soil. It rose quickly to the ground surface in the central, gently sloping part of the irrigated area. On the moderate slopes around this saturated area, such a water table either developed late in the storm, or did not occur at all where the A-horizon of the soil was deep and antecedent conditions were dry. The positions of the various wells were reflected in the order, magnitude, and rate of rise and recession of the water table. The water table reached the surface over an area of approximately 4360 square feet in the 2- and 3.5-hour-long storms, but over an area of only 1760 square feet in the one-hour-long storm (Figure 4). These areas were mapped a few minutes before the end of each storm.

Runoff in each storm (Figure 10) began within minutes of the onset of rainfall. Discharge increased rapidly, but generally at a declining rate. The peak discharge occurred 3 to 5 minutes after the end of rainfall and was followed by a sharp decline lasting 15 minutes. The first contribution of storm flow came from the channel area, which was receiving rainfall at a rate of 0.27 cfm. A perched water table developed in the A-horizon of the soil and reached the soil surface at the head of the channel within the first few minutes of rainfall. Return flow could be seen issuing from the soil surface and direct precipitation onto this saturated area also contributed to runoff. The area over which the water table reached the ground surface expanded rapidly through the gently sloping central depression during the first two hours of rainfall, but did not expand appreciably up the steeper sideslopes. As this saturated area increased, the contributions from both return flow and direct precipitation increased.

Estimates of the relative contributions from these two sources were made by calculating the rate of precipitation onto the saturated area and assuming that there was an approximate equilibrium between rainfall and runoff. Average rainfall intensities, antecedent conditions, and the behavior of the water table were almost identical in the three artificial storms. It was therefore assumed that the only major difference between runoff production in the three storms was the effect of rainfall duration. The contribution of direct precipitation to storm flow at the end of the two shorter storms could then be transferred to the longer storm, one and two hours after the beginning of rainfall (Figure 10). After the end of each storm the drainage of this direct precipitation by overland flow was assumed to have ended when the recession limb of the hydrograph abruptly changed its slope. The rough calculations gave consistent results in the various storms. The remainder of the runoff after subtraction of this direct precipitation was produced by a combination of flow that moved to the channel through the A-horizon of the soil and water that returned to the soil surface and reached the channel as overland flow. The recession limb of the separated hydrograph of subsurface and return flow was almost as steep as that of direct precipitation, suggesting that the former was also dominated by the drainage of overland flow, i.e., was mainly return flow. Rough estimates of the contribution of subsurface flow based on water table slope, and thickness and hydraulic conductivity of the topsoil placed the maximum flow rate into the channel at approximately 0.13 cfm. Peak rates of combined subsurface and return flow in the various storms varied from 3.81 cfm to 5.4 cfm, while peak rates of direct precipitation onto the saturated area varied from 2.45 cfm to 6.53 cfm. This evidence was taken to confirm the conclusion that subsurface flow could only contribute significantly to the stream hydrograph when it emerged from the soil surface as return flow and traveled to the stream at velocities characteristic of overland flow.

DISCUSSION OF RESULTS

Subsurface storm flow was not as important a contributor to the storm hydrograph as would be

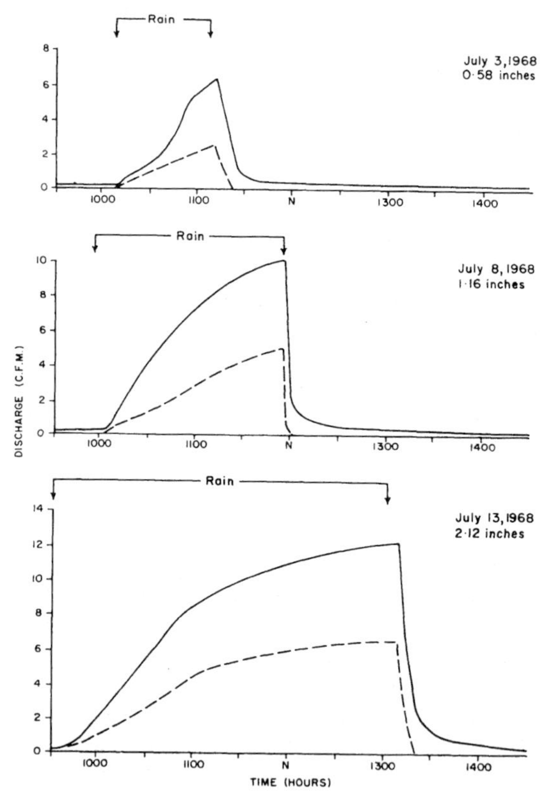

Fig. 10. Runoff from seep *B* during three artificial rainstorms. The dashed curves represent the contributions from direct precipitation onto the saturated area.

expected from much of the recent literature on runoff production in humid areas [*Roessel,* 1950; *Kirkby and Chorley,* 1967]. Large artificial storms on the watershed studied produced subsurface contributions that were measurable but small. Volumes of storm runoff from the trenched hillside, seep *B*, and the channel area were so large and responsive compared with subsurface flow that the latter did not even control the form of the recession limb of the hydrograph until discharge was almost back to its prestorm level (see Figure 10 in this paper and Figures 4, 6, 8, 9, and 10 in *Dunne and Black* [1970]). Storage and transmission beneath the soil heavily dampened the response of subsurface flow to rainfall. In storms of the kind that generally occur in northern Vermont, little subsurface storm flow was produced in this watershed. Under some rather specialized geologic conditions, or in watersheds of a markedly different size, subsurface storm flow, as defined by Kirkby and Chorley, might contribute significantly to storm flow. The conclusions outlined in this paper are based on a study in an area of well drained slopes and poorly drained valley bottoms and seeps, such as characterize much of Vermont. These conclusions were confirmed by a similar study in another basin of the Sleepers River experimental watershed. In a nearby area of Vermont, *Ragan* [1967] calculated larger subsurface contributions from a poorly drained floodplain on a highly permeable Pleistocene delta. But as a description of the major source of storm flow in humid regions, and as an important alternative to the Horton model in such regions, the subsurface storm flow model seems to be less useful than has been claimed.

The results of this study are in general agreement with the partial area concept of runoff production which can include the important factor of overland flow where soil and topographic conditions are favorable [*Betson,* 1964; *Ragan,* 1967]. Small areas in which the water table was at the surface of the ground contributed a high proportion of the rainfall applied to them. Runoff from these areas occurred mainly as overland flow. Storage within them was small and travel times out of them were short. Runoff from these sources was, therefore, largely controled by rainfall intensity; rates of rise and recession were extreme. To this extent runoff from these areas resembled that from urban watersheds.

At the beginning of a storm, the area over which the water table was at the surface of the ground was confined to the stream channel itself. Consequently, this area was the first to contribute runoff, and in short, intense storms such runoff virtually alone controlled the hydrograph from this small watershed. In longer storms, the seepy valley floor and the channel area of seep *B* also contributed most of the water supplied to them. In practically all summer storms these small areas were the only important source of the overland flow responsible for storm runoff.

Areas in which the water table was close to but not at the ground surface at the beginning of a storm also contributed large amounts of flow to the stream channel, particularly in heavy storms of long duration. The greater portion of the seep *B* catchment and the concave area of the trenched slope fall into this category. From these areas, subsurface flow was measured, but proved to be too small, too late, and too insensitive to changes of rainfall intensity to be an important contributor to the channel hydrograph. The inability of the soil to transmit most of the water supplied to it during large rainstorms was an important factor producing high rates of runoff from these areas. As the soil stored the water applied to it, the water table rose to the surface of the ground over small areas. When this occurred, water emerged from the soil surface and ran quickly to the channel as overland flow. Although this return flow had the same origin as flow that remained within the soil, its reaction to rainfall and importance to storm runoff were very different [*Dunne and Black,* 1970]. Its velocity was measured to be 100 to 500 times greater than the velocity of subsurface flow, allowing a much greater area to supply runoff to the channel during the storm. When rainfall ceased, the water table fell quickly below the soil surface and runoff rates were drastically reduced. The amount of rainfall required to bring the water table to the surface in a particular area of the watershed depended on topographic position, soil profile characteristics, depth to water table, antecedent condition of the topsoil, and the intensity and duration of rainfall. These were also the most important controls of runoff production, for the importance of an area as a contributor to the storm hydrograph depended on its ability to produce overland flow.

The small areas of hillside from which return flow occurred were in effect an expanded channel system. Direct precipitation onto them moved quickly to the stream as overland flow. Expansion of the channel area in this watershed by up to 170% during storms caused a large addition of runoff to the channel from this source. Figure 11 is a map of the areas contributing overland flow to the storm hydrograph as return flow and direct precipitation. It shows the differing partial areas in short intense summer storms and in heavy storms of longer duration, such as the artificial storms described above. The results are an experimental verification of a conceptual model used by hydrologists of the *Tennessee Valley Authority* [1965] as a basis for their mathematical model. They assume that in a humid region, the portion of a watershed contributing storm runoff can expand or contract during a rainstorm. This model is essentially the same as the 'variable source-area concept' originally proposed for the Coweeta watersheds by *Hewlett and Hibbert* [1965], although these writers do not emphasize the importance of the threshold between the surface and subsurface flow systems. Differences between their conceptual model and the present experimental findings may be the result of pedalogic, meteorologic, or other differences between Georgia and New England.

Results of the present study show that the partial area or source area varies seasonally between rainstorms and during rainstorms. The

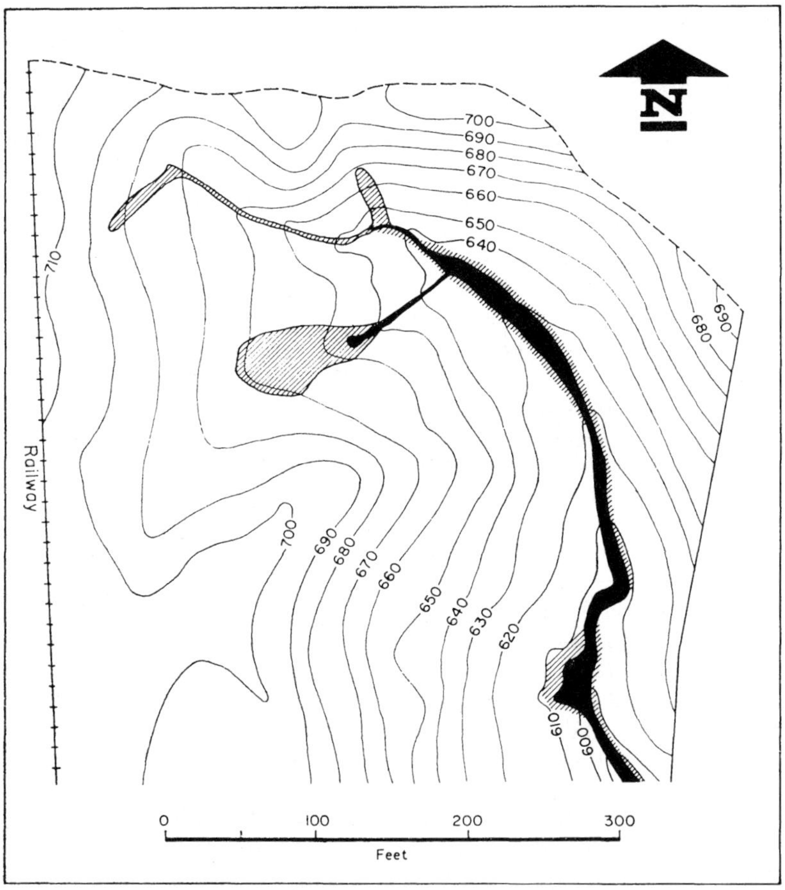

Fig. 11. Areas contributing storm flow: variation between short intense summer storms, and extreme autumn storms on wet conditions.

dynamic nature of this partial area may explain the value of an antecedent precipitation index in predicting storm runoff from humid areas, where the more familiar relationship between antecedent moisture and infiltration rate [*Linsley et al.*, 1949] seems irrelevant. The distribution and extent of the saturated areas in the present watershed and in other basins of the Sleepers River experimental watershed are closely related to the pattern of stream channels. This is probably why various workers have had some success in relating drainage density to flood discharges and water yields in humid areas [*Carlston*, 1963; *Burford and Lillard*, 1966]. Detailed examination and mapping of the pattern of small partial areas in Vermont, Maryland, England, and Kenya, however, indicates that small scale effects of geologic structure can play an important part in their localization. This agrees with *Amerman*'s [1965] description of the location of runoff producing areas in Ohio.

Within the glaciated upland region represented by the Sleepers River experimental watershed, the juxtaposition of deep permeable well-drained soils and shallow poorly drained soils is a common feature and a major control of runoff production. Within this watershed, runoff is measured from 17 basins ranging in size from 0.18 to 43 square miles. Within these basins, streams draining areas of a few acres on different soils have been gaged at various times. No detailed analysis of runoff from these areas has been made, but a preliminary survey of the data confirms the conclusions drawn from the present study. Runoff from the basins seems to have been generated on small areas of poorly drained soil, seeps, and ill-defined marshy channels close to the stream channel, which resemble the valley bottom and lower side slopes of the small watershed described in this report. Runoff rates from these areas are strongly controlled by rainfall intensity and reach the channel quickly. As this runoff proceeds downstream, channel storage in the main streams reduces the sensitivity of measured runoff to rainfall fluctuations. This change is demonstrated by a series of hydrographs from different points on the same channel of the Sleepers River (Engman, unpublished data, 1965). As storm flow from small partial areas moves downstream, individual peaks lose their identity and the resultant hydrograph is controlled by storage and translation in the channel. The hydrographs from the larger watershed have simpler shapes and long recession limbs reflecting the drainage from the channel system of water delivered to it by the small runoff producing areas that are remote from the gaging station. As the duration of this recession is so long in the larger watersheds, time is available for the contribution of some subsurface storm flow, but the results of the present and other experimental studies (for example, *Whipkey* [1965]) indicate that this contribution is relatively unimportant, especially near the peak of the hydrograph, in small watersheds.

CONCLUSION

In the upland watersheds of Vermont the major portion of storm runoff seems to be produced as overland flow on small saturated areas close to streams. The remainder of the watershed acts mainly as a reservoir during storms, and between storms it supplies base flow and maintains the wet areas that produce storm flow. Runoff from these wet areas is supplied by water escaping from the ground surface to reach the channel as overland flow and by direct precipitation onto the saturated area, which is essentially an expanded stream system. The runoff process depends on the operation of several continuous, interdependent flow systems, which though difficult to separate in the field except by detailed experiment, have very different reactions to rainfall. The most important transition between flow systems seems to be at the soil surface, when water is released from the extreme damping effect of subsurface flow. The importance of an area as a source of storm runoff, therefore, depends on its ability to produce overland flow.

The area contributing overland flow is dynamic in the sense that it may vary seasonally or throughout a storm. The fluctuations of this partial area can be rationally related to topography, soils, antecedent moisture, and rainfall characteristics. Different soils of a watershed contribute most of their storm runoff by the same process, but with differing frequencies. Thus, the channel areas below WC-1, the upper end of the channel area above WC-1, much of the seep *B* catchment, the central concave area of the trenched hillside, and the straight trenched slope can be viewed as potential contributing areas arranged along a scale of rainfall return period at intervals of approximately 10^{-2}, 10^{-1}, 10^{0}, 10^{1} to 10^{2}, and 10^{2} to 10^{3} years, respectively. Hydrographs of overland flow from these small areas reflect the hyetograph in a manner analogous to those from urban source areas. Storage and translation within the main channel system of a basin converts the effects of these hydrographs into a more familiar attenuated form.

The partial area concept provides an attractive alternative to the Horton and subsurface storm flow models as the basis for a study of storm runoff production in an area such as Vermont. It also provides a point of departure for models of catchment behavior not based on infiltration theory; for a new look at such hydrologic concepts as the unit hydrograph, time of concentration, and time–area diagrams; and for studies of nutrient losses from watersheds as well as sources of agricultural water pollution.

Acknowledgments. This paper is a contribution from the Northeast Branch, Soil and Water Conservation Research Division, Agricultural Research Service, U. S. Department of Agriculture in cooperation with the Vermont Agricultural experiment station. The project was initiated with the encouragement of M. L. Johnson. Data reduction and analysis were greatly facilitated by computer programs written by G. H. Comer, who together with D. J. Calkins, R. L. Hendrick, and S. H. Kunkle also assisted at many stages of the field work and in the discussion of results.

REFERENCES

Amerman, C. R., The use of unit-source watershed data for runoff prediction, *Water Resour. Res.*, *1*(4), 499–508, 1965.

Betson, R. P., What is watershed runoff?, *J. Geophys. Res.*, *69*(8), 1541–1552, 1964.

Burford, J. B., and J. H. Lillard, Relation of selected characteristics to hydrologic performance of two small watersheds, *Amer. Soc. Agr. Eng. Trans.*, *9*, 394–397, 1966.

Carlston, C. W., Drainage density and streamflow, 8 pp., *U. S. Geol. Surv. Prof. Pap. 422-C*, 1963.

Dunne, T., Runoff production in a humid area, Ph.D. dissertation, Johns Hopkins University, Baltimore, 1969.

Dunne, T., and R. D. Black, An experimental investigation of runoff production in permeable soils, *Water Resour. Res.*, *6*(2), 179–191, 1970.

Hewlett, J. D., and A. R. Hibbert, Moisture and energy conditions within a sloping soil mass during drainage, *J. Geophys. Res.*, *68*(4), 1081–1087, 1963.

Hewlett, J. D., and R. A. Hibbert, Factors affecting the response of small watersheds to precipitation in humid areas, paper presented at the International Symposium on Forest Hydrology, Pennsylvania State University, University Park, 1965.

Horton, R. E., The role of infiltration in the hydrologic cycle, *Trans. Amer. Geophys. Union*, *14*, 446–460, 1933.

Horton, R. E., Erosional development of streams and their drainage basins, *Bull. Geol. Soc. Amer.*, *56*, 275–370, 1945.

Kirkby, M. J., and R. J. Chorley, Overland flow, throughflow, and erosion, *Int. Ass. Sci. Hydrol. Bull.*, *12*(3), 5–21, 1967.

Linsley, R. K., M. A. Kohler, and J. H. L. Paulhus, *Applied Hydrology*, 689 pp., McGraw-Hill, New York, 1949.

Ragan, R. M., An experimental investigation of partial area contributions, *Int. Ass. Sci. Hydrol. Proc. Berne Symposium*, 1967.

Roessel, B. W. P., Hydrologic problems concerning the runoff in headwater regions, *Trans. Amer. Geophys. Union*, *31*(3), 431–442, 1950.

Tennessee Valley Authority, Area–stream factor correlation, a pilot study in the Elk River basin, *Int. Ass. Sci. Hydrol. Bull.*, *10*(2), 22–37, 1965.

Whipkey, R. Z., Subsurface stormflow from forested slopes, *Int. Ass. Sci. Hydrol. Bull.*, *10*(2), 74–85, 1965.

(Manuscript received February 18, 1970; revised June 24, 1970.)

Commentary on Paper 16

Osborn, H. B. & Renard, K. G. (1970) Thunderstorm runoff on the Walnut Gulch experimental watershed, Arizona, USA.

This paper represents one of the early published analyses of runoff from Walnut Gulch, the classic example of an ephemeral semiarid basin in the American Southwest (see http://www.tucson.ars.ag.gov/unit/Watersheds/WGEW.htm). The basin experiments have, for many years, been supported by the USDA Agricultural Research Service group in Tucson. Both Herb Osborn and Ken Renard worked with the ARS. The first raingauges were installed in Walnut Gulch in 1954 and it continues to be the subject of research into rainfall patterns (Goodrich *et al.*, 1995), runoff generation (Goodrich *et al.*, 1994), the hydraulics of surface runoff (Abrahams *et al.*, 1994), erosion and sediment transport (Abrahams & Parsons, 1991; Abrahams *et al.*, 1995; Nearing *et al.*, 1999), evapotranspiration (Menenti & Ritchie, 1994; Hultine *et al.*, 2003), and the use of remote sensing in hydrology (Kustas *et al.*, 1993; Goodrich *et al.*, 1995; Anderson *et al.*, 1997; Houser *et al.*, 1998).

Modelling the responses of Walnut Gulch continues to be a real challenge for hydrological modellers (Grayson *et al.*, 1992; Michaud & Sorooshian, 1994; Goodrich *et al.*, 1994, 1997; Faurès *et al.*, 1995; Sivapalan *et al.*, 2002). A glance through this paper will indicate why. All the channels in the catchment are normally dry; there will be only 10–15 runoff events per year on the larger catchments and, because of the variability of the rainfall, smaller catchments will have less. The larger channels in the catchment, which are dry for 99% of the time, are losing water into their beds during storm events so that transmission losses are important in modelling the storm hydrograph magnitude and shape.

Ken Renard was Director of the USDA ARS Laboratory for more than 30 years until his retirement in 1994. Herb Osborn also continued to work at Tucson until his retirement. Both still work with current ARS staff as retired collaborators, and are contributing to a volume to celebrate the 50th anniversary of Walnut Gulch that is expected to be published in 2007.

References

Abrahams, A. D. & Parsons, A. J. (1991) Relation between infiltration and stone cover on a semi-arid hillslope. *J. Hydrol.* **122**, 49–59.

Abrahams, A. D., Parsons, A. J. & Wainwright, J. (1994) Resistance to flow on semi-arid grassland and shrubland hillslopes, Walnut Gulch, southern Arizona. *J. Hydrol.* **156**, 431–446.

Abrahams, A. D., Anthony, J. P. & Wainwright, J. (1995) Effects of vegetation change on inter-rill runoff and erosion, Walnut Gulch, Southern Arizona. *Geomorph.*. **13**, 37–48.

Anderson, M. C., Norman, J. M., Diak, G. R., Kustas, W. P. & Mecikalski, J. R. (1997) A two-source time-integrated model for estimating surface fluxes using thermal infrared remote sensing, *Remote Sens. Environ.* **60**, 195–216.

Faurès, J.-M., Goodrich, D. C., Woolhiser, D. A. & Sorooshian, S. (1995) Impact of small-scale rainfall variability on runoff modelling. *J. Hydrol.* **173**, 309–326.

Goodrich, D. C., Schmugge, T. J., Jackson, T. J., Unkrich, C. L., Keefer, T. O., Parry, R., Bach, L. B. & Amer, S. A. (1994) Runoff simulation sensitivity to remotely sensed initial soil moisture content. *Water Resour. Res.* **30**, 1393–1405.

Goodrich, D. C., Faurès, J. M., Woolhiser, D. A., Lane, L. J. & Sorooshian, S. (1995) Measurement and analysis of small-scale convective storm rainfall variability. *J. Hydrol.* **173**(1–4), 283–308.

Goodrich, D. C., Lane, L. J., Shillito, R. M., Miller, S. N., Syed, K. H. & Woolhiser, D. A. (1997) Linearity of basin response as a function of scale in a semi-arid watershed. *Water Resour. Res.* **33**(12), 2951–2965.

Grayson, R. B., Moore, I. D. & McMahon, T. A. (1992) Physically-based hydrologic modelling. 2. Is the concept realistic? *Water Resour. Res.* **28**, 2659–2666.

Houser, P. R., Shuttleworth, W. J., Famiglietti, J. S., Gupta, H. V., Syed, K. H. & Goodrich, D. C. (1998) Integration of soil moisture remote sensing and hydrologic modeling using remote sensing. *Water Resour. Res.* **34**(12), 3405–3420.

Hultine, K. R., Williams, D. G., Burgess, S. S. O. & Keefer, T. O. (2003) Contrasting patterns of hydraulic redistribution in three desert phreatophytes. *Oecologia* **135**(2), 167–175

Kustas, W. P., Schmugge, T. J., Humes, K. S., Jackson, T. J., Parry, R., Weltz, M. A. & Moran, M. S. (1993) Relationships between evaporative fraction and remotely-sensed vegetation index and microwave brightness temperature for semi-arid rangelands. *J. Appl. Met.* **32**, 1781–1790.

Menenti, M. & Ritchie, J. C. (1994) Estimation of effective aerodynamic roughness of Walnut Gulch watershed with laser altimeter measurements. *Water Resour. Res.* **30**(5), 1329–1338.

Michaud, J. & Sorooshian, S. (1994) Comparison of simple versus complex distributed runoff models on a mid-sized semiarid watershed. *Water Resour. Res.* **30**, 593–606.

Nearing, M. A., Simanton, J. R., Norton, L. D., Bulygin, S. J. & Stone, J. (1999) Soil erosion by surface water flow on a stony, semiarid hillslope. *Earth Surf. Process. Landf.* **24**(8), 677–686.

Sivapalan, M., Jothityangkoon, C. & Menabde, M. (2002) Linearity and nonlinearity of basin response as a function of scale: discussion of alternative definitions. *Water Resour. Res.* **38**(2), doi: 10.1029/2001WR000482.

THUNDERSTORM RUNOFF ON THE WALNUT GULCH EXPERIMENTAL WATERSHED, ARIZONA, U.S.A. *

H. B. OSBORN and K. G. RENARD,**

SUMMARY

The Southwest Watershed Research Center of the Agricultural Research Service operates the 58-square-mile (148 km*2*) Walnut Gulch Experimental Watershed in southeastern Arizona. This watershed is fairly representative of several hundred thousand square miles of mixed grass and brush rangeland in the Southwestern United States. Rainfall and runoff records have been collected from the watershed since 1954. The present network contains 95 weighing-type recording rain gages and 21 permanent runoff measuring stations.

About 70 percent of the average annual precipitation of 11.5 inches (29.2 cm), and almost all runoff, occurs in the summer thunderstorm season from June through September. Rainfall variability in time and space dominates any rainfall-runoff relationship, especially on the smaller subwatersheds (up to several square miles in area). Flow abstractions in the normally dry ephemeral channels become increasingly important with increasing watershed size. Whereas onsite runoff may average 20 to 25 percent of summer rainfall, runoff from the 58-square-mile (148 km*2*) watershed averages only about 4 percent of the summer rainfall. Correlation between Thiessen-weighted rainfall and runoff for individual events decreases with increasing watershed size because of the limited areal extent of thunderstorm rainfall and because of the increasing magnitude of channel abstractions for the larger watersheds.

The effects of transmission loss on the channel flows, as represented by changes in hydrograph shape, height, and volume, are being studied. Outflow hydrographs have been predicted for a 4-mile (6.4 km) reach of Walnut Gulch, with no tributary input, using regression techniques and a three-parameter gamma distribution. This work is being expanded to include other channel reaches with varying physical dimensions.

The "key gage" principle is not satisfactory for the Walnut Gulch watershed for representing either storm or annual runoff. Peak discharges have been highest following relatively dry periods, which is probably not coincidence. The larger thunderstorms have followed dry periods when convective heating of the ground surface is possible because of less cloud cover and drier surface conditions.

RÉSUMÉ

ÉCOULEMENT D'ORAGE SUR LE BASSIN EXPÉRIMENTAL DE WALNUT GULCH EN ARIZONA

Le Centre de Recherches sur bassins du Sud-Ouest de l'ARS conduit les recherches sur le bassin de 58 milles carrés (148 km^2) de Walnut Gulch dans le Sud-Ouest de l'Arizona. Ce bassin est assez bien représentatif de plusieurs milliers de milles carrés de terrains couverts d'herbes ou de buissons dans le Sud Ouest des États-Unis. Les précipitations et l'écoulement ont été relevés sur ce bassin depuis 1954. Le réseau actuel comprend 95 pluviographes à pesée et 21 stations de jaugeage permanents.

Environ 70% de la précipitation annuelle de 11,5 pouces (29,2 cm) et presque tout l'écoulement se produit dans la saison des orages d'été de juin à septembre. La variabilité des précipitations dans le temps et l'espace domine toute relation pluie-écoulement, spécialement dans les plus petits sous-bassins (grands de quelques milles carrés au maximum). Les obstructions à l'écoulement dans les chenaux éphémères normalement asséchés deviennent de plus en plus

* Contribution of the Southwest Watershed Research Center, Agricultural Research Service, Soil and Water Research Division, USDA, in cooperation with Arizona Agricultural Experiment Station, University of Arizona, Tucson, Arizona.

** Research Hydraulic Engineers, Southwest Watershed Research Center, Agricultural Research Service, USDA, SWC, 442 East Seventh St., Tucson, Arisona 85705.

importantes avec l'étendue croissante du bassin. Alors que l'écoulement sur place peut valoir 20 à 25% en moyenne des pluies d'été, l'écoulement du bassin 58 milles carrés (148 km^2) n'atteint en moyenne que 4% de la pluie d'été. La corrélation entre la pluie donnée par la méthode Thiessen et l'écoulement pour des pluies individuelles décroît avec l'étendue des bassins du fait de l'étendue limitée des pluies d'orage et du fait aussi de l'accroissement des obstructions pour les grands bassins.

Les effets des pertes en transmission sur les écoulements des chenaux se traduisent par des modifications de l'hydrogramme en forme, hauteur et volume : elles ont été étudiées. Les hydrogrammes de sortie ont été prévus pour une section de 4 milles (6,4 km) de Walnut-Gulch, sans arrivée de tributaire, en utilisant une technique de régression et une distribution gamma à trois paramètres. Ce travail sera continué pour d'autres sections avec des conditions physiques différentes. A remarquer que les plus forts orages ont suivi des périodes relativement sèches quand le chauffage convectif de la surface du sol est possible du fait de la moindre nébulosité et des conditions plus sèches de surface.

INTRODUCTION

The Southwest Watershed Research Center of the Agricultural Research Service operates the 58-square-mile (148 km^2) Walnut Gulch Experimental Watershed in southeastern Arizona, U.S.A. (fig. 1). Two-thirds of the rangeland watershed is dominated by brush; one-third by grass. There is no cultivation. Most of the watershed is grazed by cattle, and a small part is urban (the town of Tombstone). The Walnut Gulch watershed is fairly representative of several hundred thousand square miles of semiarid rangeland in the Southwestern United States, as well as of other similar semiarid rangelands throughout the world.

Rain gages were first placed on the watershed in 1954, and at present precipi-

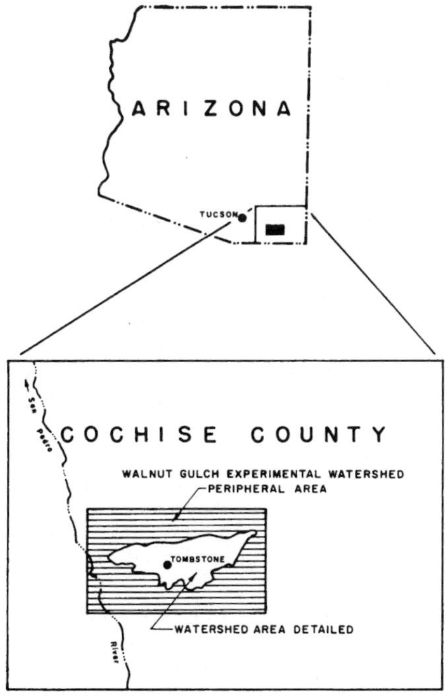

FIGURE 1. *Walnut Gulch Experimental Watershed location map.*

tation is measured with a network of 95 weighing-type recording rain gages (fig. 2). The principal watershed is subdivided into 20 subwatersheds. Surface runoff from these smaller units, as well as from the principal watershed, is determined with continuous water-level recorders at concrete control structures on the ephemeral channels. There are also several groups of plot studies, where the effects on onsite runoff of such variables as vegetation, soils, and surface conditions are studied more closely (Kincaid, Osborn, and Gardner, 1966; Schreiber and Kincaid, 1967).

WATERSHED CHARACTERISTICS

The Walnut Gulch watershed is butterfly shaped with the maximum length about 13 miles (21 km) and the maximum width about 5 miles (8 km). The watershed outlet lies at an elevation of just under 4000 feet (1200 meters) above mean sea level, while the upper reaches of the watershed rise to over 6000 feet (1800 meters). The watershed is drained by an extensive and well-defined channel system. Channels are clearly defined on even the smallest instrumented watershed (0.5 acre or 0.002 km^2). Onsite runoff moves relatively rapidly into well-defined channels.

The lower two-thirds of the watershed is dominated by native shrubs, while the upper end is primarily native grasses. The watershed is generally rolling with very little land suitable for cultivation. The major channels have about a one-percent slope. The channels are generaly filled with highly porous, relatively coarse-grained alluvial material. Infiltration rates at the beginning of flow often exceed 10 inches per hour (25.4 cm per hour) and decreases during the flow to about 1 to 2 inches per hour (2.54 to 5.08 cm per hour).

PRECIPITATION

In southeastern Arizona precipitation is seasonal. In the summer, the combination of moist air moving into the region from the Gulf of Mexico and strong convective heating produces thunderstorms. Most thunderstorms occur in the late afternoon or early evening and produce high-intensity, short-duration rains of limited areal extent. Winter storms, on the other hand, result from weak cold fronts and produce relatively low-intensity rainfall [generally less than 0.1 inch per hour (0.25 cm/hr)]. Occasionally in the late summer, a tropical storm off Baja California forces additional moist air into the region, which may increase thunderstorm potential.

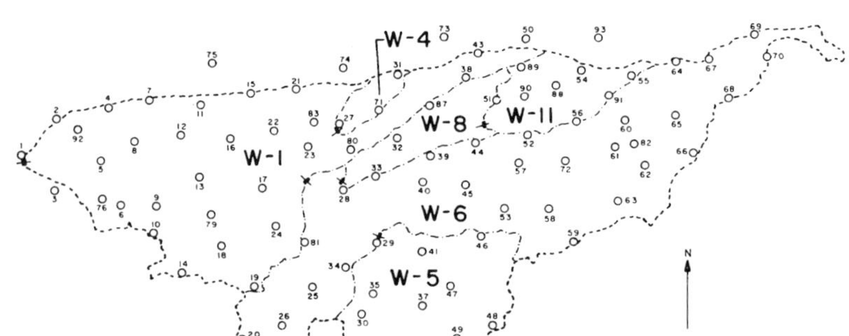

FIGURE 2. *Walnut Gulch Experimental Watershed*

On Walnut Gulch about 70 percent of the annual precipitation of 11.5 inches (29.2 cm) occurs from June through September (Osborn and Hickok, 1968). Summer rainfall is highly variable from year to year as well as over the watershed. Generally, the highest summer point rainfall is more than twice the lowest. For example, about 10.5 inches (26.7 cm) and 4.8 inches (11.9 cm) were the maximum and minimum recorded point rainfall depths for the summer of 1963. Storm rainfall varies much more, with some parts of the watersheds generally remaining dry while other parts are receiving runoff-producing rainfall. The extreme variability of thunderstorm rainfall is by far the major factor in determining rainfall-runoff relationships.

Runoff

Onsite Runoff

Because of the high intensities of rainfall, onsite runoff normally commences within minutes after the rain begins. Rainfall intensities can exceed infiltration rates by 10 to 1, and infiltration rates during the short-lived storms depend generally on the condition and type of soil in the top few surface inches. Runoff-producing rainfall (roughly, rain with intensities exceeding 0.5 inch per hour, or 1.27 cm per hour) seldom lasts for more than 30 minutes at any one point. In general, two-thirds of the total storm rainfall occurs in the first 20 minutes of the storm (Osborn and Reynolds, 1963). Onsite runoff may average as much as 20 to 25 percent of summer rainfall.

Channel Flow

The ephemeral sand channels on the Walnut Gulch watershed are dry most of the year. Surface flows are limited generally to a relatively few afternoons and evenings during the summer. Approximately 10 to 15 runoff events might be expected annually

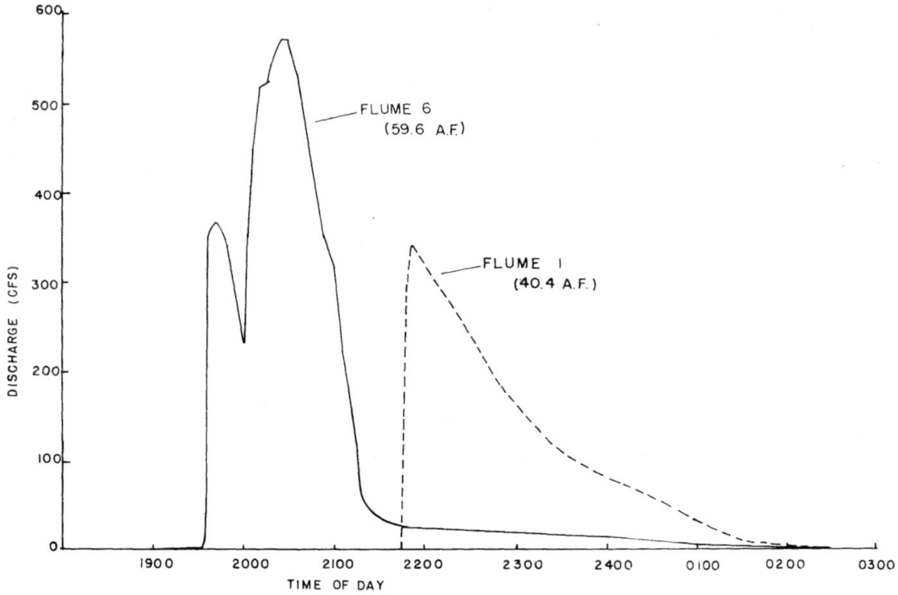

FIGURE 3. *Walnut Gulch Experimental Watershed hydrographs for August 2, 1964, Flumes 1 and 6*

on a 30-to 50-square-mile (74 to 127 km^2) watershed. Fewer events are recorded on the smaller watersheds. For the larger watersheds of 30 to 50 square miles (74 to 127 km^2), the duration of annual storm streamflow is about 2 to 3 days. In other words, the channels are dry over 99 percent of the time.

The average duration of flow for an individual storm event is relatively short on the very small watersheds—on the order of one hour. It is considerably longer on the larger watersheds—on the order of 6 hours. Variations in durations between similar sized watersheds are primarily due to differences of the alluvial channels, both in areal extent and in confining layers below the alluvium. Occasionally, local surface conditions such as urban development (Tombstone) or impervious rock

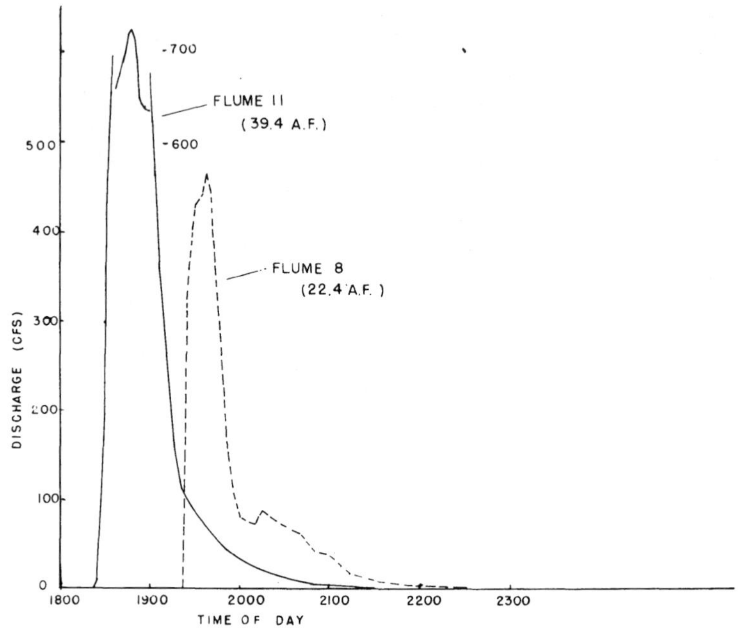

FIGURE 4. *Walnut Gulch Experimental Watershed hydrographs for August 2, 1964, Flumes 8 and 11*

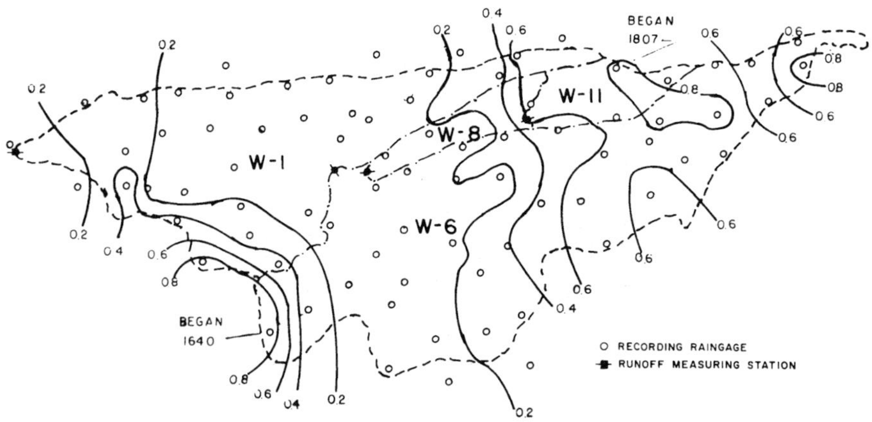

FIGURE 5. *Walnut Gulch Experimental Watershed rainfall on August 2, 1964*

barriers may influence the duration or number of events recorded at a particular station.

Because the streambeds are dry most of the year and because of the large volumes of alluvial material in the streambeds, runoff reductions are large as the water moves downstream. These flow reductions, or transmission losses, are extremely important in evaluating the hydrologic response of ephemeral streams in the Southwestern United States.

Transmission losses are measured on Walnut Gulch by comparing the inflow and outflow hydrographs for individual storms for channel reaches having little or no unmeasured tributary inflow. Because of the limited areal extent of the runoff-producing precipitation and the density of runoff-measuring sites, losses are evaluated from several storms in most years. The large critical-depth flumes measuring runoff from the larger complex watersheds [drainage areas >560 acres (220 hectares)] create seven channel segments with average bottom widths varying between 36 feet (12 meters) and 217 feet (71 meters) and channel lengths varying between 0.9 and 4.0 miles (1.4 and 6.4 km). Data show that the transmission losses are related to: (1) channel length and wetted perimeter; (2) hydrograph characteristics; (3) antecedent moisture conditions of the alluvium; (4) volume and porosity of the alluvium; (5) amount of clay in suspension in the water infiltrating into the alluvium.

Examples of transmission losses are shown in figures 3 and 4 for the storm of August 2, 1964. The rain on the lower portion of the watershed as shown on the isohyetal map (fig. 5) began at 1640 and produced only a trace of runoff at the watershed outlet (Flume 1) between 1900 and 2000 hours (fig. 3). The storm on the upper portion of the watershed began at 1807 hours and produced the runoff beginning at 1820 hours at Flume 11, as well as a significant amount of runoff in other tributaries above Flume 6. The additional drainage area between Flumes 11 and 8 probably produced very little flow. Thus, the 726-cfs (25.6 m^3/sec) peak discharge and 39.4 acre-feet (890 m^3) of runoff were reduced to the 463-cfs peak discharge and 22.4 acre-feet (505 m^3) of runoff in the 4 miles (6.4 km) of channel between the two flumes. The runoff was further reduced in the 0.9 mile (1.4 km) of channel between Flumes 8 and 6 to a peak discharge of 365 cfs (12.8 m^3/sec), as indicated by the first

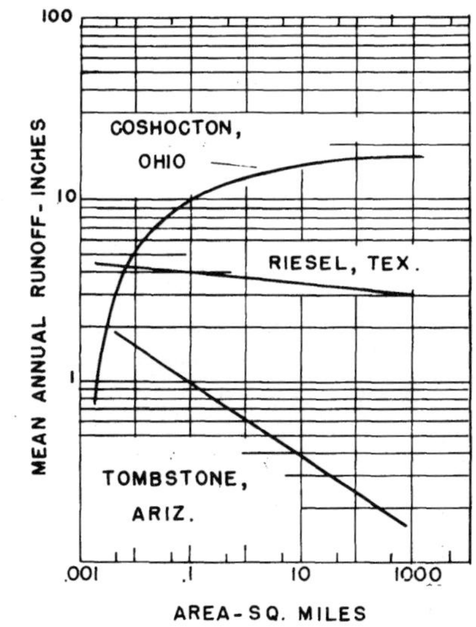

FIGURE 6. *Relationship between mean annual runoff and area*

peak on the recorded hydrograph. The double-peak hydrograph at Flume 6 resulted from runoff in the unmeasured tributaries above Flume 6 combining with the runoff from subwatershed 8.

The outflow hydrograph at Flume 1 demonstrates the effects of transmission losses and other channel hydraulic conditions on the hydrograph shape. The double-peak hydrograph at Flume 6 was changed to the triangular shape at Flume 1 as the flow traversed 6.8 miles (11 km) of relatively coarse-textured alluvium. The transmission losses, which are very large during early portions of the runoff, produce the truncation of the rising side of the hydrograph. Thus, for storm events traversing long reaches of stream channel, the rising limb of the hydrograph can be represented as a straight line, and the receding limb can be represented by an exponential decay (Renard and Keppel, 1966).

Figure 6 from a paper by Glymph and Holtan (1969) shows that the relationship between mean annual runoff and area changes for different physiographic and climatic provinces. In the semiarid Southwest, the streambeds absorb all of the runoff from many storms. There is little base flow. The transmission losses thus affect runoff per unit area, causing a decrease in runoff with increasing drainage area. In other areas such as Coshocton, Ohio, the runoff increases with increasing drainage area because of groundwater discharge.

Mathematical modeling of transmission losses has been limited. Recently, however, Lane, Diskin, and Renard (submitted for publication 1970) successfully predicted downstream hydrographs using regression techniques and a three-parameter gamma distribution to represent the inflow and outflow hydrographs for the 4-mile (6.4 km) reach of channel between Flumes 11 and 8. Using the known parameters of the inflow hydrograph at Flume 11 and two fitting methods, outflow hydrographs were obtained at Flume 8 for the storm of August 2, 1964 (fig. 7). The predicted volume of runoff at Flume 8 was 22.9 acre-feet (510 m³), which agreed very well with the measured volume of 22.3 acre-feet (500 m³).

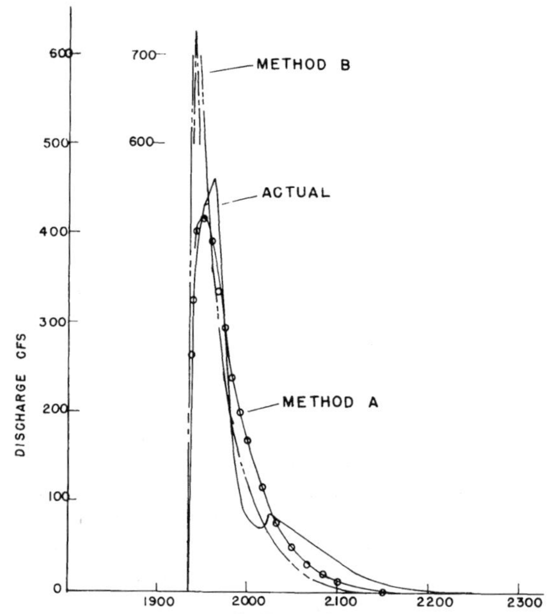

FIGURE 7. *Predicted and actual hydrographs for August 2, 1964, Flume 8*

Two methods were used to predict the gamma distribution parameters used to represent the outflow hydrograph from the properties of the inflow hydrograph. Method A involved the simultaneous solution of the equations for the mean and the mode of the gamma distribution to determine the parameters of the theoretical distribution. Method B involved the solution of the equations for the maximum ordinate and the interval between the mean and the mode to obtain the parameters of the theoretical distribution. In both methods, the third parameter of the theoretical distribution was equated to the time of start of the observed hydrograph.

In the illustration used, Method B overestimated the peak discharge, as well as displacing the peak toward the beginning of the storm (increased the skew). This was not true in all of the events used in the analysis, and the authors concluded that the two methods used to fit the gamma distribution to the outflow hydrographs were about equal in giving a good representation of the hydrograph.

Additional work is in progress using this procedure on additional channel segments of Walnut Gulch as well as on other ephemeral channels in the Southwestern United Stats. Hopefully this work will relate the coefficients of the regression equations with the physical characteristics of the streams to provide a regional description of transmission losses.

Correlation of Rainfall and Runoff

For very small watersheds, runoff can be directly correlated to point rainfall values by simple regression equations (Osborn and Lane, 1969). About 70 to 80 percent of the variability in peak discharge and runoff volume was explained by the variability in either the maximum 15-minute or total rain depths for 4 very small watersheds [0.5 to 11 acres (0.2 to 4.5 hectares)]. In general, the variability of thunderstorm rainfall so dominated the rainfall-runoff relationships that no other single variable appreciably improved the initial regression equation.

$$Q = a + bP_{15}$$

and

$$V = c + dP_{tot}$$

where

Q = peak discharge in inches per hour;
V = volume of runoff in inches;
P_{tot} = total storm rainfall in inches;
(a and c are both negative values).

For larger watersheds (over one square mile) runoff was best correlated to the maximum 30-minute rainfall for both peak discharge and total storm runoff. Simple correlations between the maximum 15- and 30-minute point rainfall values and peak (Q) and total (V) discharge are shown in table 1. As would be expected, the best correlations were for the very small watersheds.

For watersheds less than 120 acres (49 hectares), point rainfall from a single gage gave good rainfall-runoff correlations. For the 560-acre (220 hectares) watershed, W-4, with a 3-gage network, the correlations were poor, and the differences between using the central gage and average rainfall were not significant.

Correlations between point rainfall values and runoff for 4 larger watersheds, W-11, W-5, W-6, and W-1 also are shown in table 1. For each of these stations, 20 to 26 events were chosen from the period 1961-68. These events represented all but one of the major recorded runoffs for which there was good record. One event, which produced significant runoff at 3 of the 4 stations, was deleted since it occurred

TABLE 1. Coefficient of Determination of Rainfall and Runoff for Eight Walnut Gulch Watersheds.

			Coefficient of determination(R^2)					
			P15			*P30*		
Watershed	area	Runoff variable	Ave.	Max.	Central	Ave.	Max.	Central
	(acres)							
LH-5	0.5	V/A			0.81*			0.77*
		Q/A			0.85*			0.74*
LH-3	8.5	V/A			0.85*			0.85
		Q/A			0.88*			0.86*
K-1	120.	V/A	0.61	0.60	0.60			
		Q/A	0.72	0.72	0.71			
W-4	560.	V/A	0.48	0.37	0.41			
		Q/A	0.48	0.37	0.43			
W-11	2,030.	V/A				0.71	0.67	**
		Q/A				0.76	0.66	**
W-5	5,500	V/A				0.35	0.23	0.33
		Q/A				0.33	0.33	0.28
W-6	23,500	V/A				0.60	0.48	0.53
		Q/A				0.50	0.46	0.49
W-1	36,900	V/A				0.24	0.15	0.34
		Q/A				0.24	0.35	0.27

* Only one rain gage for the very small watersheds.
** No "central" gage on the W-11 watershed until 1966.

only 8 hours after another major event. All other events were separated by at least one day. In general, except for the occasional "double" event, antecedent rainfall and prestorm channel conditions account for less than 10 percent of the variability in runoff. Rainfall variability was by far the most significant variable for explaining runoff.

The results shown in table 1 probably are not conclusive, however. The records were relatively short and the range of runoff magnitude varied considerably between stations. For example, there were no "large" runoff events on W-4 between 1961 and 1968, whereas all other stations recorded a good range of storms from small to large. Also, there are three large stock ponds amounting to over 20 percent of the watershed area on W-5, which confounds the rainfall-runoff relationship there. Therefore, correlations for W-4 and W-5 would be expected to be considerably poorer than those for the other 6 watersheds, other factors remaining equal.

The correlations between rainfall and runoff for W-11, W-6, W-5, and W-1 could be rated as good, poor, poor, and nonexistent for the 4 watersheds, respectively. The most significant point here was the difference in correlation between W-6 and W-1. There was obvious correlation, although admittedly poor, between rainfall and runoff for W-6, and just as obvious a lack of correlation on W-1. The amount of difference would appear to rule out chance. Probably, the combined effect of the limited areal extent of the runoff-producing thunderstorm and the larger channel abstraction between Station W-6 and W-1 cause the difference.

Most major runoff events appear to cover from 25 to 35 square miles (65 to 91 km^2) with runoff-producing rainfall [over 0.5 inch (1.27 cm) in less than one hour]. Such events come much closer to covering the 37-square-mile (94 km^2) W-6 watershed than the 58-square-mile (147 km^2) W-1 watershed; therefore, an average rainfall value on W-6 is meaningful, whereas on W-1 it is not.

There was no "key" gage from which rainfall and runoff would be significantly correlated on the W-1 watershed. Furthermore, there was no "best" gage for rainfall-runoff correlation on the W-5 and W-6 watersheds, and the data were not sufficient to prove or disprove the "key" gage theory on W-11.

The largest runoff events with the highest peak discharges all occurred after relatively dry periods. The record is too short to determine conclusively whether this is by chance or has real meaning. However, other records, plus some conclusions concerning the mechanics, or thermodynamics, of thunderstorms, strongly suggest that it is not chance. Surface temperatures are generally lower following a period of runoff both because of the evaporative cooling at the ground level and the greater likelihood of cloud cover. The chances for thunderstorms to occur in the summer under such conditions are excellent, but there seems to be little chance of an extreme storm developing.

This, then, would eliminate the concept of the maximum flood occurring from the maximum probable rainfall on a saturated watershed. Furthermore, the "maximum" rainfall appears to be so much larger than what can occur on a saturated watershed that it would produce a significantly greater peak discharge than the largest possible rainfall on a previously saturated watershed.

REFERENCES

Kincaid, D.R., Osborn, H.B. and Gardner, J.L. (1966): Use of unit source watersheds for hydrologic investigations in the semiarid Southwest. *Water Resoursec Res.*, 2(3), 381-392.

Schreiber, H.A. and Kincaid, D.R. (1967): Regression models for predicting on-site runoff from short-duration convective storms, *Water Resources Res.*, 3(2), 389-395.

Osborn, H.B., and Hickok, R.B. (1968): Variability of rainfall affecting runoff from a semiarid rangeland watershed, *Water Resources Res.*, 4(1), 199-203.

Osborn, H.B. and Reynolds, W.N. (1963): Convective storm patterns in the Southwestern United States, *Internatl. Assoc. Sci. Hydrol. Bull.* 8(3), 71-83.

Renard, K.G. and Keppel, R.V. (1966): Hydrographs of ephemeral streams in the Southwest, *J. Hydraul. Div.*, ASCE 92 (HY-2), 33-52.

Glymph, L.M. and Holtan, H.N. (1969): Land treatment in agricultural watershed hydrology research. *In* Effects of watershed changes on streamflow, *Water Resources Symp. No. 2*, Univ. Texas Press, Austin, pp. 44-68.

Lane, L.J., Diskin, M.H. and Renard, K.G. (1970): Input-output relationships for an ephemeral stream channel system, *Submitted for publication in Water Resources Res.*

Osborn, H.B. and Lane, L.J. (1969): Precipitation-runoff relationships for very small semiarid rangeland watersheds, *Water Resources Res.*, 5(2), 419-425.

Commentary on Paper 17

Weyman, D. R. (1970) Throughflow on hillslopes and its relation to the stream hydrograph.

In the late 1960s, a rather remarkable group of PhD students were working in the Geography Department of the University of Bristol, in the UK, under the supervision of Mike Kirkby and D. Ingle (Dingle) Smith (later at the Centre for Resource and Environmental Studies at the Australian National University in Canberra). They included Malcolm Newson (later Professor at Newcastle University), Pete Smart (later Professor at Bristol University), Tim Atkinson (later Professor at University College, London), Steve Trudgill (later at Sheffield and Cambridge universities), and Darrell Weyman (who later became a priest). It was the period when the concepts of a quantitative physical geography, spreading from the work of Luna Leopold, Stan Schumm, Reds Wolman and others in the USA, through the teaching of Dick Chorley at Cambridge University, were really taking hold in the UK. Thus, many PhD students were set the task of taking measurements of processes in the field, rather than just interpreting the landscape in a qualitative way.

Malcolm Newson, Tim Atkinson and Steve Trudgill all made quantitative advances in karst hydrology and erosion processes; Darrell Weyman was looking at the hydrological processes in the small East Twin Brook catchment underlain by impermeable rocks on the Mendip Hills in Somerset, southwest England. His work resulted in three influential papers published in 1970, 1973 and 1974, any one of which could have been chosen to represent his classic work here.

Weyman adopted a similar experimental strategy to Dunne & Black (Paper 15) using nested V-notch weirs, a throughflow trough with collectors at different depths (but here only 1 m wide), and tensiometers and piezometers on the slope above the throughflow trough. He also attempted to collect the seepage from different 1-m lengths of channel bank, since one of the most interesting aspects of his study, and the reason why I have chosen the 1970 paper here, was the recognition of the dramatic variability in seepage rates along the channel (see top of p.31). Note also the references to the effect of slope convergence on the occurrence of saturation, throughflow, and runoff (p.32) and to the role of piping (p.31–32). Tim Atkinson (1978) later pointed out the potential effects of the throughflow trough on the flow on the slope and the potential for the effective contributing area to a trough to change under wet and dry conditions (especially for narrow troughs).

The East Twin proved to be a very interesting catchment with which to illustrate dynamic contributing area concepts. In the lower basin between the two discharge measurement sites, flow was dominated by subsurface throughflow and Weyman discusses the role of saturated and unsaturated flow in controlling the contributing area for throughflow under different antecedent conditions (see also Weyman, 1973,1974). In the upper basin, storm runoff was produced predominantly by saturation excess surface runoff, with a dynamic network of surface channels and saturated areas. These differences allowed Weyman (1974) to discuss a classification of runoff types for the lower basin. Type 1 was a small rapid peak thought to be due to rainfall directly into the channel between the weirs; Type 2 was a delayed, but larger peak, resulting from saturated throughflow from the hillslopes; Type 3 was interpreted as a contribution of unsaturated throughflow to the hydrograph. This was separated from the saturated contribution by a line of constant slope that intersects the recession curve close to a break in the recession slope (see Weyman, 1974, p.38). Weyman notes, however, that such a separation is arbitrary and, lacking further evidence, does not pursue this interpretation any further.

The throughflow trough and potential measurements are mostly reported in Weyman (1973), where he talks about the importance of breaks in the permeability profile in the soil during the build up of saturated conditions and downslope throughflows in the lower basin response. He suggests, however, that baseflows will still be dominated by unsaturated lateral flow on the slopes. He notes that "*A hydraulically distinct system of non-capillary spaces will require the application of other flow laws for part of the flow and renders the parameter "average hydraulic conductivity" rather meaningless*" (p.286), citing the work of Chamberlin (1972) in porous forest soils in British Columbia. My own attempts to model the East Twin data as part of my PhD thesis would have benefited from taking closer notice of this statement. Model simulations based on core-scale hydraulic conductivity measurements failed miserably to simulate the discharges or water table shapes. The story is told in Beven (2001).

Finally, Weyman does not specifically mention the extensive network of bracken rhizomes in the A horizon of the brown earth soils of the lower basin in any of the papers, though these may be important in inducing the high infiltration capacities of the soil and, possibly, downslope preferential flows (although his own comment is that "*there is no evidence to suggest that any part of the hillslope flow occurs within macro-voids of the soil which are hydraulically unrelated to the soil mass*", 1973, p.285). In the upper basin, by contrast, "*storm discharge has also been observed to occur from subsurface pipes which may be animal burrows although their origin remains obscure*" (1974, p.40).

References

Atkinson, T. C. (1978) Techniques for measuring subsurface flow on hillslopes. In: *Hillslope Hydrology* (ed. by M. J. Kirkby), 73–120. Wiley, Chichester, UK.

Beven, K. J. (2001) How far can we go in distributed hydrological modelling? *Hydrol. Earth System Sci.* **5**(1), 1–12.

Chamberlin, T. W. (1972) Interflow in the mountainous forest soils of coastal British Columbia. In: *Mountain Geomorphology* (ed. by . O. Slaymaker & H. J. McPherson), 147–150. Tantalus Research, Vancouver, Canada.

Weyman, D. R. (1973) Measurements of the downslope flow of water in a soil. *J. Hydrol.* **20**, 267–288.

Weyman, D. R. (1974) Runoff process, contributing area and streamflow in a small upland basin. In: *Small Instrumented Watersheds in the United Kingdom.* (ed. by K. J. Gregory & D. E. Walling) Inst. Brit. Geogr. Special Publication, IBG, London, UK.

THROUGHFLOW ON HILLSLOPES AND ITS RELATION TO THE STREAM HYDROGRAPH

D. R. WEYMAN
Department of Geography, Bristol University

ABSTRACT

Throughflow has been measured from three soil horizons on a 12° slope with impermeable. bedrock. Storm flow comes from the 10-45 cm horizon and is controlled by the upslope extent of saturated conditions. Base flow comes from the 45-75 cm horizon and is supplied by slow unsaturated flow from the whole soil mass to a small constant zone of saturation.

Differences between input and output stream hydrographs over 270 metres of channel are attributed to throughflow and correlate well with measured values providing a basis for separating throughflow components from the stream hydrograph. Observed stream flows contain no overland flow or ground water flow components. The main basin flood peak is not generated within this control section of channel but is produced in the headwater zone (0.1 km^2) by the faster runoff characteristics of the soils in that area and by topographic factors which lead to rapid channel extension.

INTRODUCTION

The lateral flow of water through the soil cover of hillslopes (throughflow) has been described in detail by several writers (Hewlett, 1961; Hewlett and Hibbert, 1963; Whipkey, 1965; Kirkby and Chorley, 1967). At the basin scale the effect of throughflow has been inferred from the discrepancies arising in some predictive models (Betson, 1964; T.V.A., 1965). This paper outlines the results of a study which is attempting to evaluate the direct effect of throughflow on the stream hydrograph. It is hoped that these findings will provide the foundation of a predictive model based upon the response of throughflow to input factors and its subsequent distribution through the stream channel.

EXPERIMENTAL BASIN

The East Twin Brook is a very small (0.21 km^2) upland basin in the Mendip Hills, Somerset. The basin is underlain by sandstones, shales and quartz-conglomerates of Devonian age which are known to be virtually impermeable in this locality (Whittard, 1949). The basin head is a very shallow topographic bowl in a summit plateau at 350 metres from which the channel drops steeply to its outfall at 230 metres. Most of the basin is occupied by long (300 metres) valley side slopes which are broadly convex, ending with a short steep (12°) drop into the incised channel. Rainfall is moderate (103 cm, 1969-70) and spread throughout the year. Intensities are low (0.5 cm/hr max.) from frontal rainfall but occasional summer convective storms may produce intensities up to 1.7 cm/hr. On the valley side slopes a brown earth soil is developed showing typically a loose organic *A* horizon (0-10 cms), a sandy loam *B* horizon with stones (10-45 cm) and a transition *B/C* horizon (45-75 cm) which starts with a distinctive stone layer. Vegetation on the brown earth is dominated by bracken *(Pteridium aquilinum)* and bramble *(Rubus sp.)* with a grass mat (*Festuca-Agrostis* community). A peaty podsol is developed in the headwater area. This shows a thin peat (20 cm) overlying a sandy loam podsol (60 cm) supporting a broken cover of bilberry *(Vaccinium myrtillum)*, heather *(Calluna vulgaris)* and grass *(Molinia caerulea)*.

Total runoff from the basin is measured by a 90° 'V-notch' weir with associated continuous stage float recorder. A duplicate gauging station with ½ 90° weir is located 270 metres upstream at the junction between the brown earth and podsol. The basin is thus divided into two parts: a "control section" 270 metres long where contributions to stream discharge from the valley side brown earth system can be isolated, and a "headwater zone" (0.1 km^2) where discharge is from the peaty podsols of the summit plateau. Interest is concentrated in the control section where two soil plots have been established for detailed investigation. These plots are assumed to be metre-wide strips running upslope in the direction of maximum surface slope from the stream channel. Discharge at the base of the plots is measured by troughs, one metre long, inserted into the natural stream bank. (Whipkey, 1965). Discharge is measured separately for the three horizons of the soil profile in the brown earth. In this paper throughflow discharges will be referred to *A*, *B* or *B/C* horizon discharge (0-10, 10-45 and 45-75 cm respectively). Soil moisture equipment, duplicated as far as possible, has been established at set distances upslope measuring tension at each of the three horizons. Manometer-type tensiometers, capable of recording positive piezometric pressures as well as tension, have been included (Webster, 1966).

THROUGHFLOW HYDROGRAPHS

During the period March 1969-February 1970 a total of 16 separately identifiable storms occurred which produced significant discharge from the soil plots in excess of base flow which

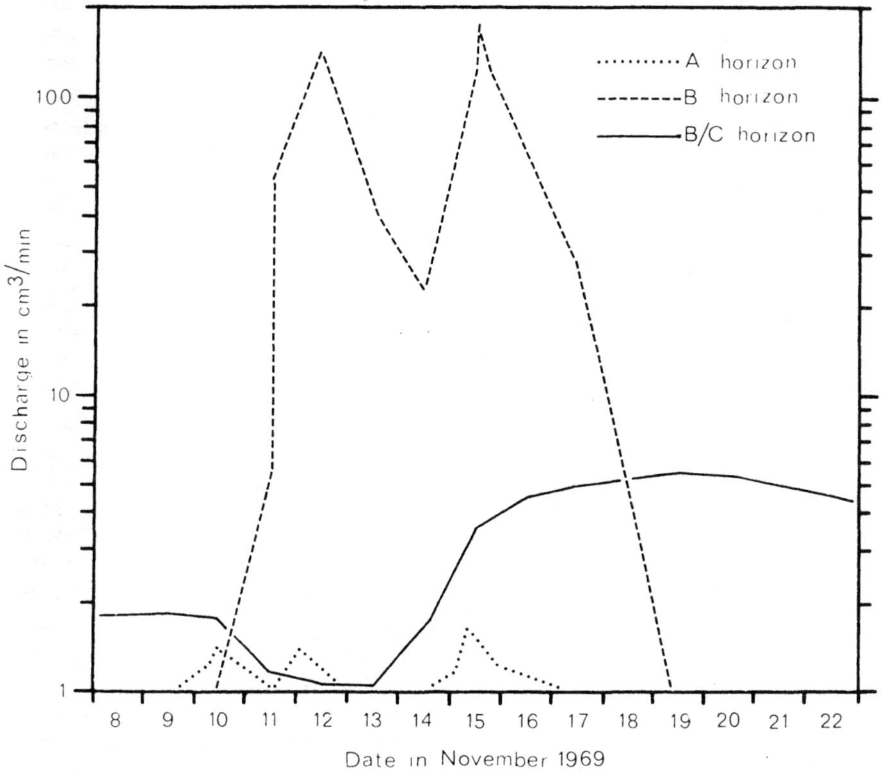

Fig. 1 — Throughflow hydrographs for Soil plot 1, 8-22 November, 1969. *A* horizon 0-10 cm; *B* horizon 10-45 cm; *B/C* horizon 45-75 cm.

26

is continuous under natural recharge. Figure 1 illustrates the throughflow hydrograph resulting from a storm of 7.0 cm (9-15 November 1969). The general relationships are true for all 16 storms observed, both summer and winter, although input-output relationships vary seasonally with evapotranspiration changes. The first part of figure 1 shows the response to 4.0 cm rain falling in 40 hours which may be compared with results obtained by Whipkey (Whipkey, 1965, fig. 3) for throughflow response to 3.1 cm of rain in 100 mins. An important difference between the graphs is considered to be the time of concentration. In Whipkey's case the time lag from start of rain to throughflow peak is 3 hours, in comparison with a lag of 26 hours between rainfall peak and throughflow peak in the East Twin example. This would seem largely explicable in terms of the vastly greater rainfall intensity generated by Whipkey (1.8 cm/hr against 0.3 cm/hr natural rainfall maximum). There is a corresponding difference in the peak throughflow discharges (680 cm^3/min against 140 cm^3/min). In terms of general relationships, however, several similarities may be observed. Both hydrographs show only minor discharge from the organic (*A*) horizons and no overland flow. The main discharge in each case is from a horizon within the profile lying above a boundary of vertical permeability decrease. The 90 cm boundary in the Whipkey profile is associated with a bulk density change from 1.41 to 1.78 gms/cm^3. At the base of the *B* horizon in the East Twin profile there is a bulk density change from 1.6 to 1.9 gms/cm^3 and a sharp increase in stoniness. Base flow for Whipkey's example and the East Twin is from the lower profile and has a much smaller peak discharge, longer lag to peak and greater continuity of flow than the main discharge. Base flow in the East Twin brown earth is from the *B/C* horizon, varies from 1.0 to 10 cm^3/min and has not ceased during the recording period of one year which included 40 days with no appreciable rainfall.

Nature of water flow on the hillslope

The general model of slope flow considered appropriate for this situation is a modification of the "dynamic watershed concept" (T.V.A., 1965). Water moves vertically and laterally in the soil mass under unsaturated conditions. The lateral component of flow may be increased by a drop in vertical permeability. This may occur at the soil/bedrock interface or at some boundary within the soil. Saturated flow conditions must exist at the base of the slope for discharge from the slope to occur. This saturated zone is supplied by unsaturated flow from upslope. Instantaneous discharge from the profile is directly related to the hydraulic head in the saturated zone, which in turn is a function of the upslope extent of saturated conditions. Assuming that saturated permeability of the soil is constant upslope, the extent of the saturated zone is dependent only upon unsaturated supply rate. When supply exceeds lateral permeability the saturated zone grows upslope and discharge from the slope base increases.

In the brown earth of the East Twin catchment there are two significant changes in vertical permeability. The first is at the *B-B/C* junction (45 cm), which becomes effective only during rapid infiltration, and the second is at the soil/bedrock interface (base of *B/C* horizon −75 cm). The effect of these permeability variations is that the *B* and *B/C* horizons may operate as separate discrete water transmission systems according to the prevailing conditions.

Following heavy rain (intensities in excess of 0.2 cm/hr) unsaturated flow is partially diverted at the *B-B/C* junction and flows downslope at that level to create a saturated zone in the *B* horizon at the slope base. Discharge from the horizon starts once saturated conditions have spread at least one metre upslope (from tensiometer readings). Thereafter discharge increases as the saturated zone grows. For the 9-15 November storm the maximum *B* horizon discharge of 180 cm^3/min was associated with saturated conditions in the *B* horizon 7 metres upslope. Tension values of soil moisture upslope of the saturated zone fall during zone extension suggesting the dependence of zone growth on increasing unsaturated supply from upslope. The lag of about 24 hours between rainfall peak and *B* horizon discharge peak is related to percolation and lateral flow time to the slope base. Once rainfall has stopped the horizon drains rapidly, flow normally ceasing after about five days.

Discharge from the *B/C* horizon is continuous throughout the year and shows only small variation in rate (1.0-10 cm^3/min). Discharge is maintained by slow unsaturated flow from the

27

entire soil mass which is transmitted through a small stable zone of saturation in the horizon at the base of the slope. Small discharge fluctuations are a reflection of small fluctuations in the extent of saturated conditions. A maximum upslope extent of between 1 and 2 metres is normal for peak *B*/*C* discharge.

During rapid infiltration following a storm most of the water on the lower part of the slope is transmitted through the *B* horizon saturation zone. The *B*/*C* horizon thus loses part of its supply and discharge may actually decline (fig. 1). After a lag of about 5 days storm water supplied to the slope outside the *B* horizon "capture" zone arrives at the slope base of the *B*/*C* horizon as a diffused wave. The *B*/*C* saturation zone then extends slowly to accommodate this increased supply and a small discharge peak results.

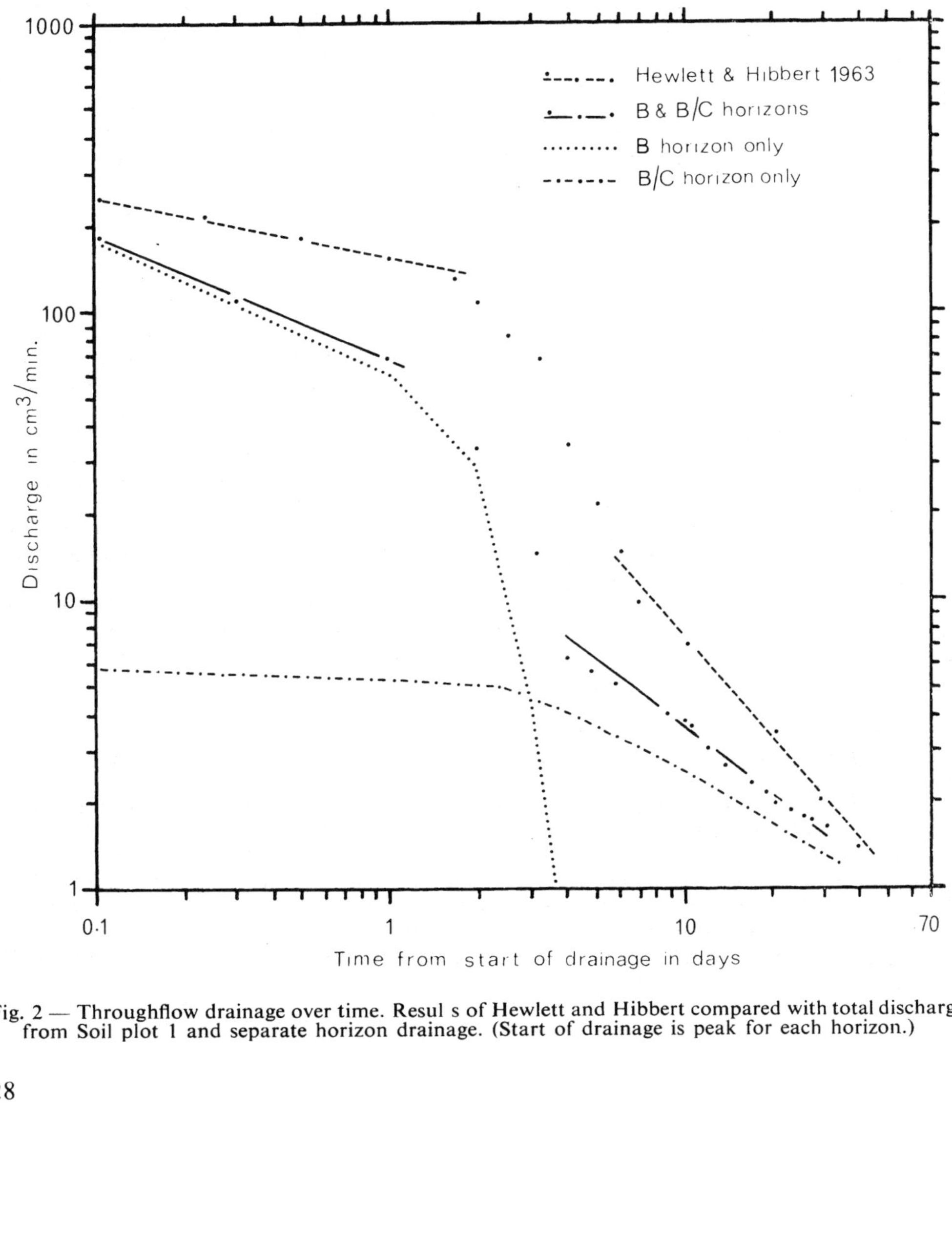

Fig. 2 — Throughflow drainage over time. Resul s of Hewlett and Hibbert compared with total discharge from Soil plot 1 and separate horizon drainage. (Start of drainage is peak for each horizon.)

28

No precise relationships are yet available for extent of saturation zone and horizon discharge, but significant relationships have been established between horizon discharge and upslope tension (the presumed control on unsaturated supply). Due to the hysteresis effect in wetting and drying there is a series of logarithmic relationships each applicable to an initial tension state. For a drainage cycle in the *B*/*C* horizon starting at 2.5 cm Hg the relationship is:

$$q_2 = 19.6\,\psi^{-1.51} \qquad r = -.97 \qquad (1)$$

An equivalent infiltration cycle starting at 6.5 cm Hg is given by:

$$q_2 = 4.39\,\psi^{-.4} \qquad r = -.78 \qquad (2)$$

In each case q_2 is *B*/*C* horizon discharge in cm^3/min and ψ is the soil moisture tension in cm Hg in the *B*/*C* horizon measured by a tensiometer located 1 metre upslope.

Flow in each horizon is thus a form of unsaturated supply to a dynamic saturated zone which reaches steady state when upslope supply is equal to discharge from the base of the slope. Lag time from storm to the start of saturation extension is different for each horizon with the result that each will operate as a discrete system for a given storm. The drainage of both horizons approximate to similar forms:

$$q_1 = 59.9 - 98.5 \log t \qquad r = -.99 \qquad (3)$$

$$q_2 = 5.53 - 2.89 \log t \qquad r = -.99 \qquad (4)$$

where q_1, q_2 are discharge from *B* and *B*/*C* horizons respectively in cm^3/min and t is the time from start of drainage for each horizon (i.e. from peak discharge) in days.

Equations 3 and 4 may relate to two parts of a composite form of drainage covering all stages in the decay of a saturated zone. Hewlett and Hibbert, argue in favour of a change from dominantly saturated to unsaturated flow during drainage on the basis of the complex drainage forms obtained from total soil plot discharge over time (Hewlett and Hibbert, 1963). By totalling discharge from both horizons, a very similar drainage form can be obtained for the East Twin plots (fig. 2). Total discharge decay over time approximates to a logarithmic form up to 1 day and after 5 days from start of drainage. Hewlett and Hibbert considered drainage up to 1 day to be dominated by saturated flow conditions, and after 5 days by unsaturated flow with a transition period from 1 to 5 days. As shown above, this complex drainage form may be simplified by considering the horizons as discrete systems with the same flow processes acting over different time scales. The main point raised by Hewlett and Hibbert is probably still true in as far as equation 3 relates to saturated conditions from 1 to 7 metres upslope while equation 4 relates largely to saturation less than 1 metre upslope, although the two are still parts of one continuum.

Stream hydrographs and throughflow

Hydrographs for the control section of the stream are produced from the difference between hydrographs at the upper and lower weirs. A constant time lag of 1½ hours is used on the basis of peak concordance which is adequately definitive for rapid discharge rises. The variability of time lag with discharge and the spatial distribution of inputs to the stream have been ignored at this level of analysis. This simplification is felt to be justified because travel time between the weirs is fast in comparison with input changes through time.

A derived hydrograph for the storm of 9-15 November is given in figure 3. Again, the general statements made here are applicable to all 16 storms in the period considered.

In assessing the contribution of throughflow to the hydrograph of flow generated in the control section the first difficulty is met in comparing the slope output of one metre width with a total input to the stream of 540 such units. This crude comparison is made in figure 3. The similarity of the throughflow and control section hydrographs over time would seem to indicate a very close relationship between the two. One problem in this respect is the storage effect of infill material in the valley which gives rise to higher stream discharges on the recession limb of

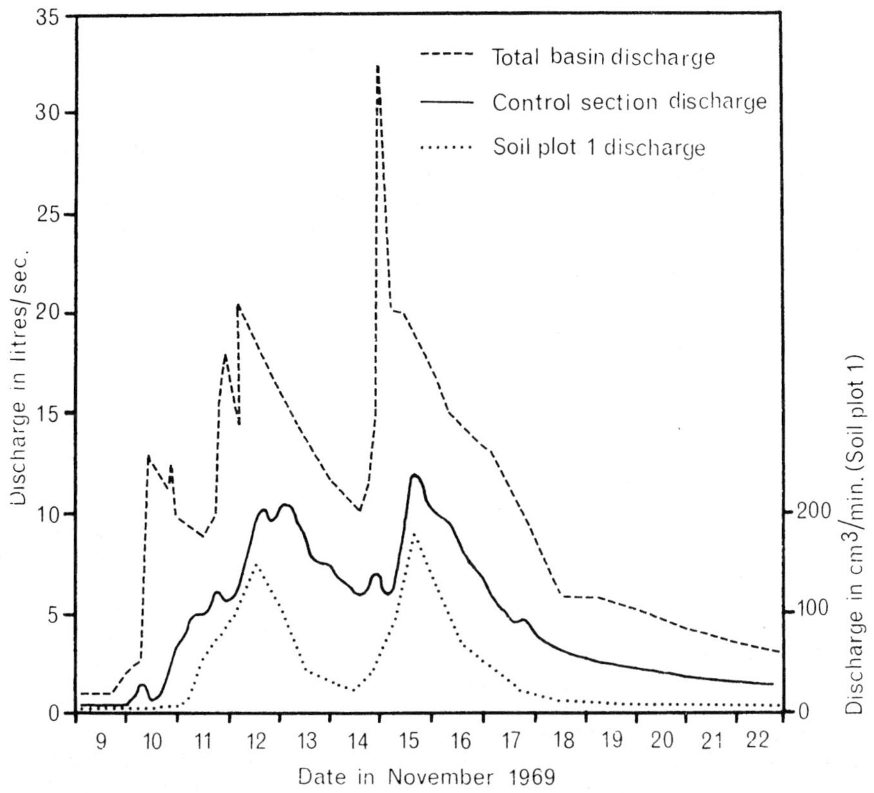

Fig. 3 — Total basin discharge, control section discharge and throughflow discharge for Soil plot 1 (different scale) for 9-22 November, 1969.

the hydrograph than would be expected if throughflow accounted for the total instantaneous discharge of the control section. This effect may be further specified by relating throughflow output directly to control section discharge. Separate relationships exist for rising and falling stage. For the storm in question:

Rising:

$$Q_c = 6.81 \log q - 3.59 \qquad r = .99 \tag{5}$$

Falling:

$$Q_c = 7.02 \log q - 3.86 \qquad r = .90 \tag{6}$$

where Q_c is control section output in litres/sec and q is total throughflow discharge from one soil plot in cm^3/min.

30

A strong similarity thus exists between throughflow and the control section discharge in a time sequence sense. No contributory process is operating on a very different time scale to throughflow—(for example, overland flow or ground water discharge) with the exception of short term infill storage of small amounts of water. In a non-time sense, however, there is still the problem of the spatial variability of input rates. To produce a peak control section discharge of 12 litres/sec by throughflow alone from 540 metres of bank requires a mean peak throughflow discharge of 1320 cm^3/min/metre. In contrast the peak discharge from the soil plots was only 185 cm^3/min/metre. On the other hand measured seeps from the soil at other locations on the channel give peak discharges for this storm of up to 7800 cm^3/min. The supply area for these inputs is indeterminate but in terms of bank length is certainly not more than one metre as seep spacing is often less than that distance. The seeps also show the same variation through time as the soil plot outputs and control section hydrograph.

This evidence together strongly implies that the entire discharge generated within the control section is the result of throughflow inputs from the adjacent slopes. The main rise in discharge is related to unsaturated percolation supplying a rapidly expanding zone of saturation in the *B* horizon of the soil and rapid discharge increase from that horizon into the stream. Rate of stream discharge rise is reduced initially due to the small absorption effect of the valley infill. The recession limb is produced by the rapid drainage of *B* horizons, the peak and slow drainage of the *B/C* horizons and the release of water stored temporarily in the valley infill. Stream flow after 5 or more days of no rain is produced by *B/C* horizon base flow contributed by the whole hillslope. To some extent it is possible to equate *B* horizon discharge with "quick flow" and *B/C* horizon discharge with "delayed flow" in the simple response system suggested by Hewlett and Hibbert (1965).

The drainage of the control section through time gives very similar results to the time decay of discharge from the soil plots. Discrete systems cannot be isolated for the control section as was achieved in equations 3 and 4, but total flow decay approximates very closely to two logarithmic relationships:

$$Q_c = 9.2t^{-.15} \qquad r = -.95 \qquad (7)$$

$$Q_c = 8.2t^{-.9} \qquad r = -.99 \qquad (8)$$

where Q_c is control section discharge in litres/sec and t is time in days from start of drainage to 1.0 days in equation 7 and from 1.0 to 11 days in equation 8.

Generation of the flood peak

A significant point emerging at this stage is that a flood peak in the normal sense is not generated within the control section. Time from peak rainfall to control section discharge peak is up to 36 hours, which for a basin this size (0.11 km^2) is very slow. Total basin output (fig. 3) does, however, show a true flood peak occurring 3-4 hours after peak rainfall. It is clear from the analysis that this peak is generated within the headwater zone and is transmitted through the control section. Discharge generated within the control section has the effect of extending the recession limb of the main flood.

There are two ways of looking at the problem of flood peak generation. Firstly, it is possible to argue that the flood peak is generated within the headwater zone as a result of the faster runoff characteristics of the peaty podsols in that area. Limited throughflow measurements on the podsol have indicated that *A* horizon (peat) discharge is more important than on the brown earths. The main discharge from the podsol, however, is through an extensive system of natural pipes (5 cm diameter) at the base of the peat. Overland flow in the accepted sense does not occur extensively but very rapid pipe transmission has a similar discharge effect, allowing a short time lag for percolation and saturation of the base of the peat. The result is that each unit width of podsol slope produces a hydrograph which is an exaggerated form of the brown earth hydrograph, with main discharge occurring from the 20 cm level, 2-3 hours after peak rainfall. This alone will account for a flood peak generation in the headwater zone.

Secondly, it is probable that a flood peak would be generated in the headwater zone if brown earths covered that area. The bowl-shaped topography of the valley head concentrates slope flow in such a way that channel extension occurs. Whereas along a stream channel each point on the channel receives water from an area proportional to twice the slope length, a point at the head of a channel in a topographic bowl receives water from an area equal to half a circle whose radius is equal to slope length. It is to be expected that the lateral permeability of a soil in the latter case would be exceeded easily during heavy throughflow discharge and the surface stream would extend rapidly upslope as throughflow discharge increased. Discharge from the system will thus increase as the number of contributing units grows.

In the East Twin headwater zone both exaggerated unit throughflow discharge and channel extension combine to produce a sharp flood peak. A good relationship has been established between stream discharge and channel length, as measured against reference pegs.

$$Q_u = 0.014L\,1.09 \qquad r = 0.90 \tag{9}$$

where Q_u is discharge from the headwater zone in litres/sec and L is the length of channel above the gauging point in metres.

Equation 9 agrees in general with the work of Gregory and Walling (1968) who found that discharge varied according to L^2. The greater exponent values may be due to ground water flow contributing to stream discharge in their case. Flow at all stages and at all times of the year has been taken into account in equation 9. There is consequently no support for Carlston's suggestion of base flow varying inversely to the second power of drainage density (Carlston, 1963)– a conclusion in agreement with Gregory and Walling.

Throughflow discharge increase and channel extension are closely related processes which together dominate the nature of headwater zone discharge in the East Twin catchment. They result in one simple form of discharge decay through time:

$$Q_u = 6.4t^{-.5} \qquad r = -.99 \tag{10}$$

Conclusions

In an impermeable catchment with a well-developed soil cover throughflow is found to be the only process of water transmission effectively contributing to stream flow. The "quick response" of stream discharge to rainfall is a result of rapid increases in throughflow output as saturated conditions extend upslope. The "delayed response" of stream discharge and subsequent base flow are both produced by stages in the slow unsaturated drainage of the whole slope, a process which is sufficient to maintain stream flow throughout the year. Overland flow does not contribute to "quick response" in the stream, probably as a result of high *A* horizon permeability and low rainfall intensities. Ground water flow is not contributory to "delayed flow" although temporary storage in valley infill material may supplement unsaturated throughflow.

Owing to the time lag involved in producing saturated throughflow conditions, "quick response" stream discharge in a section of channel does not generate a true flood peak. In this instance the basin flood peak is generated in the headwater zone by a combination of faster response per unit width of slope (due to piping) and an increase in the number of contributing units through time by channel extension (due to topographic concentration of flow). It would seem possible to create a flood peak in a channel section with rapid throughflow response, or in a headwater zone with fairly slow response.

Acknowledgements

The author is grateful to Dr. M.J. Kirkby, Mr. M.D. Newson and Mr. T.C. Atkinson of the Department of Geography, Bristol University for their helpful critical review and discussion of the manuscript of this paper.

32

REFERENCES

Betson, R. P. (1964), What is watershed runoff?, *Journal of Geophysical Research*, v. 69, pp. 1541-1552.
Carlston, C. W. (1963), Drainage density and streamflow; *U. S. Geol. Survey Prof. Paper*, 422-C, 8 p.
Gregory, K. J. and Walling, D. E. (1968), The variation of drainage density within a catchment, *Int. Assoc. Sci. Hydrology Bull.*, v. 13, no. 2, pp. 61-68.
Hewlett, J. D. (1961), Soil moisture as a source of base flow from steep mountain watersheds: Southeastern Forest Experiment Station, Asheville, North Carolina, U.S. Dept. of Agriculture-Forest Service, Station Paper no. 132, 11 p.
Hewlett, J. D. and Hibbert, A. R. (1963), Moisture and energy conditions within a sloping soil mass during drainage, *Journal of Geophysical Research*, v. 68, pp. 1081-1087.
Hewlett, J. D. and Hibbert, A. R. (1965), Factors affecting the response of small watersheds to precipitation in humid areas, *Int. Symp. on Forest Hydrology*, Pennsylvania State University.
Kirkby, M. J. and Chorley, R. J. (1967), Throughflow, overland flow and erosion, *Int. Assoc. Sci. Hydrology Bull.*, v. 12, no. 3, pp. 5-21.
Tennessee Valley Authority, (1965), Area-stream factor correlation, *Int. Assoc. Sci. Hydrology Bull.*, v. 10, no. 2, pp. 22-37.
Webster, R. (1966), The measurement of soil water tension in the field, *New Phytologist*, v. 65, pp. 249-258.
Whipkey, R. Z. (1965), Subsurface stormflow from forested slopes, *Int. Assoc. Sci. Hydrology Bull.*, v. 10, no. 3, pp. 74-85.
Whittard, W. F. (1949), Temporary exposures and borehole records in the Bristol area. IV-Boreholes on Mendip, *Proc. Bristol Naturalists' Soc.*, v. 27.

Commentary on Paper 18

Jones, J. A. A. (1971) Soil piping and stream channel initiation.

An extreme case of macroporosity in soils is that of soil piping. In 1971, as Tony Jones notes in the very first sentence of the abstract, "*the hydrological significance of soil piping has been generally ignored*" (p.602). He comments that pipes may occur widely in English soils, may have a variety of origins, may include "pseudo-pipes" associated with root channels and soil cracks, and may be important in soil erosion. Following the tendency of the time for classification, he suggests that a drainage basin is made up of seven conceptual subsystems: the channel net; the pipe net; the pseudo-pipe net; the zone of diffuse seepage; the zone of overland flow; the zone of diffuse groundwater flow; and groundwater springs (p.604). He also, however, provides data on the occurrence of pipes in the banks of the Burbage Brook catchment in the English Peak District, noting that they are associated with seepage "percolines" in the sense of Bunting (1961), and provides a discussion of the potential origins in different circumstances.

One of the locations where Jones notes that piping has been observed was in the experimental catchments at Plynlimon, mid-Wales, run by the Institute of Hydrology at Wallingford. Kevin Gilman and Malcolm Newson, who worked at Plynlimon, carried out extensive studies of the pipes that occur in the peats and peaty gleyed podsols there (Newson, 1976; Gilman & Newson, 1980). These pipes seem to be of two types: those formed by peat growing over small channels eroded to bedrock, and hill-slope pipes formed as a result of occasional dessication cracking of the pipes and preferentially maintained by channelled flow after rewetting. Gilman & Newson produced a monograph on their results in 1980, which was closely followed by a general review of pipeflows by Tony Jones (1981) in the same series. Pipe tracing was described by Smart & Wilson (1984) in the Cwm Llwch catchment in the Brecon Beacons in Wales, not far to the south, though there they suggest that mole burrows might be important in pipe initiation (Wilson & Smart, 1984). McCaig (1983) also provides evidence of pipeflows from peaty podsols in Yorkshire, England. Further work on the pipes at Plynlimon was described by Sklash *et al.* (1996) who, using information provided by environmental isotopes, suggested that the predominant source of water in the hydrograph from an ephemeral pipe with a lag time of about 15 minutes, was pre-event water.

Tony Jones still works as a Professor in the Department of Geography at the University College of Wales, Aberystwyth.

References

Bunting, B. T. (1961) The role of seepage moisture in soil formation, slope development and stream initiation. *Am. J. Sci.* **259**, 503–518.

Gilman, K. A. & Newson, M. D. (1980) Soil pipes and pipeflow – a hydrological study in upland Wales. British Geomorphology Research Group Monograph 1. GEO Books, Norwich, UK.

Jones, J. A. A. (1981) The nature of soil piping: a review of research, British Geomorphology Research Group Monograph 3. GEO Books, Norwich, UK.

Newson, M. D. (1976) Soil piping in upland Wales: a call for more information. *Cambria* **3**(1), 33–39.

McCaig, M. (1983) Contributions to storm quickflow in a small headwater catchment – the role of natural pipes and soil macropores. *Earth Surf. Process. Landf.* **8**, 239–252.

Sklash, M. G., Beven, K. J., Gilman, K. & Darling, W. G. (1996) Isotope studies of pipeflow at Plynlimon, Wales, UK. *Hydrol. Processes* **10**, 921–944.

Smart, P. L. & Wilson, C. M. (1984) Methods for the tracing of pipeflow on hillslopes. *Catena* **11**, 159–168.

Wilson, C. M. & Smart, P. L. (1984) Pipes and pipe flow process in an upland catchment, Wales. *Catena* **11**, 145–158.

Soil Piping and Stream Channel Initiation

ANTHONY JONES

Experimental Cartography Unit, London SW 7, England

Abstract. The hydrologic significance of soil piping has been generally ignored. Piping has been associated primarily with drylands, yet evidence of piping is available from a large range of climatic regions. In particular, soil piping is found to be widespread in the United Kingdom. Preferred locations for piping are either just above or within a horizon of low relative permeability and low aggregate stability. Chemical environment may range from acidic moorland soils to saline marshes. There are significant trends to lower aggregate stability and coarser grain size (particularly in the range above 250 μm) in the bed of a pipe than in the roof. Many pipes in the areas studied appear to be dormant or relatively inactive and may well be in approximate equilibrium with the soil pore and channel subsystems. However, when equilibrium is destroyed (e.g., by stream incision) pipes can form loci for channel extension. Studies of the spatial distribution of outlets show that to create normal channel networks, pipe clusters within the ensemble and, similarly, individual pipes within those clusters must be selected on an unequal basis. A low density random selection from pipes located on percolines would fulfill the requirements. The presence of piping may have a significant effect on the form of the hydrograph.

PREVIOUS WORK

The phenomenon of piping or tunneling in the soil has been known for some years, particularly to engineers [*Terzaghi,* 1936]; yet piping has been generally overlooked by hydrologists and geomorphologists. Engineers have been mainly concerned with piping failure in earth dams, and a considerable amount of research has been done on this aspect [*Aitchison and Wood,* 1965; *Kassiff and Henkin,* 1967]. However, the importance of piping in the natural landscape has only recently been brought before a wide international audience. First, *Parker* [1963] stressed the significance of piping as a form of erosion in the arid southwest of the United States, and second *Dalrymple et al.* [1968] set up a hypothetical model for land surface soil and slope units that recognized the importance of piping erosion as a form of lateral soil movement in four out of the nine units. The latter group drew principally on their experience in North Island, New Zealand, a humid temperate climate, and implicitly stressed for the first time in a major journal that piping can be expected in humid as well as arid environments. This point is significant, since many theories of pipe development have relied on severe desiccation cracking to begin the process. A reinterpretation by *Smith* [1968] of certain periglacial features in the United States as the result of piping completes the range of climatic environments.

Recent investigations indicate that the phenomenon not only is present in the United Kingdom but widespread. The author has observed piping in brown earths (sols lessivés) and peats on the Carboniferous Millstone Grit series of the Peak District; in recent marine deposits near Scolt Head Island and Blakeney Point, Norfolk; in the Fal estuary, Cornwall; in podzol A_1 horizons on the Devonian Old Red sandstone of the Mendips, Somerset; in solifluction materials over Old Red sandstone in the Brecon Beacons, Brecknock; and also in brown earths overlying igneous rocks along the Afon Cerist, Merionethshire. Piping has also been observed in lessivé podzols on the Bagshot Sands of Berkshire (J. B. Dalrymple, personal communication, 1968); in the Atherfield Clay, Isle of Wight (C. Jeans, personal communication, 1970); in peaty podzols on Plynlimon, Cardiganshire [*Knapp,* 1970]; and in peat in the Lake District, Westmorland (J. Pethick, personal communication, 1970) (Figure 1).

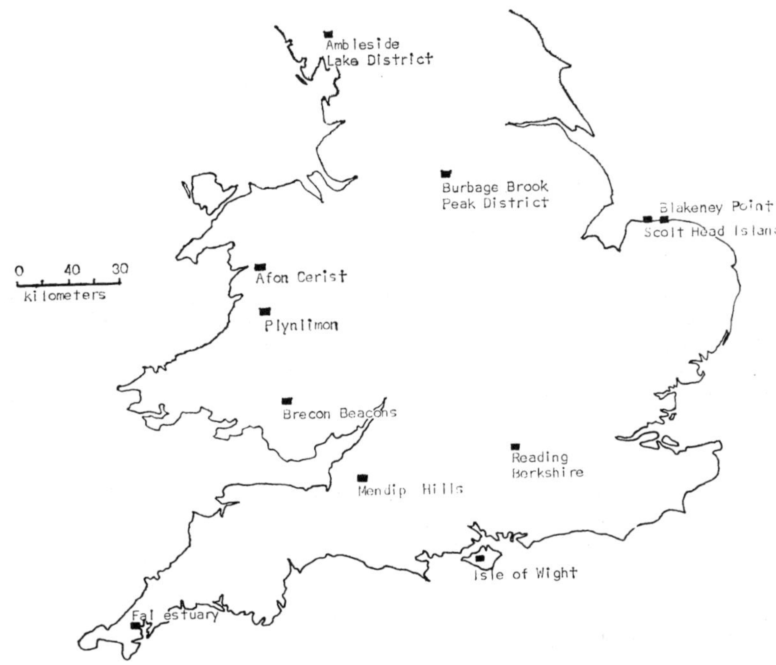

Fig. 1. Location of piping sites in England and Wales.

HYDROLOGIC SIGNIFICANCE

Whereas a number of writers have linked piping with gully development by roof collapse [*Buckham and Cockfield,* 1950], the theory of stream channel initiation has been inordinately dominated by Horton's theory of critical distance of overland flow [*Horton,* 1945]. Only recently has subsurface flow of any type been generally considered, for example by *Bunting* [1961] and *Kirkby and Chorley* [1967]. Yet seepage, throughflow, or shallow interflow is probably a more effective agent of erosion than overland flow in humid, vegetated drainage basins, where surface flow is impeded by plant growth and where infiltration capacity is increased by biotic activity. Piping may prove to be a significant form of stream initiation in humid lands such as the British Isles and New Zealand. Piping may also prove a significant form of soil erosion in Britain, as it has in Australia [*Downes,* 1949], the United States [*Parker,* 1963], and New Zealand [*Blong,* 1965]. Piping is a form that has been completely ignored in studies of rates of erosion on slopes, from both the theoretical [*Culling,* 1965] and practical [*Kirkby,* 1967] points of view.

The immediate hydrologic importance of piping has never been analyzed. Yet by increasing the transmissibility of the soil mass, piping increases the probability of a given runoff exceedance in proportion to the capacity of the pipe network. This possibility is ignored in current theoretical work [*Liao and Scheidegger,* 1969; *Childs,* 1969], although an indication of the potential importance has been given in recent work by the U.S. Department of Agriculture at Columbus, Ohio [*Whipkey,* 1969], notwithstanding the conflicting evidence of *Dunne and Black* [1970], who do not appear to be aware of Whipkey's latest work. Whipkey found that subsurface flow, not overland flow, was the major source of floodwater from forest soils on the Allegheny-Cumberland plateau. He further observed that this flow did not satisfy any established law but often flowed through unsaturated soil along 'pipe-like' channels. We might term this feature 'pseudopiping,' in which soil cracks or root channels rather than hydraulic laws almost completely determine pipe geometry. In this hydrologic role, piping and pseudopiping may well achieve a quasi-equilibrium state, in which piping is not merely a precursor to surface channels but is a relatively stable component of the drainage system.

Therefore, our conceptual model of a drainage basin should be revised to comprise seven subsystems: (1) channel net, (2) pipe net, (3) pseudopipe net, (4) zone of diffuse seepage, (5) zone of overland flow, (6) zone of diffuse groundwater flow, and (7) groundwater spring net. Subsystems 6 and 7 are the more stable counterparts of subsystems 2–4; in general, the former are not found in the soil cover. Soil pipes correspond to *Bryan*'s [1919] 'minor tubular springs.' The other five subsystems show marked expansion or contraction with mutual interference in response to rainfall or snowmelt. Naturally there are problems of operational definition, and the hydrologic and geomorphologic boundaries may not coincide. The stream net may expand during storms well outside the channel net [*Gregory and Walling,* 1968] so that the fingertips of the net become concentrated linear elements of overland flow; conversely, the stream net may contract to be smaller than the channel network. *Kirkby and Chorley* [1967] suggested that the heads of the channel net reflect a complex relationship between soil properties and the expansion and contraction of the stream net over time. The outer links of the pseudo-pipe and pipe networks may well also expand and contract in relation to storm history, but the magnitude of soil creep and infilling processes along the pipe walls will be greater in relationship to channel size. Hence the network is inherently less stable. This process may be aggravated by biological, especially bacterial, activity. The writer has observed a pipe 10 cm in diameter blocked by ferric iron bacteria in a gleyed podzol in the Burbage basin (Figure 1). Similar blockages have proved hydrologically significant in agricultural drains in the Netherlands [*Wiklander,* 1945]. Furthermore, the relationship of both channel and pipe nets to storm incidence may often be less direct than the proportional scouring and infilling sequence envisaged by Kirkby and Chorley. As shown by *Heede* [1967] in gullies of the Colorado front ranges, major extensions may occur in minor storms after periods of desiccation have weakened the soil structure.

Observations suggest that there is a considerable delay in contraction after scouring. In the Peak District catchment (Burbage Brook), the pipe network rarely carries any discharge (for example, 5% flow in a storm of 2.5 cm), yet the net is maintained despite processes of soil creep. Indeed many of the outlets appear to be relict features, cut off from their catchment volume by more competitive neighbors.

SPACING OF OUTLETS

The possibility of relict features poses a problem in studying the spatial distribution of the process, because they add an unwanted historic element. Nevertheless, the spatial distribution is an important property of piping, and its relationship to stream channels may be important from the genetic viewpoint. Distances along the stream channel from one pipe outlet to another might be expected to follow the observed frequency distribution of internal stream links, if the pipe ensemble is a nascent channel ensemble.

An independent events or random model provides a good fit to the distribution of stream link lengths [*Dacey,* 1970] with but a few reservations [*James and Krumbein,* 1969]. If all pipes are potential stream channels, their linear distribution along the stream banks should be random. Both reflexive neighbor and quadrat analyses show, however, that distribution is not random along Burbage Brook and Afon Cerist (Figure 3; Table 1). Instead the pipes are clustered. Moreover, the mean distance between clusters is shorter than would be expected for stream links. If pipes can initiate stream channels, we must therefore postulate unequal opportunity, i.e., the growth of a few at the expense of the others. It is, indeed, not uncommon to find one or two pipe outlets more developed than others in the cluster and some clusters with significantly more runoff events in the 'master' pipe (Figure 3). In general, the clustering probably reflects an area attraction, a zone of deeper soil, more susceptible soil, or surface hollows. In particular, despite the scatter of possible relict pipes, a two by two contingency test shows a significant association between pipe outlets and percolines or seepage surfaces. However, at least one case that might be termed point attraction was found on the Burbage Brook, where a partially collapsed pipe containing an apparently perennial flow debouches into a system of deltaic distributaries with about 30 outlets in the stream bank.

Clearly, piping associated with percolines will have a better chance of development, and given the general tendency for drainage capture, our hypothesis requires analysis of the distribution of percolines. The distribution has proved to be random on the Burbage Brook (Table 1). Hence, it is possible that a random selection from the percolines could extend the channel net in accordance with a major 'requirement' of the network structure. This extension along percolines or seepage surfaces could occur without piping in the manner suggested by *Bunting* [1964] in Rossendale, but piping should increase both the likelihood of extension and the rate of extension. The association between piping and percolines is significant at the 0.1% level for Burbage, despite many presently inactive pipes.

CAUSES OF PIPING

Piping does not appear in all soils and cannot be the ubiquitous cause of stream initiation, although it is clearly important in many cases. Review of the literature indicates a certain consensus of opinion on the causes of piping. The overall ranking in descending order of

TABLE 1. Measures of the Distribution of Pipe Outlets and Percolines of a 1500-Meter Sample on the Banks of the Burbage Brook, Derbyshire, England

Test	Type of Distribution: More Clustered	Random	More Regular
Neighbor and distance analyses using stream distance combining both banks			
Reflexive neighbor index of spacing*	0.56		
Transformed statistic of spacing†	$z = -2.22$		
Cell frequency analyses on separate banks‡			
Pipe outlets, 10-cm units	$R(0.001)$ $L(0.001)$		
Pipe outlets, 5-meter units	$R(0.002)$ $L(0.02)$		
Pipe outlets in altered sequence, 5-meter units§		R L	
Percolines, 10-meter units	$R(0.025)$ (interference from extra large percoline)	L	
Percolines with piping, 10-meter units	$R(0.025)$	L	
Pipes on percolines, 5-meter units	$R(0.025)$ $L(0.025)$		
Pipes in sections not recognized as percolines, 5-meter units	$R(0.025)$ $L(0.025)$		
Association			
Contingency test between pipes and percolines, 10-meter units	coefficient 0.27	significance level 0.001	

Vertical location of pipe outlets in stream banks: $\langle x \rangle = 44\%$ bank-full height; s.d. $= 20\%$ normally distributed; standard error of $\langle x \rangle = 1.8\%$; standard error of s.d. $= 1.1\%$.

* Proportion of reflexives observed to reflexives expected in a random case [*Dacey*, 1960].

† Normal deviate measure based on spacing of events [*Barton and David*, 1956].

‡ The Poisson and Runs tests were both used to establish nonrandomness. Ten cm was chosen as being approximately the mean diameter of the outlets, i.e., the closest horizontal packing distance. Ten meters was chosen as approximately the average width of seepage lines at the stream bank. Significance level is in parentheses. R is the right bank; L is the left bank.

§ Where two or more adjacent cells contain pipe outlets only the first is recorded as containing any (i.e., 1) and the rest are considered empty (i.e., 0). The result suggests that pipe clusters are randomly distributed.

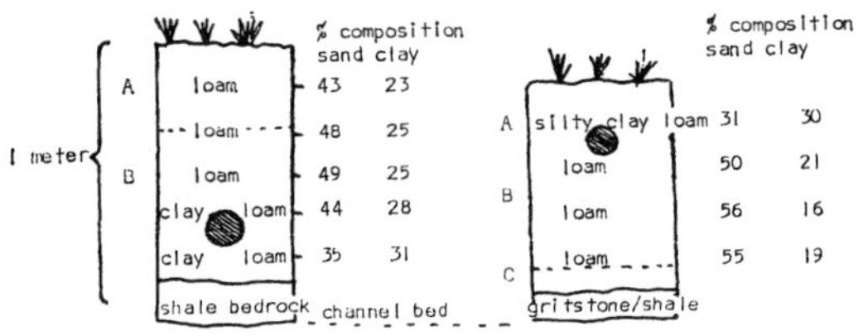

Fig. 2. Textural profiles of piped banks on Burbage Brook (shaded areas are pipe outlets).

emphasis is: (1) susceptibility to cracking in dry periods, high silt–clay content, and high percentage of swelling clays (especially montmorillonite); (2) periodic high intensity rainfall and devegetation; (3) biotic breakup of soil and a relatively impermeable basal horizon; (4) an erodible layer above this base, high exchangeable sodium and high base exchange capacity or high soluble salts (alkaline soils); and (5) steep hydraulic gradient due, for example, to active downcutting of baselevel.

Analyses of soil samples from the two main field areas suggest that whereas the average silt–clay content is approximately 40%, there is a large variation, and soils with similar content can be found without piping. However, mechanical analysis of soil profiles does suggest that the piping tends to occur within horizons containing higher percentages of silt–clay (Figure 2). Within these horizons there is a tendency for the roof of the pipe to have higher fines content than the base (Figure 3). The floors of the Burbage soil pipes show significantly higher percentages of material in the range 9 mm to 250 μm and significantly lower percentages of coarse silt, i.e., 75–20 μm (confidence level α = 0.10, samples from 30 pipes). This relationship is accompanied by an extremely significant trend (α = 0.001) to higher organic content in the roof. The latter results reflect the general gradients to be expected in the soil profile and may not be important to pipe location for pipes of the size observed (mean 9 cm), although high silt–clay and root contents could play an important part in maintaining the roof in more developed pipes. However, the apparent association between piping and layers of finer material may be more critical. The need for a less permeable layer to impede vertical drainage and to cause lateral seepage has been suggested by many writers [*Carroll,* 1949]. However, these workers suggest piping above the layer, whereas pipes in Burbage are often within the layer. Observations show that these layers may vary from the base of the soil profile to the base of the A horizon over very short distances (20 meters between profiles in Figure 2). The layers can often be described as a Bt horizon, and a number of physical reasons can be suggested for the piping. First, the layer tends to have a rather blocky or prismatic structure and can dry out sufficiently in the average summer to produce cracks of the order of 10 cm long between the prismatic peds. This cracking could provide a pseudopipe basis from which piping may develop. Second, the relative impermeability of the layer holds up a greater head of water, which is then capable of greater erosion when it is concentrated on a crack or fracture. Third, it is possible that the fines content preserves the open tunnels, whereas elsewhere tunnels tend to be ephemeral, that is, they develop

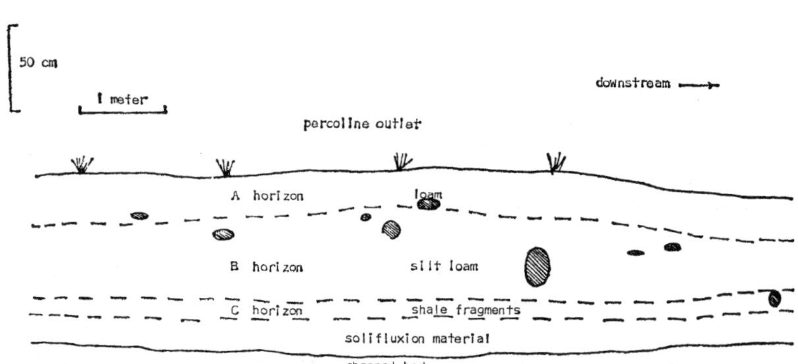

Fig. 3. Section of part of the bank of the Burbage Brook, showing a cluster of pipe outlets (vertical exaggeration × 2).

TABLE 2. Physical and Chemical Properties of Piped Soils from the Burbage Brook, Derbyshire, England

Property	Location of Sample at Pipe Outlet			Number of Samples for Each of the Three Locations
	Top	Side	Base	
Clay finer than 2 μm, %	18	18	16	21
Silt–clay finer than 75 μm, %	46	38	40	24
Finer than 250 μm, %	69	63	62	24
D_{50}, μm	100	150	150	24
Textural type	loam–sandy loam			82 (total)
Organic material, %	6.8	5.3	5.4	24
Clay minerals*	kaolinite–chlorite (dominant), illite, and chlorite–vermiculite–montmorillonite complex			15 (total)
Aggregate stability, %	73	60	55	20
Mean weight diameter†	3.69	3.12	2.67	20
Saturation percentage	58	56	55	24
*p*H	3.9	4.0	4.2	15
Electrical conductivity, mmho cm^{-1} at 25°C	0.602	0.724	0.667	24
Salt concentration, mg l^{-1}	385.4	463.4	426.7	24
Total cation concentration, meq l^{-1}	6.0	7.2	6.7	24

All methods of analysis followed the routines suggested in *Black et al.* [1965].
* X-ray diffraction on untreated and glycerolated samples by Dr. Chris Jeans, Department of Geology, Cambridge.
† Van Bavel's index calculated by the Youker-McGuinness approximation [*Black et al.*, 1965, p. 505].

entrainment piping as described by *Parker and Jenne* [1967]. These soils are capable of cracking despite the lack of severe droughts and despite the lack of any significant amounts of highly expansive clay minerals (Table 2).

In the light of much discussion on the importance of the erodibility of piping soils, it is surprising that hardly any investigator has measured erodibility except *Brown* [1961]. A series of tests of water stable aggregates on the Burbage soils (Table 2) suggests that the base of the pipes tends to be located at the level of lowest aggregate stability and mean weight diameter in the profiles. Thus the idea that piping is linked with an erodible layer in the soil is confirmed. Within the piped layer, there are more aggregates greater than 2 mm in diameter in the roof and more smaller aggregates in the sides and floor ($\alpha = 0.01$, $N = 20$). The lower degree of flocculation in the base may be partially secondary owing to either deflocculating salts in the pipe water or deltaic deposition when flow ceases. The lower degree of flocculation is a much more marked feature than the graduation in grain size distribution and appears to be more abrupt than the overall gradient in aggregate size through the soil profile. The importance of deflocculating salts cannot at present be substantiated. However, some evidence has been obtained of a thin (less than 2 mm) lining of finer material on the base of some pipes, which could slightly lower the apparent aggregate stability. *Brown* [1961] found that piped soils in Colorado were highly erosive to extremely erosive, although he did not sample for variations within the profile.

Despite the contentions of various workers, *p*H and salinity do not appear to be critical (Table 2). The literature records *p*H as varying from 9.8 to 4.1, and analyses on the Derbyshire soils indicate a variation of 3.8–4.3. Again, published data suggest high soluble salts content as the most common piping environment, yet the Derbyshire samples indicate values between one and two orders of magnitude lower than *Parker's* [1963] values. These differences probably reflect the difference between humid and arid environments and have no clear genetic relationship to piping. In Arizona, *Jones* [1968] concluded that soil chemistry was largely irrelevant, although his sampling was inadequate. Nevertheless, although his conclusion may be true of the 'static' chemistry measured after the initiation, 'dynamic' chemistry during initiation is probably very important to deflocculation [*Ingles,* 1964]. Unfortunately, observations of the process, like Ingles' observations on earth dams, are very difficult to make in the natural landscape.

Hydraulic gradient is plainly of paramount importance. There is strong evidence that piping is associated with stream incision [*Jones,* 1968], and there is evidence of such rejuvenation in the River Derwent system in Derbyshire, of which Burbage Brook is part. A common cause of stream incision has been overgrazing [*Fletcher and Carroll,* 1948]. Similarly, there is circumstantial evidence that piping is more common in upland areas than in lowlands. The disparity between observations of piping on steep, valley side slopes [*Zeitlinger,* 1959; *Knapp,* 1970] and on shallow floodplains [*Ward,* 1966; *Dalrymple et al.,* 1968] may be explained by the fact that the controlling factor is hydraulic gradient and not physiographic gradient. Hence, to develop on low slopes, piping requires a gully wall or stream bank nearby, whereas on steep slopes pipe outlets may be found on the surface of the same slope segment as the inlet. The latter situation can lead to a discontinuous pipe network analogous to the discontinuous gully networks described by *Leopold and Miller* [1956] and *Heede* [1970], and fusion may take place by both upslope and downslope extension. Examples of this type of pipe network have been found in the Mendips (D. Weyman and M. Stagg, personal communication, 1970), and on the Bagshot Sands (J. B. Dalrymple, personal communication, 1968). Weyman has observed linear elements of saturated overland flow extending from these pipe outlets and integrating with the expanded stream net.

A number of authors have suggested the possibility of biotic origin for some pipes. Whereas biotic initiation (e.g., by dead root holes or animal burrows) is possible, the evidence clearly indicates that it is of very secondary importance compared to physical factors [*Bond,* 1941]. An interesting case was found on salt marshes on the Norfolk coast, in which burrowing crabs appeared to be aiding pipe drainage of the salt pans.

VERTICAL LOCATION OF PIPE OUTLETS

The vertical location of pipe outlets along the Burbage Brook was found to be almost normally distributed about a mean of 44% bank-full height (Table 1). This finding compares with the observation by *Buckham and Cockfield* [1950] that the outlets 'often' occur at one-half to two-thirds of the way down gully walls. However, *Parker* [1963] notes that 'many outlets' occur at the base of the walls and explains that this is the lowest point above the local baselevel. It seems reasonable to suppose that the baselevel for piping is determined by the flood recession curve. Although the writer has observed pipes flowing or trickling at levels 30–40 cm above the stream stage during flood recession, presumably these pipes have not yet eroded down to their baselevel. Lower outlets suggest either (1) more developed piping with larger cross-sectional area; (2) older soils, in which a Bt horizon has migrated down the profile, in which case the cross-sectional area may remain small; (3) more flashy river regimes with lower base flow stage; (4) deeper desiccation and pseudopipe development; or (5) ephemeral streams with sudden drawdown.

CONCLUSIONS

Piping is clearly a widespread feature in the British Isles and seems likely to be as characteristic of humid temperate regions as of semiarid areas, if not more so. The location and development of piping seem to be more dependent on physical than on chemical or biological factors. Analyses suggest that pipes cluster areally around favorable soils and steep hydraulic gradients and cluster vertically around fine, more erodible soil horizons. The pipes analyzed above are probably mostly in a youthful or arrested youthful stage. Observations of partially collapsed piping in Derbyshire and the Brecon Beacons suggest that in a favorable location pipes develop beyond the control of the original horizon. This favorable horizon is commonly Bt in the Burbage area, but observations show piping developed at many other levels, for example, at peat–sandy loam interfaces on podzols or silt–gravel interfaces on salt marshes.

Hydrologically, piping speeds throughflow and will result in a quicker hydrograph response than might otherwise be expected. During stream incision, selected pipes may become important as sites of active headward erosion, as in the San Pedro Valley, Arizona [*Jones,* 1968], and probably in a wide variety of climatic regions.

Pipes appear to be the first of a series of fluvial landforms that show increasing symmetry and regularity downstream with increasing discharge. As discharge increases, hydraulic laws become dominant over soils, geology, or land slope. This dominance appears to be reflected in spatial distribution as well as in landform geometry, although the transition is not smooth. There is a marked discontinuity at the stream bank, with, for example, a bifurcation ratio of approximately 150 or more for part of the Burbage Brook. The continued increase in the relative proportion of discharge in larger channels downstream may, at least in part, account for the increasing sparseness of channels downstream reflected in *Schumm*'s [1956] observation that bifurcation ratio decreases with stream order.

Acknowledgments. I am particularly indebted to Mr. R. J. Chorley for criticism of the manuscript, to my numerous field assistants, and to the technical staff of the Department of Geography, Cambridge. I am also extremely grateful for the assistance of Dr. Chris Jeans of the Department of Geology, Cambridge, who performed the clay mineral analyses; Dr. John B. Dalrymple, Department of Soil Science, Reading, and Mike Garth, University of East Anglia, for describing field observations; and Messrs. Mike Stagg and Darrell Weyman of the University of Bristol and John Pethick, Brathey Field Study Centre, Ambleside, Westmorland, for leading me to field examples in the Mendips and Norfolk, respectively.

REFERENCES

Aitchison, G. D., and C. C. Wood, Some interactions of compaction, permeability, and post-construction deflocculation affecting the probability of piping failure in small earth dams, *Proc. 6th Int. Conf. Soil Mech. Found. Eng.*, Montreal, Canada, 442–446, 1965. (Also *Div. Soil Mech. Res. Pap. 61,* Commonwealth Scientific and Industrial Research Organisation, Canberra, Australia, 1965.)

Barton, D. E., and F. N. David, Tests for randomness of points on a line, *Biometrika, 43,* 104–112, 1956.

Black, C. A., et al., Editors, *Methods of Soil Analysis,* 2 vols., American Society of Agronomy, Madison, Wisconsin, 1965.

Blong, R. J., Subsurface water as a geomorphic agent with special reference to the Mangakowhiriwiri catchment, *Auckland Stud. Geogr. (N.Z.), 1*(2), 82–95, 1965.

Bond, R. M., Rodentless rodent erosion, *Soil Conserv., 10,* 269, 1941.

Brown, G. W., Some physical and chemical soil properties associated with piping erosion in Colorado, M.Sc. thesis, Colorado State University, Fort Collins, Colorado, 1961.

Bryan, K., Classification of springs, *J. Geol., 27,* 522–561, 1919.

Buckham, A. F., and W. E. Cockfield, Gullies formed by sinking of the ground, *Amer. J. Sci., 248,* 137–141, 1950.

Bunting, B. T., The role of seepage moisture in soil formation, slope development and stream initiation, *Amer. J. Sci., 259,* 503–518, 1961.

Bunting, B. T., Slope development and soil formation on some British sandstones, *Geogr. J., 130,* 73–79, 1964.

Carroll, P. H., Soil piping in southeastern Arizona, 21 pp., *Reg. Bull. 110,* Soil Series 13, U.S. Department of Agriculture Soil Conservation Service, Region 6, Albuquerque, New Mexico, 1949.

Childs, E. C., *An Introduction to the Physical Basis of Soil Water Phenomena,* 320 pp., John Wiley, London, 1969.

Culling, W. E. H., Theory of erosion on soil-covered slopes, *J. Geol., 73,* 230–254, 1965.

Dacey, M. F., The spacing of river towns, *Ann. Ass. Amer. Geogr., 50,* 59–61, 1960.

Dacey, M. F., Length of stream for an independent events model, *Water Resour. Res., 6* (1), 341–344, 1970.

Dalrymple, J. B., R. J. Blong, and A. J. Conacher, An hypothetical nine unit landsurface model, *Z. Geomorphol., 12*(1), 60–76, 1968.

Downes, R. G., A soil, land-use, and erosion survey of parts of the counties of Moira and Delatite, Victoria, 89 pp., *Bull. 243,* Commonwealth Scientific and Industrial Research Organisation, Canberra, Australia, 1949.

Dunne, T., and R. D. Black, An experimental investigation of runoff production in permeable soils, *Water Resour. Res., 6*(2), 478–490, 1970.

Fletcher, J. E., and P. H. Carroll, Some properties of soils that are subject to piping in southern Arizona, *Soil Sci. Soc. Amer. Proc., 13,* 545–547, 1948.

Gregory, K. J., and D. E. Walling, The variation of drainage density within a catchment, *Int. Ass. Sci. Hydrol. Bull., 13*(2), 61–68, 1968.

Heede, B. H., The fusion of discontinuous gullies, *Int. Ass. Sci. Hydrol. Bull., 12*(4), 42–50, 1967.

Heede, B. H., Morphology of gullies in the Colorado Rocky Mountains, *Int. Ass. Sci. Hydrol. Bull., 15*(2), 79–90, 1970.

Horton, R. E., Erosional development of streams and their drainage basins: hydrophysical approach to quantitative morphology, *Bull. Geol. Soc. Amer., 56,* 275–370, 1945.

Ingles, O. G., The water-soil regime in three dams sampled during failure, in Colloquium on failure of small earth dams, 4 pp., *Water Res. Ass. Austr. Ltd. Soil Mech. Sect. Pap. 31,* Commonwealth Scientific and Industrial Research Organisation, Canberra, Australia, 1964.

James, W. R., and W. C. Krumbein, Frequency distributions of stream link lengths, *J. Geol., 77*(5), 544–565, 1969.

Jones, N. O., The development of piping erosion, Ph.D. thesis, University of Arizona, Tucson, Arizona, 1968.
Kassiff, G., and E. N. Henkin, Engineering and physico-chemical properties affecting piping failure of low loess dams in the Negev, pp. 13–16, *Proc. 3rd Reg. Conf. Soil Mech. Found. Eng., 1,* Haifa, Israel, 1967.
Kirkby, M. J., Measurement and theory of soil creep, *J. Geol., 75*(4), 359–378, 1967.
Kirkby, M. J., and R. J. Chorley, Throughflow, overland flow and erosion, *Int. Ass. Sci. Hydrol. Bull., 12*(3), 5–21, 1967.
Knapp, B. J., A note on throughflow and overland flow in steep mountain watersheds, *Reading Geogr., 1,* 40–43, 1970.
Leopold, L. B., and J. P. Miller, Ephemeral streams: Hydraulic factors and their relation to the drainage net, 37 pp., *U.S. Geol. Surv. Prof. Pap. 282-A,* 1956.
Liao, K. H., and A. E. Scheidegger, Theoretical stream lengths and drainage areas in Horton nets of various orders, *Water Resour. Res., 5*(3), 744–746, 1969.
Parker, G. G., Piping, a geomorphic agent in landform development of the drylands, *Int. Ass. Sci. Hydrol. Publ. 65,* 103–113, 1963.
Parker, G. G., and E. A. Jenne, Structural failure of western U.S. highways caused by piping, 27 pp., U.S. Geological Survey Water Resources Division, Washington, D. C., 1967.
Schumm, S. A., Evolution of drainage systems and slopes in badlands at Perth Amboy, N. Y., *Bull. Geol. Soc. Amer., 67,* 597–646, 1956.
Smith, H. T. U., Piping in relation to periglacial boulder concentrations, *Biul. Peryglacjalny, 17,* 195–204, 1968.
Terzaghi, K., Earth slips and subsidences from underground erosion, *Eng. News Rec., 107,* 90–92, 1936.
Ward, A. J., Pipe/shaft phenomena in Northland, *J. Hydrol. (N.Z.), 5*(2), 64–72, 1966.
Whipkey, R. Z., Storm runoff from forested catchments by subsurface routes, in Floods and their computation, *Int. Ass. Sci. Hydrol. Publ. 85, 2,* 773–779, 1969.
Wiklander, L., Recipitation of iron hydroxide in drain pipes, *Kgl. Lautbruksakad. Tidskr., 84,* 358–367, 1945.
Zeitlinger, J., Observations of sub-surface erosion in loam (in German), *Mitt. Oesterr. Geogr. Ges., 101*(1), 9–95, 1959. (English abstract in *Geomorphol. Abstr.* 61/6, 1961.)

(Manuscript received October 5, 1970;
revised November 16, 1970.)

Commentary on Paper 19

Trafford, B. D. & Rycroft, D. W. (1973) Observations on the soil-water regimes in a drained clay soil.

Wherever the hillslope hydrologist works, s/he has to remember that every hillslope has a history, and very many of those histories involve modification by man in a way that affects the hydrology. In many cases the modifications may have been long in the past (and often forgotten; I have known several streamflow generation studies in which the research discovered the presence of artificial drains only after instrumentation had been installed). In many cases those modifications are on-going. This volume does not consider the impacts resulting from urbanization, nor those resulting from afforestation and deforestation, but agricultural practices can also have important effects on the hydrology, including field drainage, tillage/no-tillage practices, direction of ploughing, etc.

The issue of whether field drainage increases or decreases flood peaks has been a contentious issue in the UK for a long time. Nicholson (1953) reviewed the discussion of a paper given at the Institution of Civil Engineers, London, in 1861 by J. Bailey Denton, in which it was suggested that underdrainage in the catchment area upstream during the preceding 20 or 30 years (1830–1860) had greatly reduced the time to peak in the Thames near Abingdon. Note that this is some 130 years before the peak of the government grant-aid supported drainage schemes of the 1960s and 1970s in the UK. Nicholson (1953) concludes that field drainage should in general reduce peak flows by increasing the available storage between events and reducing the likelihood of saturation causing fast runoff

As an example of such studies, this paper by Brian Trafford and David Rycroft (1973) is an investigation of the effects of field drainage on discharges from clay soils. There is an interesting history of such investigations in the UK which, for a period at least, even had a Field Drainage Experimental Unit (FDEU), near Cambridge, that was a research institute of the UK Ministry of Agriculture, Fisheries and Food (MAFF). MAFF provided subsidies for field drainage, by both tile drains and mole drains, in the UK in the 1970s and 1980s. Analysis for the design of field drainage systems was important in many other countries, particularly The Netherlands. However, the FDEU made measurements of how drainage systems were actually working at a number of different sites, some of which are still in use today (see for example, Cannell *et al.*, 1984; Harris *et al.*, 1984, 1993; Harris & Catt, 1999), and this paper is an example of the types of response found for discharges, soil moisture potentials, water table levels and profiles. Trafford & Rycroft also provide calculations of effective hydraulic conductivities based on drainage curves, resulting in very high values for the mole drained plot when the water table is close to the surface.

In the UK, there were a number of other studies at different sites including at Grendon Underwood (Beven, 1980); Enfield Chase (Reid & Parkinson, 1984); North Wyke (Scholefield *et al.*, 1993; Armstrong & Garwood, 1991); Wytham Wood (Morecroft *et al.,* 2000). The Grendon Underwood experiment carried out by the Institute of Hydrology (see Robinson & Beven, 1983; Robinson, 1990) reveals some of these complexities. This experiment, in the River Ray catchment, Oxfordshire, was based on a pair of field plots (2500 m^2) on heavy clay soil (Denchworth series) and under permanent pasture. The plots had a low amplitude ridge and furrow topography. One plot was mole drained in the furrows; the other was left undrained such that the furrows saturated easily in the winter period.

Robinson (1986, 1990; Robinson & Rycroft, 1999), summarizing the results from many different studies, suggests that drainage tends to reduce peak flows from clay catchments by increasing antecedent storage deficits, but that drainage may increase peak flows for more permeable soils. He notes that the type of drainage scheme might also be important, with moling and subsoiling giving higher peaks than pipe drains alone, and open ditches giving higher peaks than subsurface drains. Ground condition, due to both agricultural practices and cracking in heavy clay soil, may also be important in controlling the responses, but in general drainage tended to modify the timing of runoff and the peak flow rather than the volume of runoff from a given storm.

Brian Trafford retired as Director of the Field Drainage Experimental Unit; David Rycroft later moved to teach in Civil and Environmental Engineering at the University of Southampton, becoming Director of the Institute of Irrigation Studies there.

References

Armstrong, A. C. & Garwood, E. A. (1991) Hydrological consequences of artificial drainage of grassland. *Hydrol. Processes* **5**, 157–174.

Beven, K. J. (1980) The Grendon Underwood field drainage experiment. *Institute of Hydrology Report 65*. Wallingford, UK.

Cannell, R. Q., Goss, M. J., Harris, G. L., Jarvis, M. G., Douglas, J. T., Howse, K. R. & LeGrice, S. (1984) A study of mole drainage with simplified cultivation for autumn-sown crops on a clay soil. 1. Background, experiment and site details, drainage systems, measurement of drainflow and summary of results, 1978–1980. *J. Agric. Sci.* **102**, 539–560.

Harris, G. L., Goss, M. J., Dowdell, R. J., Howse, K. R. & Morgan, P. (1984) A study of mole drainage with simplified cultivation for autumn-sown crops on a clay soil. 2. Soil water regimes, water balances and nutrient loss in drain water, 1978–1980. *J. Agric. Sci.* **102**, 561–581.

Harris, G. L. & Catt, J. A. (1999) Overview of studies on the cracking clay soil at Brimstone Farm, UK. *Soil Use Manage.* **15**, 233–239.

Morecroft, M. D., Burt, T. P., Taylor, M. E. & Rowland, A. P. (2000) Effects of the 1995–1997 drought on nitrate leaching in lowland England. *Soil Use & Manage.* **16**, 117–123.

Nicholson, H. H. (1953) *The Principles of Land Drainage.* Cambridge University Press, London, UK.

Reid, I. & Parkinson, R. J. (1984) The nature of the tile-drain outfall hydrograph in heavy clay soils. *J. Hydrol.* **72**, 289–305.

Robinson, M. (1986) Changes in catchment runoff following drainage and afforestation. *J. Hydrol.* **86**, 71–84.

Robinson, M. (1990) Impact of improved land drainage on river flows. *Institute of Hydrology Report no. 113.* Wallingford, UK.

Robinson, M. & Beven, K. J. (1983) The impact of mole drainage on the hydrological response of a swelling clay soil. *J. Hydrol.* **64**, 205–223.

Robinson, M. & Rycroft, D. W. (1999) The effect of drainage on streamflow. In: *Agricultural Drainage* (ed. by R. W. Skaggs & J. van Schifgaarde), 767–800. Agronomy Monograph no. 38. ASA, CSSA and SSSA, Madison, Wisconsin, USA.

Scholefield, D., Tyson, K. C., Garwood, E. A., Armstrong, A. C., Hawkins, J. & Stone, A. C. (1993) Nitrate leaching from grazed grassland lysimeters: effects of fertilizer input, field drainage, age of sward and pattern of weather. *J. Soil Sci.* **44**, 601–614.

OBSERVATIONS ON THE SOIL-WATER REGIMES IN A DRAINED CLAY SOIL

B. D. TRAFFORD AND D. W. RYCROFT

(*Field Drainage Experimental Unit, Anstey Hall, Trumpington, Cambridge*)

Summary

Details of soil-water profiles between drains and the associated drain-flow data are presented for the winter period on a clay soil of the Ragdale series, formed in drift over Lias clay in the Midlands of England. A conventional water-table is shown to exist and the rate of drain flow is shown to depend on whether the water-table is high enough to be within layers having a better hydraulic conductivity than the normal subsoil. Mole drainage is shown to modify the drainage performance considerably by providing a quicker disposal of water, and significantly lower water-table subject to much smaller fluctuations in level. Where drains have been laid in the natural subsoil, without this being disturbed by deep cultivations, the water-table level is frequently close to cultivation level and there appears to be little water movement in the subsoil. It is concluded that for effective drainage, soils of this type require deep cultivations to modify the subsoil and allow water movement at greater depth.

Introduction

THE importance of the cultivated layer to water movement in the drainage of a clay soil has long been recognized (Childs, 1943; Nicholson, 1953). Some workers have reported investigations which suggested that the subsoil took no effective part in the drainage of such soils (Schleifer and Feichtinger, 1954; Steinhardt, *et al.*, 1972; Coulbois, 1972). The Field Drainage Experimental Unit operate three experimental sites on clay soils; in all cases present results suggest that simple pipe drainage has had little effect on either the soil water regime or cropping. Significant changes have been achieved only where the subsoil has been modified by either mole drainage or subsoiling (Annual Reports of the Field Drainage Experimental Unit, 1969, 1970, and 1971). The purpose of this paper is to present details of one winter's work at one site (Brooksby) which aimed at elucidating the relationship between observed drainflow and soil water profiles.

Experimental site and methods

The experimental site at Brooksby Hall has been fully described in the Field Drainage Experimental Unit Annual Reports for 1970 and 1971 and only brief details will be given here. The general layout of the experiment is shown in Fig. 1.

The soil is classified as Ragdale series formed in drift over Lias clay. This soil is in general decalcified down to approximately 30 cm and the most marked distinction in the profile on this site is the fairly sharp transition between the cultivated layer and the subsoil approximately 26 cm below ground level. A detailed soil profile description appears in

Journal of Soil Science, Vol. 24, No. 3, 1973

the F.D.E.U. Report for 1970. Prior to this investigation the site had been ploughed and was still in furrow awaiting spring cultivation.

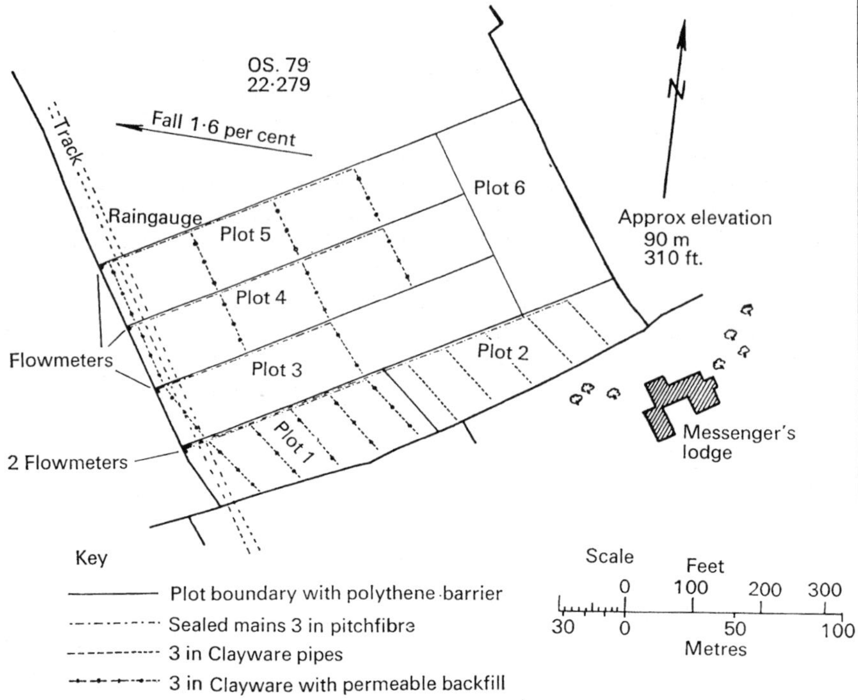

Plot No	Drainage treatment	Cost on field scale (1970)
1	1 Chain spacing, permeable fill, no secondary treatment	£72 per acre
2	1 Chain spacing no P.F. no secondary treatment	£50 per acre
3	4 Chain spacing permeable fill & moled	£29 per acre
4	2 Chain spacing, permeable fill & moled	£47 per acre
5	2 Chain spacing permeable fill & sub-soiled	£47 per acre
6	None-control plot	

FIG. 1. Site plan showing experimental layout at Brooksby Hall.

The five drainage treatments correspond to the range of designs found on similar soils in the area. Results in earlier years have shown that the effect of the two simple piped drainage treatments in plots 1 and 2 have been similar in terms of drainflow hydrographs. Plot 4, drains at 40 m with moling, has generally given the best drainage performance.

The similarly spaced drains with subsoiling (plot 5) and the 80 m spacing with moling (plot 3) gave results intermediate between the records for plots 1 and 4. Because of this it was decided to investigate the soil-water regime in plot 1 intensively in an attempt to elucidate the role of the various soil layers in the water movement. To obtain comparative data from what appeared to be the most effectively drained plot some observations were made on plot 4.

As direct methods for measuring the water movement in any given layer of the soil are not available it was decided to use the relationship between the soil-water profile, on a cross section between two drains, and the drain outflow to give some indications of the respective contributions of the cultivated layer and the subsoil to the total flow. Isherwood and Pillsbury (1958), Kirkham and Gaskell (1950), together with French and O'Callaghan (1966), indicate that the drain outflow may display differing characteristics during recession depending upon whether the water-table is in a zone of high hydraulic conductivity where the water may be transmitted easily to the trench zone which may have appreciably greater conductivity than the surrounding soil. Goins and Taylor (1959) reported a linear relationship between mid-plane water-table height and drain flow whilst the water-table was in the upper layers of the soil (0–15 cm). Hoffman and Schwab (1964) reported a curvilinear relationship for the same two variables in a silty clay soil whilst the water-table was below the cultivated layer. In earlier work on the data from the Brooksby site Rycroft (1971) had shown that the recession limbs of the winter-period hydrographs for the individual plots could in general be fitted to a standard curve. During the winter, some factors normally affecting the hydrograph, such as absorption of water by the soil and evaporation following storms, are negligible and hence the recession limb of the hydrograph, occurring after rainfall has ceased, is primarily a function of the water-table height, the drain spacing, and the soil characteristics. Large differences have been shown to occur in the recessions on plots 1 and 4. In previous seasons continuous drainflow data had been recorded but only a few soil-water levels measured at weekly intervals.

A considerable problem in measuring soil-water profiles in a clay soil is that most methods cause some disturbance to the soil-water system in the vicinity and have a time lag. This time lag can be minimized by reducing the volume of water entering or leaving the instrument for a given level change. In a clay soil water movement tends to be confined to the fissures and hence there may be significant differences in pore-water pressures in short distances. Tensiometers have a small volume change and hence minimum lag times but are somewhat troublesome in use and only contact a limited area of soil. Simple auger holes or dipwells sample a large volume of soil but suffer large lag times. In this experiment both tensiometers and dip wells were used. However, the dip wells were modified to retain the large contact area yet minimize the volume change by filling up a large part of the hole with a very thick walled tube. The recording water-table meters used sense the water head by means of flexible bellows, the movement of which is transmitted to a pen on a chart. This instrument has a lag time sufficiently short for fluctuations occurring during storms to be satisfactorily recorded.

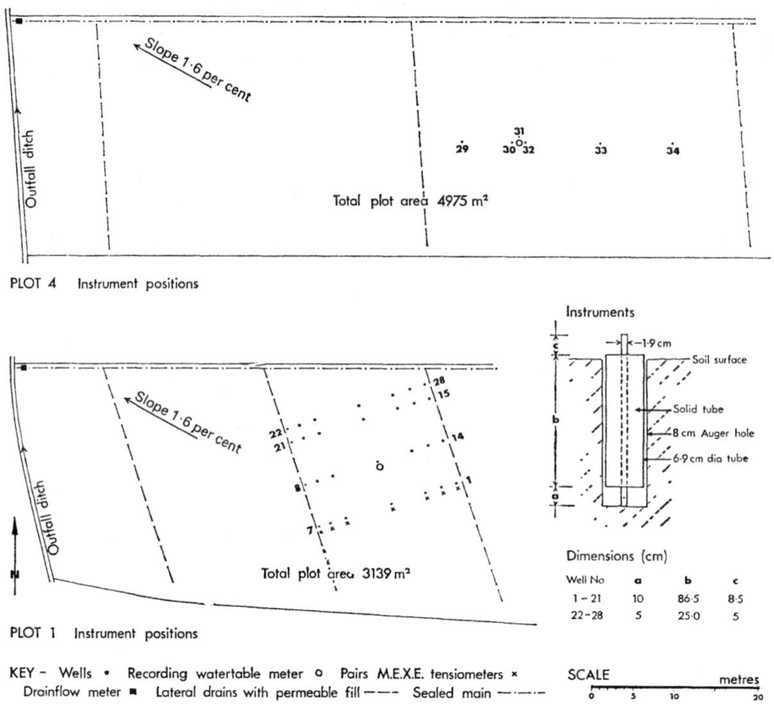

FIG. 2. Instrument positions. Brooksby Hall 1971–2.

Fig. 2 indicates the positions and types of instruments used. The shallow wells Nos. 22–8 were installed almost entirely within the plough layer on plot 1 in order to reduce the response time which affected the deeper wells 1–21. Tensiometers were also installed with the same purpose and also to yield data on the hydraulic gradients. The continuous recording water-table meter was sited at the central position, adjacent to well 11.

For comparative data on plot 4 a line of wells 29–34 was installed. Three of these (30 to 32) were control wells around the recording water-table meter.

Measurements of water-table height, tension, and drainflow were recorded throughout the winter at intervals, varying between 2 days during dry periods and 4 hours during storms. Hydraulic conductivity and infiltration measurements were also made on plots 1 and 4. The purpose of these on plot 1 was to establish a general figure for the plot and help to explain differences of water-table profile on the three lines of wells. The measurements were made by sampling the soil immediately below plough depth with 10 cm diam. and 12 cm long samplers. After saturation, the cores were fitted into a falling head permeameter. Measurements were also made *in situ* at the existing wells. These were done by removing the loose fitting core and recording the subsequent recovery of water level in the well, using the formula derived by Ernst (1950) in the bulletin by van der Beers (1963) to convert the recovery to values of hydraulic conductivity. On plot 4 hydraulic conductivity measurements were made by the auger-hole method only at positions 30, 31, and 32.

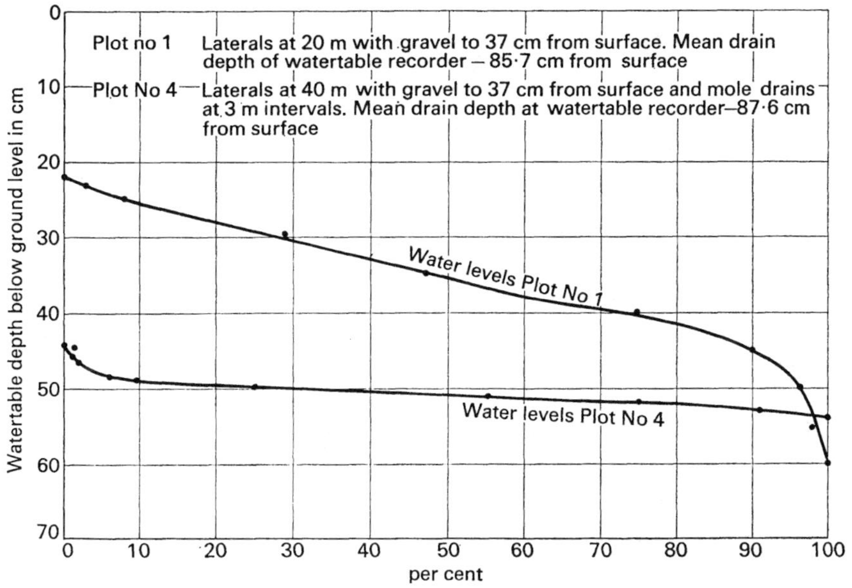

FIG. 3. Percentage number of daily water levels equalling or exceeding the values stated.

Results and discussion

Water-table heights

The details of the water-table levels as indicated on the recording water-table meters are given in Fig. 3. These results are plotted as percentage frequencies for the winter period and show that, on plot 1 water levels were in the topsoil (top 26 cm) for a total of 15 per cent of the time. Water levels were higher than 40 cm for the major part of the time. By contrast on the moled treatment, plot 4, the water level was fairly constant at around 50 to 55 cm. This is very close to the depth of the moling.

Drainflow and water-table profiles

Fig. 4*a* shows the hydrographs for a winter storm and Fig. 4*b* the simultaneous continuous water-table meter records. The total drainflow expressed as a percentage of rainfall was 69 per cent on plot 1 and 90 per cent on plot 4. The hydrograph evidence taken with the water-table levels at mid-spacing (the recording meter on plot 4 was only approximately at mid-spacing due to an initial measurement error)

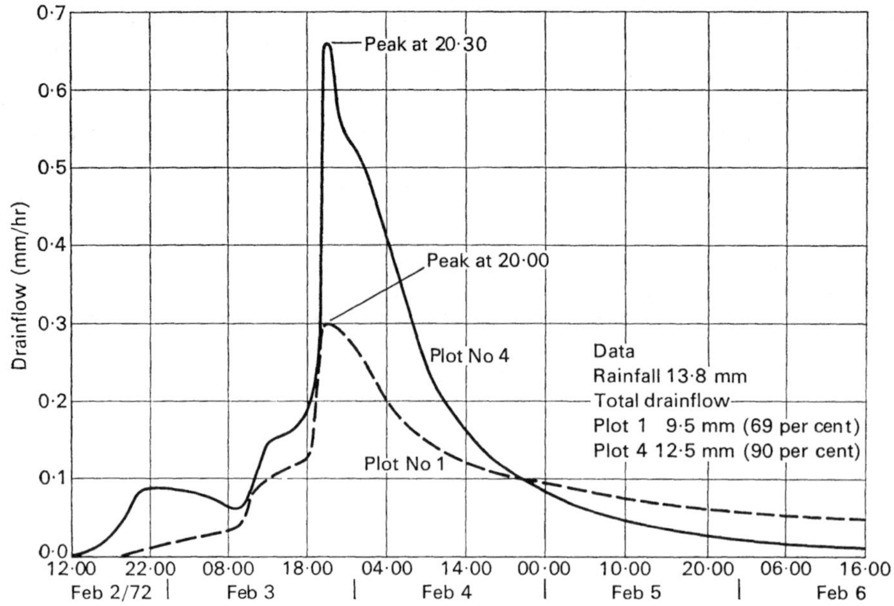

FIG. 4 (*a*). Hydrographs. Plots 1 and 4.

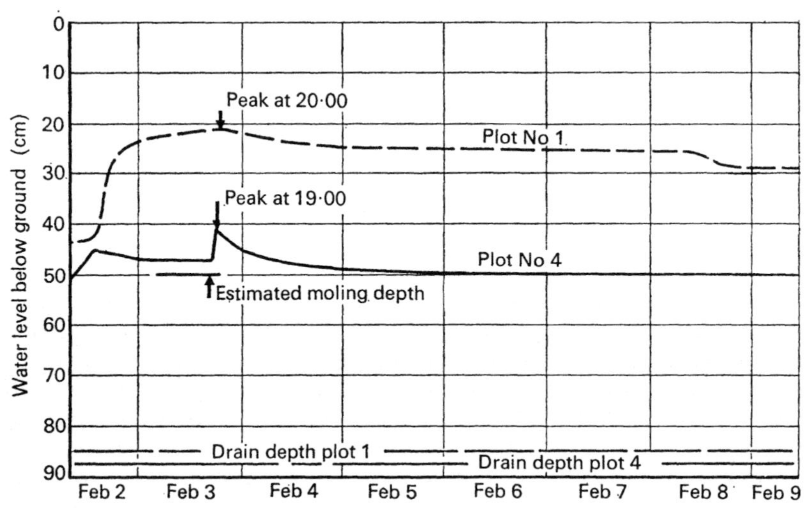

FIG. 4 (*b*). Water-tables.

suggest very different hydrologic behaviour in plots 1 and 4. There is evidence of rapid movement of water to the drains in plot 4 (moled plot) with little rise in the water-table. The water-table profile on two sections on plot 1 are shown on Fig. 5. Line 8–14 is the central line of 1 m deep wells. It is similar to the variation occurring on the lines 1–7 and 15–21 except that the residual and maximum levels increased from lines 1–7 to 15–21 by 10 cm. The line of wells installed mainly in the plough layer (22–8) shows a further increase which is a logical extension of the general increase down the plot.

The soil-water profiles shown in Fig. 5 differ from profiles usually associated with two-layered soils. If the subsoil had a uniform and significant hydraulic conductivity, the water-table level would become horizontal at drain level at zero flow. However, as is shown by the data

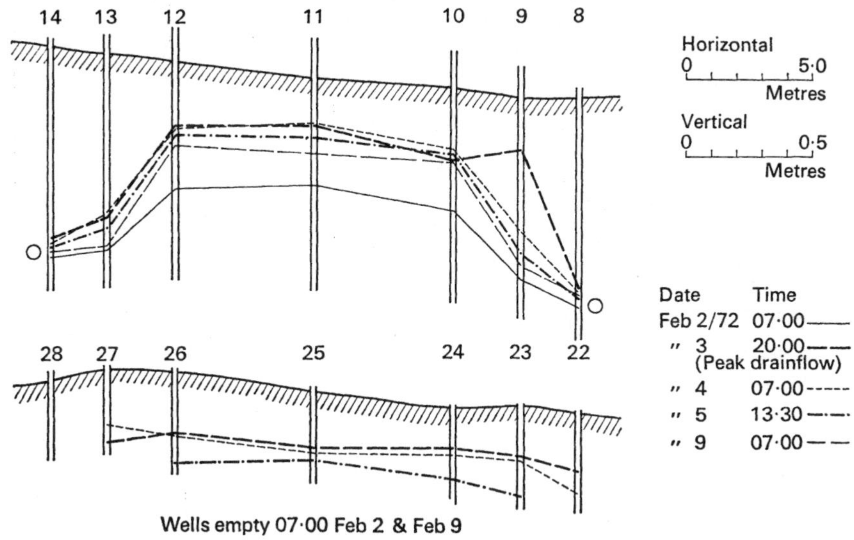

FIG. 5. Plot No. 1. Water-table profiles.

for 2 February 1972 07.00 h, this is not the case by a large margin. In fact at no time during the observations did the water-table on plot 1 become horizontal. On the other hand if the subsoil were impermeable the water-table could be virtually flat topped with no draw down until the trench zone was reached. (Water-tables of this type have been recorded on other F.D.E.U. clay sites). Generally speaking, the results from plot 1 are an amalgam of both modes of water movement. One hypothesis which might explain these water profiles would be that the soil structure has improved near the trench. Some workers, Fauser (1935) and Hakansson (1971), have shown structural improvements proceeding outwards from the trench. However, soil structure is very difficult to measure in quantitative terms and we have as yet no firm evidence that soil structure improvements do account for the shape of the water-table profiles.

On the moled plot (plot 4) there were insufficient data to plot water-table profiles due largely to the fact that replication was restricted to wells 30, 31, and 32 surrounding the water-table meter. The decision not to replicate was felt to be justified at the time of installation by the fact that wells could not be sited in known relation to the positions of the mole channels, and by the assumption that moling, by creating an artificially high hydraulic conductivity above a subsoil of much lower conductivity, resulted in fairly flat water-tables between permanent drains. Most of the evidence from water-table data on plot 4 would suggest that this was reasonable, as very small fluctuations (up to about 12 cm) were recorded during storms and the water-table appeared to follow the ground slope (which is also the gradient on the mole channels) closely. However, one well on plot 4 showed fluctuations in water level of up to 20 cm during storms. Whether this was because it was more remote from a mole channel is not known.

Tensiometers

The original intention was to use the tensiometers to monitor changes during and immediately after storms but this was not possible partly because of insufficient data during storms and partly because of large variations between adjacent instruments. The most likely cause of the latter problem is the heterogeneous nature of the soil and the small volume sampled by the tensiometers. Under quiescent conditions it was possible to analyse the data (Table 1).

TABLE 1

Mean soil-moisture tensions in Brooksby topsoil, plot 1 (10 cm), 4 February 1972 to 11 March 1972

Tensions in cm of water at cup (10 cm depth)		
Near western drain	*Mid-spacing*	*Near eastern drain*
19·7 S.D. ± 10·7	6·5 S.D. ± 15·9	22·1 S.D. ± 11·7

The tensiometer data show a curvilinear water-table with mid-spacing levels close to or into the cultivated layer (top 26 cm or so). These data support the dip-well data in general, and in particular curvilinear water-tables at very low drainflows.

Hydraulic conductivity

The values of hydraulic conductivity obtained from both the laboratory measurements on 'undisturbed' cores and by field measurements in auger holes are given in Table 2. With the exception of the reading for well 31 all the hydraulic conductivity values are low. The values for the 50–100 cm layer decrease from cross section 1–7 to 8–14 and further to 15–21. This is in agreement with the water-table profiles which increased in height over the same three cross sections. These sections are only 5 metres apart and hence there is a hydraulic conductivity reduction of roughly 20:1 in 10 metres. The values for the lower part of the topsoil on wells 8–14 do not suggest high hydraulic conductivities.

However, the present ploughed layer was somewhat more shallow and evidence obtained during attempts to measure the infiltration rate suggested that its hydraulic conductivity was of the order of 8×10^3 mm/day and that there was rapid water movement at plough sole level down the natural slope of the field (about 1·6 per cent).

TABLE 2

Position		Measurement depth (cm)	Hydraulic conductivity mm/day		Method used
			Mean	Standard deviation	
1–7	Plot 1	50–100	78·0		Auger hole
8–14		50–100	37·0	20·5	,, ,,
8–14		20–30	16·0	13·0	10 cm lab cores
15–21		50–100	4·0	2·5	Auger hole
29	Plot 4	50–100	30·0		,, ,,
30		50–100	10·0		,, ,,
31		50–100	400·0		,, ,,

The difficulties of measuring hydraulic conductivities and the variations in results in quite small areas are well known. However, knowing the drainflow, the boundary conditions, and the water-table profile it is possible to calculate an effective hydraulic conductivity. At Brooksby there is no precise impermeable layer, in general the hydraulic conductivity appears to become less with depth, and thus we did not know the boundary conditions sufficiently well to make other than checks on the values obtained by alternative techniques. The following equation (van Schilfgaarde, 1963) was used in estimating the hydraulic conductivity (K):

$$K = \frac{S^2}{9A^2} \cdot \frac{2n(Ho - H)}{t(de + H)(de + Ho)}$$

where:

S is the drain spacing.

H is the water-table height above the tile drains at the water-table meters. In the equation H is the mid-plane water-table height. This condition was met on plot 1 but not on plot 4 where the meter was 5 m from the mid-plane position.

Ho is the initial water-table height.

d is the depth of the impermeable layer below the tile drains. This was assumed to be 1 m, coinciding with the permanently saturated soil.

de is the equivalent depth defined by Hooghoudt, the numerical values of which are given by van Schilfgaarde. It is dependent on d.

A is a constant dependent upon de and Ho.

n is the effective drainable porosity calculated from the following equations:

$$n = \frac{3Qt}{2(Ho-H)}$$ for plot 1 where the water-table profile was assumed to be parabolic, and

$$n = \frac{Qt}{(Ho-H)}$$ for plot 4 where the water-table profile was assumed flat.

Q is the mean drainflow in mm/day over the time period considered.

t is the period over which the drainflow Q occurred.

The equation was applied in a number of situations where the rate of fall of the water-table was known. Each of the resulting values of hydraulic conductivity applied to a soil which was assumed to be homogeneous between the initial water-table and the assumed impermeable barrier. This is an assumption which clearly becomes difficult to justify as the water-table rises, on plot 1 into the ploughed zone, and on plot 4 above the mole drains. The values are included, however, on Fig. 6, as an indicator of the importance of these zones to water movement.

Fig. 6 contrasts the relationships between water-table height, and drainflow as recorded on the respective meters and also values of hydraulic conductivity, K, and effective drainable porosity, n. These have been associated with particular depth intervals within the soil which coincide with the intervals of water-table recession from which they were calculated. The results obtained for effective hydraulic conductivity by calculation (as in Fig. 6) and those obtained by measurement (Table 2) are of the same order.

General conclusions

The evidence in general and the shape of the water-table height/drain-discharge curves in particular suggest that, on both plots, drainage is a two-stage process. On plot 1 there is a linear relationship between water-table height and drainflow for mid-plane water tables nearer the surface than 27 cm, which is the approximate depth of the ploughed layer. The attempts to measure the infiltration rate of the ploughed layer demonstrated that it had a hydraulic conductivity greatly exceeding that of the subsoil, which formed a barrier, restricting downward movement of water and enhancing lateral flow about it. It is suggested that this ease for lateral flow is one possible reason why total drainflow on plot 1 is substantially less than the total rainfall during winter storms.

On plot 4 there is a linear relationship between water-table height and drainflow for water-tables between 47 cm and 40 cm below ground level, the 47 cm depth, approximating to moling depth. During storms the water-table rose slightly relative to the rise on plot 1, and fell much more rapidly to its original level than did the water-table on plot 1. At the end of a storm the drains on both plots continued to trickle; this

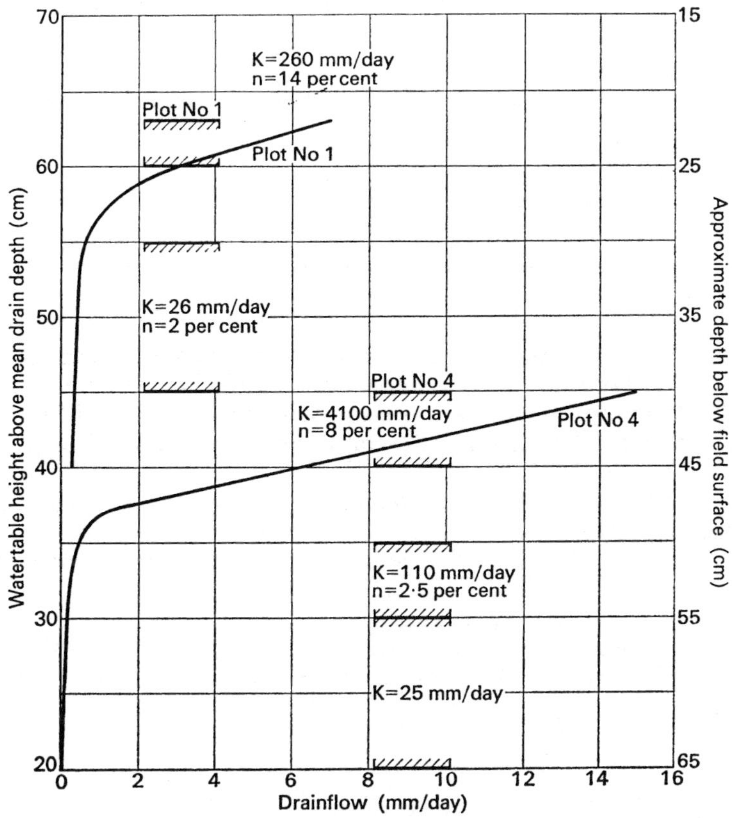

FIG. 6. Water-table/drainflow. Plots 1 and 4.

resulted in a water-table recession of approximately 2 cm/day on plot 1 and less than 0·5 cm/day on plot 4.

All the evidence suggests that the ploughed layer on plot 1 is most important in drainage terms, subsoil taking a very secondary part in the drainage process. On plot 4, the moled zone, down to approximately 50 cm, appears to have very different soil-water characteristics from the unmodified subsoil which dominate the drainage performance. One must conclude from these results that moling improves the drainage efficiency both in the amount of surplus water removed and in the control of the water-table at desirable levels.

Acknowledgement

Thanks are due to the Principal, staff, and students of the Brooksby Agricultural College, Brooksby, Melton Mowbray. Their co-operation and help made this work possible.

REFERENCES

CHILDS, E. C. 1943. Studies in mole-drainage—interim report on an experimental drainage field. J. agric. Sci. **33,** 136–46.

COULBOIS, C. 1972. Le drainage des sols argileux dans l'estuaire de la Charente. Trans. 8th Congress I.C.I.D. Varna. Report 19, 287–96.
ERNST, L. F. 1950. A new formula for the calculation of the permeability factor with the auger hole method. T. N. O. Groningen. Cornell Univ. Ithaca, N.Y. 1955.
FAUSER, O. 1935. Die Wirkung der Dränung auf das Gefüge des Bodens. Trans. 3rd int. Congr. Soil Sci. **1**, 391–7.
Field Drainage Experimental Unit Annual Reports 1969, 1970, and 1971. Ministry of Agriculture, Fisheries and Food, London.
FRENCH, B. E., and O'CALLAGHAN, J. R. 1966. A field test of drain spacing equations for agricultural land. J. Agric. Engng. Res. **11**, 282–95.
GOINS, T., and TAYLOR, G. S. 1959. Drain depth and spacing effects on water removal from a grey brown podzolic soil. Soil Sci. Soc. Am. Proc. **23**, 413–18.
HAKANSSON, A. 1971. Private communication.
HOFFMAN, G. J., and SCHWAB, G. O. 1964. Tile spacing prediction based on drain outflow. Trans. Am. Soc. Agric. Eng. 444–7.
ISHERWOOD, J. D., and PILLSBURY, A. F. 1958. Shallow ground water and tile drainage in the Oxnard plain. Trans. Am. Geophys. Union **39**, 1101–10.
KIRKHAM, D., and GASKELL, R. E. 1950. The falling water table in tile and ditch drainage. Soil Sci. Soc. Am. Proc. **15**, 37–42.
NICHOLSON, H. H. 1953. The principles of field drainage. Cambridge University Press.
RYCROFT, D. W. 1971. Drainflow Hydrographs. Annual Report Field Drainage Experimental Unit, Ministry of Agriculture, Fisheries and Food.
SCHLEIFER, H., and FEICHTINGER, F. 1954. Die Dränung tagwasservernässter Böden. Oesterreichische Wasserwirtschaft. 237–49.
STEINHARDT, R., SHALHEVET, J., and STERNBAUM, B. 1972. Field Drainage in the Western Jezreel Valley. Volcani Institute Prelim. Rep. No. 711.
VAN DER BEERS, W. F. J. 1963. The auger hole method. International Institute of Land Reclamation and Improvement. Bulletin 1.
VAN SCHILFGAARDE, J. 1963. Design of tile drainage for falling water tables. J. Irrig. and Drain Div. Proc. Am. Soc. Civil Engng. 1–11.

Discussion

E. W. Russell (Reading): How far does improving the drainage of heavy soil increase soil temperature at depths 5–10 cm during March and April? An effective drainage system in such soils seems to me to allow crops to start growth earlier in the spring because drainage increases the air space rather than the temperature in the surface soil.

B. D. Trafford: The evidence for increased soil temperature due to drainage is mainly in the older literature and the soil type is often unclear. Certainly, in the F.D.E. 11 experiments on heavy clay soils, no significant temperature differences have been recorded to date, despite considerable hydrologic differences between plots. In previous seasons, weekly manual temperature recordings have been taken at Brooksby. However, in 1972–3 automatic hourly recordings will be taken at several depths on a site at Layer Breton in Essex. We hope this work will demonstrate whether or not a temperature effect occurs on heavy soils.

(*Received 2 October 1972*)

Commentary on Paper 20

Beasley, R. S. (1976) Contribution of subsurface flow from the upper slopes of forested watersheds to channel flow.

Another set of trench experiments to measure surface and subsurface flow rates from hillslope plots in Northern Mississippi were reported by Scott Beasley (1976), working at the USDA Forest Hydrology Laboratory in Oxford, Mississippi. Beasley reports subsurface flows starting within 20 minutes of the onset of rain and peaking at the same time as the channel flow. He then discusses the mechanisms by which this might be achieved. He notes that, despite highly permeable surface soils, saturated hydraulic conductivities "*are not sufficient to account for peaks occurring <1 hour from the onset of rainfall.*" (p.957). He suggests that the Hewlett concept of translatory flow also cannot explain the fast responses "*because it requires a saturated zone near the point of outflow*" (p.957) and tensiometers above the trench at no time recorded saturated conditions during the storms following their installation. Beasley suggests that the rapid peaks in the subsurface flows would have required velocities of 800 m day^{-1} and concludes that the only mechanism that might explain such velocities is flow through macro-channels "*formed by decayed roots*" (p.957). Beasley notes that before the trench was backfilled with gravel he observed water gushing from such openings and cites the earlier studies of Gaiser (1952), Whipkey (1967), and Aubertin (1971) in support of this conclusion.

Beasley seems to think that the water flowing through the macropores is rainwater in that, "*before subsurface flow can occur in macrochannels, the channels must be open to the atmosphere and a positive head of water must exist at the openings. Such conditions are common in a forest environment. Depressions formed by uprooted trees or decayed stumps provide a place for water to concentrate. Decayed roots spreading outwards from the depressions provide natural pathways for water to flow freely through the soil towards the stream channels. Stemflow accumulating at the bases of trees is another possible source of significant quantities of subsurface flow*" (p.957). He does not discuss how such root openings might form continuous pathways and retain water through unsaturated soil.

Both Gaiser (1952) and Aubertin (1971) were working on forested catchments in Ohio. Aubertin carried out tracing experiments using artificial rainfall above 3.3-m trenches and also intensively sampled a large soil block that represented the lower part of the plot used by Whipkey (Paper 8) to map all roots and macropore features. The plots also showed water moving through macropores without the surrounding soil being saturated. He suggests that the source of water is infiltration into a spongy organic topsoil that collects in depressions in the surface of the mineral soil associated with decayed tree taproots. To explain the continuity of macropore flow pathways, Aubertin suggests that: "*In many, if not most, forested areas the soil mass is honeycombed with a complex network of intricately connected old root channels, animal passages and other microvoids. Strong arguments can be advanced to support the contention that, under these conditions, the complex network is always open to the atmosphere at numerous points and that truly dead-end isolated channels, or macrovoids, are rare and of little consequence*" (p.11).

In his summary, Aubertin notes: "*That free water can move rapidly substantial distances through relatively dry soil was demonstrated by an observation of seepage occurring from a pit face some 30 feet* [~9 m] *obliquely downslope from the pit under study. The seepage commenced between 30 and 45 minutes after starting to wet the forest floor above the observation pit*" (p. 27). He also notes evidence of new roots following the channels left by old roots, and that whether a root leaves an open channel will depend on the tree species.

Another country where flows through macropores have been extensively studied in a humid temperate environment is Japan, particularly in the studies of Kitahara and co-workers (e.g. Kitihara, 1993) on Hokkaido. These pipes, which the authors suggest were mostly due to decayed root channels, were found in the lower part of the soil profile at the bottom of the weathered regolith. Relationships between velocity and discharge were found for a number of pipes measured individually. Other trench studies of macropore flows in Japan, in the same period, were reported by Tsuboyama *et al.* (1994), Tanaka *et al.* (1988) and Tsukamoto & Ohta (1988), while more recently Uchida *et al.* (2001, 2002) also inferred that macropore flows were important in the response of basins dominated by subsurface runoff.

Scott Beasley left the Forest Hydrology Laboratory in 1976 to join the Department of Forest Resources at the University of Arkansas and in 1992 joined the Stephen F. Austin State University at Nacogdoches, Texas, as Dean of the Faculty of Forestry.

References

Aubertin, G. M. (1971) Nature and extent of macropores in forest soils and their influence on subsurface water movement. *USDA Forest Service Research Paper, NE192*. US Dept Agriculture, New York, USA.

Gaiser, R. N. (1952) Root channels and roots in forest soils. *Soil Sci. Soc. Am. Proc.* **16**, 62–65.

Kitihara, H. (1993) Characteristics of pipe flow in forested slopes. In: *Exchange Processes at the Land Surface for a Range of Space and Time Scales* (ed. by H.-J. Bolle, R. A. Feddes & J. D. Kalma), 235–242. IAHS Publ. 212. IAHS Press, Wallingford, UK.

Tanaka, T., Yasuhara, M., Sakai, H. & Marui, A. (1988) The Hachioji Experimental Basin Study – storm runoff processes and the mechanisms of its generation. *J. Hydrol.* **102**, 139–164.

Tsuboyama, Y., Sidle, R. C., Noguchi, S. & Hosada, I. (1994) Flow and solute transport through the soil matrix and macropores of a hillslope segment. *Water Resour. Res.* **30**(4), 879–890.

Tsukamoto, Y. & Ohta, T. (1988) Runoff processes on a steep forested slope. *J. Hydrol.* **102**, 165–178.

Uchida, T., Kosugi, K. & Mizuyama, T. (2001) Effects of pipeflow on hydrological processes and its relation to landslides: a review of pipeflow studies in forested headwater catchments. *Hydrol. Processes* **15**(11), 2151–2174.

Uchida, T., Kosugi, K. & Mizuyama, T. (2002) Effects of pipeflow and bedrock groundwater on runoff generation in a steep headwater catchment in Ashiu, central Japan. *Water Resour. Res.* **38**(7), doi: 10.1029/2001WR000261.

Whipkey, R. Z. (1967) Theory and mechanics of subsurface stormflow. In: *Forest Hydrology* (ed. by W. E. Sopper & H. W. Lull) (Proc. NSF Advanced Science Seminar), 255–260. Pergamon, New York, USA.

Contribution of Subsurface Flow from the Upper Slopes of Forested Watersheds to Channel Flow[1]

R. S. Beasley[2]

ABSTRACT

Overland flow and subsurface flow from 0.054- and 0.089-ha plots on the upper slopes of forested watersheds were intercepted and measured in trenches located at the bases of the plots. Channel flow was measured in flumes at the watershed outlets and intensity, duration, and amount of rainfall were measured by gages on the watersheds. Over a 3-year period, 115 storms produced little overland and subsurface flow above the B-horizon. Subsurface flow over an impermeable clay layer was as much as 90% of rainfall during a calendar quarter. Thirty-six of the storms produced subsurface flow on one or both plots. For these storms, subsurface flow from the upper third of the watersheds was equivalent to 26% of the channel flow on one watershed and to 27% on the other. For 27 of the 36 storms, subsurface flow from the plots peaked sooner than channel flow. Twelve storms produced subsurface flow within 20 min of the onset of rainfall, and 15 storms produced channel flow within the same time. Subsurface flow often began shortly after rainfall began even when there was neither saturation at the point of outflow nor high antecedent soil moisture. Interconnected channels through the soil formed by decayed roots and animal burrows are the best explanation for the quick subsurface flow responses. The similarity in timing of subsurface and channel flows indicates that subsurface flow from upper slopes can contribute significantly to storm hydrographs in forested areas where permeable surface soils overlie an impermeable stratum.

Additional Index Words: **lateral flow, soil moisture, hydrology, runoff, streamflow.**

Investigators disagree about the timing, mechanics, and magnitude of subsurface flows from upper slopes. Some deny that subsurface flow contributes significantly to runoff (Dunne and Black, 1970; Freeze, 1972); others maintain that only subsurface flow from near the stream channel contributes to storm hydrographs (Hewlett and Hibbert, 1967; Hewlett and Nutter, 1970; Nutter 1973). Gaiser (1952), Whipkey (1967), and Aubertin (1971), however, assert that interconnected macrochannels formed by roots and animal burrows can provide the means for rapid subsurface flow from upper slopes to stream channels. This paper examines the contribution of subsurface flow from upper slopes to channel flow and discusses flow timing in relation to rainfall and antecedent soil moisture for two forested watersheds in northern Mississippi.

METHODS

Study Area

The study area is typical of much of the terrain in the upper Coastal Plain. The two watersheds studied support mature shortleaf pines (*Pinus echinata*) and mixed hardwoods. The upper slopes have average gradients of 30% and are composed of two highly permeable Coastal Plain soils, McLaurin and Smithdale sandy loams. Wilcox sandy loam, a poorly drained soil with very low permeability, dominates the lower slopes. A dense stratum of kaolinite clay lies 6 m below the surface at the watershed divides but converges to within < 1 m of the surface at midslope.

Mean annual precipitation on the study area is 135 cm; 80 cm fall from November through April. The lowest monthly rainfall average is 8 cm in October; the highest is 16 cm in March.

Installations and Measurements

Study plots were established on the upper slopes of the adjacent 1.86– and 1.62–ha watersheds. The surface drainage area of plot 1, located on the larger watershed, was 0.054 ha; that for plot 2 was 0.089 ha. The subsurface drainage areas, defined by augering to the clay layer and preparing a contour map of its surface, were 0.045 ha on plot 1 and 0.068 ha on plot 2. Average depth to the clay is 1.8 m on plot 1 and 2.1 m on plot 2.

Trenches, located about 75 m below the watershed divides and 125 m above the watershed outlets, served as the lower boundaries of the plots. They were 12.2 m long and extended into the clay; two gravel-filled troughs were constructed inside of each (Fig. 1). One trough intercepted subsurface flow from the A-horizon; the other intercepted flow from between the A/B-horizon interface and the clay. Subsurface flow at the two levels was diverted through sheet metal troughs into 0.305 m HS-flumes equipped with stage recorders. A strip of sheet metal was inserted between the litter layer and mineral soil at the lip of each trench to divert overland flow to 0.153 m H-flumes for measurement. The collection troughs were driven into the trench faces to prevent water from seeping around one trough to the one below. The trenches and diversion troughs were covered to prevent the direct inflow of precipitation. Before the trenches were constructed, subsurface flow emerged through holes about 5 to 10 cm in diameter that were 15 to 30 m below the plots. From that point flow continued downslope in well-defined intermittent stream channels. The trenches interrupted the natural pathway of subsurface flow, but the water they intercepted moved readily through the collection system and became part of channel flow. Channel flow, water from all sources (including the plots) that reached the watershed outlets, was measured in 0.915 m H-flumes constructed in the natural channels. Concrete cut-off walls beneath the flume approach sections extend into clay to minimize underflow.

Rainfall on the two watersheds was measured by a network of five gages. A centrally located gage gave rainfall durations and intensities.

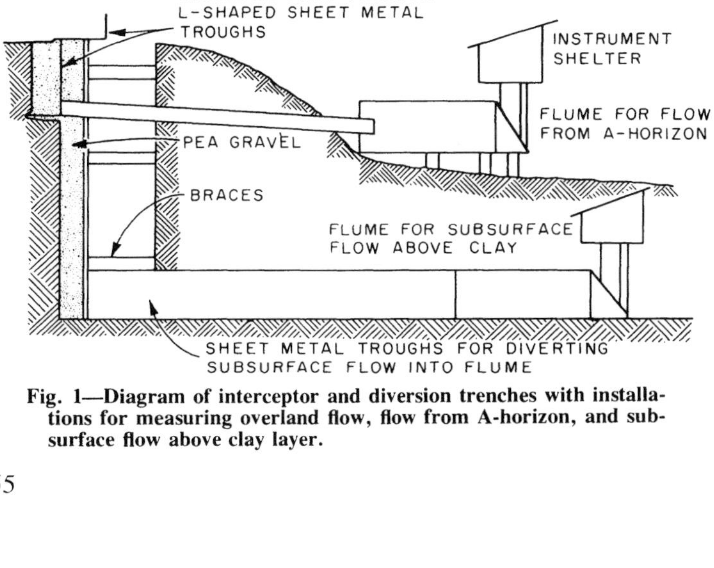

Fig. 1—Diagram of interceptor and diversion trenches with installations for measuring overland flow, flow from A-horizon, and subsurface flow above clay layer.

[1]Contribution from the For. Service—USDA. Paper presented at the Annual Meeting of the Soil Science Society of America, 10–15 Nov. 1974 Chicago, Ill. Received 5 April 1976. Approved 5 Aug. 1976.

[2]Formerly Associate Hydrologist at the Forest Hydrology Laboratory which is maintained at Oxford, MS. 38655, by the South. For. Exp. Stn., For. Service—USDA, in cooperation with the Univ. of Mississippi. Now Assistant Professor, Dep. of Forestry, Univ. of Arkansas, Monticello, AR 71655.

Analyses

Rainfall and subsurface flow volumes above the clay layer were totaled by calendar quarter for the 115 storms (rainfall events that produced subsurface flow on one or both plots) that occurred from 1971 through 1973. The storms were used to compare subsurface flow volumes with total rainfall by calendar quarters during the three years. Thirty-six storms that produced subsurface flow on both plots were used to compare: (i) subsurface flow volumes measured at the plots with channel flow volumes measured at the watershed outlets, (ii) time to beginning of subsurface flow with time to beginning of channel flow, and (iii) time to subsurface flow peaks with time to channel flow peaks. Regressions were developed using a combinatorial screening technique to relate the beginning and peaks of subsurface and channel flows to (i) maximum 10-min rainfall intensity, (ii) time to midpoint of maximum rainfall intensity, (iii) rainfall amount, and (iv) antecedent rainfall. An additional variable, antecedent soil moisture, was included along with the rainfall variables in a regression analysis of 26 storms for which soil moisture data were available.

RESULTS AND DISCUSSION

Subsurface Flow as a Percentage of Rainfall

Overland flow and subsurface flow above the B-horizon were negligible during the study period; infiltration rates on the upper slopes were high enough to accommodate even the most intense storms. Subsurface flow between the A/B-horizon interface and the clay layer, however, often was a large percentage of rainfall, especially during wet periods (Table 1). During quarter 1 of 1973, 88% of the 75.4 cm of total rain emerged as subsurface flow on plot 1, and 92% emerged on plot 2. A storm that delivered over 25 cm of rain in 42 hours produced about one-third of the subsurface flow measured during that quarter. During quarter 2 of 1973, subsurface flow accounted for 51% of the total rainfall on plot 1 and 37% on plot 2. During periods of lower rainfall a smaller percentage emerged from the plots as subsurface flow because most of the water went to evapotranspiration and soil storage. One hundred thirty-two cm of rain fell in 1971, which is slightly less than average; only 5.7% of the rain emerged as subsurface flow. In 1972, rainfall was 20 cm above average and 11.5% emerged as subsurface flow. In 1973, rainfall exceeded the average by 44 cm and 47.7% of the rain emerged as subsurface flow.

Table 1—Subsurface flow as a percentage of total rainfall for 115 storms by calendar quarters and years.

Year	Total rainfall	Subsurface flow Plot 1	Subsurface flow Plot 2
	cm	%	%
1971			
Quarter 1	45.87	11.3	14.0
2	22.35	2.1	1.6
3	43.18	1.8	1.6
4	20.32	1.6	0.4
Annual	131.57	5.1	5.7
1972			
Quarter 1	31.75	12.0	12.2
2	36.32	0.7	1.3
3	33.27	0.9	3.2
4	52.58	17.8	23.4
Annual	153.92	8.8	11.5
1973			
Quarter 1	75.44	88.2	91.6
2	39.12	50.8	36.9
3	16.51	0.4	0.4
4	47.75	3.2	3.8
Annual	178.82	49.2	47.7

Comparing Subsurface and Channel Flow Volumes

Subsurface flow from the upper slopes often was a significant proportion of total flow from the catchments. Total subsurface flow from plot 1 for 36 storms over the 3-year period was 82.9 area cm compared with 91.0 area cm of channel flow. Subsurface flow from plot 2 was 82.6 area cm compared with 92.7 area cm of channel flow. Assuming the plots represent the upper one-third of the watersheds, subsurface flow from the upper slopes is equivalent to 27% of channel flow on watershed 1 and 26% on watershed 2 for these 36 storms. Betson (1964) and Hewlett (1961) contend that most channel flow comes from small areas adjacent to the stream, and Freeze (1972) concludes that overland flow from near the stream channel dominates the storm hydrograph on well-vegetated basins. The results of the current study, however, demonstrate that subsurface flows from upper slopes can be important contributors to channel flow volumes.

Timing of Flows in Response to Rainfall

The elapsed time from the onset of rainfall to the starts and peaks of flows showed, as Kirkby and Chorley (1967) found, that subsurface flow often begins soon enough to contribute to channel flow peaks. A distribution of flow initiation times by 20-min intervals shows a high proportion of storms produced subsurface flow at the plots soon after rain began. Twelve of the 36 storms produced subsurface flow within 20 min of the beginning of rainfall (Fig. 2). Fifteen storms, including 11 of the 12 that had quick subsurface flow, caused a rise in channel flow within the same time.

The average time lag between the onset of rainfall and a rise in the channel hydrograph was 93 min for the 36 storms. The delay was caused by the few storms that began gradually and continued at low intensities, sometimes for hours, before producing stormflows.

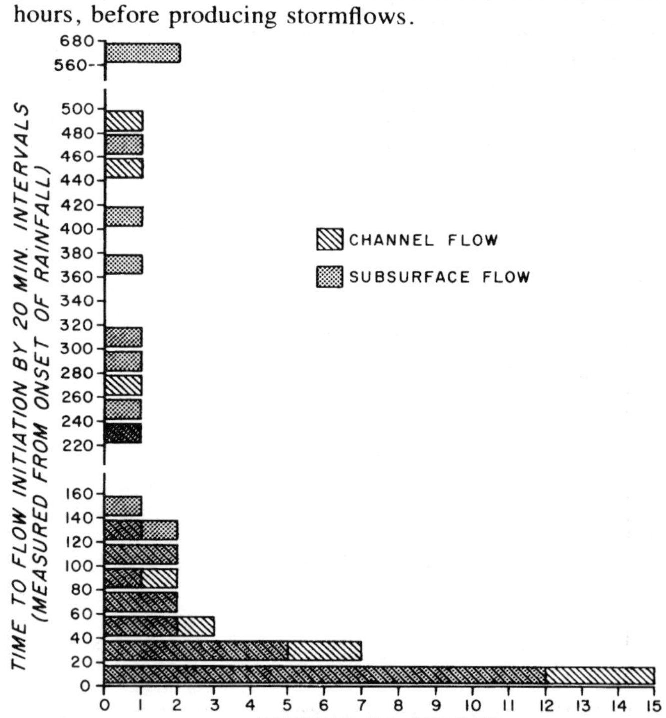

Fig. 2—Distribution of times to initiation of subsurface and channel flows for 36 storms.

Fourteen storms produced subsurface flow peaks in the trenches < 1 hour after rainfall began, and 11 of those produced peaks at the watershed outlets within 1 hour. For the 36 storms, subsurface flows peaked an average of 5 min later than channel flows (Fig. 3); however, subsurface flow on one or both plots peaked sooner than channel flow during 27 of the storms. Many storms produced quick rises on the subsurface flow hydrographs even when flow from preceding storms had stopped.

Of the variables considered, only one, time to maximum 10-min rainfall intensities, was included in the regressions of best fit. The regressions accounted for about 85% of the variation in time to peaks for both types of flow on both watersheds. This similarity in behavior indicates that the storm hydrographs probably are dominated by subsurface flow from the hillslopes. A close relationship between time to channel peaks and rainfall intensity has been shown by other workers (Hewlett and Helvey, 1970, Linsley et al., 1949); however, the close association between subsurface flow peaks and the occurrence of maximum intensity was unexpected in light of the usual assumption that subsurface flow velocities are governed primarily by hydraulic properties of the soil mass and not by rainfall characteristics.

The quick response of subsurface flow to rainfall cannot be explained solely on the basis of saturated flow through the main soil mass. Measured saturated hydraulic conductivities of the highly permeable surface soils ranged up to 64 cm per hour, but even these fast rates are not sufficient to account for peaks occurring < 1 hour from the onset of rainfall. Translatory flow (Hewlett and Hibbert, 1967) the process by which soil water near the stream channel is pushed out of the soil by rain infiltrating upslope, does not explain the quick responses because it requires a saturated zone near the point of outflow. On 13 Aug. 1974, 10 tensiometers, 2 at each 0.305 m interval down to the clay, were installed immediately above the trench on plot 1. Five storms produced quick subsurface flows during the following 3 months. Three of the storms produced a peak within 1 hour of the onset of rainfall. At no time during this period was there a saturated zone above the trench. Timing of subsurface flow peaks was also unrelated to antecedent soil moisture.

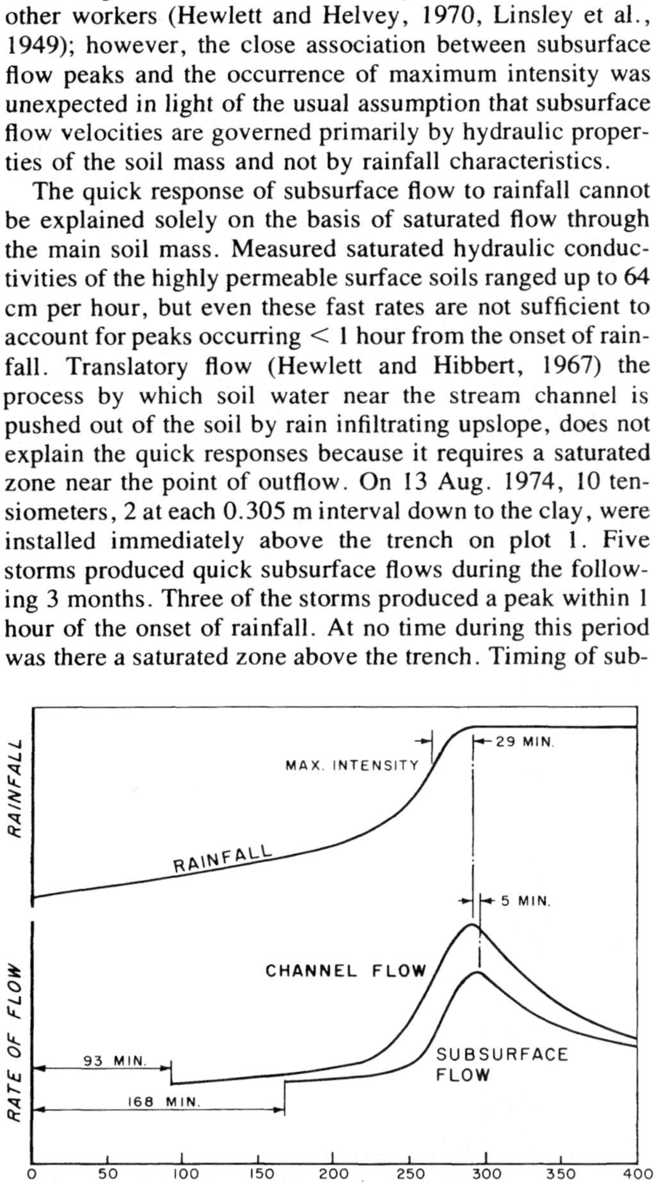

Fig. 3—Hydrographs showing average time from onset of rainfall (time zero) to maximum rainfall and times from beginning of flows to flow peaks for 36 storms. (The vertical scales are not quantitative.)

Hewlett and Nutter (1970) state that subsurface flow velocities seldom exceed 3 to 5 m per day. However, for subsurface flows from my study plots to peak within an hour after rainfall began, flow velocities must have exceeded 800 m per day. The best explanation for the fast rates is that the water travels through macrochannels formed by decayed roots. During several storms that occured before the trenches were filled with gravel, I observed water gushing from such openings. Others, including Aubertin (1971) and Whipkey (1967), recognize the importance of flow through macrochannels in the transport of water through forest soils. Gaiser (1952) studied vertical and lateral channels formed by decayed root systems in an Ohio hardwood forest and asserted the root channels are pathways for the rapid movement of free water because they contain materials more permeable to water than the surrounding soil mass.

Before subsurface flow can occur in macrochannels, the channels must be open to the atmosphere and a positive head of water must exist at the openings. Such conditions are common in a forest environment. Depressions formed by uprooted trees or decayed stumps provide a place for water to concentrate. Decayed roots spreading outward from the depressions provide natural pathways for water to flow freely through the soil toward the stream channels. Stemflow accumulating at the bases of trees is another possible source of significant quantities of subsurface flow.

LITERATURE CITED

1. Aubertin, G. M. 1971. Nature and extent of macropores in forest soils and their influence on subsurface water movement. USDA For. Serv. Res. Pap. NE-192, 33 p. Northeast. For. Exp. Stn., Upper Darby, Penn.
2. Betson, R. P. 1964. What is watershed runoff? J. Geophys. Res. 69: 1541–1552.
3. Dunne, T., and R. D. Black. 1970. Partial area contributions to storm runoff in a small New England watershed. Water Resour. Res. 6: 1296–1311.
4. Freeze, R. A. 1972. Role of subsurface flow in generating surface runoff. II. Upstream source areas. Water Resour. Res. 8: 1272–1283.
5. Gaiser, R. N. 1952. Root channels and roots in forest soils. Soil Sci. Soc. Am. Proc. 16: 62–65.
6. Hewlett, J. D. 1961. Soil moisture as a source of base flow from steep mountain watersheds. USDA For. Serv., Southeast. For. Exp. Stn. Pap. 132,11 p. Asheville, N. C.
7. Hewlett, J. D., and J. D. Helvey. 1970. Effects of forest clearfelling on the storm hydrograph. Water Resour. Res. 6: 768–782.
8. Hewlett, J. D., and A. R. Hibbert. 1967. Factors affecting the response of small watersheds to precipitation in humid areas. p. 275–290 *In* W. E. Sopper and H. W. Lull (eds.) Forest hydrology: proceedings of a National Science Foundation advanced science seminar. Pergamon Press, New York.
9. Hewlett, J. D., and W. L. Nutter. 1970. The varying source area of streamflow from upland basins. p. 65–83. *In* Proc., Symp. interdisciplinary aspects of watershed manage. Am. Soc. Civ. Eng., New York.
10. Kirkby, M. J., and R. J. Chorley. 1967. Throughflow, overland flow and erosion. Bull. Int. Assoc. Sci. Hydrol. 12(3): 5–21.
11. Linsley, R. K., M. A. Kohler, and J. L. H. Paulhus. 1949, Applied hydrology. McGraw-Hill Book Co., Inc., New York. 689 p.
12. Nutter, W. L. 1973. The role of soil water in the hydrologic behavior of upland basins. P. 181–193. *In* R. R. Bruce, K. W. Flach, and H. M. Taylor (eds.) Field soil water regime. Spec. Publ. 5 Soil Sci. Soc. Am., Madison, Wis.
13. Whipkey, R. Z. 1967. Theory and mechanics of subsurface stormflow. p. 255–290. *In* W. E. Sopper and H. W. Lull (eds.) Forest hydrology: proceedings of a National Science Foundation advanced science seminar. Pergamon Press, New York.

Commentary on Paper 21

Yair, A. & Lavee, H. (1976) Runoff generative processes and runoff yield from arid talus mantled slopes.

This paper is a classic of arid hydrology (see also Paper 16 by Osborn & Renard from the semiarid Walnut Gulch experiment). There is a long history of experimental work on rainfall patterns and runoff generation in Israel by Yair, Lavee, Schick, Sharon and others (see summary of Yair & Lavee, 1985), in part investigating ancient water harvesting techniques (e.g. Shanan & Schick, 1980). This paper reports on plot experiments carried out in an area where the average annual rainfall is only 25 mm. The authors therefore resorted to using artificial rainfall to generate surface runoff. Rainfall intensities (if not raindrop sizes and kinetic energy) were based on records from the area, and runs were made at three different intensities. As happens so often in experimental studies, nature intervened and a rainstorm during the series of experiments created difficulties by modifying the initial conditions for some of the plots.

One of the most interesting aspects of the work at this site and at Asher Schick's Yael watershed, is that runoff coefficients tend to decrease with increasing scale, from runoff plot to hillslope to increasing catchment scales. This appears to be common, if not characteristic, of runoff in arid and semiarid areas (see also Dunne, 1978, for data from the Southwest USA, and Puigdefàbregas & Sánchez, 1996, for data from Spain), implying that surface runoff generated on slopes with either low infiltration rates or limited storage capacities then infiltrates on areas having higher infiltration rates or greater storage capacities downslope or in valley bottoms. At larger scales the patchiness of rainfalls and transmission losses from the channel itself will also be an important factor in reducing catchment average runoff coefficients (e.g. Sharon, 1972, 1981; Goodrich *et al.*, 1997).

Aaron Yair became Professor (now Professor Emeritus) and Chair of the Department of Geography at the Hebrew University, Jerusalem; Hanoch Lavee became Professor of Geography and Dean of the Faculty of Social Science at the Bar-Ilan University, Israel.

References

Dunne, T. (1978) Field studies of hillslope flow processes. In: *Hillslope Hydrology* (ed. by M. J. Kirkby), 227–294. Wiley, Chichester, UK.

Goodrich, D. C., Lane, L. J., Shillito, R. M., Miller, S. N., Syed, K. H. & Woolhiser, D. A. (1997) Linearity of basin response as a function of scale in a semi-arid watershed. *Water Resour. Res.* **33**(12), 2951–2965.

Puigdefàbregas, J. & Sánchez, G. (1996) Geomorphological implications of vegetation patchiness on semi-arid slopes. In: *Advances in Hillslope Processes* (ed. by M. G. Anderson & S. M. Brooks), 1027–1060. Wiley, Chichester, UK.

Shanan, L. & Schick, A. P. (1980) A hydrological model for the Negev Desert Highlands: effects of infiltration, runoff and ancient agriculture. *Hydrol. Sci. Bull.* **25**(3), 269–282.

Sharon, D. (1972) The spottiness of rainfall in a desert area. *J. Hydrol.* **17**, 161–175.

Sharon, D. (1981) The distribution in space of local rainfall in the Namib Desert. *J. Climatol.* **1**, 69–75.

Yair, A. & Lavee, H. (1985) Runoff generation in arid and semi-arid zones. In: *Hydrological Forecasting* (ed. by M. G. Anderson & T. P. Burt), 183–220. Wiley, Chichester, UK.

EARTH SURFACE PROCESSES, VOL. 1, 235–247 (1976)

RUNOFF GENERATIVE PROCESS AND RUNOFF YIELD FROM ARID TALUS MANTLED SLOPES

A. YAIR AND H. LAVEE

Department of Geography, Hebrew University of Jerusalem, Jerusalem, Israel.

SUMMARY

Previous works dealing with the influence of a stone cover on runoff yield showed that runoff, attributed to the sealing effect of the topsoil by raindrops impact, was negatively related to the per cent of stone cover and stone size. These works were conducted on gentle slopes (3–11·5 degree) with a per cent of stone cover generally lower than 50 per cent, and composed of gravels. The present study deals with the runoff yield of steep talus slopes (26–36 degree) whose per cent of stone cover is very high (90–100 per cent), composed of cobbles and boulders. Three stimulated rainstorms were performed at various rainfall intensities and durations on each one of six plots representative of the North eastern sector of Sinai. Although the contiguous stony cover prevented surface sealing by raindrops impact, runoff developed quite quickly on most slopes, and reached at peak discharge, after approximately ten minutes, up to 56 per cent of the rainfall. Differences between plots, in time lag, peak discharge and other hydrological variables, could be attributed mainly to the properties of the upper stony layer, with stone size as the most influential factor. Contrary to previous works, a positive relationship, was obtained between stone size to runoff yield. The result is explained by the process of water concentration. Each cobble and boulder behaves, on a smaller scale, like a bare rocky surface and yields per unit area water amounts beyond the infiltration rate of the limited uncovered areas. For a given stony cover the effect of water concentration is quicker with the big blocks than with gravels. A series of graphs trying to relate theoretically the relative importance of sealing and water concentration processes in runoff generation, at various conditions of stone cover and stone size, is proposed. The graphs enable to reconcile the results of the present study with those of previous works.

KEY WORDS Arid, Sinai Talus mantled slopes Experimental plots Simulated rainstorms Stone cover, stone size Runoff generation Runoff yield Hydrological variables

INTRODUCTION

The influence of the properties of a stone cover on the runoff yield of slopes was studied in several agricultural research works. Grant and Struchtemeyer (1959) found, in a laboratory experiment, that the clearing of stones, 0·24–1·27 cm in size, increased the amount of runoff. Similar results were obtained by Jung (1960), on a 6 degree slope with a stone cover of 39 per cent; by Epstein *et al* (1966) on a 5 degree slope with a stone cover of 31 per cent, which included few stones larger than 3·8 cm, and by Evenari *et al* (1958-1967) on slopes between 6 and 11·5 degrees with a per cent of stone cover up to 75 per cent. Agassi (1970) experimented in the laboratory, with a 3 degree slope, using high rainfall intensities. He found a positive relationship between the infiltration rate and the size of the gravels, 1–5 cm in size, and the per cent of stone cover (0–50 per cent). All authors cited suggest that the stone cover prevents the impact of raindrops which, by the sealing of the upper pore space in the soil, contributes to runoff generation.

Received 8 December 1975
Revised 3 March 1976

Figure 1. View of an upper layer. The material is loose and devoid of fine particles

Talus mantled slopes in arid areas represent a specific and extreme case of a stone cover, characterized by steep slopes and a high per cent of stone cover (90-100 per cent) composed mainly of cobbles and boulders (Figure 1). Considering the contiguous stony cover of such slopes, runoff generation by the sealing process is unlikely to occur and one has to look for a different mechanism to account for the occurrence of quite frequent runoff events on such slopes (Yair and Klein (1973), Yair and Lavee (1974)).

AIM OF THE WORK

The present study has two main objectives:

(a) To obtain quantitative data on the runoff yield from arid talus mantled slopes, which differ in their lithology and grain size distribution of the coarse surficial material.

(b) To identify the mechanisms of runoff generation under conditions of a dense stony cover and try to relate to it runoff data obtained from the studied slopes.

DESCRIPTION OF THE STUDIED SLOPES

Six slopes were selected in the north eastern sector of the Sinai Peninsula (Figure 2). The average annual rainfall in the area is estimated to be 25 mm. A concise description of the surface properties of the slopes is given in Table I. Five of the slopes are scree slopes and the sixth, a bare bedrock surface. The slopes differ in lithology and structure, which control the size of the blocks composing the scree mantle. Slopes numbered 1, 2 and 6 are on crystalline and metamorphic rocks. They have a very steep rocky escarpment in their upper part and a scree slope below. Slope 3 is on bare crystalline bedrock densely jointed. Its profile is very irregular and stepped. Slopes 4 and 5 are on sedimentary rocks consisting of coarse sandstone and limestone respectively. The scree mantle covers the whole slope, up to the top.

Pits dug in the scree mantle, where it exists, show that the talus materials is built of two distinct layers. An upper layer composed of loose gravels, cobbles and boulders 5 to 100 cm thick (Figure 1), and a lower layer, overlying the bedrock, which is built of gravels and boulders buried and slightly cemented by a fine grained matrix of sand silt and clay (Figure 3). Its thickness may reach several meters.

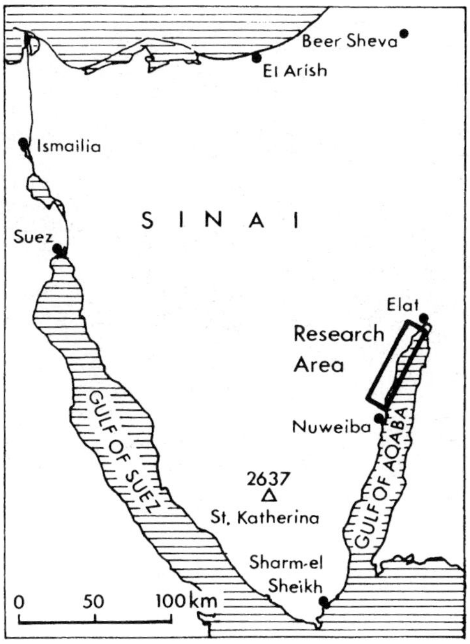

Figure 2. Location map

EXPERIMENTAL DESIGN

As rainfall is scarce in the area, and in order to work with uniform rainfall conditions, simulated rainstroms were produced on experimental plots located at the lower part of the slopes (Figures 4 and 5).

Water was supplied by a 5 m^3 tank with the aid of a generator operating a pump. The water distribution assembly (Figure 5) included two main pipes connected to the pump and located on both sides

Table I. Surface characteristics of the slopes

		Escarpment		Scree slope				Sprinkled and drained area		
Plot No.	Lithology	Length (m)	Mean gradient (degrees)	Length (m)	Mean gradient (degrees)	Size of big blocks (cm)	Topographic description	Area (m^2)	Length (m)	Gradient (degrees)
1.	Schist and granite	15	75	70	34	15–25	Linear profile. A central gully appears in the lower 20 m of the slope. Bare bedrock outcrops in small patches.	100	14	30°
2.	Granite and schist	30	60	150	34	30–40	Profile slightly concave. Several gullies transect the slope. Heaps of very big blocks are spread over the slope.	100	14	26°
3.	Schist and granite	Bare bedrock is exposed all over the slope. The rock is densely jointed. Joints are filled with fine material. The slope profile is irregular and stepped.								
4.	Coarse sandstone	none	none	230	28	20–40	Linear profile; no gullies, bedrock outcrops in small patches.	140	14	36°
5.	Limestone with flint concretions	none	none	85	31	15–20	Linear profile. Gullies are spaced and very shallow.	110	14	26°
6.	Magmatic dyke	15	52	20	30	3–7	Linear profile. Gullies are very shallow.	70	14	30°

Figure 3. View of the lower layer. Compacted mixture of boulders, gravels and fine particles

of the sprinkler plot. Smaller pipes, to which sprinklers were attached, were laid on the ground across the plot, at 3 m intervals and connected to the main pipes. The sprinklers were located so as to produce a quite uniform rainfall distribution all over the sprinkled area, at each of the runs performed. Rainfall intensity was controlled by regulation of the water pressure and the number of operating sprinklers. Rainfall amount and distribution were checked with the aid of twenty metal cans scattered over the drained area (Figure 5).

As the central thesis of this paper is that, in the stony environment studied, runoff is not generated by the sealing process, due to the impact of raindrops, drop size distribution and terminal velocities of raindrops were not measured as they were assumed to be minor factors in runoff generation.

The sprinkled area covered some 200 m^2. In order to present lateral flow of water from the drained area to its vicinity, only the central part of the sprinkled area was drained (Figure 4). Low walls were built at the base of the plots which directed runoff water to collectors (Figure 5) emptied through volumetric water meters read at a time interval of 30 sec.

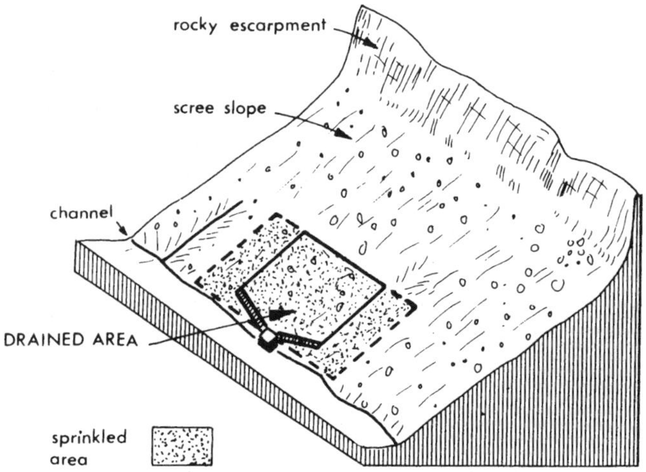

Figure 4. Location of an experimental plot

Figure 5. View of an experimental plot

Three successive rainstorms were simulated on each plot at the following rain intensities and durations:

First run—1 mm/min during 10 min.

Second run—0·5 mm/min during 10 min.

Third run—0·25 mm/min during 20 min.

The sprinkling intensities and durations selected are based on an analysis of rainfall events recorded in the area during the last fifteen years (Yair and Lavee (1974)), and are of quite frequent occurrence.

The first run on each plot started under relatively dry surface conditions. As the time interval between the sequence of three runs varied from 30 to 60 min, the second and third runs were always performed under wet surface conditions. Soil moisture increased therefore parallel with the decrease in the rainfall intensities used, and allowed a relatively quick development of runoff even under the lowest rainfall intensity performed.

After the performance of the experiments on plots No. 1, 3 and 4, a natural rainfall event, amounting to 14–28·5 mm occurred in the area. The first runs on plots No. 2, 5 and 6 were therefore performed on previously wetted surfaces. For this reason, as well as due to the fact that for technical reasons rainfall intensities and durations obtained during the experiments slightly differed from planned values (Table II), any comparative analysis of hydrologic data recorded or computed at the various plots cannot be considered very accurate.

The following data were recorded during the experiments: rainfall amounts over the sprinkled area (P); time elapsed from beginning of rainfall to the beginning of runoff (T); and cumulative volume of runoff measured at an interval time of 30 sec (Q). On the basis of data recorded the following variables, allowing to draw a hydrograph for each of the runs, were computed: rainfall intensity (I); initial losses (F); runoff coefficient (Q/P); peak discharge (q max) and minimum infiltration (f min) as was obtained at the end of the rainfall.

All measured and computed hydrologic parameters are given in Table II. They are related to a unit area and expressed in millimeters or millimeters per minute. A graphic presentation of the most significant parameters is given in Figure 6.

DATA ANALYSIS

In spite of the non-uniformity in initial soil moisture and rainfall conditions during the experiments, some general trends pertaining to differences in the hydrological behaviour between plots and within

Table II. Hydrological variables obtained at the experiments

Plot No.	P Rainfall amount (mm)	I Rainfall Intensity (mm/min)	T Lag time (min)	F Initial losses (mm)	Q Total runoff (mm)	Q/P Runoff coefficient (%)	q max Peak discharge (mm/min)	f min Minimum infiltration (mm/min)
	7·2	0·72	3·5	2·52	2·20	30·6	0·37	0·37
1	4·4	0·44	4·0	1·76	0·86	19·5	0·20	0·26
	4·4	0·22	6·5	1·43	0·87	19·8	0·07	0·15
	9·3	0·93	2·5	2·33	2·73	29·4	0·42	0·52
2	4·0	0·40	3·5	1·40	1·23	30·8	0·21	0·20
	6·0	0·30	4·5	1·35	1·68	28·0	0·12	0·18
	10·5	1·05	2·0	2·10	4·15	39·5	0·54	0·53
3	5·5	0·55	2·5	1·38	2·25	40·9	0·31	0·25
	5·4	0·27	3·5	0·95	1·93	35·7	0·13	0·14
	9·4	0·94	10·5	9·87	0·00	00·0	0·00	0·94
4	6·0	0·60	7·0	4·20	0·30	05·0	0·10	0·50
	5·4	0·27	6·5	1·76	0·94	17·4	0·09	0·19
	9·3	0·93	3·5	3·26	2·90	31·2	0·46	0·48
5	5·5	0·55	4·0	2·20	0·96	17·5	0·19	0·38
	6·0	0·30	7·5	2·25	0·73	12·2	0·09	0·21
	12·1	1·21	4·0	4·84	2·56	21·2	0·47	0·77
6	6·3	0·63	4·5	2·84	1·14	18·1	0·22	0·50
	8·4	0·42	5·5	2·31	2·22	26·4	0·14	0·28

a single plot during the sequence of three runs, appear quite clearly and deserve to be mentioned and analyzed.

TIME LAG AND INITIAL LOSSES

Runoff developed quite quickly on all sprinkled plots. The time lag recorded at all runs performed varied between 2 to 10·5 min. The quickest response to rainfall was always obtained on the bare bedrock slope (plot 3) and the slowest on the sandstone slope (plot 4). The ranking of the other plots, with increasing time lag was as follows: plot 2, 1, 5 and 6.

The values of the initial losses are generally quite low and vary from 0·95 mm to 9·9 mm. The ranking of the slopes is similar to that described for the time lag on which the initial losses greatly depend.

In spite of the increasing soil moisture the time lag increased from the first to the third run, parallel with the decrease in rainfall intensity. The only exception was obtained at the sandstone slope in which the trend is reversed (Table II, Figure 6).

The initial losses decrease in all plots from the first to the third run, in spite of the increase in the time lag. Such a result is explained by the fact that in the equation $F = T.I$ (1) the rate of increase in T is lower than the rate of decrease of I (Table II).

PEAK DISCHARGE AND RUNOFF COEFFICIENT

Values of peak discharge, for all runs, vary from 0·07 mm/min (plot 1, run 3) to 0·54 mm/min (plot 3, run 1). Runoff coefficients vary from 0 per cent (plot 4, run 1) to 40·9 per cent (plot 3, run 2). Once more the highest values at the sequence of three runs were obtained on the bare bedrock slope and the lowest on the sandstone slope (run 3 excepted). The remaining slopes are ranked similarly, to the ranking described for the time lag (Figure 6). The relatively high figures obtained at plot 6 cannot be considered as indicative of a better ranking when compared to plots 1, 2 and 5, as the rainfall intensities simulated on plot 6 were relatively high.

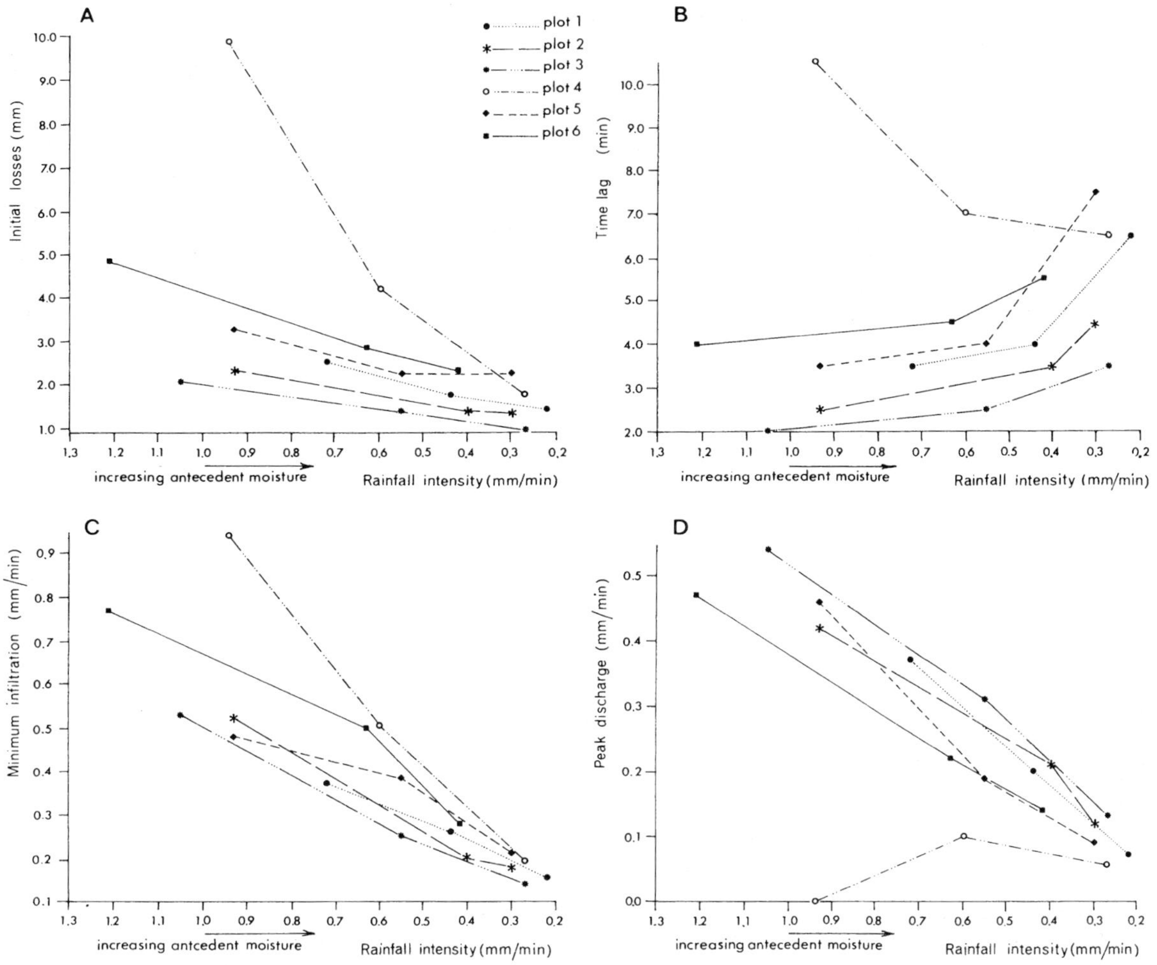

Figure 6. Initial losses, time lag, minimum infiltration and peak discharge as related to rainfall intensity and antecedent moisture

It is worthwhile noting that equilibrium was almost reached in plots 3 and 2 at the first run, and at all other plots at the second or third run (Figure 7).

Differences in values of peak discharge, between plots, diminished with increasing soil moisture. A similar trend was obtained for values of minimum infiltration, which reached nearly uniform values, around 0·19 mm/min, in all plots at the third run, sandstone slope included (Figure 6).

Values of peak discharge and minimum infiltration decrease with decreasing rainfall intensity and increasing soil moisture. The positive relationship between rainfall intensity and minimum infiltration is probably related to the phenomena of partial area contribution detected on the sprinkled plots (Yair and Lavee (1974)). Even at the end of rainfall, with the highest intensity performed, large areas were still able to absorb all the rain. Less water, in absolute values, was absorbed at the same time, by the same areas at lower rain intensities, resulting thus in a decrease in the absolute values of f.

$$\text{As: } q \max = I - f \min \qquad (2)$$

the parallel decrease in q max and f min, is due to a higher rate of decrease in the values of I as compared to those of f min (Table II, Figure 6).

A positive relationship between infiltration and rainfall intensity was previously obtained by Modenhauer *et al* (1960), Vatson (1965), Morin *et al* (1968), and Agassi (1970). Because their works deal with the infiltration process of ponded areas, the explanations they propose differ from that given above and pertain to various physical processes which occur in the soil profile at different rainfall intensities.

THE RUNOFF GENERATIVE PROCESS

The performance of simulated rainfalls at similar intensities and durations enables to relate the hydrological data obtained to the surface properties of the sprinkled areas. Field observations during natural and simulated rainfall events showed that runoff on scree slopes occurs at the contact zone between the upper gravelly layer and the compacted lower layer rich in fine grained material. The runoff process

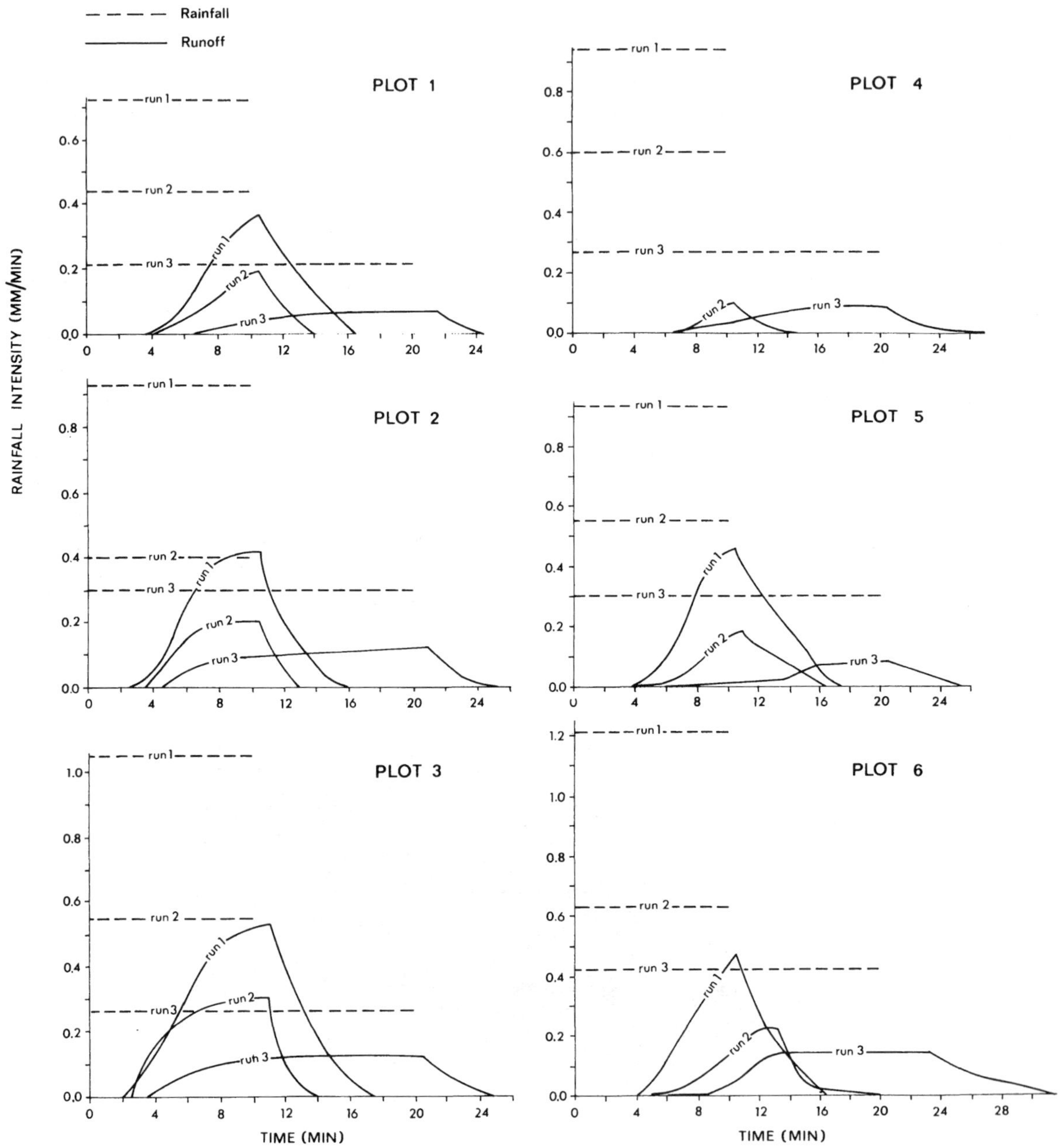

Figure 7. Hydrographs obtained at simulated rainfalls

Table III. Particle size composition of the matrix in the lower talus layer

Plot No.	% of clay	% of silt	% of sand	% < 2 mm
1	3·9	12·1	14·0	30·0
2	4·3	10·1	12·0	25·5
3	10·2	17·8	42·0	70·0
4	3·0	10·0	15·0	28·0
5	5·4	5·6	7·5	18·5
6	3·6	13·4	14·0	31·0

is therefore due to the fact that the infiltration rate of the lower layer is lower than the rate of water delivery by the cobbles and boulders composing the upper coarse layer. The infiltration capacity of the lower compacted layer is assumed to depend mostly on the composition of its fine grained matrix. Ring infiltrometers inserted in the stony material of the lower layer caused major disturbances that seriously affected the infiltration results. Possible differences between infiltration rates were therefore related to the particle size composition of the matrix. Several samples were taken from each plot and the figures obtained were averaged (Table III). At the bedrock slope, samples were taken from joints filled with fine material.

On the whole the grain size composition of the matrix is quite similar on all slopes. The differences fall within the various possible sources of error of a standard granulometric analysis. The only exception is the bedrock slope which is richer in fines. Such results suggested that the differences in the infiltration rates of the lower layer, although probably existing, are unable to account for the differences in the hydrological data obtained. These differences are therefore assumed to be mostly related to the properties of the upper gravelly layer.

The latter assumption was checked by performing, with the aid of a small rainfall simulator (Morin *et al* (1970)), an experiment over a plot 1·5 m^2 in area, from which all the stones, 8–20 cm in size, were cleared (Figure 8). The slope was previously wetted by natural rainfall. Intensity of the rainfall simulated was 0·5 mm/min. Runoff began 7 min after the rain started. This time lag is twice as long as those obtained on all sprinkled plots at the second run (Table II). The result points clearly at the decisive role that should be attributed to the properties of the upper talus layer in runoff production.

Figure 8. View of the small rainfall simulator

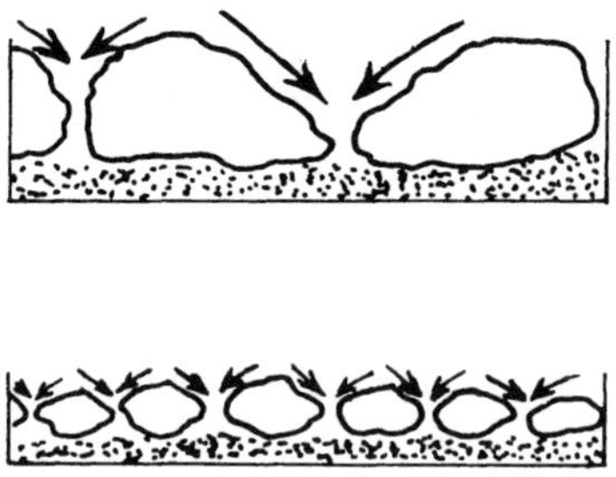

Figure 9. The influence of stone size on water concentration and runoff yield

The most influential factors affecting water delivery by the gravelly layer are considered to be:
1. Thickness of the layer.
2. Its absorption capacity.
3. The size distribution of the cobbles and boulders.

The first two factors are assumed to be negatively related to runoff production, whilst the third one is assumed to be positively related to it. A positive relationship between runoff yield and grain size composition of coarse fragments can be explained by the assumption that, over a uniform lower layer, big blocks allow a higher concentration of water above limited patches of the lower layer than do small gravels (Figure 9). In addition, for a given volume of rocky material, water losses due to absorption and wetting of particles surface are higher in numerous small gravels as compared to a single boulder. The ranking of the sprinkled plots, according to their runoff yield, can therefore be expressed by the following equation

$$R = D_{50}/V \, . \, L \tag{3}$$

where the value of R represents the location of each plot in the ranking of sprinkled plots. D_{50} is the median of blocks' size composition of the gravelly layer. The thickness of the latter layer is expressed by V which represents the volume of gravels per unit area. V was measured by clearing, at each plot, several unit areas of 0·25 m^2 from their stony cover. The stones cleared were placed in a receptacle containing water, whose volume was measured before and after the introduction of the stones. L is the absorption capacity of the stones. It was measured by placing stones previously dried in an oven in water for 10 min. The amount of water absorbed was obtained by weighing the stones before and after wetting. L is expressed in percent of the dry weight. Results are given in Table IV.

The ranking obtained in equation (3) is similar to that previously described for the various hydrological variables (Table II). Data analysis shows that the differences in R values are mainly related to D.50. The only exception was found at the coarse sandstone slope whose absorption capacity is six times

Table IV. The ranking of the sprinkled plots according to the properties of the upper coarse layer

Plot No.	D_{50} Median by particles size (mm)	V Volume, per unit area, of particles size (dcm^3)	L Absorption capacity (%)	R Ranking coefficient
1	152	12·7	0·94	12·7
2	168	13·2	0·80	15·9
3	very big	very small	0·86	very big
4	115	38·1	5·67	0·5
5	82	9·1	1·10	8·2
6	27	12·2	0·79	2·8

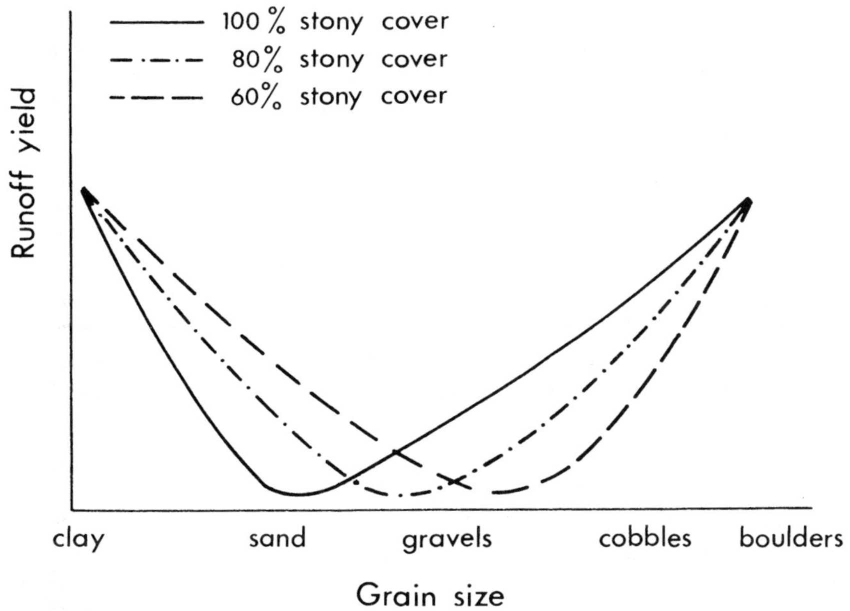

Figure 10. The relationship between runoff yield and particle size composition of a uniform cover overlying a compacted layer rich in fine grained material

higher when compared to the crystalline and limestone material of other slopes. Such a result explains clearly the high water losses characteristic of the sandstone slope, mainly at the first run.

A positive relationship between the size distribution of stones and runoff production seems to contradict the results obtained in previous works (Grant and Struchtemeyer (1959), Jung (1960), Epstein *et al* (1966), Evenari *et al* (1958), (1967), Agassi (1970)). The contradiction is apparent and pertains to the differences in the surface properties of the studied areas. The works mentioned above dealt with a relatively low per cent of stone cover and small gravels. The process of water concentration due to a high rate of water delivery by the stones did not occur. Runoff was therefore attributed to the sealing effect of the soil surface by the raindrops impact. At such conditions, an increase in the stone cover and stone size limited the area submitted to raindrops impact and resulted in a decrease in runoff production. Nevertheless, it is likely that such a trend will not continue beyond some combined threshold value of stone cover and stone size. The present study clearly shows that, under conditions of a high per cent of stone cover composed of cobbles and boulders, in which raindrops impact is negligible, runoff developed quickly due to the delivery of water by the stones in excess of the infiltration rate.

The relative importance of sealing and water concentration processes in runoff generation at various conditions of stone cover and stone size can be expressed by the theoretical graphs shown in Figure 10. For each of the graphs a trend of decrease in runoff production when particle size increases, exists for small particles. At such conditions runoff is mainly attributed to the sealing effect of the topsoil by raindrops impact. An opposite trend begins when the particle size is beyond some threshold value able to yield and locally concentrate water amounts higher than the infiltration rate of a lower fine grained layer. This value is represented, on each graph, by the inflection point in which water delivery by the stones gains progressively in importance over the sealing effect, as runoff generator. As shown in Figure 10, the inflection point is not supposed to be fixed. Its location depends mainly on the per cent of stone cover, the stones size and rainfall intensity. Assuming uniform rainfall conditions, the more the stone cover is low, the more the stone size should be high in order to generate runoff by the process of water concentration. Obviously, such a process will not occur below some threshold value of stone cover. It is likely that for a stone cover of 5–10 per cent the contribution to runoff by the water concentration process can be considered as negligible.

CONCLUSIONS

The data obtained in this study lead to the following conclusions and geomorphic implications, some of which need further confirmation by laboratory experimental methods.

(a) In spite of the high infiltration rate characteristic of arid talus slopes runoff develops on such slopes quite quickly under natural rainfall conditions. This conclusion accords with results previously obtained by Yair and Klein (1973) and suggests that runoff erosion contributes to the modelling of arid slopes, as proved in fact by the occurrence of numerous gullies incised in the scree mantle.

(b) Runoff is generated due to a high rate of water delivery by the stony cover in excess of the infiltration rate of the scree mantle. Although this mantle is composed vertically of two layers which differ by their properties, it is possible to explain the differences observed in the values of runoff yield from the studied slopes, by the properties of the upper gravelly loose layer. Under a nearly uniform high per cent of stone cover, the most influential factor was found to be the stone size composition to which runoff yield is positively related.

(c) Data obtained enabled us to rank the runoff yield of the studied slopes as follows.

Bare bedrock slope: highest runoff yield.

Various crystalline and limestone talus slopes: medium runoff yield.

Sandstone talus slope: lowest runoff yield.

(d) Differences in values of the hydrological variables are quite important at the first run performed under relatively dry conditions. These differences decrease seriously by the end of the third run, when all slopes behave quite similarly. It seems therefore that differences in runoff yield from arid slopes, which differ by their lithology and surface properties, should be expected mainly during short storms. Prolonged storms, amounting to more than 20 mm per day, quite rare in the area, will probably result in a more uniform water yield.

(e) The very high runoff coefficients obtained suggest that overland flow, generated on talus slopes, should be considered as the most important contributor to channel flow in arid areas, thus explaining the high values of peak discharge characteristic of arid watersheds. Moreover the very short lag time and concentration time obtained are helpful in explaining the steepness of the rising limb of hydrographs observed during channel flow events in small arid watersheds (Schick (1970)).

ACKNOWLEDGEMENTS

Financial support from the Ministry of Defence of Israel and the Hebrew University of Jerusalem, is gratefully acknowledged. Special thanks are accorded to Dr. Y. Morin and Mr. M. Agassi of the Ministry of Agriculture station for soil erosion research, for their help and advice in the use of the sprinkling equipment; to Mr. H. Levy, Mr. Y. Rutin and Mr. Y. Trostler for their technical help in the installation of the instrumentation; and to Mrs. T. Sofer for drawing the illustrations.

REFERENCES

Agassi, M. (1970). 'The influence of stone cover on infiltration and erosion rates laboratory experiments on a loessial soil', M.Sc. thesis, faculty of Agriculture, Hebrew University of Jerusalem. Mimeog. (In Hebrew).

Epstein, E., Grant, W. J., and Struchtemeyer, R. A. (1966). 'Effects of stones on runoff, erosion and soil moisture', *Soil Science Society of America Proceedings*, **30**, 638 640.

Evenari, M., Shanan, L., and Tadmor, N. H. (1958 1967). 'Runoff farming in the Negev desert of Israel'; I (1958–1962), II (1962 3), III (1963 4), IV (1964 7), Special Publications, National and University Institute of Agriculture, Rehovot, Israel.

Grant, J. W., and Struchtemeyer, R. A. (1959). 'Influence of the coarse fraction in two main potato soils on infiltration, runoff and erosion', *Soil Science Society of America Proceedings*, **23**, 391 394.

Jung, L. (1960). 'The influence of the stone cover on runoff and erosion on slate soil. International Association of Scientific Hydrology', *General Assembly of Helsinki, Commission of Land Erosion*, publication No. 53, 143–153.

Moldenhauer, W. C., Burrows, W. C., and Swartzendruber, D. (1960). 'Influence of rainstorm characteristics on infiltration measurements', *7th Internation Congress of Soil Science, Madison*, **I**, 426.

Morin, J., Goldberg, D., and Seginer, I. (1968). 'The kinetics of raindrop impact and its influence on infiltration capacity and erosion on soils (Hebrew with English summary). Research report (R-19). Soil Conservation Department, Ministry of Agriculture, Israel.

Morin, J., Cluff, C. B., and Powers, W. K. (1970). 'Realistic rainfall simulation for field testing', Water resources research center, University of Arizona. Paper presented at the 51st ann. meeting of the American Geophysics Union in Washington.

Schick, A. P. (1970). 'Desert floods. Interim results of observations in the Nahal Yael watershed, southern Israel', 1965–1970. Symposium on the results of research on representative and experimental basins, Wellington (N.Z.) *IASH* **96,** 463–478.

Vatson, K. K. (1965). 'A statistical treatment of the factors affecting the infiltration capacity of a field soil', *Journal of Hydrology*, **3,** 58.

Yair, A., and Klein, M. (1973). 'The influence of surface properties on flow and erosion processes on debris covered slopes in an arid area', *Catena*, **1,** 1–18.

Yair, A., and Lavee, H. (1974). 'Areal contribution to runoff on scree slopes in an extreme arid environment—a simulated rainstorm experiment', *Zeitschrift für Geomorphologie*, supplementband 21, 106–121.

Commentary on Paper 22

Harr, R. D. (1977) Water flux in soil and subsoil on a steep forested slope.

There are many examples of progress in hydrology being made by the careful, and sometimes long term, work of researchers supported by government institutes. Dennis Harr worked for the USDA Forest Service in Corvallis, Oregon, with a field site in the H. J. Andrews experimental forest close to the Blue River in the Western Cascades. The Andrews forest study was originally set up in 1950 to study the effects of forest management practices on the hydrology, geomorphology, ecology and hydroecology of the region. There are now more than 50 years of data for several experimental watersheds, some of which have been felled and treated in different ways (see http://www.fsl.orst.edu/lter/).

The study by Harr was carried out in Watershed 10, a small (10.23 ha) steep basin that had been essentially undisturbed for a long period. The study slope was steep (ranging from slopes of 50% at the divide to 110% at the stream channel). On this slope, covered by soil and sapprolite, Harr installed an intensive network of piezometers and tensiometers to examine the nature of the water flows in a situation where overland flow was expected to occur only in the channel itself. Some of this instrumentation is still to be seen on the slope. There was also an intensive study of the hydraulic characteristics of the soils, based on 450 samples and showing an exponential decline of saturated hydraulic conductivity with increasing depth into the soil, from very high values (3.5 m h^{-1}) close to the surface.

Harr's study is one of the examples in the literature of a pure subsurface stormflow response. The tensiometer and piezometer measurements suggest that a dynamic saturation zone at the very base of the slope is maintained by vertical (under drier conditions) and downslope (under wet conditions) unsaturated flows. Temporary perched water tables also sometimes occurred at the break between soil and subsoil. Harr puts his work in the context of the earlier work of Hewlett & Hibbert (Paper 6) and Weyman (Paper 17) and the concepts of displacement of stored water and "translatory flow". Similar to the earlier work, Harr found that drainage rates (and piezometer levels) showed a break in the rate of decline, in the case of this slope after about 10 hours of drainage. This he attributed to the emptying of the largest pores in the soil that had been filled during a rainfall event.

Hydrological studies on Harr's slope have recently been revitalized in the thesis work of Kevin McGuire under the supervision of Jeff McDonnell (McGuire, 2004). There are now even more anthropogenically induced holes in the hillslope. With such a steep slope it is a wonder that it has not yet failed (as happened in 1997 under a large natural rainstorm on the steep—and also heavily punctured—Coos Bay slope in the Coast Range of Oregon, see Anderson *et al.*, 1997; Montgomery *et al.*, 1997).

Dennis Harr retired from the Forest Service in the early 1990s and went into business.

References

Anderson, S. P., Dietrich, W. E., Montgomery, D. R., Torres, R., Conrad, M. E. & Loague, K. (1997) Subsurface flow paths in a steep, unchannelled catchment. *Water Resour. Res.* **33**(12), 2637–2653.

McGuire, K. (2004) Water residence time and runoff generation in the Western Cascades of Oregon. PhD Thesis, Oregon State University, USA. (http://www.ce.gatech.edu/~km216/research.htm).

Montgomery. D. R., Dietrich, W. E., Torres, R., Anderson, S. P. & Loague, K. (1997) Hydrologic response of a steep unchanneled valley to natural and applied rainfall. *Water Resour. Res.* **33**(1), 91–109.

Journal of Hydrology, 33 (1977) 37—58 37

WATER FLUX IN SOIL AND SUBSOIL ON A STEEP FORESTED SLOPE

R.D. HARR

Forestry Sciences Laboratory, Pacific Northwest Forest and Range Experiment Station, Forest Service, U.S. Department of Agriculture, Corvallis, Oreg. 97331 (U.S.A.)

(Received December 11, 1975; revised and accepted May 6, 1976)

ABSTRACT

Harr, R.D., 1977. Water flux in soil and subsoil on a steep forested slope. J. Hydrol., 33: 37—58.

Soil analyses, tensiometer, piezometer and precipitation data were used to evaluate water flux in soil and subsoil on a steep forested slope. Although unsaturated flow predominated in the soil during fourteen moderate-size winter storms, discontinuous saturation of upslope subsoil did occur at depths of 110—150 cm after at least 30 mm of rain fell in 12 h or less. This saturation persisted less than 20 h. Very slight changes in soil-water pressure head during storms effected relatively large changes in both magnitude and direction of water flux. Maximum water fluxes in the top meter of soil were 3—4.5 mm/h and maximum flux in the subsoil was 0.5 mm/h, slightly less than the minimum flux in the soil. Between storms, the vertical component of flux at the 10-cm depth was much less than the downslope component but equaled the downslope component during storms. Conversely, vertical components of flux at 70- and 130-cm depths were much less than downslope components during storms but equaled downslope components between storms. Analysis of piezometric levels, discharge from a seep, and streamflow revealed an abrupt decrease in rate of water flux to the lower part of the slope about 10 h after the end of rainfall. This decrease corresponded with nearly complete draining of larger pores that had filled with water during storms. Subsurface flow and channel interception averaged 97 and 3% of storm flow which, in turn, averaged 38% of total storm precipitation. No overland flow was observed.

INTRODUCTION

Overland flow rarely occurs on undisturbed forest soils in the western Cascade Range in Oregon, but streams respond quickly to precipitation. Steep slopes and high permeabilities of surface soils (Dyrness, 1969) are conducive to rapid, shallow subsurface flow which would account for quick response of streams. Additional information about this important hydrologic process has been lacking for this region.

That overland flow is a rare occurrence in forested watersheds in humid regions is well documented, and the importance of subsurface flow to streamflow has been stressed repeatedly (Whipkey, 1965, 1969; Hewlett and Hibbert, 1967; Weyman, 1970; Hewlett, 1974). Only a small part of a watershed produces storm runoff. This part, which appears to expand and contract according to changes in rainfall intensity and soil conductivity, was the basis for the variable source area concept (U.S. Forest Service, 1961). Reports of field studies of the interaction between unsaturated and saturated flow and streamflow have generally supported this concept (Ragan, 1968; Betson and Marius, 1969; Dunne and Black, 1970a, b; Rawitz et al., 1970), although these reports have tended to replace the word "variable" with "partial" and imply source area as being relatively fixed (Hewlett, 1974). Recently, mathematical models have been developed to couple transient unsaturated—saturated flow to streamflow (Freeze, 1972a, b; Stephenson and Freeze, 1974; Hewlett and Troendle, 1975) in attempts to supplement these field studies. Other field investigations have shown that subsurface water may move rapidly through piping channels (Jones, 1971) or through biologically created non-capillary channels (Whipkey, 1965, 1969; Patric and Swanston, 1968; Chamberlin, 1972) in otherwise unsaturated soils. Other studies have sought to describe subsurface flow on sloping ground in other areas (van 't Woudt, 1954; Minshall and Jamison, 1965; Takeda and Ishii, 1968; Megahan, 1972; Wilson and Ligon, 1973; Rogowski et al., 1974).

This study was undertaken to obtain additional information about the lateral subsurface flow phenomenon in western Oregon. Specific questions to be answered were:

(1) What is the magnitude and direction of water flux in soil during and after typical rainstorms?

(2) What soil properties and other site factors influence magnitude and direction of water flux in the soil?

(3) What are the relative importances of water fluxes under saturated and unsaturated soil conditions?

(4) How is the timing of subsurface flow related to rainfall and to streamflow?

STUDY AREA AND METHODS

The study was conducted in watershed *10*, a 10.23-ha watershed in the U.S. Forest Service's H.J. Andrews Experimental Forest about 72 km east of Eugene, Oregon. Watershed *10* is the location of one of the intensive study sites of the Coniferous Forest Biome's analysis of ecosystems and lies within the western hemlock zone (Franklin and Dyrness, 1969). Overstory vegetation is primarily 450-year-old Douglas-fir with younger western hemlock and Douglas-fir. Annual precipitation averages 235 cm, about 80% of which falls between October and April during frequent, long duration, low to moderate intensity frontal storms. Although occasional accumulations of snow are generally short lived, snowmelt concurrent with large amounts of rain has produced the maximum runoff events of record.

The study area is located on a stream-to-ridge portion of the south aspect of watershed *10* (Fig. 1). The slope is slightly convex and its gradient averages

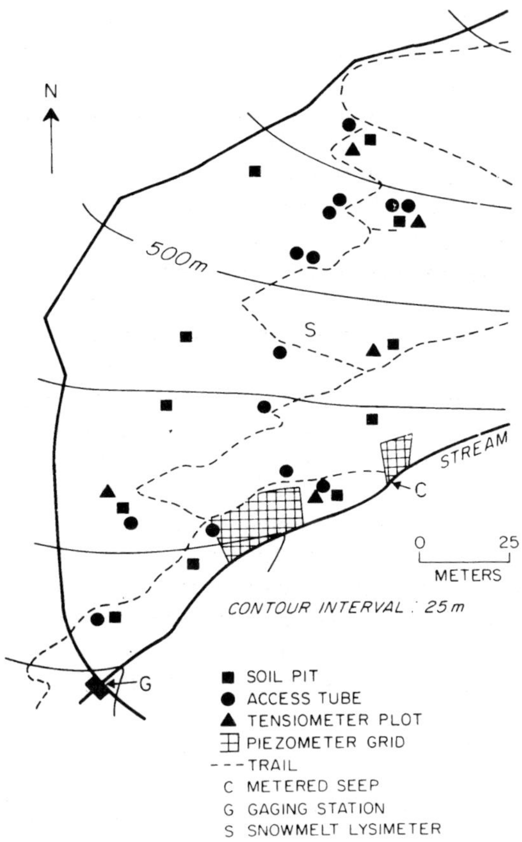

Fig. 1. Map of the study area.

75%, ranging from 50% near the ridge to 110% adjacent to the stream. Elevation ranges from 425 to 530 m. Generally, slightly more than 1 m of relatively poorly developed soil is underlain by 2—7 m of subsoil (saprolite) consisting of highly-weathered coarse volcanic breccias which formed as a result of mudflows and pyroclastic flows (Swanson and James, 1975). In the upper half of the main stream channel and the entire length of the tributary stream, water flows over this saprolite. The lower half of the main channel was scoured to bedrock by a debris avalanche that originated in the upper portion of the south aspect of the watershed. Thus, water in the lower half of the main channel flows over relatively unweathered bedrock and could be collected and its rate of flow measured. This ease of flow measurement was one criterion for selection of the study slope.

Soil and subsoil properties

During initial phases of study, hydrologic properties of the soil and subsoil were determined. A total of 450 undisturbed soil samples were taken from eleven soil pits throughout the study area (Fig. 1). A bulk-density sampler (Blake, 1965) was used to remove at least six samples from depths of 10, 30, 70 and 110 cm in the soil, and 130 and 150 cm in the subsoil as soil pits were dug as deeply as possible. Generally, visible discontinuities in physical properties of the soil at the time of soil sampling were located between these depths. In several pits, samples were also taken at depths of 200 and 250 cm. Soil samples were analyzed for saturated hydraulic conductivity with a constant-head permeameter that accommodated undisturbed samples (Ranken, 1974); pore-size distribution and drainage characteristics by the tension-table method (Vomocil, 1965); and particle-size distribution by the hydrometer method (Day, 1965). Total porosity and bulk density were also obtained from the tension-table data. Tensions applied equaled 10, 20, 30, 40, 60 and 100 cm of water. Additional details of soil analyses are given elsewhere (Ranken, 1974).

In soil and subsoil in the study area, structural characteristics affect hydrologic properties more than do textural characteristics. Texture changes little with depth (Table I). Most surface soils are gravelly clay loams, lower soil layers are gravelly silty clay loams or clay loams, and subsoils are gravelly clays or clay loams. Surface soils are well aggregated, but lower depths (70—110 cm) show subangular blocky structure with less aggregation than surface soils. Subsoil exhibits a massive structure reflecting many characteristics of the weathered pyroclastic parent material.

TABLE I

Mean values of selected soil and subsoil properties for the soil pit uppermost on the study slopes

Depth (cm)	Bulk density (g/cm^3)	Particle-size distribution (%)			Porosity (%)	Amount of porosity (%) in pores with diameter		Saturated hydraulic conductivity (cm/h)
		sand	silt	clay		> 0.30 mm	< 0.030 mm	
10	0.807	22.5	38.8	38.7	60.8	38.4	46.0	352
30	0.897	29.8	34.1	36.1	63.8	40.7	47.2	412
70	1.015	28.5	33.8	37.7	60.3	23.4	65.2	163
110	0.981	28.8	33.0	38.2	63.1	21.4	69.3	175
130	1.080	30.2	31.4	38.4	55.4	11.3	80.6	16
150	1.053	34.0	31.3	34.7	57.6	7.9	83.2	22

Differences in soil aggregation from depth to depth most likely account for differences in pore-size distribution and hydraulic conductivity. Although total porosity changes only slightly with depth, pore-size distribution and hydraulic conductivity change markedly (Table I). Significant shifts occur between 30 and 70 cm and between 110 and 130 cm involving primarily pores 0.03 and 0.3 mm in diameter. Lower depths have less porosity in the 0.3-mm diameter class and more porosity in the 0.03-mm diameter class than do upper depths. Differences in pore-size distribution among depths are further reflected in the drainage curves for soil (Fig. 2). At the bottom of the slope the more massive structure characteristic of the saprolite is located closer to the ground surface. Most likely, past landslides in this area of over-steepened slopes adjacent to the stream removed the more developed surface soils.

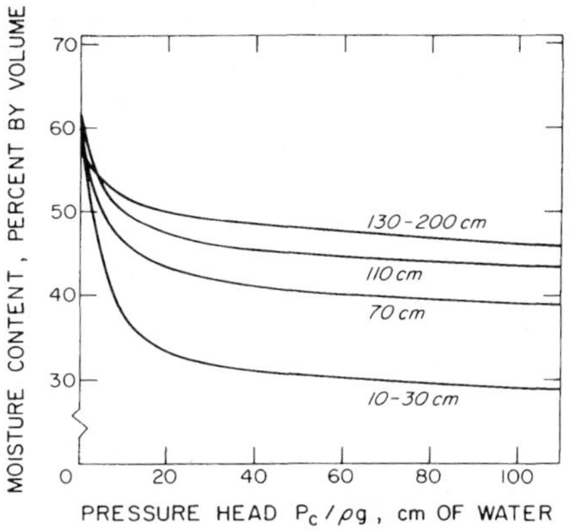

Fig. 2. Soil-water characteristics for soil and subsoil.

Instrumentation

Magnitude and direction of water flux at several depths in the soil and subsoil were estimated by 72 tensiometers in five plots within 5 m of soil pits (Fig. 1). A tensiometer plot consisted of two or four groups of tensiometers. Half of a four-group plot is shown in Fig. 3. Individual tensiometers of a group were spaced 20 cm apart and were located at depths of 10, 30, 70, 110, 130 and 150 cm, the soil sampling depths at nearby soil pits.

Water input to the study slope was measured continuously with a modification of the snowmelt lysimeter described by Haupt (1969). Although the lysimeter could measure both rainfall and snowmelt, water input consisted of only rainfall for the period described in this paper.

Water was observed flowing from the base of the study slope and from other slopes in the lower half of watershed *10*, in relatively well-defined seeps of various sizes and shapes and spaced various distances apart (Brown et al., 1972). These seeps are highly localized zones of saturated soil rather than pipelike channels described by Jones (1971) and Chamberlin (1972) and are apparently related to the microrelief of the relatively unweathered bedrock near the stream or to the presence of vertical, andesitic dikes (Swanson and James, 1975). Some saturated zones have been traced 12 m upslope.

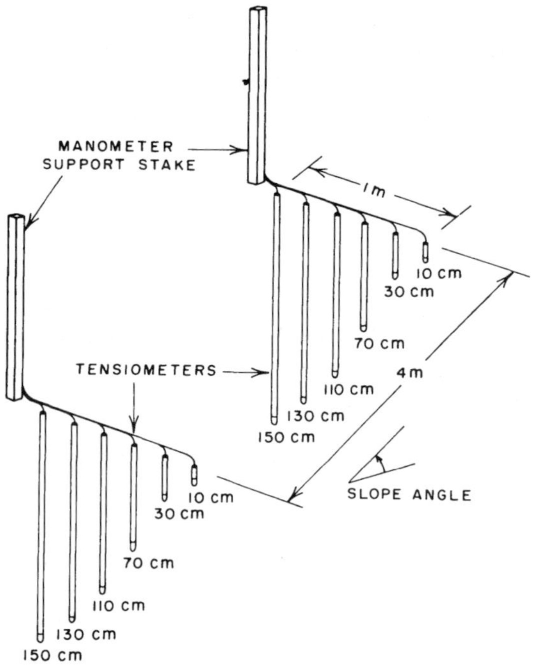

Fig. 3. Relative location of tensiometers within a tensiometer plot.

During and between major storm events, virtually no water was observed entering the stream from soil located between outlets of the seeps.

The nature of water flow in the region immediately adjacent to the stream was determined by two grids of piezometers installed in the vicinity of two seeps (Fig. 1). Each piezometer consisted of 1.9-cm PVC pipe with a perforated lower end covered with fine-mesh nylon screen. After piezometers were placed in holes power-augered to bedrock, holes were backfilled with sand and bentonite in a manner similar to that described by Swanston (1967). The smaller grid contained ten piezometers placed 40—90 cm deep and spaced 1—2 m apart, and the larger grid contained 34 piezometers at a depth of from 48 cm to over 4 m and spaced 2—4 m apart.

Preliminary observations indicated considerable fluctuation in piezometric levels over relatively short periods of time and among piezometers. Because of this variation, the difficulty in visiting each piezometer frequently enough, and the damage that would be done to the fragile, highly porous surface soils during frequent visits, an electronic water-level measuring and recording system was devised (Holbo et al., 1975). This system, which operates under battery power, was capable of measuring and recording water level in sixteen piezometers simultaneously every half hour.

In addition, flow from one large seep downslope from the small grid of piezometers (Fig. 1) was caught in a metal collector and routed to a large tipping-bucket measuring device. This seep is located in or upslope from an andesitic dike that trends in a north-northwesterly direction from the stream. Attempts to trace this dike upslope and to locate other similar dikes were unsuccessful. Thus the size of the source area of this seep is unknown.

RESULTS AND DISCUSSION

Soil-water pressure head

Measurements of soil- and subsoil-water pressure head were made at 4-h intervals during fourteen winter storms in the 1973/1974 winter rainy season, and at 8-h intervals between these storms. Storm events were defined as periods of major rainfall separated by at least 6 h of rainfall intensities averaging less than 0.1 mm/h. Storm sizes ranged from 16 to 127 mm. Average intensity of storms was about 3 mm/h, ranging from less than 0.1 mm/h to 22 mm/h.

An example of fluctuation in soil- and subsoil-water pressure head is shown in Fig. 4. These data are from one pair of tensiometer groups in the tensiometer plot uppermost on the study slope (Fig. 1) and represent only two storm events. Fluctuations of similar magnitude were observed at other locations on the study slope during other storms, although relative differences in pressure head with depth varied somewhat from place to place. Minimum pressure head observed at any depth during the study period was −28

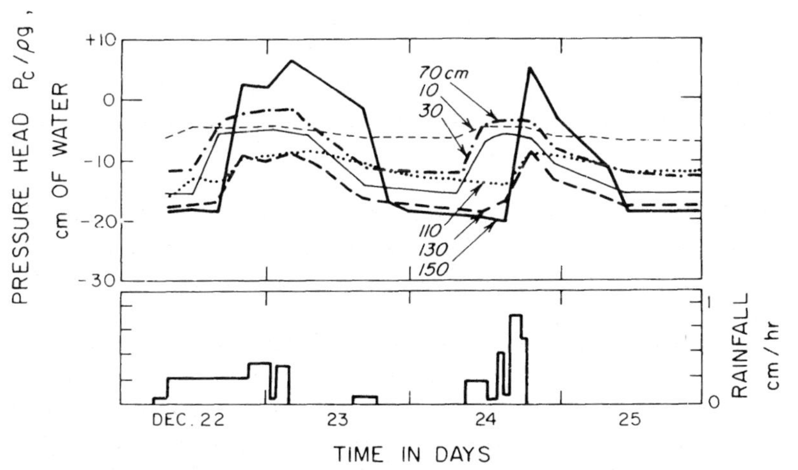

Fig. 4. Fluctuation in soil-water pressure head at highest elevation tensiometer plot for two storms.

cm of water. Subsequent discussion will center largely on two groups of tensiometers, data from one of which are shown in Fig. 4, but it also applies to most other groups of tensiometers located elsewhere on the study slope.

An important feature of Fig. 4 is the positive pressure head detected at the 150-cm depth near the end of the storm event on December 22 and again after the storm event on December 24. Results of the soil analyses had suggested that saturation would most likely occur at 130 cm, the level at which saturated hydraulic conductivity shows the greatest decrease (Table I) because of an abrupt change in pore-size distribution. The positive pressure head shown in Fig. 4 indicates a saturated zone at 150 cm with a maximum depth of 6.6 cm above the tensiometer level. A saturated layer was not detected at any other depth at this location nor at any depth in the downslope group of tensiometers. However, saturated subsoil was detected frequently at the 110-cm level in the other pair of tensiometer groups in this plot and at the 110- or 130-cm levels in other plots. Local variation in the location of the soil—subsoil boundary and the critical change in pore-size distribution probably account for the variation in the depth at which saturation occurred. Non-uniform saturation occurred only four times during the 3-month study period, after at least 30 mm of rain fell in less than 12 h. Saturation did not persist longer than 20 h at any location.

Tensiometer data were used in a two-dimensional analysis of water flux. Total flux consisted of a vertical component, e.g., between the 10- and 30-cm depths in one tensiometer group, and a downslope (parallel to slope or lateral) component, e.g., between the 10-cm depths of a pair of tensiometer groups. In either case, flux was determined by the general equation:

$$q = K(P_c)i \tag{1}$$

where q = water flux, K = hydraulic conductivity as a function of capillary pressure P_c, and i = hydraulic gradient. Hydraulic gradient is described by:

$$i = \Delta H/L$$

where H = hydraulic head = $Z + P$ at each tensiometer, and L = distance between tensiometers in the direction of the flux. Z is gravitational head (elevation of a tensiometer cup), and P is the pressure head of water.

Unsaturated hydraulic conductivity

Conductivity of soil and subsoil were determined for transient unsaturated conditions following an empirical method described by Brooks and Corey (1966). This method uses saturated hydraulic conductivity and the shape of the moisture release curve of an undisturbed soil sample to estimate unsaturated conductivity as a function of capillary pressure. Unsaturated hydraulic conductivity was estimated for each tensiometer depth by:

$$K(P_c) = K_s(P_b/P_c)^{\eta} \tag{2}$$

where $K(P_c)$ = unsaturated conductivity, K_s = saturated hydraulic conductivity, P_b = the bubbling pressure of the soil or subsoil (approximately the minimum P_c on the drainage cycle at which air is continuous in the soil), P_c = capillary pressure, and $\eta = 2 + 3\lambda$. Both P_b and λ are derived from soil and subsoil desorption data. Brooks and Corey (1966) refer to λ as an index of pore-size distribution: λ is large for porous media having a narrow range of pore sizes and small for media having a wide range of pore sizes. Sands, for example typically exhibit λ values around 3.5 indicating a narrow range of pore size. Mean values of λ and $P_b/\rho g$ are given in Table II (ρ= density of water; g = acceleration due to gravity). The relatively low values of λ shown here indicate a wide range of pore sizes at all depths.

TABLE II

Mean values of minimum and maximum unsaturated conductivity at various depths for the soil pit uppermost on the study slope as determined by Brooks and Corey's (1966) method

Depth (cm)	λ	η	$P_b/\rho g$	Saturated hydraulic conductivity, K_s (cm/h)	Unsaturated hydraulic conductivity, $K(P_c)$ (cm/h)	
					minimum	maximum
10	0.386	3.16	0.38	352	0.05	0.46
30	0.396	3.19	0.22	412	0.01	0.36
70	0.199	2.60	0.57	163	0.06	11.4
110	0.171	2.51	0.36	165	0.01	0.06
130	0.097	2.29	0.93	16	0.02	0.09
150	0.104	2.31	1.2	22	0.03	22*

*Saturation occurred at this depth. At $P_c/\rho g = 0$, $K(P_c) = K_s$.

Water flux

Water fluxes in the vertical and downslope directions were summed to obtain resultant flux. The magnitude of the resultant flux is shown graphically in Fig. 5 and can be written:

$$q_R = [(q_D + q_V \sin\alpha)^2 + (q_V \cos\alpha)^2]^{0.5} \quad (3)$$

where q = water flux; R, D, V denote resultant downslope and vertical directions; and α = slope angle. Flux angle γ becomes:

$$\gamma = \sin^{-1}(q_D \cos\alpha/q_R) \quad (4)$$

Theoretically, γ can very from 0° (vertical, downward) through 180° (vertical, upward) to 360° (vertical, downward).

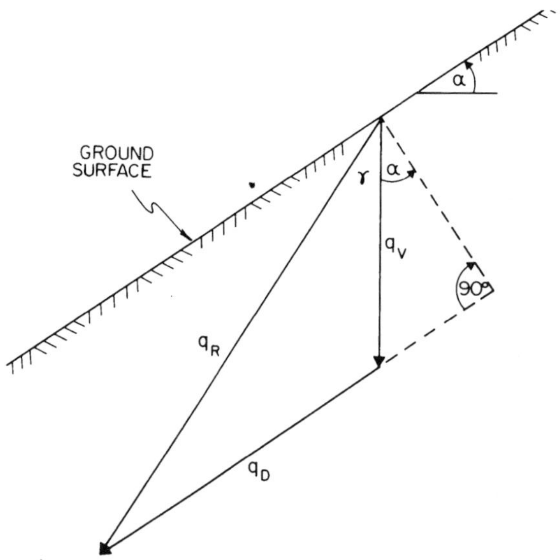

Fig. 5. Vertical, downslope, and resultant water fluxes.

The magnitude and direction of resultant water fluxes during and between two storm events are shown in Fig. 6 for three depths. At this uppermost slope position, slope angle is 28° (53% slope). Flux parallel to the slope would therefore have a flux angle of 62° (90°—28° = 62°), and downward vertical flux would have a flux angle of 0°. Fig. 6 is based partly on the data shown in Fig. 4.

Fig. 6A shows fluctuations in magnitude and direction of water flux between 10 and 30 cm. Here, flux varied from about 0.9 mm/h between storms to 2—4 mm/h during storms. As expected, response to rainfall was most rapid at this shallow depth. Flow direction at 10 cm showed a fairly consistent pattern of fluctuation during the experimental period (Fig. 6A). Between storms, when flux was minimum, flow was directed mostly downslope. Conversely, during storms, when water flux was considerably greater than between storms, the vertical and downslope components of flow were about equal. Thus, water moving from 10 to 30 cm during or shortly after a storm acquired a substantial downslope component.

Water flux between 70 and 110 cm is shown in Fig. 6B. Note that the inverse relationship between flux and flux angle did not occur at this depth. Generally when flux was low, flow was more downward. As flux increased during storms, it also became directed more downslope. Flux at this depth varied between about 0.5 and about 4.5 mm/h.

Water flux in the subsoil is shown in Fig. 6C. This figure differs substantially from either Figs. 6A or 6B in two ways. First, the magnitude of water flux was considerably less in the subsoil than in the top meter of soil. Al-

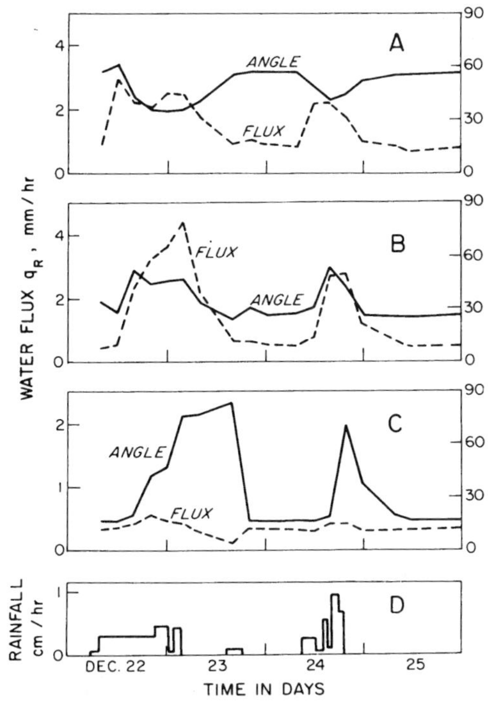

Fig. 6. Water flux at (A) 10 cm, (B) 70 cm and (C) 130 cm, in relation to rainfall (D).

though magnitude of flux in the subsoil more than doubled during storms, maximum fluxes were still less than minimum fluxes in the top 110 cm of soil. Thus, the top 110 cm of soil at this slope position appears to be the most active zone of downslope movement of water. The second difference in flux in the subsoil concerns flux angle. The flux angle at 130 cm at this location frequently exceeded 62°, the angle of flow parallel to the slope. This large flux angle was caused by the saturated zone at 150 cm in the upslope group of tensiometers. Because of this saturation, hydraulic head was greater at 150 cm than at 130 cm, where subsoil was still unsaturated, causing q_V in Fig.5 to be vertical upward and the resulting flux angle γ to exceed 62°.

That flux angles were always greater than 0°, indicating a substantial lateral component of flow, agrees with Zaslavsky and Rogowski's (1969) theoretical analysis of lateral flow. In this analysis, a lateral component was predicted to occur in sloping soil wherever a vertical decrease in permeability exists.

Malfunctions of one or more tensiometers in other tensiometer plots precluded flux analyses for other slope positions; thus, comparisons of water flux at different slope positions are not possible. In subsequent discussion, soil-water fluxes described above will be assumed to be reasonably applicable to all but the lower 10—15 m of the study slope. Similarity of properties exhibited by soil and subsoil, from midslope to upslope positions (Ranken, 1974) tends to support the validity of this assumption. Additionally, intermittent tensiometer readings indicate that capillary pressures were similar to those measured at the plot uppermost on the slope. Further support is added by an analysis of η, P_b and K_s for the entire study slope. These parameters vary with depth just as do those of the soil pit uppermost on the slope (Table II) and for which similar analyses were made.

Although computed fluxes between tensiometers were very small if one or more tensiometers were located in unsaturated soil, computed water fluxes in the subsoil would have been much greater if both upslope and downslope tensiometers at that depth had been in saturated subsoil. If a saturated zone were parallel to the slope of the ground surface, the hydraulic gradient in this saturated zone could be approximated by the sine of the slope angle, about 0.47 at this plot. Laboratory measurements show saturated hydraulic conductivities for the subsoil here could be 20—50 cm/h. Thus, fluxes of 10—25 cm/h directed entirely downslope (no component vertically downward) could have been possible if the mid- to upslope saturated zones were continuous and connected with the seeps at the bottom of the slope. Such a network of saturated zones could comprise a subsurface drainage system similar to those described for other areas (Whipkey, 1969; Chamberlin, 1972) and would partly account for the rapid hydrologic response of watershed *10* to precipitation. But if the midslope to upslope saturated zones were isolated as measurements of soil-water pressure head suggest, then water movement through these zones would be limited by the relative hydraulic conductivity of the unsaturated subsoil immediately downslope from the saturated zones, a situation that is approximated in Fig. 6C.

Saturated zones do occur elsewhere on watershed *10*. Access tubes installed throughout the watershed for soil-moisture measurements with a neutron soil-moisture meter consist of 3.66-m sections of tubing connected by a non-watertight joint. During many storm events, water has entered through this joint in some access tubes located on mid- to upslope positions on the north aspect of the watershed. Water entry could occur only if the subsoil surrounding these joints became saturated and soil-water pressures were positive. The 1.70—3.0-m depth of these joints suggests considerably more variation in the vertical location of saturated zones than was observed on the study slope. No estimates of water fluxes in these saturated zones are available.

Slope—stream linkage

At some point in a sloping soil which supplies water to a stream, saturation must occur so that a hydraulic gradient exists from the soil to the stream. Earlier, transient saturated zones at some midslope to upslope locations on the study slope were described. But saturated zones, because of their spatial and temporal limits, probably had little effect on the overall hydraulic behavior of the slope. Additional saturated zones did occur in certain locations along the base of the slope. The conductivity of saturated zones and the supply of water to them by unsaturated drainage from above determined their upslope extent. Measurements of water level in piezometers indicate the saturated zones extended both upslope and laterally during runoff events. However, because of discontinuity of the saturated zones and problems associated with installing piezometers in grids of small enough spacing, exact upslope and lateral extents are unknown.

Some insight into the hydraulic linkage between the slope and the stream can be obtained, however, by analysis of streamflow and saturated flow conditions in soil at the bottom of the slope. Streamflow, discharge from the metered seep, and piezometric level upslope from the seep all responded relatively quickly to precipitation (Fig. 7). Because of channel interception, the effects of rainfall were first evident in streamflow. For example, prior to the storm on December 22, streamflow had been decreasing after the peak of the December 20 storm. At 05^h30^m, rainfall began, and channel interception noticeably slowed the rate of decrease in streamflow. However, streamflow did not increase until about 11^h00^m when flow rate increased from the seep. Rainfall essentially ceased at 03^h30^m on December 23, but stream-

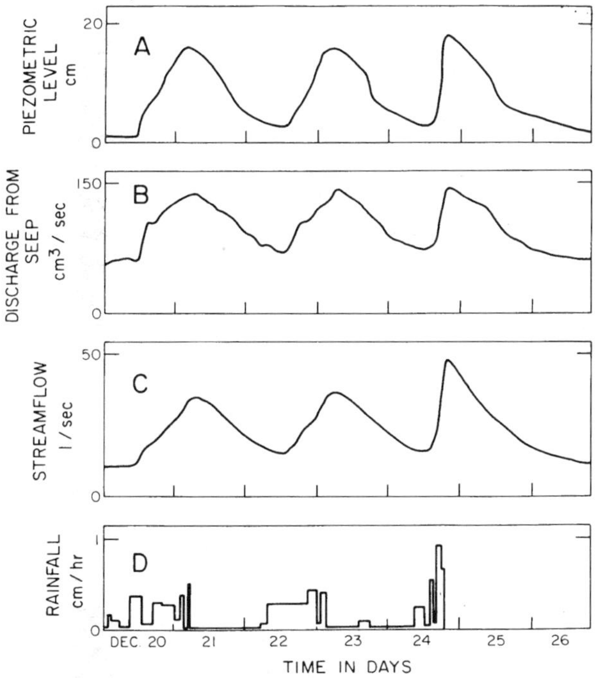

Fig. 7. Fluctuation of (A) piezometric level at a piezometer 1 m upslope from the metered seep, (B) discharge from the seep, (C) streamflow, and (D) rainfall from three storms.

flow didn't peak until $05^h 00^m$; discharge from the seep didn't peak until about $07^h 00^m$, reflecting continued drainage of the slope above this seep.

The water level in piezometer *35* (Fig. 7A) peaked some 4 h before flow from the seep. Analysis of data from piezometer *35* and other piezometers in the same grid suggests at least two sources of water for the seep downslope from this grid. One source appears to be shallow and close to the piezometer grid so that flow from this source peaks almost immediately after the end of rainfall as indicated by piezometric level in Fig. 7A. Water from the larger and probably more distant source continuous to enter the saturated zone upslope from the seep after rainfall has ceased, causing the discharge from the seep to peak several hours after the end of a storm.

Analysis of recession rates of the plots in Fig. 7 revealed additional similarities among piezometric level, discharge from the seep, and streamflow. In Fig. 8, values of each of these are plotted over time after the maximum value of each. Logarithmic scales are used for both axes of all plots because the logarithm of unsaturated drainage tends to be linearly related to the logarithm of time (cf. Wilcox, 1959; Nixon and Lawless, 1960; Hewlett and Hibbert, 1963; Weyman, 1970). Overland flow was never observed in the

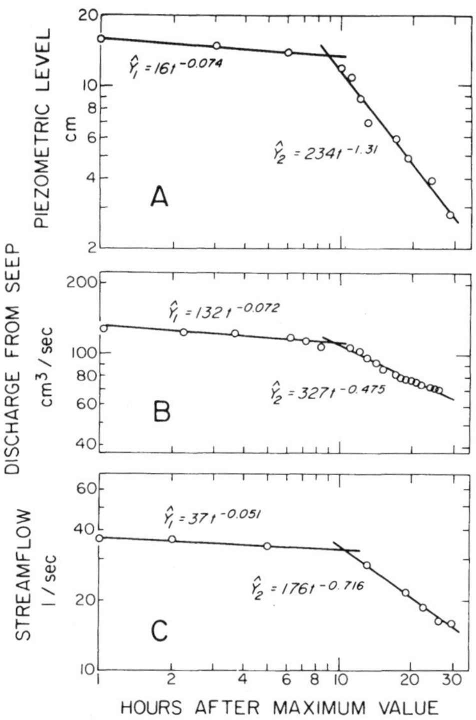

Fig. 8 Relationship of (A) piezometric level at a piezometer 1 m upslope from the metered seep, (B) discharge from the seep and (C) streamflow with time after maximum value of each for the storm of December 22, 1973.

watershed; subsurface flow predominated and ultimately controlled the rate of delivery of water to the saturated zone at the bottom of the study slope and other slopes in the watershed. Consequently, streamflow was largely a function of the rate of unsaturated drainage of sloping soil masses in the watershed, so that linear relationships between log of time and log of streamflow were expected.

In each plot in Fig. 8 there are two separate periods during which linear relationships exist between log of the Y variable and log of time. During each period, the Y variable is related to time by:

$$Y = at^{-b} \quad (5)$$

where Y = piezometric level, discharge from the seep, or streamflow; t = time after maximum value of piezometric level, discharge from the seep,or streamflow; and a and b are positive constants peculiar to each drainage period for each Y variable. The two periods are separated by a change in slope at t = 10 h. Differentiating eq. 5 for the two drainage periods in Fig. 8 shows the rate of decrease in piezometric level, discharge from the subsurface channel, and that streamflow increased about 10 h after the peak value of each. Streamflow plots are similar to plots of discharge from the seep; this suggests that the remainder of watershed *10* responded to storms in much the same way as did the study slope. Most likely, streamflow is the sum of a series of components of flow similar to that from the seep (Fig. 7B).

The forms of the curves in Fig. 8 are similar in some respects to those reported from two other studies of discharge of subsurface water from sloping soil masses. Hewlett and Hibbert (1963) described a two-phase process of drainage from a repacked, sloping soil mass 1 m deep. During the initial 1.5 days, drainage of large pores occurred; and after a 5-day transition period, drainage was from small pores. Weyman (1970, 1973) found that a 1-day first stage of drainage from a natural slope was dominated by saturated flow through primarily noncapillary pores. After a 4-day transition period, drainage from the soil was dominated by unsaturated lateral flow in small pores to a narrow saturated zone at the base of the slope. In both these studies, log-log plots of discharge from the slope over elapsed time showed that slopes of the discharge curves during the transition periods were steeper than for the first phase of drainage. The curves of the second phase of drainage were less steep than for the transition period but steeper than for the first phase of drainage.

In the study area, moderate-size winter storms associated with fronts moving in a wavelike fashion from the Pacific Ocean were spaced 24—48 h apart during the initial part of the study period. Therefore, in the examples shown in Fig. 8, piezometric level, discharge from the seep, and streamflow could recede for only about 30 h before beginning to respond to the next storm (Fig. 7). A third phase of drainage was observed, however, during the latter part of a virtually rainless 18-day period following the 7-cm storm on December 28—30. At $t \geqslant 60$ h, plots of both piezometric level and discharge from the seep showed another increasing rate of decrease. At $t = 62$ h, piezometer *35* emptied although the seep discharged water continuously at a very low rate. Other piezometers showed that a shallow saturated zone still existed, presumably because of drainage of water from the more extensive and remote source described earlier. At $t = 90$ h, streamflow decrease finally showed a similar increase in rate. Adding this third phase of drainage described above made the pattern of drainage of the study slope similar to that of the B horizon described in Weyman's (1970) study. Most of the discharge from Weyman's slope occurred from the B horizon while unsaturated percolation supplied water to a rapidly expanding zone of saturation.

In any event, drainage reported here appears to be much faster than that described by either Hewlett and Hibbert (1963) or Weyman (1970, 1973). Presumably, differences in pore-size distributions and in slope account for these differences in drainage rates.

Because unsaturated flow controls rate of water supply to the base of the slope, dynamics of water flux in the vicinity of the saturated zones as well as elsewhere on the slope should account for the change in rate of decreasing drainage at $t = 10$ h. This does appear to be the case. During the first stage of drainage, i.e., up to about 10 h, water moved out of the larger pores. This drainage can be illustrated by inserting transient values of $P_c/\rho g$ from Fig. 4 into the equation for capillary rise which is used to relate size of pores filled with water to $P_c/\rho g$. This analysis, ignoring the effects of hysteresis, indicates that pores with diameters <0.25 mm in the lower soil (70-cm depth) were water filled when $P_c/\rho g = -12$ cm at the start of the storm on December 22. At this time and depth, water flux was about 0.5 mm/h (Fig. 6B). As rainfall continued, larger pores began filling with water until pores with diameters up to about 1.5 mm (as indicated by $P_c/\rho g = -2$ cm) became filled. During this period, which lasted 10—12 hr, water flux at 70 cm was at a maximum of 3—4.5 mm/h. When rainfall ceased at 03^h30^m on December 23, these larger pores began draining as water was no longer being supplied from above by rainfall. As soil water flux decreased, flux to the lower slope decreased but initially at a relatively low rate. Within about 10 h after the end of rainfall, $P_c/\rho g = -11$ cm, which was nearly equal to pre-storm $P_c/\rho g$, and nearly all larger pores that had filled during the storm had drained. At this point, water flux had decreased to about 0.7 mm/h. Decreasing water flux to the lower slope was reflected by an increasing rate of decrease in discharge from the metered seep and in streamflow as the saturated zones contracted. After this 10-h period, soil continued to drain slowly for an additional 20 h during which time water flux gradually decreased to less than 0.5 mm/h. Then the soil began to respond to a new storm, and the cycle was repeated. Similar sequences of pores filling and draining can be observed for other soil depths although sizes of pores differed somewhat among depths.

During the 18-day rainless period described earlier, the soil undoubtedly drained considerably more than that described above. But, because temperatures during this period were subfreezing, tensiometers could not be used to measure capillary pressures. Also, because of the severe freezing weather, precipitation and soil-water fluxes could not be determined for the study period's maximum storm event which occurred immediately after this period of freezing.

Contributions to storm flow

As stated before, overland flow has not been observed in watershed *10*. Subsequently, streamflow has only two components: (1) channel interception, i.e., rain falling directly onto the water and rock surfaces between streambanks; and (2) subsurface flow.

The absence of overland flow can be explained by the values of unsaturated hydraulic conductivity $K(P_c)$ for the top 70 cm of the soil mantle (Table II) and low rainfall intensities. The highly porous, well-aggregated surface soils can readily accept all precipitation and easily transmit water to lower soil depths. When pressure heads were least prior to a storm, values of $K(P_c)$ throughout the soil were 0.01—0.06 cm/h. As water was added to any depth, $K(P_c)$ at that depth increased as pressure head increased and as water moved into larger pores. During storms, soil at the 70-cm depth apparently came closest to saturation. Before a storm, saturation here was 75%, increasing to 90% during storms. Similarly, before storms, saturations at 10- and 30-cm depths, respectively, were 67 and 55%, both increasing to 75% during storms. Considerable storage was available during storms for additional water had it occurred. However, the wetter the soil became, the better able it would have been to transmit water, and up to a point, overland flow perhaps would have been no more likely to have occurred than it was during storms in this study.

Hewlett and Hibbert (1967) have described a concept of rapid watershed response to precipitation without appreciable overland flow. They discriminate between two types of streamflow: quick flow and delayed flow. An arbitrary separation line with a constant slope of 0.55 l sec^{-1} km^{-2} h^{-1} separates quick flow from delayed flow. Each unit of rainfall on a slope contributions to both temporary storage, which eventually becomes delayed flow during some later period, and quick flow. Upslope from a stream, a unit of rainfall contributes less to quick flow and more to retention—detention storage. Hewlett and Hibbert further theorize that only part of the water contributed to quick flow is actually new water, i.e., water from the current storm; the remainder is called "translatory flow" or flow by displacement. This is water already stored in the soil prior to the storm and will be released in large quantities only where the soil is extremely wet. They regard this movement as the result of thickening of water films around soil particles and a resulting pulse of water as the saturated zone is approached. This translatory process has been supported by laboratory studies (Horton and Hawkins, 1965) in which 2.5 cm of tritiated water was added to 1.2-m soil drained to equilibrium. Then 2.5 cm of pure water was added each day, and effluent was analyzed daily for tritium. Only after 87% of the original water flowed from the columns did the tritiated water emerge. Horton and Hawkins concluded that in deep soil profiles rainwater likewise tends to displace water as it moves primarily through capillary pores. More recently, Martinec (1975), using environmental tritium to trace the movement of snowmelt water, concluded that infiltration of meltwater causes a corresponding increase in outflow of subsurface water to streams. Thus, infiltrated water, although moving slowly, appears "in substitution" in the hydrograph, represented by water previously stored in the soil mantle.

Table III shows relative contributions of channel interception and subsurface storm flow to total storm flow for seven storm events in the study reported here. Quick flow here is synonymous with that defined by Hewlett and Hibbert (1967) and is composed of channel interception and subsurface storm flow. This subsurface storm flow is roughly synonymous with that described by van 't Woudt (1954), Whipkey (1965, 1969) and Freeze (1972b). Strict comparisons with subsurface storm flow as described by these authors or with "interflow" (Chamberlin, 1972) and "throughflow" (Kirkby and Chorley, 1967; Weyman, 1973) would, or course, require uniformity of methods used to determine duration of storm flow and to separate quick flow from delayed flow.

A relatively high proportion of gross precipitation appeared as quick flow. Quick flow averaged 38% of gross precipitation, ranging from 23 to 51% for the storms analyzed. Proportion of quick flow contributed by channel interception was estimated from measurements of stream channel length and

TABLE III

Relative magnitude of storm flow, channel interception, and subsurface storm flow for watershed *10* during seven storm events

Date	Gross precipitation (cm)	Quick flow as percent of gross precipitation	Channel interception as percent of quick flow	Subsurface storm flow as percent of quick flow
Dec. 19—21	7.0	34	3	97
Dec. 22/23	6.7	23	3	97
Dec. 24	4.2	40	2	98
Dec. 26—30*	18.3	38	3	97
Jan. 30—Feb. 1	10.7	42	2	97
Feb. 3/4	4.3	38	4	96
Feb. 15—20	17.8	51	2	98
	Mean	38	3	97

*Complex storm and hydrograph occurred during this period.

width during the rainy season combined with storm precipitation after the time streamflow began to increase. As can be seen in Table III, channel interception was only 2—4% of quick flow. (The channel interception which served to slow streamflow recession as described earlier is not included here.) The remaining 96—98% was contributed by subsurface storm flow. Thus, subsurface streamflow is by far the major contributor to storm runoff in this watershed. Most likely this is also true for most other headwater basins in the western Cascade Range of Oregon.

If the ratio of storm flow to gross precipitation can be taken as a rough estimate of the percentage of a watershed contributing to storm flow (Hewlett, 1974), then, according to Table III, on the average about 38% of watershed *10* contributed to storm flow during the study period. If watershed *10* were idealized somewhat (total channel length = 550 m and slope gradient for the lower half of all slopes = 75%), a storm flow-contributing area equal to 38% of the watershed corresponds to an area extending about 42 m upslope on either side of stream channels. Clearly this area would have to be 100% saturated for a given volume of rainfall to completely displace water stored in it. Judged by pressure and piezometric levels, however, these zones were at best only about 65% saturated. Thus, the area contributing to storm flow must have extended, on the average, more than 42 m upslope to allow for incomplete displacement of soil water stored in the contributing area of the watershed. The exact upslope extent of this contributing area remains unknown, however. Most likely it is highly irregular owing to the presence of the discontinuous saturated zones on lower slopes.

Quick flow appears to be a larger percentage of storm precipitation in headwater basins in the western Cascade Range in Oregon than in other areas for which similar analyses have been reported. In the eastern United States, only the exceptional storm produces quick flow in excess of 25% of gross precipiation and the largest flows have rarely exceeded 50% of the rainfall that produced them (Hewlett and Hibbert, 1967). In contrast, quick flow from watershed *10* in Oregon averaged 38% for seven winter storms that are low to moderate size at best; quick flow from a 17.1-cm storm of 5-day duration amounted to 51% of rainfall. Presumably, during the largest storm events during which 24-h rainfall has exceeded 16 cm (Rothacher et al., 1967) and more than 32 cm of rain has fallen in 3 days, even a greater proportion of the watershed can contribute to storm flow.

Results of storm flow analyses are in agreement with the concept of variable source area of runoff production (cf. Hewlett, 1974). Piezometer data show that saturated zones whose outlets are visible at the toe of the study slope expand gradually upslope and laterally as rainfall continues. However, overland flow on wet zones near streams, which can be an important factor in runoff production in other areas (Ragan, 1968; Betson and Marius, 1969; Dunne and Black, 1970a, b), was notably absent and has not appeared in similar small watersheds in this region. Stream channels apparently do not expand or contract appreciably over the course of frequent winter storms. That soils have high conductivities, slopes are steep, and slopes are slightly convex is very conducive to subsurface storm flow and precludes overland flow on any segment of a slope.

SUMMARY

(1) Both magnitude and direction of water flux in soil and subsoil varied temporally. Maximum fluxes of 3—4 mm/h in the top meter of the soil mantle were 6—9 times greater than those in the subsoil. Water flux below 30 cm was directed mostly downslope during storms and more vertically downward between storms. Water flux in surface soils was directed downslope between storms and more vertically downward during storms.

(2) A gradual change in pore-size distribution throughout the soil and an abrupt change at the soil—subsoil boundary influenced values of relative hydraulic conductivity estimated from measurements of capillary pressures. Because of steep slopes, the gravitational component of hydraulic head overshadowed the small changes in pressure head measured during the course of winter storms. However, these slight changes in pressure head effected large changes in both magnitude and direction of water flux at various depths.

(3) Unsaturated flow dominated over all but the bottom 12—15 m of the study slope. Temporary saturated zones frequently occurred at the soil—subsoil boundary at mid- to upslope locations. Potential fluxes in these saturated zones were high if these zones were connected to the permanently saturated zones at the base of the slope. However, these upslope saturated zones appeared to be isolated, and flux in them was limited by the unsaturated conductivity of subsoil immediately downslope.

(4) Saturated flow conditions at the bottom of the slopes in the watershed appeared to be related to the drainage of soil pores which filled with water during a storm. About 10 h after rainfall had ceased, these pores had essentially drained. Discharge from a seep and streamflow exhibited a marked increase in their rates of decrease at this time.

(5) Subsurface storm flow dominated storm flow from the study watershed. Overland flow did not occur. Total storm flow averaged 38% of storm precipitation. Subsurface storm flow and channel interception, respectively, averaged 97 and 3% of total storm flow.

ACKNOWLEDGMENTS

Winter fieldwork by John S. Fekete, computer programing by Tom C. Corbett, and helpful comments by Royal H. Brooks are gratefully acknowledged. This study was partially funded by National Science Foundation Grant No. GB-38610X to the Coniferous Forest Biome, U.S. Analysis of Ecosystems, International Biological Program while the author was Assistant Professor of Forest Hydrology, School of Forestry, Oregon State University, Corvallis. This is Contribution No. 178 from the Coniferous Forest Biome.

REFERENCES

Betson, R.P. and Marius, J.B., 1969. Source areas of storm runoff. Water Resour. Res., 5: 574--582.
Blake, G.R., 1965. Bulk density. In: C.A. Black (Editor), Methods of Soil Analysis, Am. Soc. Agron. Madison, Wisc., pp. 374—390.
Brooks, R.H. and Corey, A.T., 1966. Properties of porous media affecting fluid flow. J. Irrig. Drain. Proc. ASCE, 4855 (IR2): 61—88.
Brown, G.W., Burgy, R.H., Harr, R.D. and Riley, J.P., 1972. Hydrologic modeling in the Coniferous Forest Biome. In: J.F. Franklin, L.J. Dempster and R.H. Waring (Editors), Proceedings—Research on Coniferous Forest Ecosystems—A Symposium, Pac. Northwest For. Range Exp. Stn., For. Serv. U.S. Dep. Agric., Portland, Oreg., pp. 49—70.
Chamberlin, T.W., 1972. Interflow in the mountainous forest soils of coastal British Columbia. In: O. Slaymaker and H.J. McPherson (Editors), Mountain Geomorphology, Tantalus Res., Vancouver, B.C., pp. 121—127.
Day, P.R., 1965. Particle fractionation and particle-size analysis. In: C.A. Black (Editor), Methods of Soil Analysis, Am. Soc. Agron., Madison, Wisc., pp. 545—567.
Dunne, T. and Black, R.D., 1970a. An experimental investigation of runoff production in permeable soils. Water Resour. Res., 6: 478—490.
Dunne, T. and Black, R.D., 1970b. Partial area contributions to storm runoff in a small New England watershed. Water Resour. Res., 6: 1296—1311.
Dyrness, C.T., 1969. Hydrologic properties of soils on three small watersheds in the western Cascades of Oregon. USDA For. Serv., Res. Note PNW-111, Pac. Northwest For. Range Exp. Stn., Portland, Oreg., 17 pp.
Franklin, J.F. and Dyrness, C.T., 1969. Vegetation of Oregon and Washington. USDA For Serv. Res. Pap. PNW-80, Pac. Northwest For. Range Exp. Stn., Portland, Oreg., 216 pp.
Freeze, R.A., 1972a. Role of subsurface flow in generating surface runoff, 1. Base flow contributions to channel flow. Water Resour. Res., 8: 609—623.
Freeze, R.A., 1972b. Role of subsurface flow in generating surface runoff, 2. Upstream source areas. Water Resour. Res., 8: 1272—1283.
Haupt, H.F., 1969. A simple snowmelt lysimeter. Water Resour. Res., 5: 714—718.
Hewlett, J.D., 1974. Comments on letters relating to *Role of subsurface flow in generating surface runoff, 2. Upstream source areas* by R. Allan Freeze. Water Resour. Res., 10: 605—607.
Hewlett, J.D. and Hibbert, A.R., 1963. Moisture and energy conditions within a sloping soil mass during drainage. J. Geophys. Res., 68: 1081—1087.
Hewlett, J.D. and Hibbert, A.R., 1967. Factors affecting the response of small watersheds to precipitation in humid areas. In: W.E. Sopper and H.W. Lull (Editors), Forest Hydrology, Pergamon, New York, N.Y., pp. 275—290.
Hewlett, J.D. and Troendle, C.A., 1975. Non-point and diffused water sources: a variable source area problem. Watershed Management, Am. Soc. Civ. Eng., New York, N.Y., pp. 21—46.
Holbo, H.R., Harr, R.D. and Hyde, J.D., 1975. A multiple-well, water-level measuring and recording system. J. Hydrol., 27: 199—206.
Horton, J.H. and Hawkins, R.H., 1965. Flow path of rain from the soil surface to the water table. Soil Sci., 100: 377—383.
Jones, A., 1971. Soil piping and stream channel initiation. Water Resour. Res., 7: 602—610.
Kirkby, M.J. and Chorley, R.J., 1967. Throughflow, overland flow, and erosion. Int. Assoc. Sci. Hydrol. Bull., 12: 5—21.
Martinec, J., 1975. Subsurface flow from snowmelt traced by tritium. Water Resour. Res., 11: 496—498.

Megahan, W.F., 1972. Subsurface flow interception by a logging road in mountains of central Idaho. In: Proc. Nat. Symp. on Watersheds in Transition. Am. Water Resour. Assoc., Urbana, Ill., pp. 350—356.

Minshall, N.E. and Jamison, V.C., 1965. Interflow in claypan soils. Water Resour. Res. 1: 381—390.

Nixon, P.R. and Lawless, G.P., 1960. Translocation of moisture with time in unsaturated soil profiles. J. Geophys. Res., 65: 665—661.

Patric, J.H. and Swanston, D.N., 1968. Hydrology of a slide-prone glacial till soil in southeastern Alaska. J. For., 66: 62—66.

Ragan, R.M., 1968. An experimental investigation of partial-area contributions. Int. Assoc. Sci. Hydrol. Publ., 76: 241—249.

Ranken, D.W., 1974. Hydrologic properties of soil on a steep forested slope. M. Sc. Thesis, Oregon State University, Corvallis, Oreg. (unpublished).

Rawitz, E., Engman, E.T. and Cline, G.D., 1970. Use of the mass balance method for examining the role of soils in controlling watershed performance. Water Resour. Res., 6: 1115—1123.

Rogowski, A.S., Engman, E.T., and Jacoby, Jr., E.L., 1974. Transient response of a layered, sloping soil to natural rainfall in the presence of a shallow water table: experimental results. Agric. Res. Serv. Res. Pap. ARS-NE-30.

Rothacher, J., Dyrness, C.T. and Fredriksen, R.L., 1967. Hydrologic and Related Characteristics of Three Small Watersheds in the Oregon Cascades. Pac. Northwest For. Range Exp. Stn., For. Serv., U.S. Dep. Agric., Portland, Oreg., 54 pp.

Stephenson, G.R. and Freeze, R.A., 1974. Mathematical simulation of subsurface flow contributions to snowmelt runoff, Reynolds Creek watershed, Idaho. Water Resour. Res., 10: 284—294.

Swanson, F.J. and James, M.E., 1975. Geology and geomorphology of the H.J. Andrews Experimental Forest, western Cascades, Oregon. USDA For. Serv. Res. Pap. PNW-188, Pac. Northwest For. Range Exp. Stn. Portland, Oreg., 14 pp.

Swanston, D.N., 1967. Soil-water piezometry in a southeast Alaska landslide area. USDA For. Serv. Res. Note, PNW-68, Pac. Northwest For. Range Exp. Stn., Portland, Oreg., 17 pp.

Takeda, S. and Ishii, M., 1968. Movement of subsurface flow from forest area and mechanism that affects soil erosion. Mt. Land Res. Lab., Bull., No. 1, Iwate Univ. Morioka.

U.S. Forest Service, 1961. Watershed management: Some ideas about storm runoff and base flow. Annu. Rep., Southeastern For. Exp. Stn., Asheville, N.C., pp. 61—66.

Van 't Woudt, B.D., 1954. On factors governing the subsurface storm flow in volcanic ash soils, New Zealand. Trans., Amer. Geophys. Union, 35: 136—144.

Vomocil, J.A., 1965. Porosity. In: C.A. Black (Editor), Methods of Soil Analysis, Am. Soc. Agron., Madison, Wisc., pp. 299—314.

Weyman, D.R., 1970. Throughflow on hillslopes and its relation to the stream hydrograph. Int. Assoc. Sci. Hydrol. Bull., 15: 25—33.

Weyman, D.R., 1973. Measurements of the downslope flow of water in a soil. J. Hydrol., 20: 267—288.

Whipkey, R.Z., 1965. Subsurface stormflow from forested slopes. Int. Assoc. Sci. Hydrol. Bull., 10: 74—85.

Whipkey, R.Z., 1969. Storm runoff from forested catchments by subsurface routes. Int. Assoc. Sci. Hydrol. Publ. 85: 773—779.

Wilcox, J.C., 1959. Rate of soil drainage following an irrigation,I. Nature of soil drainage curves. Can. J. Soil. Sci., 39: 107—119.

Wilson, T.V. and Ligon, J.T., 1973. The interflow process on sloping watershed areas. Water Resour. Res. Inst. Rep., No. 38, Clemson Univ., Clemson, S.C.

Zaslavsky, D. and Rogowski, A.S., 1969. Hydrologic and morphologic implications of anisotropy and infiltration in soil profile development. Soil Sci. Soc. Am. Proc., 33: 594—599.

Commentary on Paper 23

Anderson, M. G. & Burt, T. P. (1978) The role of topography in controlling throughflow generation.

The role of the University of Bristol in UK studies of streamflow generation, that had started with the studies in the East Twin Brook catchment of Mike Kirkby and Darrell Weyman (Paper 17), was continued, when Mike Kirkby moved to the University of Leeds, by the appointment of Malcolm Anderson (later one of the founding editors of *Hydrological Processes*, editor in chief of the *Encyclopaedia of Hydrological Sciences*, and still a professor at Bristol). Malcolm Anderson instigated a number of further field studies, including the work in the Bicknoller Combe catchment by his then PhD student, Tim Burt who went on to work at the universities of Huddersfield and Oxford before becoming a professor at Durham University.

This was one of the first studies in the UK to use the "scanivalve" system of logging tensiometers. Each tensiometer was connected to a port on the valve at a central point in the system. The ports were then accessed in turn by a single pressure transducer and the resulting pressures logged. The scanivalve systems were notoriously difficult to keep free of air bubbles, but were the only way of obtaining multiple, continuously logged soil water potential measurements for a reasonable cost at the time. The system allowed one of the first extensive studies of the pattern of soil moisture potentials within a topographic hollow (as opposed to the earlier studies looking at conditions only in the soil profile or upstream of a throughflow trough).

The experiments at Bicknoller were designed to examine the role of topography on the patterns of saturation in a small hillslope hollow. The measurements suggested, not surprisingly, that the flow pathways were not quite as simple as the hydrological models of the time were theorizing. The flow lines did not necessarily follow the lines of greatest slope, but were also influenced by the down-valley slope and the local soil variability.

Tim Burt later became closely involved with experiments at the Slapton Wood experimental catchment in Devon. This is one of the longest running experimental catchments in the UK. It was originally set up to improve understanding of the movement of nutrients into the streams contributing to Slapton Ley, the largest natural freshwater lake in the south of England. Slapton Ley is formed by a coastal gravel bar that developed in the post-glacial period and which blocks the valleys that were overdeepened during periods when sea level was much lower. The Ley has suffered from severe eutrophication problems during successive summer periods because of the high nutrient levels resulting from the intensive agricultural use of parts of the catchments (Trudgill *et al.*, 1991; Coles & Trudgill, 1985).

The Slapton catchments have been the subject of a number of PhD research studies on both flow and transport processes, one of which was that of Dave Butcher. This led to another benchmark paper, Burt & Butcher (1985) that extended the understanding of how connectivity on hillslopes controls catchment responses. They chose to study the flow processes in one of the three hillslope hollows of the Slapton Wood catchment, in the pasture fields on the northern side of Eastergrounds. Again, they made use of a spatial network of tensiometers to indicate patterns of saturation on the hillslopes. Their results demonstrate the importance of connectivity of saturation, which here builds up from both the valley bottom and the flat upper hillslopes, on the catchment response (Figs 23.1, 23.2). The importance of connectivity of surface and subsurface saturated areas has also been stressed by, amongst others, Engman & Rogowski (1974) and Bazemore *et al.* (1994).

Later work has revealed that the hydrological response of the Slapton hillslopes is more complex than presented here. Although the hollows do saturate, there are old field drains made of stone slabs (probably dating back to 19th century), there are places where surface runoff re-infiltrates into periglacial head deposits, and there are also springs where fault lines in the Devonian shales intersect the surface that are unrelated to any surface expression of the topography. The catchment hydrographs in winter, once subsurface connectivity has been established, typically show a double peak. There is a fast early peak, resulting from saturated contributing areas (and possibly surface runoff from agricultural fields in some storms), and a slow (often higher) peak 2–3 days later resulting from the response of a significant subsurface storage (Burt *et al.*, 1983; Burt & Heathwaite, 1986).

Data sets from Slapton have been used to test hydrological models, including versions of TOPMODEL (Fisher & Beven, 1996; Beven & Freer, 2001) and the SHE model (Bathurst *et al.*, 2004). The latter is a particularly interesting study as it was formulated as a "blind validation" test of the SHE model.

References

Bathurst, J. C., Ewen, J., Parkin, G., O'Connell, P. E. & Cooper, J. D. (2004) Validation of catchment models for predicting land-use and climate change impacts. 3. Blind validation for internal and outlet responses. *J. Hydrol.* **287**, 74–95.

Bazemore, D. E., Eshleman, K. N. & Hollenbeck, K. (1994) The role of soil water in stormflow generation in a forested headwater catchment: synthesis of natural tracer and hydrometric evidence. *J. Hydrol.* **162**, 47–75.

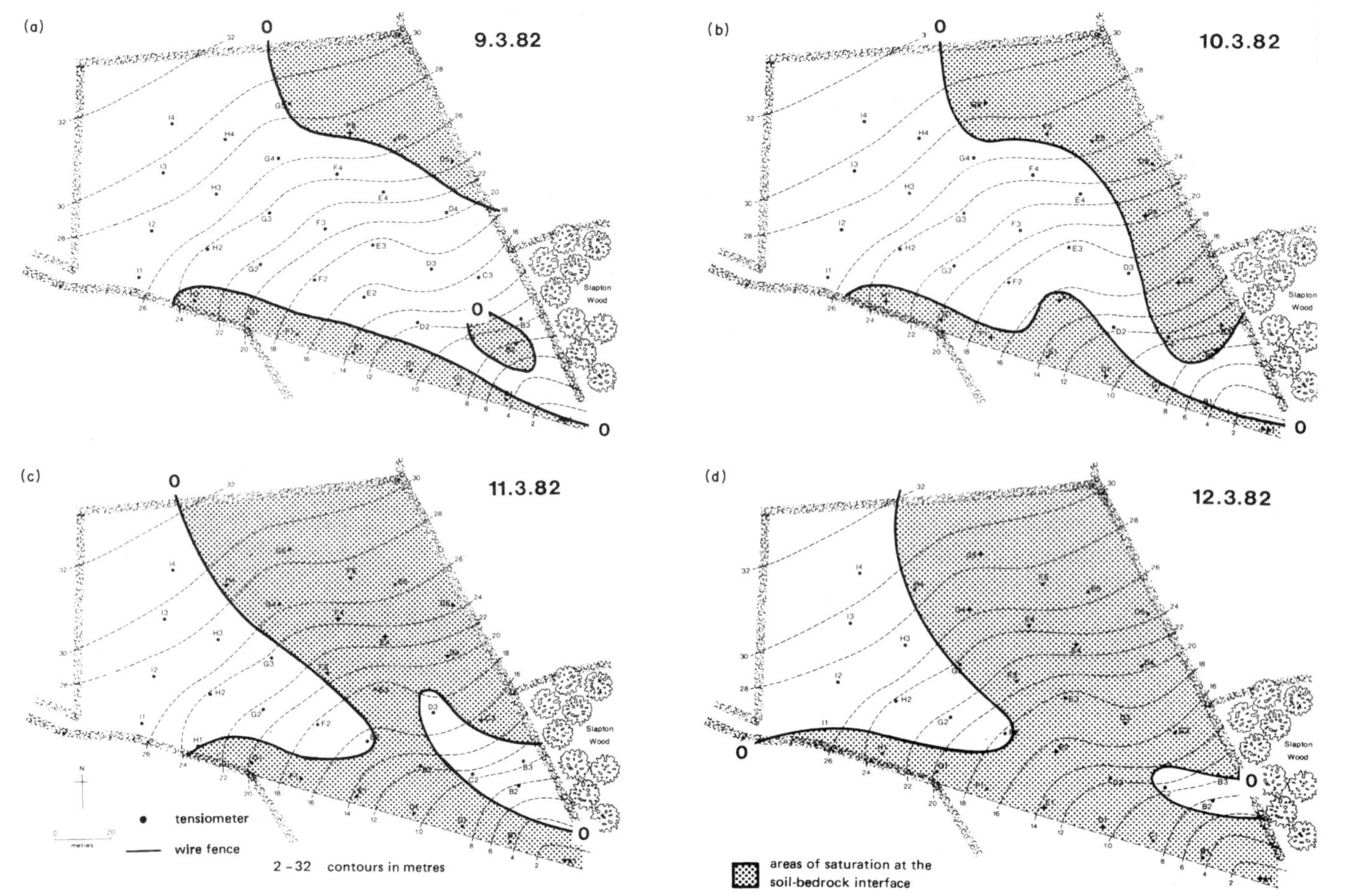

Fig. 23.1 Saturated zone responses in March 1986 in a field of the Slapton Wood catchment, Devon, UK. The southern boundary of the field lies along the axis of the Eastergrounds hollow. Contour spacing is 2 m. Shaded area represents saturation at the soil bedrock boundary as inferred from the manually read tensiometers (reproduced with permission from Burt & Butcher, 1985, © Blackwell Publishing).

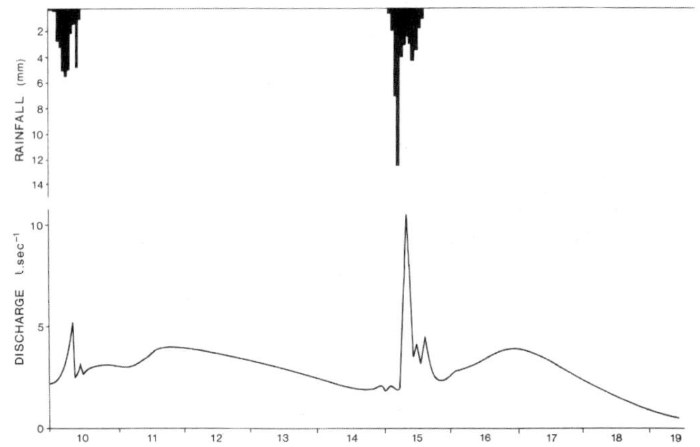

Fig. 23.2 Storm hydrographs for the Slapton Wood catchment, Devin, UK for the period 10–19 March 1982 showing the double peak response typical of this catchment (reproduced with permission from Burt & Butcher, 1985, © Blackwell Publishing).

Beven, K. & Freer, J. (2001) A dynamic TOPMODEL. *Hydrol. Processes* **15**, 1993–2011.

Burt, T. P. & Butcher, D. P. (1985) Topographic controls of soil moisture distributions. *J. Soil Sci.* **36**, 469–486.

Burt, T. P. & Heathwaite, A. L. (1986) The hydrology of the Slapton catchments. *Field Studies* **8**, 543–557.

Burt, T. P., Butcher, D. P., Coles, N. & Thomas, A. D. (1983) The natural history of the Slapton Ley Nature Reserve. XV. Hydrological processes in the Slapton Wood catchment. *Field Studies* **5**, 731–752.

Coles, N. & Trudgill, S. (1985) The movement of nitrate fertilizer from the soil surface to drainage waters by preferential flow in weakly structured soils, Slapton, South Devon. *Agriculture, Ecosystems & Environment* **13**(3–4), 241–259.

Engman, E. T. & Rogowski, A. S. (1974) A partial area model for storm flow synthesis. *Water Resour. Res.* **10**(3), 464–472.

Fisher, J. I. & Beven, K. J. (1996) Modelling of streamflow at Slapton Wood using TOPMODEL within an uncertainty estimation framework. *Field Studies* **8**, 577–584.

Trudgill, S. T., Burt, T. P., Heathwaite, A. L. & Arkell, B. P. (1991) Soil nitrate sources and nitrate leaching losses, Slapton, South Devon. *Soil Use Manage.* **7**(4), 200–206.

EARTH SURFACE PROCESSES, VOL. 3, 331–344 (1978)

THE ROLE OF TOPOGRAPHY IN CONTROLLING THROUGHFLOW GENERATION

M. G. ANDERSON AND T. P. BURT*

Department of Geography, University of Bristol, Bristol, England

Received 28 March 1977
Revised 26 September 1977

SUMMARY

Previous studies of the downslope movement of soil moisture have been largely confined to two-dimensional slope profiles—the influence of contour curvature (three-dimensional) on moisture movement has received little attention. An automatic tensiometer system has been used to monitor soil moisture status continuously in a single hollow and adjacent spurs. Maximum saturated hillslope flow is shown to coincide with the stream discharge 'throughflow peak'. Relative to the hollow, little saturation is evident at the base of the spurs. This subcatchment model is applied to a major portion of the catchment, showing the general importance of topography in the control of stream discharge.

KEY WORDS Throughflow Soil moisture Topography

INTRODUCTION

Models of hillslope hydrology in Humid Temperate regions have rightly concentrated on the throughflow process in recent years. Following the work of Hewlett and Hibbert (1963), studies have been undertaken which have sought to monitor the growth and diminution of the saturated wedge at the slope base (for example Weyman (1973) and Dunne, Moore and Taylor (1975)). Although the size of the saturated wedge has been shown to be highly correlated with hillslope discharge (Anderson and Burt (1977a)), an explanation of the moisture movement within the soil requires a relatively sophisticated data base if the controls on the saturated wedge are to be identified and modelled.

The purpose of this paper is to examine the control of topography upon throughflow discharge and the stream hydrograph. Such an examination, in necessitating the application of Darcy's Law, demands a significant improvement in data coverage in three major areas if theory and observation are to be matched.

Measurement of key soil parameters

Darcy's Law (1856) enables the prediction of outflow discharge at a slope base to be made. In order to use Darcy's Law successfully detailed knowledge is required of the size and slope of the saturated wedge. More important is the accurate determination of hydraulic conductivity for the soil in question. This requires measurement either by a field test or in a laboratory permeameter, since calculations using moisture content can easily lead to large errors. Nor is the determination of hydraulic conductivity restricted to the saturated case, since unsaturated flow beyond the wedge should also be evaluated; this requires establishing the hydraulic conductivity–soil moisture tension curve. The consequences of ignoring hydraulic conductivity in such investigations can lead to markedly inaccurate process inference. Hewlett and Hibbert (1963) postulated that the saturated wedge is maintained by unsaturated flow, but calculations using unsaturated hydraulic conductivity show that input to the saturated wedge is

* Present address: Geography Department, The Polytechnic, Huddersfield, England.

0360–1269/78/0403–0331$01.00

much less than the outflow; indeed, the outflow discharge of the wedge at all stages of drainage is entirely explained by the size and slope of the wedge itself (Anderson and Burt (1977b)). Other workers have been similarly handicapped by lack of detailed information of hydraulic conductivity. Weyman (1973) inferred a decrease in hydraulic conductivity at depth to explain slope outflow discharge; although no measurements of hydraulic conductivity were actually carried out. Kirkby and Chorley (1967) calculated rates of throughflow using data from Hewlett and Hibbert's laboratory slope experiment. However, their results are based upon a calculated relationship of hydraulic conductivity and tension, and had measured values been available their conclusions as to the effectiveness of unsaturated flow above the saturated zone might have been revised.

Spatial variation of soil moisture

Kirkby and Chorley (1967) enumerated four zones where maximum hillslope water flow might occur; at the base of slopes, in hollows, in slope profile concavities and in areas of thin or less permeable soils. However most detailed studies of soil moisture have been confined to the first case, examining saturated conditions at the base of a single slope profile in an area of straight contours (N.E.R.C. (1976)). Other workers have noted the importance of convergent flow in slope hollows causing large saturated areas to develop (Dunne and Black (1970)). In addition to the flow concentration effect, Kirkby and Chorley note that hollows are associated with concave slope profiles which cause the development of saturated conditions at the slope base (relative to the straight slope case). The development of large saturated zones in slope hollows is thus doubly affected.

Although Dunne and Black noted the importance of the saturated wedge in the hollow, the spatial variation of soil moisture content was not recorded. The upslope extension of saturation was measured using a line of piezometers up the centre of the hollow—the lateral extension of saturation was not considered. Obviously a spatial grid of moisture sensors is required to allow the complete size and shape of the saturated wedge to be assessed.

Lack of adequate temporal data

The lack of adequate temporal data in the study of hillslope soil moisture involves two considerations: the general preoccupation of researchers with stormflow rather than baseflow, and the absence of continuous data recording in moist studies.

Previous research into throughflow has been concerned mainly with stormflow rather than baseflow. Dunne and Black show the major result of hollow saturation is the development of saturated overland flow, which produces rapid and large discharge outflow from the hollow. In contrast, the straight slope section, lacking a saturated zone produced no saturated overland stormflow. Baseflow from all slope sections produced a small delayed response to the precipitation input (Figure 1). This delayed baseflow peak, because of its size, is given little attention by Dunne and Black, but we have noted elsewhere that, under different topographical and pedological conditions, the delayed throughflow component can produce a stream discharge peak larger than the stormflow peak (Anderson and Burt (1977a)). Compared to the Sleepers River Catchment, the soils in the catchment used in this study are deeper, and the slopes are three times as long, producing a much enhanced convergent effect in the hollows. The nett result is that the size and depth of the saturated wedge is not greatly affected by the precipitation input, but is controlled largely by the convergence of throughflow into the hollows.

Most previous studies of soil moisture movement on slopes have ignored completely the possibilities of automatic monitoring of field data. Manual operation of equipment requires the continued presence of the operator in the field, especially during storm conditions. Moreover, the occurrence of a storm may be relatively simple to anticipate, and the operator can act accordingly, but the timing of the throughflow peak cannot be accurately predicted so the need for automatic monitoring is doubly important in this case.

The instrumentation for this paper has been installed in a small west-facing valley at Bicknoller, on the Quantock Hills in Somerset, England (Figure 2). The valley-side slopes vary from 20° to 30°. The freely drained Brown Earth soils are one to two metres deep, being developed on impermeable

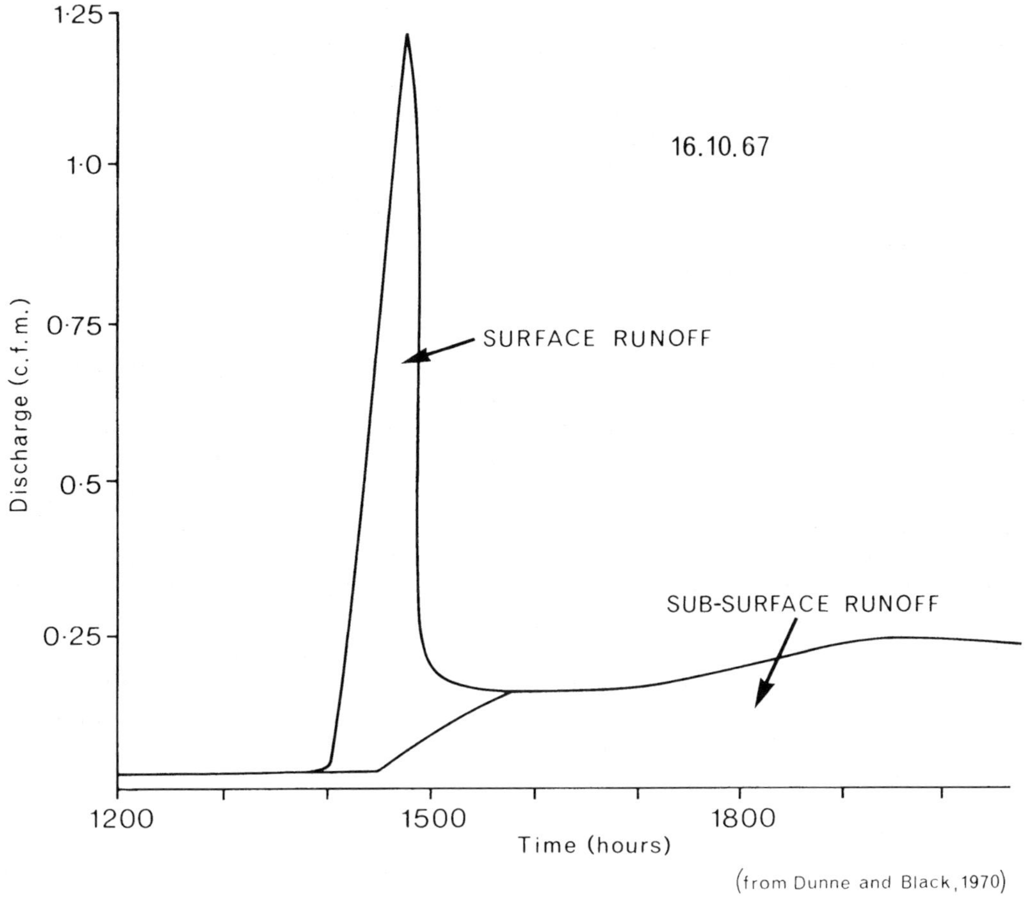

Figure 1. Surface and subsurface runoff as recorded by Dunne and Black (1970) for a hillslope hollow

Devonian Old Red Sandstone. An automatic tensiometer system has been designed and installed in a hillslope hollow and adjacent spurs to study the potential development of saturated conditions in the hollow. The automatic tensiometer system is based on a Scanivalve fluid scanning switch which allows 24 tensiometers to be converted to one pressure transducer. Two such systems are used at Bicknoller, and since one cycle of 24 readings takes one hour, the moisture conditions over the whole slope are automatically recorded every hour. The tensiometer grid is shown in Figure 2 and provides a map of soil moisture on the slope, whilst banks of tensiometers at each point provide vertical information on soil moisture conditions within a single soil profile. The Scanivalve tensiometer system itself is fully described in Burt (1977). The tensiometer system is accompanied by an autographic Casella raingauge, and by Munro stage recorders on 90° v-notch weirs in the stream. Weirs above and below the hollow allow the hollow discharge input into the stream to be calculated.

SUBCATCHMENT THROUGHFLOW MODEL

Introduction

Recent laboratory investigations have shown the significance of the saturated wedge in controlling the slope discharge hydrograph at all stages of drainage (Anderson and Burt (1977b)). This aspect has received scant attention in field monitoring of hillslope hydrology processes, due in some instances to an insufficient density of tensiometers. This is a particularly important criterion for any tensiometer layout, and although the authors are unaware of previous empirical guidelines the density adopted

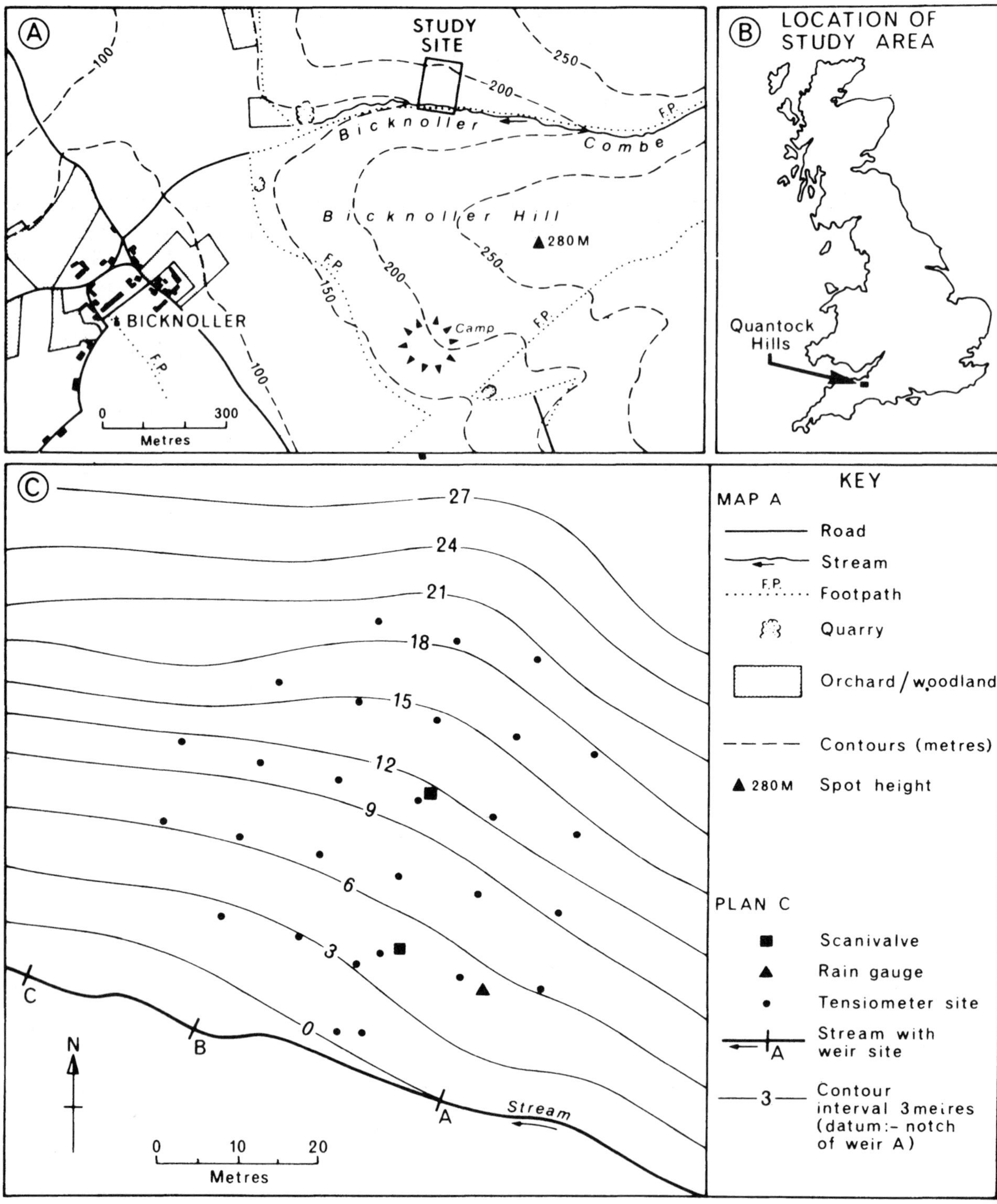

Figure 2. Location of study catchment and instrumented hillslope

in this study (10 metre grid points) has shown quite adequate resolution in this respect. In addition hydraulic conductivity determinations were made in the field using dye tracing methods to establish soil moisture flow velocities; infiltration tests on saturated soils (Hills (1970)) yielding hydraulic conductivity directly, were made across the hollow-spur configuration, and at different depths. This data allows slope discharge hydrograph prediction by Darcy's Law.

Flow nets for hillslope moisture movement

Figure 3 provides examples of tension data for the hollow (Figure 2) during a storm of October 1976. The stream hydrograph for the storm is given in Figure 4. The storm peak in stream discharge during 14 October results mainly from overland flow with additional contributions from channel precipi-

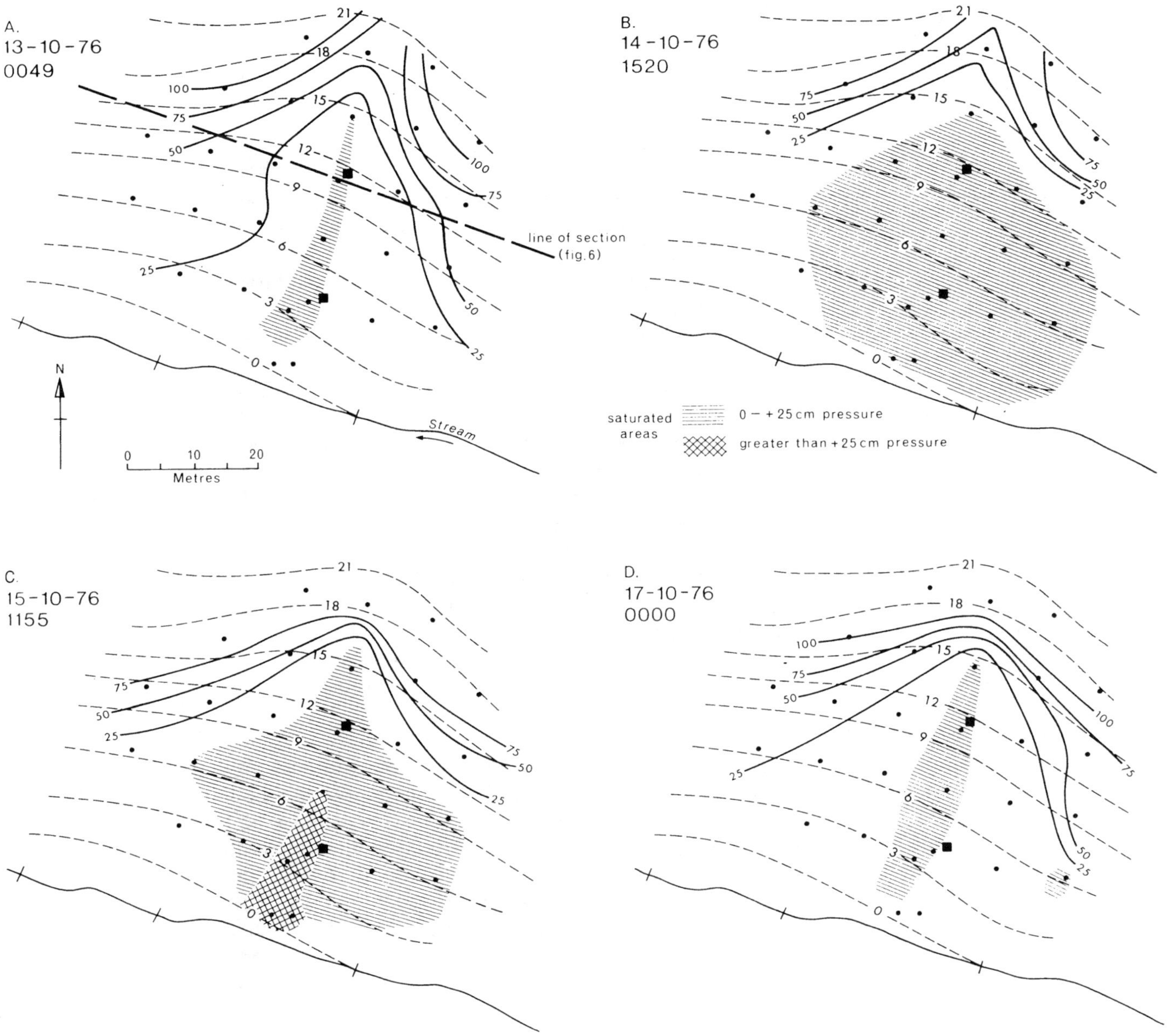

Figure 3. Maps of soil moisture tension and pressure at 60 cm depth, for one storm on the instrumented hillslope

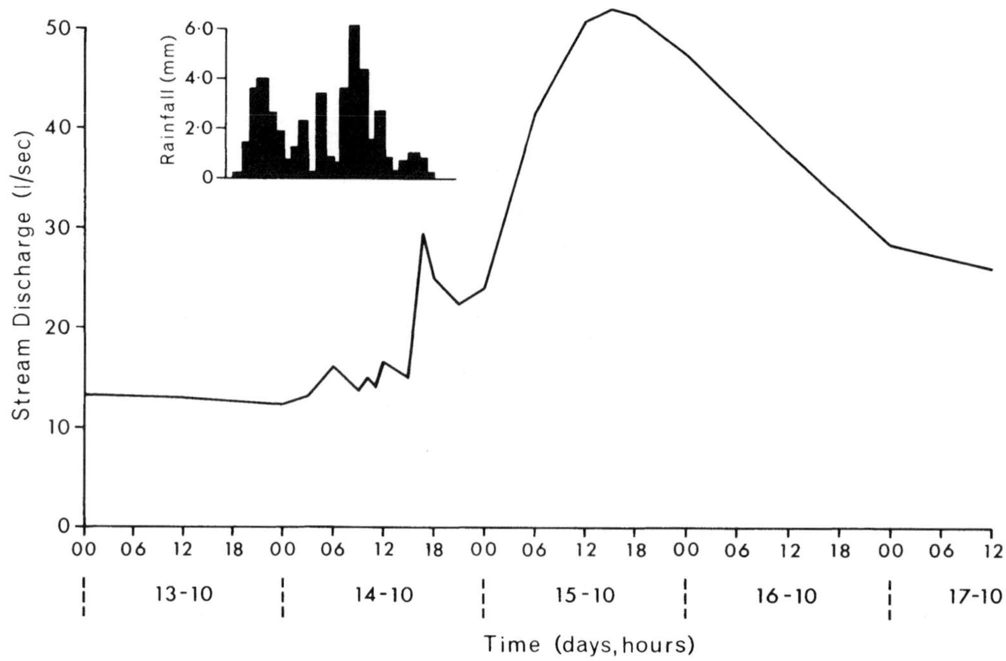

Figure 4. Stream hydrograph for the storm associated with the tension maps in Figure 3

tation, and from a small marshy zone in the headwater area (100 m^2). The delayed peak in stream discharge (15 October) reflects the hillslope throughflow contribution—the peak occurring in all parts of the catchment almost simultaneously (Anderson and Burt (1978)). The tension maps illustrate clearly the spatial extent of the saturated wedge during the storm, and in addition show the close temporal correspondence between soil water pressure ($\geqslant$25 cm) and the delayed throughflow peak. Flow nets, constructed from the tension maps, based on total potential (ϕ), Figure 5, show clearly the convergence of soil water flow in the hollow. Comparison of flow nets during a storm reveal little substantive change in the *degree* of convergence. This is attributable to the dominance of the elevation potential in the total potential summation. The maximum tension changes at a point during the storm of 14 October 1976 (Figure 3) were only of the order of 50 cm. On a comparatively steep hillslope, elevation potentials will be almost everywhere an order of magnitude more than the largest tension change. Only on low slopes, such as the lower base of the hollow, are tension changes capable of modifying the flow net. This confirms and explains the dominant role of topography in controlling flow direction thereby emphasizing the need for hillslope hydrology models to be three dimensional (Anderson (1973)).

Prediction of hollow throughflow discharge

This qualitative association depicted in Figures 3 and 4 and observed elsewhere (Anderson and Burt (1977a)) can be improved by attempting a prediction of hollow discharge. Darcy's Law requires knowledge of cross-sectional area of flow and hydraulic conductivity if discharge is to be predicted. Depth to bedrock over the hollow and spur was determined by a portable refraction seismograph, and confirmed by augering. Generally soil depths were 1–1·5 m on the spurs and up to 2 m in the hollow. Despite spatial variations in hydraulic conductivity, previous work has largely assumed a single value to be representative. Because of the wide ranging variation in hydraulic conductivity (commonly four orders of magnitude) then it must be important to take due account of this for accurate discharge prediction. Accordingly hydraulic conductivity values were determined across the hollow and spur using saturated infiltration capacity as a measure of saturated hydraulic conductivity. Figure 7 shows the variation across the section indicated in Figure 3. From salt dilution gauging downstream across the hollow reach, an estimate of discharge input from the hollow reach was obtained. The total discharge increment

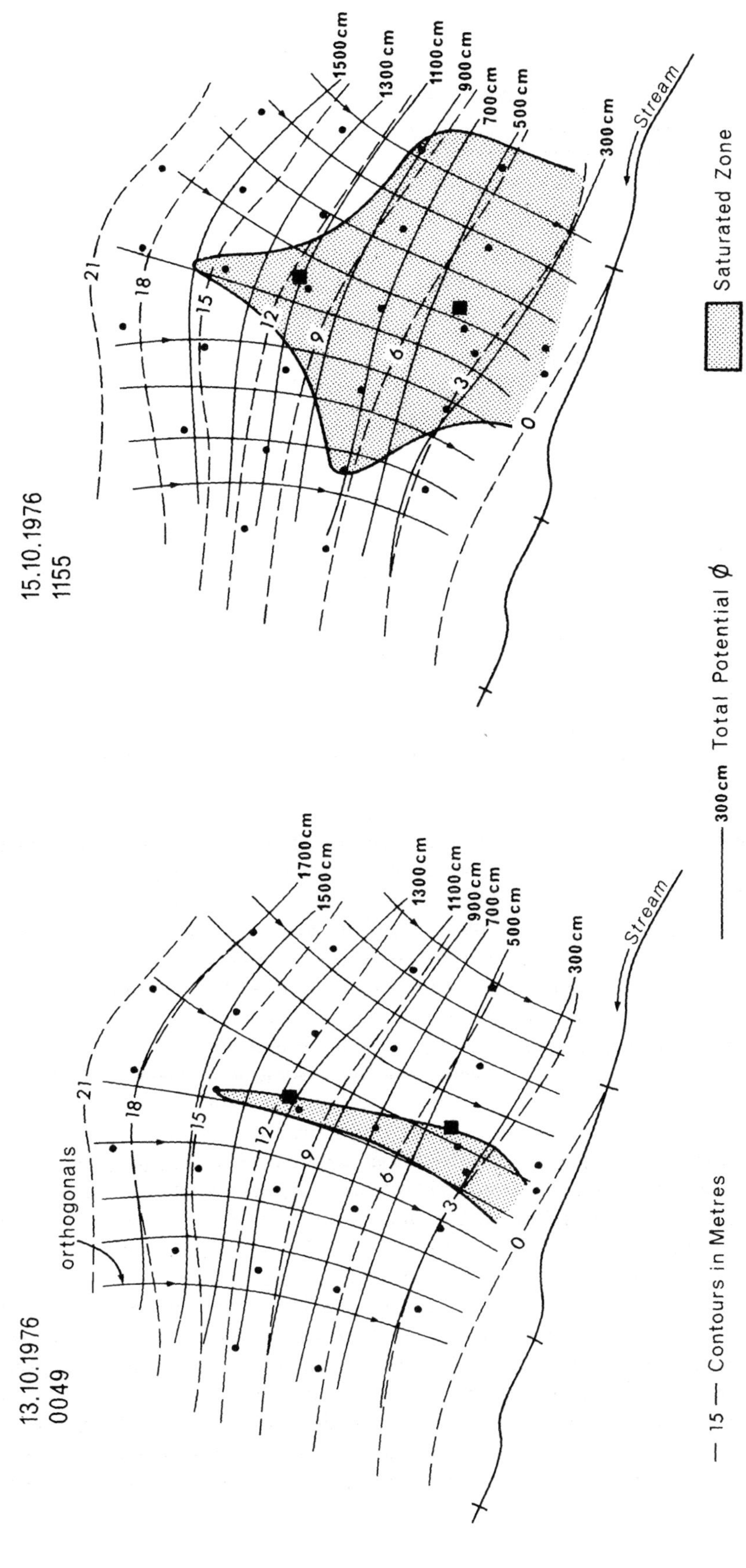

Figure 5. Flow nets and total potential (ϕ – cm) for two occasions over the October storm event

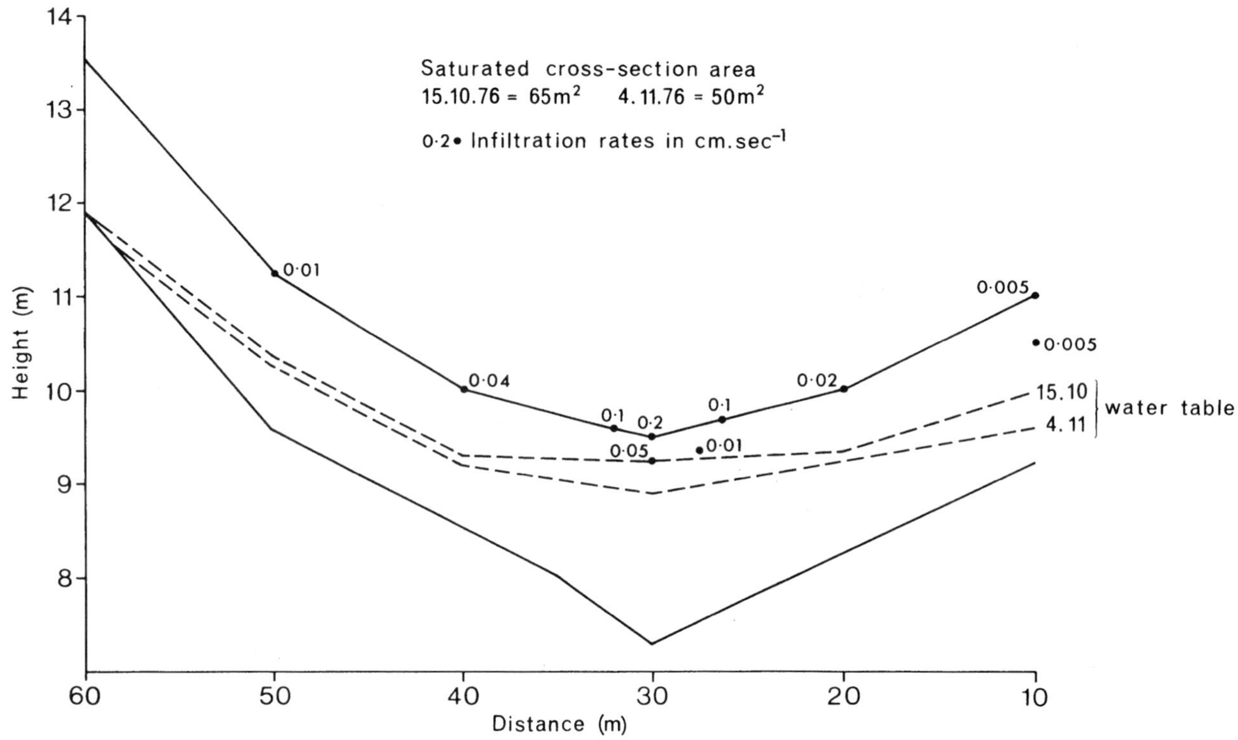

Figure 6. Cross-section of hillslope hollow (line of section shown in Figure 3a) showing the saturated zone on two occasions. Saturation infiltration rates are also shown at selected points

thus obtained was reduced to take account of the spur slope input on the opposite bank. This was achieved by dilution gauging over a reach with spur input on both banks, the spur input per metre of channel length calculated and subtracted from the hollow reach increment. Table I shows hollow discharge contribution, saturated area in the hollow cross-section and calculated hydraulic conductivity for the conditions during the storms of 15 October and 4 November 1976. Hydraulic conductivity calculated in this manner is an average figure for the whole of the saturated area in the hollow. Using the *measured* hydraulic conductivity values from the field it is possible to predict hollow discharge input. This relies on accurate apportionment of zones of differing hydraulic conductivity within the hole (Figure 6). Table II shows the predicted hollow discharge figures for the two occasions referred to in Table I. From Table III it can be seen that both the low discharge and field hydraulic conductivity values have been accurately predicted. Slight error occurs in peak hollow discharge prediction—in part perhaps due to errors in defining the exact saturated area and the total variation in hydraulic conductivity. Nevertheless, the results do show that hollow discharge can be reasonably predicted from knowledge of the saturated wedge and field hydraulic conductivity—and more importantly thereby confirms the dominance of saturated flow in hillslope hydrograph generation.

The timing of the delayed throughflow peak

The timing of the delayed throughflow peak in stream discharge has already been related to the general development of saturation conditions in the hollow following a storm. In the hollow where soil moisture was monitored continuously, the throughflow peak seems to coincide with the occurrence of maximum positive pore water pressures in the centre of the hollow (Figure 3). The fact that the overall basin discharge produces a single throughflow peak suggests that all the hollows may produce peak discharge at more or less the same time in a basin of this size. Figure 7 shows the stream discharge as measured at the lowest weir for a single storm and associated delayed throughflow peak. Also shown

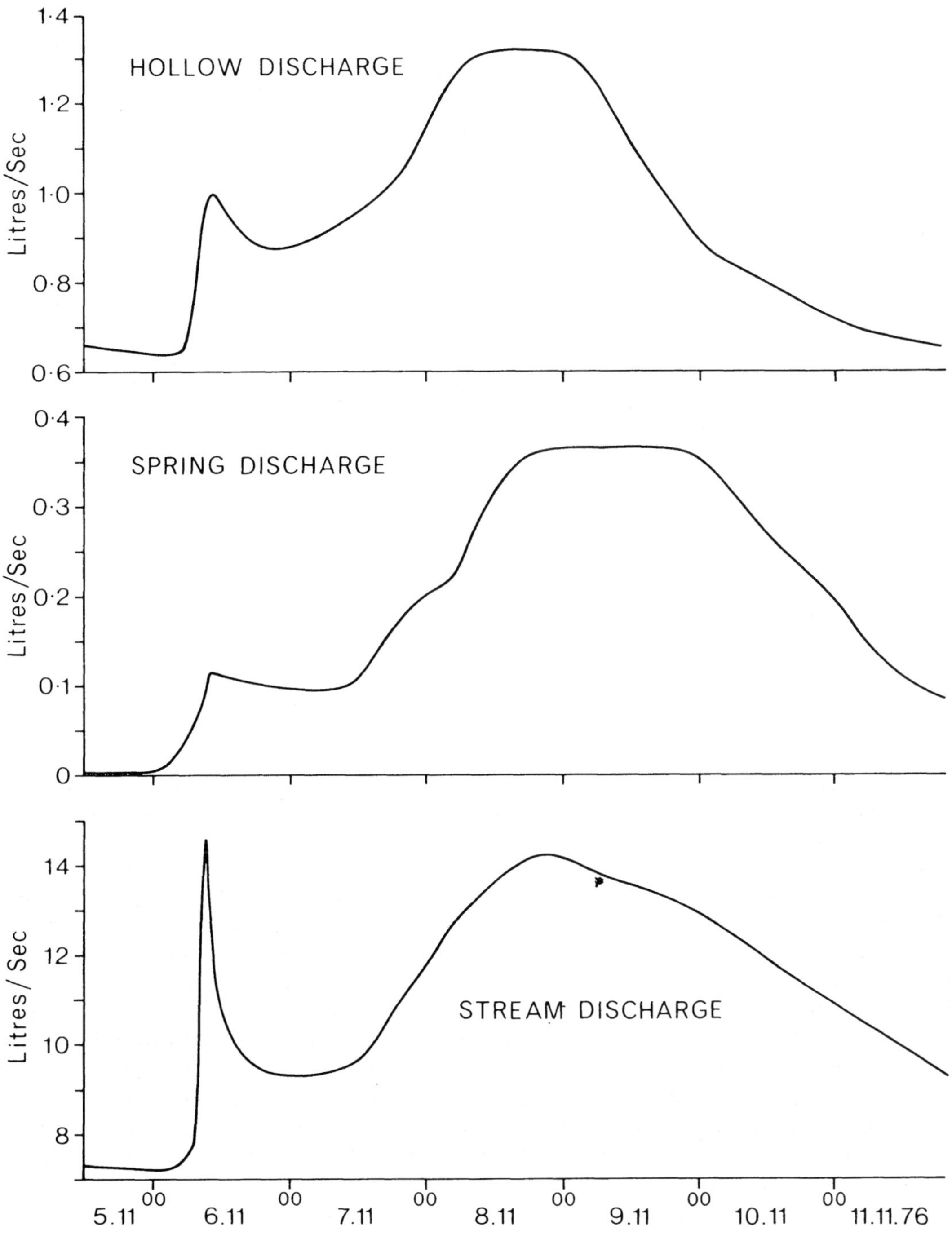

Figure 7. Total stream discharge response, hillslope hollow discharge and 'spring' discharge (see Figure 9) for a single storm event

are two discharge–time curves for different hollows in the catchment. The 'spring' discharge–time curve represents part of the throughflow contribution for the hollow immediately below the headwater weir (see Figure 9 for spring location). The slope outflow produces a spring which was monitored using a v-notch weir and stage recorder. The main hollow studied produced a general seepage rather than a single spring outflow and continuous monitoring of discharge proved impracticable.

Table I. Estimation of hydraulic conductivity (K) for two storms

	15.10.76	14.11.76
Stream discharge ($l\ s^{-1}$)	51	9
Hollow discharge-Q ($l\ s^{-1}$)	4·1	0·73
Saturated area-A (m^2—see Figure 6)	65	50 n
Water table slope-I (dimensionless—from tensiometer data)	0·30	0·30
Hydraulic conductivity-K ($cm\ s^{-1}$) from Darcy's Law K = Q/IA)	0·0210	0·0048

Table II. Estimation of hollow discharge input for two storms (notation as Table I)

(i) Storm of 15 October 1976	
(a) for area in which	$K = 0{\cdot}1\ cm\ s^{-1}$
	$A = 6\ m^2$
	$I = 0{\cdot}3$
since Q = KIA	$Q = 1{\cdot}8\ l\ s^{-1}$
(b) for area in which	$K = 0{\cdot}005\ cm\ s^{-1}$
	$A = 59\ m^2$
	$I = 0{\cdot}3$
therefore	$Q = 0{\cdot}89$
Total discharge input from hollow = $1{\cdot}8 + 0{\cdot}89 = 2{\cdot}69\ l\ s^{-1}$	
(ii) Storm of 4 November 1976	
	$K = 0{\cdot}005\ cm\ s^{-1}$
	$A = 50\ m^2$
	$I = 0{\cdot}3$
Total discharge input from hollow = $0{\cdot}75\ l\ s^{-1}$	

Table III. Predicted and measured hydraulic conductivity (K) and hillslope hollow discharge (Q) for two storms

	Q $l\ s^{-1}$		K $cm\ s^{-1}$	
Storm date	15.10.76	4.11.76	15.10.76	4.11.76
Measured	4·1	0·73	0·005 and 0·1	0·005
Predicted	2·69	0·75	0·0210	0·0048

The pore water variation measured in the centre of the hollow over the storm is taken to be a direct indication of the form of the complete hollow hydrograph. Actual hollow discharge was determined for a variety of stream flow conditions using dye dilution methods, thus facilitating the establishment of a relationship between hollow discharge and stream discharge. Use of this relationship for the 'stream discharge' hydrograph shown in Figure 7, allowed calibration of the pore water variation curve to hollow discharge (also shown in Figure 7).

The results shown in Figure 7 suggest that hollows produce a peak in throughflow discharge almost simultaneously, resulting in a single delayed peak in total stream discharge. The timing pattern as shown in Figure 7 for the storm of 6 November 1976, has been reproduced for several other storms, showing the 'spring' peak to lag consistently behind the main basin maximum.

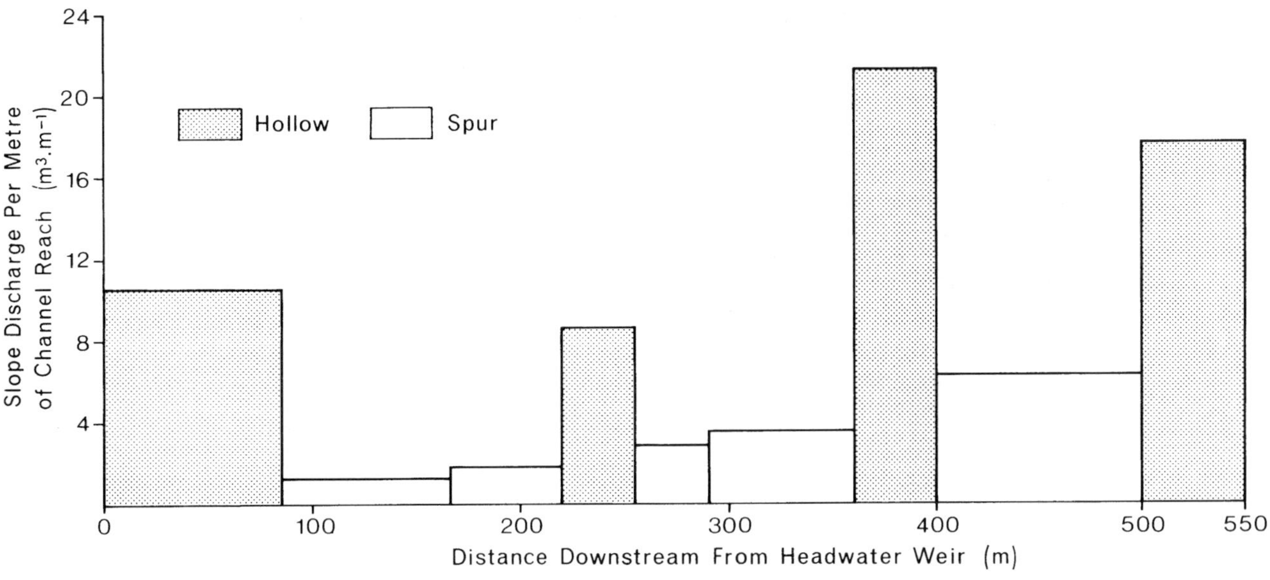

Figure 8. Hillslope input discharge over 550 m channel length—discharge per metre of channel length

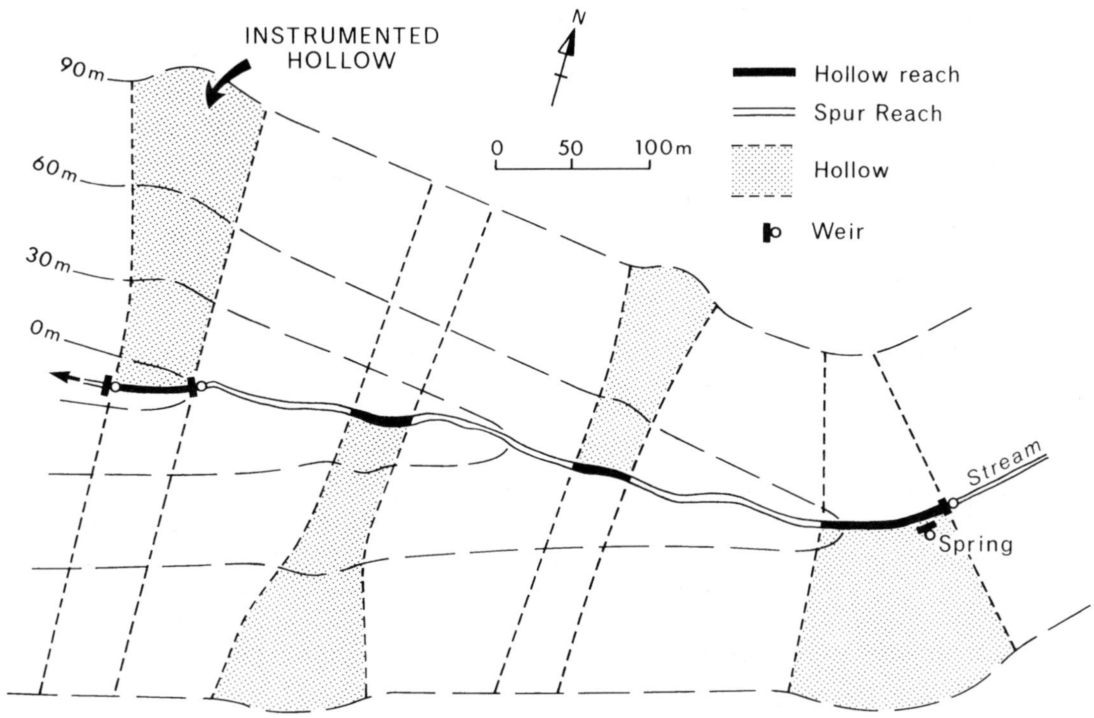

Figure 9. Map of catchment subunits of hollow and spur reaches

TOTAL CATCHMENT THROUGHFLOW APPORTIONMENT

Dye dilution gauging was employed to determine discharge increments down the entire channel length below the complex headwater area. Over this distance of 550 m fourteen determinations were made. Figure 8 shows the discharge input per metre of channel length, given in terms of hollow and spur reaches. Summations of discharge input over the total stream lengths, based upon togography are shown in Table IV, and are derived from the stream sections labelled 'hollow' and 'spur' in Figure 9. For

Table IV. Total hollow and spur reach inputs for two storms

	15.10.76	4.11.76
Total hollow reach input (l s^{-1})	16·34	2·78
Total hollow reach length (m)	190	190
Hollow discharge input per metre of channel length (cc min^{-1})	86	14·6
Total spur reach input (l s^{-1})	11·66	1·47
Total spur reach length (m)	360	360
Spur discharge input per metre of channel length (cc min^{-1})	32·4	4·1

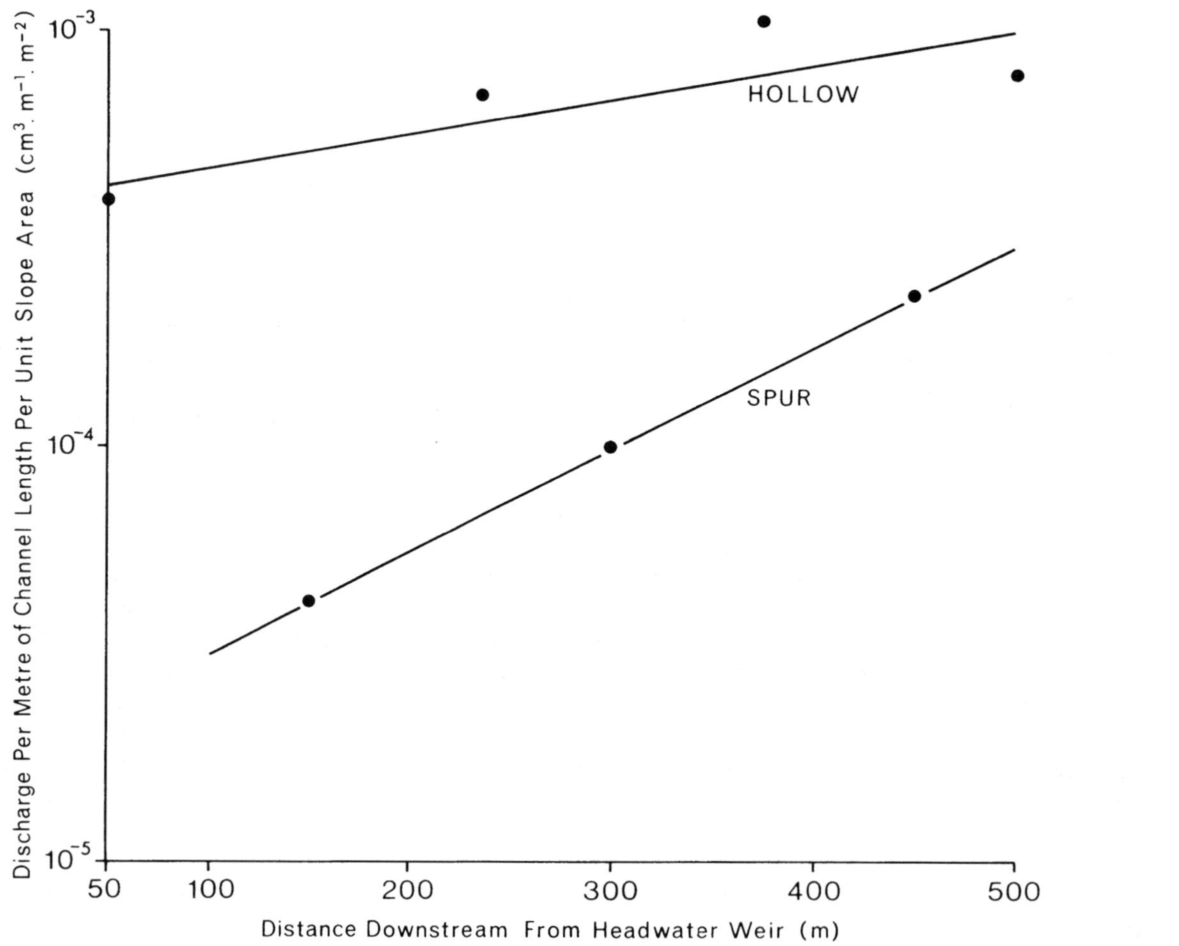

Figure 10. Hollow and spur reach discharge inputs over 550 m channel length—discharge per metre of channel length per square metre of drained area

these purposes a hollow reach is defined as the channel length across a hollow base and includes discharge input from the opposite slope. A spur reach contains no hollows on either bank. These results (Table IV) show that hollows form the dominant input sources to the stream at both flow levels for the two storms studied, but that at higher flows significant saturation occurs on the spurs as well as in the hollows.

Of additional interest, Figure 8 shows both spur discharge and hollow discharge per metre of channel length to be increasing downstream; undoubtedly due to the increase in slope lengths from stream to divide. Slope length increases from 100 m at the top of the channel reach to 200 m at a distance of 0·5 km downstream. A division of these figures by area drained for the hollow reaches reveals only slightly increased efficiency of drainage for the larger slope lengths, with hollow reach slope input increasing from 4×10^{-4} cc/m of channel length/m^2 of drained area to approximately 1×10^{-3} cc/min/m^2. Similar calculations for spur contributions show significant influence of increased slope length (Figure 10). This graph shows that where convergent flow occurs (hollows) then efficiency of flow is almost independent of hillslope length. Without convergence (spurs) hillslope length exerts a highly significant control on spur discharge. These results accord with the suggestions of Calver *et al.* (1972) with respect to increased storage of soil moisture downslope. Implied in their suggestion is that greater slope lengths (i.e. downstream) will yield proportionately more throughflow.

CONCLUSIONS

An understanding of hillslope and channel hydrology relationships necessitates a well defined temporal and spatial data base for soil moisture and accurate determinations of hydraulic conductivity over different types of hillslope topography. Although some progress can be made using distributed models relying upon 'representative' parameter values (for example Calver, Kirkby and Weyman (1972)), the results presented in this study strongly suggest that ideas of adequate 'representative values' (for example of hydraulic conductivity) must be discounted in favour of detailed field determinations of hydraulic conductivity (Kirkby and Weyman (1973)) water table slope and the saturated wedge dimensions. Nevertheless in attempting to identify and model the throughflow process, approximations *must* be made. Laboratory evidence has been presented elsewhere which suggests that, in hillslopes, saturated flow alone need be considered (Anderson and Burt (1977b))—empirical evidence presented here concurs with this earlier finding. Thus it is in the *spatial* context that approximations can most reasonably be made. Hillslope hollows, through convergence, give rise to areas of saturated flow and calculation of total potential in hillslope hollows reveal the dominance of elevation potential on a slope angle of 25°. The relationships between topography and hillslope discharge have been evaluated for one hollow (Table III) and considered in general terms for the entire stream (Figure 8). Table IV has suggested that at high flows spur base saturation may occur, but that hollow reach contributions exceed spur conditions by a factor of at least two for the throughflow peak investigated. Moreover this is quite probably stating the least attractive case since what these calculations likely reflect during such flow conditions, is an increased *lateral* hydraulic gradient at the hollow base occasioned by a vertical growth in the saturated wedge at the hollow centre. At the downstream side of the hollow this hydraulic gradient is of course enhanced by the falling stream bed elevation. Thus spur base saturation may well be initiated and maintained by this hollow spill-over mechanism. This is an area to which attention is currently being paid. Moreover the storm conditions of 15 October which have been examined, being the highest recorded throughflow peak (also the highest recorded streamflow during 2 years of records), represent the most favourable conditions for spur contributions. For the two flow conditions examined spur reach contributions/m of channel length/m^2 drained area are shown to be almost an order of magnitude less than hollow contributions, although increasing downstream under the control of hillslope length (Figure 10). It is also interesting to note that the spur reaches comprise 60 per cent of the basin area, but only contribute 40 per cent of the total discharge under the most favourable of conditions.

REFERENCES

Anderson, M. G., (1973). 'Measures of three dimensional drainage basin form', *Water Resources Research*, **9**, 378–383.

Anderson, M. G. and Burt, T. P., (1977a). 'Automatic monitoring of soil moisture conditions in a hillslope spur and hollow', *Journal of Hydrology*, **33**, 27–36.

Anderson, M. G. and Burt, T. P., (1977b), 'A laboratory model to investigate the soil moisture conditions on a draining slope', *Journal of Hydrology*, **33**, 383–390.

Anderson, M. G. and Burt, T. P., (1978). 'Time synchronised stage recorders for the monitoring of incremental discharge inputs in small streams', *Journal of Hydrology* (in press).

Burt, T. P. (1977). 'An automatic fluid scanning switch tensiometer system', *Technical Bulletin, Geomorphological Research Group* (in press).

Calver, A., Kirkby, M. J. and Weyman, D. R. (1972). 'Modelling hillslope and channel flows', in *Spatial Geomorphology*, Chorley, R. J. (ed), London, Methuen, pp. 197–218.

Darcy, H. (1856). *Les Fontaines Publiques de la Ville de Dijon*, Paris, Dalmont.

Dunne, T. and Black, R. D. (1970). 'An experimental investigation of runoff production in permeable soils', *Water Resources Research*, **6**, 478–490.

Dunne, T., Moore, T. R. and Taylor, C. H. (1975). 'Recognition and prediction of runoff-producing zones in humid regions', *Hydrological Sciences Bulletin*, **20**, 305–327.

Hewlett, J. D. and Hibbert, A. R. (1963). 'Moisture and energy considerations within a sloping soil mass during drainage', *Journal of Geophysical Research*, **64**, 1081–1087.

Hills, R. C. (1970). 'The determination of the infiltration capacity of field soils using the cylinder infiltrometer', *Technical Bulletin No. 3, British Geomorphological Research Group.*

Kirkby, M. J. and Weyman, D. R. (1973). 'Measurements of contributing area in very small drainage basins', *University of Bristol Geography Department Seminar Series B, No. 3.*

Kirkby, M. J. and Chorley, R. J. (1967). 'Throughflow, overland flow and erosion', *International Association for Scientific Hydrology Bulletin*, **12**, 5–21.

N.E.R.C. (1976). 'Research on geomorphology of water-produced landforms', *Publication Series B, No. 16.*

Weyman, D. R. (1973). 'Measurements of the downslope flow of water in a soil', *Journal of Hydrology*, **3**, 267–288.

Commentary on Paper 24

Bonell, M. & Gilmour, D. A. (1978) The development of overland flow in a tropical rainforest catchment.

Mike Bonell and Don Gilmour provided one of the very first detailed studies of hillslope processes in a forested tropical catchment. Their work suggested that a very rapid decline in hydraulic conductivity in the soil profile, coupled with the high rainfall intensities of tropical regions, would result in the generation of a perched saturated zone and, where the volume of rainfall and flow from upslope was sufficient to saturate the soil to the surface, in saturation excess surface runoff even given the high topsoil conductivities measured. They also, like Hewlett (1974), Gregory & Walling (Paper 10) and others, pointed to the rapid increase in drainage density by the expansion of ephemeral channels to "tap" this overland flow and carried out some pioneering tracer experiments in tropical soils (Bonell *et al.*, 1984).

While this is certainly a benchmark paper, there has been a tendency (as with some other papers included in this volume) to assume that the type of response described is characteristic of a very wide range of tropical catchments. There is still not enough evidence available to suggest that this is necessarily the case. An alternative type of response, involving deeper subsurface flow pathways has been described, for example, by Nortcliff & Thornes (1981) in the Amazon basin, while other tracer experiments reported by Chappell & Sherlock (2005) in Singapore and Borneo suggest a complexity of local responses that may be difficult to generalize.

Mike Bonell, who had completed his PhD on the Catchwater Drain under the supervision of Roy Ward at the University of Hull in England, worked for a long time for James Cook University in Townsville, Australia, before becoming Chief of the Hydrological Processes and Climate Section within the UNESCO Division of Water Sciences in Paris, France. He has published several excellent reviews of streamflow generation in tropical forests (Bonell, 1993, 1998, 2005). Don Gilmour was a forest hydrologist with the Queensland Department of Forestry for nearly 30 years. After a short period with the IUCN in Switzerland, he now works as a consultant hydrologist based in Brisbane, Australia.

References

Bonell, M. (1993) Progress in the understanding of runoff generation dynamics in forests. *J. Hydrol.* **150**, 217–275.

Bonell, M. (1998) Selected challenges in runoff generation research in forests from the hillslope to headwater drainage basin scale. *J. Am. Water Resour. Assoc.* **34**(4), 765–785.

Bonell, M. (2005) Runoff generation in tropical forests. In: *Forests, Water and People in the Humid Tropics* (ed. by M. Bonell & L. A. Bruijnzeel), 314–406. UNESCO International Hydrology Series, Cambridge University Press, Cambridge, UK.

Bonell, M., Casells, D. S. & Gilmour, D. A. (1984) Tritiated water movement in clay soils of a small catchment under tropical rainforest in North East Queensland. In: *Int. Soil Sci. Soc. Symposium on Water and Solute Movement in Heavy Clay Soils* (ed. by J. Bouma & P. A. C. Raats), 197–201. Int. Inst. Land Reclamation and Improvement, Publ. no. 37, Wageningen, The Netherlands.

Chappell, N. A. & Sherlock, M. D. (2005) Contrasting flow pathways within tropical forest slopes of Ultisol soils. *Earth Surf. Process. Landf.* **30**, 735–753.

Hewlett, J. D. (1974) Comments on letters relating to 'Role of subsurface flow in generating surface runoff. 2. Upstream source areas' by R. Allen Freeze. *Water Resour. Res.* **10**(3), 605–607.

Nortcliff, S. & Thornes, J. B. (1981) Seasonal variations in the hydrology of a small forested catchment near Manaus, Amazonia, and the implications for its management. In: *Tropical Agricultural Hydrology – Watershed Management and Land Use* (ed. by R. Lal & E. W. Russell), 37–57. Wiley, Chichester, UK.

Journal of Hydrology, 39(1978)365—382

[1]

THE DEVELOPMENT OF OVERLAND FLOW IN A TROPICAL RAIN-FOREST CATCHMENT

M. BONELL and D.A. GILMOUR

Department of Geography, James Cook University of North Queensland, Townsville, Qld. 4811 (Australia)
Queensland Forestry Department, Atherton, Qld. 4883 (Australia)

(Received September 27, 1977; revised and accepted April 1, 1978)

ABSTRACT

Bonell, M. and Gilmour, D.A., 1978. The development of overland flow in a tropical rain-forest catchment. J. Hydrol., 39: 365—382.

A preliminary assessment is made of the factors contributing to the rapid response of the discharge hydrograph in a tropical rainforest catchment. Overland and subsurface flows were measured by means of a trough system constructed at the lower end of three sites. Water was led from each trough into a tipping-bucket device with the number of tips being recorded mechanically using a hay-bale counter and electronically by a digital event recorder. Substantial overland flow occurred at all sites despite the exceedingly high average saturated hydraulic conductivity, k_s values in the top 20 cm of the soil profile. The highest ratio of overland flow to subsurface flow was associated with a plot having an upslope position. Subsurface flow volumes, however, were higher at the two remaining plots of lower-slope position. This was attributed to the presence of residual rocks in the deeply weathered clays. The occurrence of widespread overland flow was explained by the occurrence of high-intensity rainfalls (particularly during the summer months). Such rainfalls frequently exceed the average k_s values below 20 cm depth and therefore generate a widespread perched water table within the top 20 cm. This causes subsurface flow (particularly on steep slopes) through the highly permeable surface horizon. Additional rain causes the perched water table to emerge at the soil surface and hence widespread saturation overland flow results for the remainder of the storm. This overland flow is tapped by a temporary dense drainage network, thus accounting for the rapid quickflow response. Spearman's rank correlation coefficient analysis emphasised the significance of rainfall intensity despite some detailed differences between "summer" and "winter" rainfall events. It is concluded that in this particular environment the variable source area concept of storm flow generation, associated with humid temperate areas, is not applicable mainly due to the characteristics of rainfall intensity.

INTRODUCTION

The aim of this paper is to make a preliminary assessment of the factors contributing to the rapid response of the discharge hydrograph within a tropical rainforest catchment in northeast Queensland. The majority of the discussion will concentrate on the origin of overland flow, supported by the use of Spearman's rank correlation analysis, using data collected during the 1976 wet season. Some general comparisons are made with the drainage processes in humid temperate environments.

There has been a considerable amount of work accomplished in the humid temperate latitudes evaluating the origin of quickflow. Freeze (1972b), Dunne et al. (1975) and Ward (1975) have given detailed summaries of this work. The general consensus is that widespread overland flow does not occur in forested areas but is more locally confined to areas having thin soils, swales and the lower reaches of concave slopes adjacent to streams. However, in terms of contribution to storm runoff the areas of thin soils and swales are only of significance if they occur at the foot of the concave slopes in the presence of a shallow water table. The surface emergence of this water table resulting from recharge by subsurface flow and rainfall leads to what Dunne and Black (1970a, b) termed saturation overland flow which expands upslope during the course of a storm. This forms the basis of the variable source area concept. The reason for this very localised nature can be attributed to the very high infiltration capacities which, under undisturbed conditions, occur in forest soils resulting from the continued incorporation of organic matter plus a high root density. Therefore there is a consequential improvement in soil structure which is further assisted by the presence of well developed A-horizons. Consequently, rainfall intensities are usually less than the soil infiltration capacities, and normal storm durations are insufficient to create ponding at the soil surface. This is in marked contrast with the more widespread overland flow found in certain semi-arid areas following Horton's (1933, 1945) philosophy.

Thus for temperate humid areas most writers favour some version of the Hewlett variable source area concept (Hewlett, 1961; Hewlett and Hibbert, 1967) of overland flow generation, despite few detailed attempts at field measurement (Ward, 1975). This has led to increasing attention being paid to the recognition and statistical prediction of such areas so that the hydrograph components can be modelled (Dunne et al., 1975; Hewlett and Troendle, 1975). The only contentious issue is the principal mode of recharge of the water table whether it be interflow (e.g., Whipkey, 1965, 1969; Hewlett and Hibbert, 1967) or vertical infiltration (e.g., Dunne and Black, 1970a, b). But as Ward (1975) noted:

> "... neither ... is likely to be rapid enough to make a significant contribution to storm runoff except through the medium of translatory flow or through pipe flow in saturated conditions."

Part of this debate depends on the topographical—hydrological circumstances of slopes. The most favourable circumstances for the variable source area concept to apply are regarded to be where there are extensive saturated or near saturated valley bottoms at the foot of concave slopes. Saturation overland flow is regarded as being less significant in steep, well drained and permeable hillslopes bordering narrow valley floors. This increases the significance of subsurface storm flow contribution to the hydrograph although volumetrically it is still regarded to be small. This was demonstrated by Freeze (1972b) using a mathematical model despite rigid assumptions.

By comparison the humid tropics have received only passing attention during a consideration primarily of denudation problems. Consequently, there is no certainty that humid temperate processes are as valid in this environment. Perhaps the most commonly held belief is that expressed by Thomas (1973) when he stated that:

"... rapid infiltration of heavy rain into the deep tropical soils prevents much surface flow taking place except on steep slopes ..."

and attributes this to be due to the kaolin clays being:

"highly permeable, and the intensity of precipitation at ground level is much less than outside the forest."

Thus Thomas favours slower storm recharge to streams:

"... by means of lateral movement or throughflow within the soil and regolith."

Whilst acknowledging the significance of subsurface flow Douglas (1973), however, noted that:

"... under undisturbed forest cover there is little vegetation on the ground surface, and surface wash, derived from stemflow, may add to the transport of sediment towards stream channels by subsurface flow."

The surface wash component was particularly emphasised by Ruxton (1967) during a qualitative survey of the denudation processes occurring on slopes under tropical rainforest in Northern Papua. He pointed out that open rather than closed canopies occurred in this environment due to tree fall caused by monsoonal gales and mass movement. This created less protection from rainfall on the forest floor (c.f. Thomas, 1973), which made raindrop and waterdrop erosion more effective. He suggested that overland flow commenced as soon as the soil surface was sealed by fine particles disrupted by raindrop impact. Furthermore Ruxton did not regard overland flow as only operative when the soil was saturated from heavy, prolonged rain but like Douglas (1973), supported the role of stem flow. As he commented:

"... in this way water is commonly supplied to the soil at the tree base faster than it can infiltrate, and runoff may therefore occur after only moderate rainfalls."

The significant feature is that none of these writers is able to support their views with quantitative evidence. Furthermore the matter of spatial organization and magnitude of overland flow remains inconclusive.

DESCRIPTION OF STUDY AREA

The site chosen for field experimentation was in an existing experimental catchment ca. 6 km east of Babinda (Fig.1). The climate of the area is typical of that experienced on the wet tropical coastal belt with high annual rainfall totals being an outstanding feature. The average annual rainfall on the site for the 6-yr. period 1970—1975 was 4175 mm, and about half of this total falls during the wet-season months of January to March. The regular occurrence of cyclones and tropical lows associated with the inter-tropical convergence zone in the summer months produces rainfall intersities which are among the highest recorded in Australia — daily falls in excess of 250 mm are fairly frequent. Later in the year the majority of rainfall originates from moist on-shore SE winds associated with a firm high-pressure coastal ridge and the occasional eastward progression of upper atmospheric lows. In both cases rainfall intensities are usually much lower than in the wet

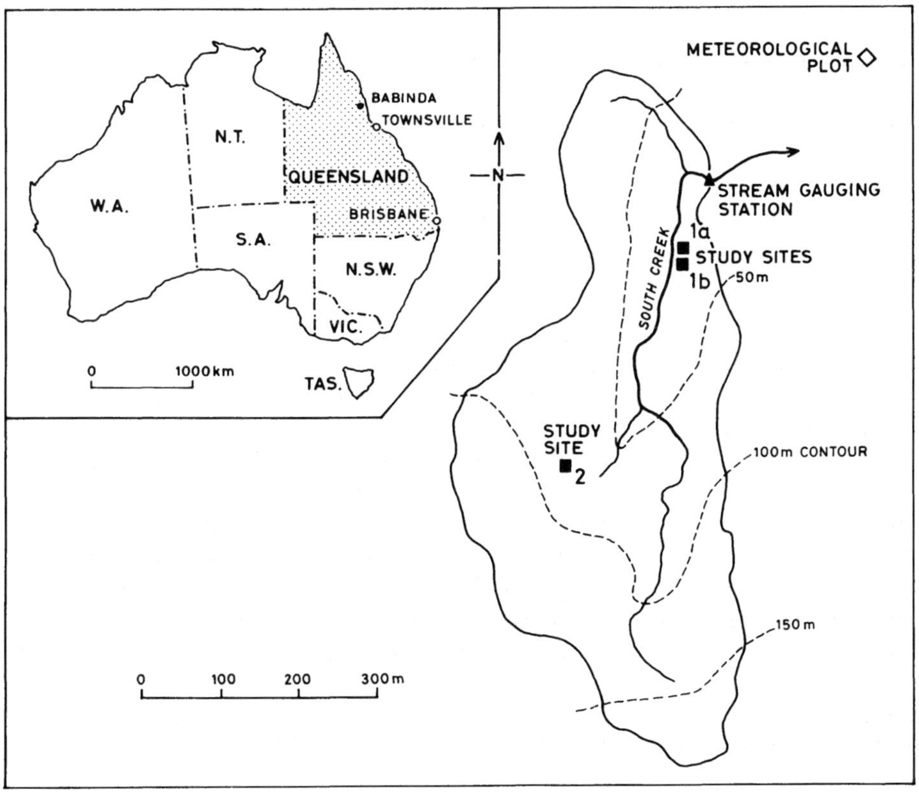

Fig. 1. The location of study area and experimental plots in South Creek catchment.

season proper, although daily totals can still be quite high with recordings occasionally in excess of 100 mm.

The experimental catchment viz. South Creek, in which the study sites are located is 0.257 km^2 in area and has steep slopes (mean catchment slope of 34%). The soils are yellow-brown clays derived from deeply weathered basic metamorphic rocks. Under the Northcote (1971) system, they are classified as Gn 3.74 and Gn 3.11/3.14, respectively (R.F. Isbell and G.G. Murtha, pers. commun., 1976).

The catchment is covered by tropical rainforest — classified Mesophyll Vine Forest according to Webb's (1959) classification. The forest has an uneven canopy resulting from disturbance by tropical cyclones. A portion of the catchment (about 37%) was logged between 1967 and 1969. However, disturbance was slight as the area was only lightly timbered. The detailed study plots were located in areas undisturbed by logging. The forest floor exhibits many of the features described by Ruxton (1967) and Douglas (1973) which suggest the occurrence of widespread overland flow. These include undermining of lateral surface roots causing root steps and the exposure and alignment of finer roots in a downslope direction. Subsurface flow was also indicated by examples of caving beneath the root mat at the edge of streams in non-meandering situations.

The stream gauging station is located at the lower end of the catchment and consists of a V notch weir and a water level recorder. Meteorological data are recorded in an adjacent clearing.

EXPERIMENTAL DESIGN

Three sites were chosen for detailed study — two sites (sites *1a* and *1b*) were located adjacent to each other on the lower slopes of the catchment and the third (site *2*) was located in an upper-slope situation in an area where first-order streams are prevalent (Fig.1). The slopes of sites *1a*, *1b* and *2* are in respective order 62, 50 and 42%. A trough system was constructed at the lower end of each site to intercept lateral flow from different depths in the soil profile. A pit was dug, and 2 m long troughs were installed on steps at the soil surface and at depths of 0.25, 0.50 and 1.00 m below the surface. The troughs were let into the vertical face of the pit and sealed with mortar to prevent leakage. Water from each trough is led into a tipping-bucket device and the number of tips of each bucket is recorded mechanically using a hay-bale counter. Mercoid® switches attached to the tipping buckets are connected to digital event recorders at each site so that the runoff is recorded with a time base of 6 min. However, problems with the recorders resulted in an incomplete record during the 1976 wet season. A raingauge at the adjacent meteorological site also provides rainfall information with a 6-min time base.

In addition, a network of tensiometers, piezometers and deep wells was established upslope from two of the sites (*1a* and *2*) to provide information on soil-moisture conditions and saturated zones. The tensiometers were set so that a zero tension was equivalent to a dial reading of 200 cm. In this way small positive pressures could be determined.

A soil field testing programme was carried out to determine the pertinent soil parameters to aid in the interpretation of the observed lateral flow and stream flow data. A ring infiltrometer was used to determine sorptivity and saturated hydraulic conductivity, k_s using a portable constant-head permeameter (Talsma, 1969, 1974) at locations close to the study plots. The k_s factor was also determined by the auger hole pumping test method (e.g., Kirkham and Van Bavel, 1948; Van Beers, 1958) where the permanent saturated zone could be located at site *2*, otherwise the shallow-well pump-in method (Boersma, 1965) was applied at the same sites previously used by the ring infiltrometers. This gave k_s values for the 0—10 and 10—20 cm depths by the constant-head permeameter, and 20—100 cm (approximately) by the shallow-well pump-in method. The piezometers located at depths of 0.5, 1.00, 2.00 and 3.00 m only had water in them for short periods during and immediately after heavy rainstorms. However, the unsteady nature of the potentiometric head prevented any reliable estimates of k_s using the piezometer pumping test method (Luthin and Kirkham, 1949).

RESULTS

In comparing the k_s results from the soil field testing programme it should be noted that the ring infiltrometers measure primarily the vertical component as against the horizontal and vertical components combined by the remaining methods. Nevertheless Table I shows the expected decline of k_s with depth. However, more notable is the rate of decline and the exceedingly high magnitude of k_s in the 0—10 cm layer. The coefficient of variation and absolute range demonstrate the high variability of these surface soils.

A mean k_s value of 32.45 m/day for the top 10 cm of the soil implies that the infiltration capacity would be far in excess of any possible rainfall intensity which could occur. The mean k_s value of the 10—20 cm layer of 1.54 m/day (64 mm/h) means that only the higher rainfall intensities would exceed the percolation rate. Thus the prospect of widespread overland flow originating from the surface 20 cm would seem to be ruled out. Even the much lower k_s below 20 cm is still sufficient to accommodate moderate rainfall intensities ($\leqslant$ 13 mm/h).

The surface configuration and pedological—geological structure of all three slopes associated with these sites would favour predominantly subsurface storm flow on the basis of temperate latitude experience. That is, they would be categorised as steep, well drained and permeable, and bordering on a narrow valley floor in the case of sites *1a* and *1b*. However, this postulation is quickly dismissed when a sample storm is examined using information from the continuous recorders at site *2*. Fig.2 shows that overland flow rather than subsurface flow is prevalent despite the upslope location. Several

TABLE I

The variation in saturated hydraulic conductivity with depth

Depth (m)	Mean k_s (m/day)	Absolute range of k_s (m/day)	Coefficient of variation (untransformed data)	Coefficient of variation (square root transformation)	Sample size, n
0—0.1	32.454^P	2.819—99.398	84.6	45.13	37
0.1—0.2	1.545^P	0.079—5.394	73.81	39.82	33
0.2—1.0 (approx.)	0.318^S	0.053—0.625	55.83	30.3	27
5.52—5.88 (Well *1*) (Site *2*)	0.435^A	0.409, 0.461	—	—	2
3.72—4.48 (Well *2*) (Site *2*)	0.106^A	0.104, 0.108	—	—	2

(a) k_s derived from: P = constant-head permeameter; S = shallow-well pump-in method; and A = auger hole pumping test method (Ernst, 1950).
(b) All k_s values corrected for viscosity at a temperature of 20° C. This is about the average ground water temperature for South Creek.
(c) Two sets of coefficient of variation are given because the coefficient of skewness was substantially positive in the case of 0—0.1 m (+1.03) and 0.1—0.2 m (+1.2). This will have some effect on the standard deviation estimate. The square root transformation was found to be the most suitable and the corresponding results are also shown. However, this does not change the rank order.

other factors are evident in this diagram, notably the rapid commencement of overland flow soon after the commencement of the storm and the very short lag response (6—18 min) between the peaks of rainfall and overland flow.

Even more notable is the close relationship between the overland flow graph and the discharge hydrograph despite the difference in spatial location between the plot site and the gauging station. Previous work (Gilmour, 1975) has noted the rapid response of this catchment to rainfall events and the importance of the quickflow (or stormflow) portion of the hydrograph in contributing to the total runoff. It was calculated that 47% of the total runoff occurred as quickflow. This compares with values of 15—20% for perennial streams in the temperate areas of the U.S.A. (Maki and Hafley, 1972). It is evident from Fig.2 that the quickflow component of the hydrograph mirrors very closely the overland flow graph thus highlighting the importance of overland flow as a major contributor to total runoff.

Fig.2 is a typical example of the situation during that portion of the wet season when heavy rainfall predominates. It is interesting to note that this storm produced an overland flow volume of 3242.9 l compared with 105.6 l from the 0.25 m depth. Therefore the ratio of contribution was 30.71 : 1 in favour of overland flow. It will also be noted that no reference has so far been made to the contributions from 0.5 and 1.00 m depths at this site. In

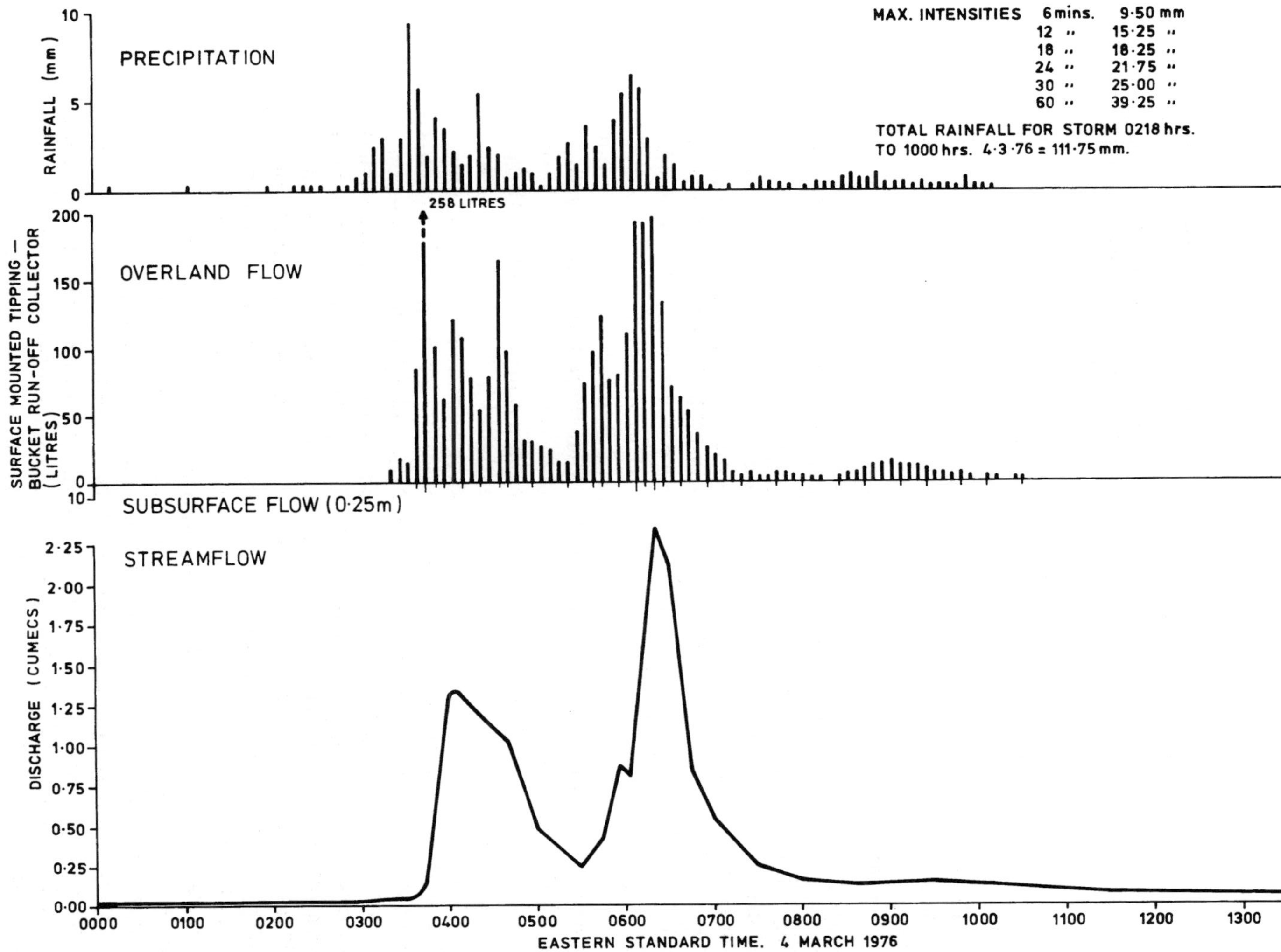

Fig. 2. The continuous record for rainfall, overland flow, subsurface flow (0.25 m) and discharge for a sample storm at site 2.

fact the 0.5-m bucket tipped only once during this storm yielding 2.8 l. There was no response from the 1-m trough.

Sites *1a* and *1b* cannot be compared on a continual recording basis with site *2* because of instrumentation problems. However, Table II showing the volume of flow per week from each depth indicates that substantial overland flow occurs at all sites. Subsurface flow also makes a major contribution to the total volume of lateral flow, particularly at sites *1a* and *1b*. However, the magnitude of overland flow is surprisingly large (considering the previously accepted explanation of drainage processes through forest soils) and it is this facet of flow which is the major subject of investigation in this paper. Certainly our visual observations during rainstorms do not indicate that stemflow is the main source of overland flow as was suggested by Ruxton (1967) and Douglas (1973). The nature of the uneven forest canopy permits a very high percentage of the rainfall to attain the forest floor as throughfall especially during heavy storms. Perhaps the most significant fact is that at none of the sites is there a very shallow permanent water table which is able to emerge at the surface to create saturation overland flow on the basis of the variable source area concept. All wells and piezometers were predominantly dry at site *1a* except during rainstorms, and this also applied to the piezometers at site *2*. Contact was made with the permanent saturation zone at site *2* but as Fig.3 shows for well *2/2*, the water table did not come within 2.6 m of the surface and for the most part was below 3.5 m. This well was situated about 2 m upslope from the troughs. The maximum rise recorded by a second well, located further upslope, was 3.8 m below the surface but remaining records for the wet season showed that the water level stayed below 5 m.

Thus it is evident that an alternative process must be considered. The principal differences between this environment and that in humid temperate areas are the characteristics of rainfall intensity. For example, in this environment maximum 6-min intensities for each week were commonly between 6 and 10 mm during the 1976 wet season. If the soil k_s properties are considered it is evident that the top 20 cm provide little obstruction to vertical percolation for most rainfall events. However, at lower levels a mean k_s value of 0.318 m/day can also be interpreted as 13.25 mm/h or 1.325 mm per 6 min. A similar k_s value in humid temperate areas, where the rainfall intensities are commonly less than 13.25 mm/h, could be expected to encourage deeper vertical percolation. When occasional rainfall intensities exceeding this value occur for short periods, lateral subsurface flow could be expected. But Fig.2 demonstrates that the majority of 6-min rainfall values exceeds 1.325 mm. Thus it is postulated that a temporary perched water table occurs soon after the onset of rain, and subsurface flow rapidly develops in the top 20 cm recharged by "pipe flow" infiltration due to innumerable macropores. The perched water table quickly emerges at the soil surface and subsequent rainfall moves either over the surface as overland flow or recharges the perched water table through losses by vertical percolation and subsurface

TABLE II

Weekly tipping-bucket volumes (l) per week at the three study plots

Date	Throughfall (mm)	Site *1a*			
		surface	0.25 m	0.5 m	1.0 m
21.1.76	81.54 (*1a* + *1b*) 70.3 (2)	299.93	*	43.55	*
28.1.76	56.58	380.14	520.65	107.2	83.85
4.2.76	156.34	826.54	1076.4	549.4	*
11.2.76	27.58	184.84	178.43	6.7	9.68
18.2.76	98.99	767.25	491.4	63.65	45.15
25.2.76	65.86	641.7	327.6	60.3	22.58
3.3.76	52.37	571.95	269.1	53.6	25.8
10.3.76	153.73 (*1a* + *1b*) 154.81 (2)	2197.13	1325.02	475.7	238.65
19.3.76	59.84 (*1a* + *1b*) 58.75 (2)	345.26	429.98	140.7	35.48
24.3.76	96.89	488.25	976.95	331.65	16.13
1.4.76	145.03	850.95	1427.4	*	*
7.4.76	244.35	7149.38	1064.7	402	548.25
14.4.76	113.13 (*1a* + *1b*) 89.93 (2)	1541.48	239.85	30.15	16.13
21.4.76	17.57	0	2.93	0	0
27.4.76	60.93	41.85	137.48	3.35	3.23
5.5.76	58.75 (*1a* + *1b*) 65.13 (2)	*	277.88	10.05	12.9
12.5.76	99.71 (*1a* + *1b*) 93.33 (2)	2734.2	359.78	20.1	70.95
19.5.76	18.88	94.16	46.8	0	0
26.5.76	10.18	0	0	0	0
2.6.76	71.59	428.96	462.15	10.05	19.35
9.6.76	15.55	6.98	49.73	0	0
16.6.76	28.3	13.95	81.9	0	0

(a) Dates quoted in most cases represent 7-day periods.
(b) Throughfall calculated from total rainfall for appropriate period by equation y -- $0.7 + 0.725x$ where y = throughfall (mm), x = rainfall (mm) (Gilmour, 1975). Two throughfall values are shown for the appropriate sites when readings were taken at different times during rainfall.
(c) Asterisks denote unreliable records.

flow. The mechanism has strong parallels with that put forward by Dunne and Black (1970a, b) who described it as saturation overland flow. The major difference is that in this study a temporary perched water table over virtually the whole catchment is the important factor in initiating overland flow whereas Dunne and Black postulate the rise of a permanent saturated zone adjacent to the stream channels. Thus the main implication is that overland flow can occur over widespread areas of the catchment once the top 20 cm store has been filled. Furthermore, the temporary high drainage den-

Site *1b*				Site *2*			
surface	0.25 m	0.5 m	1.0 m	surface	0.25 m	0.5 m	1.0 m
42.5	116.61	3.17	0	861.3	112	0	0
245	167.3	0	0	1650.83	195.2	5.58	0
412.5	415.26	9.5	5.39	*	*	13.94	0
35	32.86	0	0	*	*	2.79	0
255	125.48	6.34	0	1513.8	185.6	0	0
195	71.7	6.34	0	991.8	160	0	0
240	56.76	3.17	0	809.1	60.8	0	0
1340	236	19	2.69	4600.13	201.6	8.36	5.18
355	41.83	6.34	0	1428.98	60.8	2.79	0
517.5	176.26	22.17	0	2205.45	*	2.79	0
1337.5	265.89	34.84	*	3830.18	723.2	19.51	*
2295	313.69	41.18	21.54	*	*	*	*
585	44.81	6.34	2.69	1631.25	115.2	2.79	0
2.5	2.99	0	0	61.99	12.8	0	0
67.5	29.88	3.17	0	512.21	76.8	0	0
145	47.8	9.5	0	694.91	67.2	0	0
995	131.45	0	0	2381.63	108.8	2.79	0
22.5	17.93	0	2.69	270.79	28.8	5.58	0
0	0	0	0	3.26	0	0	0
142.5	203.15	12.67	0	499.16	83.2	11.15	0
7.5	14.94	0	0	127.24	12.8	2.79	0
7.5	32.86	6.34	0	270.79	22.4	0	0

sity created by the addition of numerous ephemeral streams during rainstorms enables a substantial amount of this overland flow, and to a lesser extent subsurface flow, to be tapped by the main stream. This accounts for the very high magnitude of quickflow response.

To investigate this hypothesis further, weekly surface tipping-bucket volumes were correlated using Spearman's rank correlation coefficient, r_s (Siegel, 1956), with 14 weekly rainfall variables with special emphasis on intensity. They were the weekly maximum intensities for 0.1, 0.2, 0.3, 0.4,

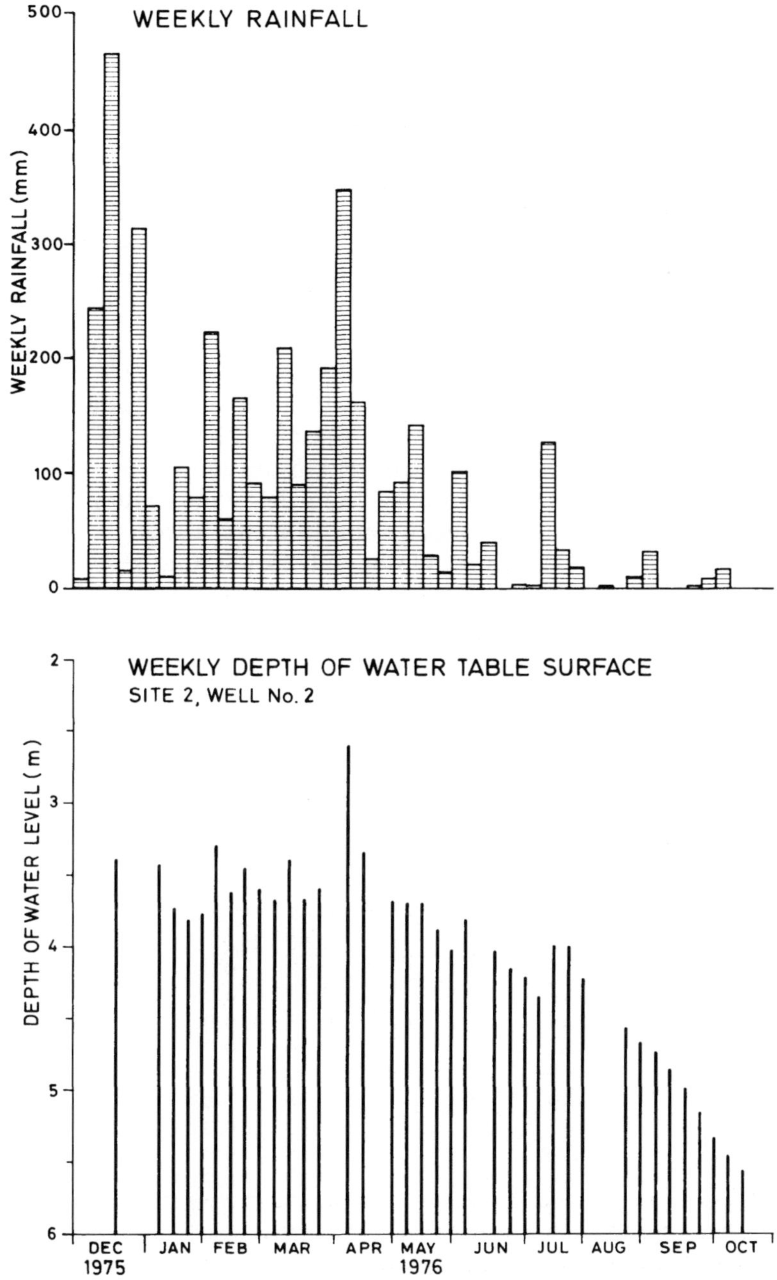

Fig. 3. The well hydrograph for well No. *2* located about 2 m upslope from the troughs at site *2*.

0.5, 1, 2, 3, 4, 5, 6 and 12 h. In addition, the total weekly precipitation and the amount for the maximum storm were incorporated into the analysis after using a correction for interception (Gilmour, 1975). Spearman's r_s method was used because its distribution-free property was considered more suitable when dealing with small samples. The period for this particular investigation was 14.1.76—16.6.76. The tensiometers indicated that the soil-moisture conditions were similar throughout this time interval with the suction in most cases lying between 0 and 100 cm. In addition, small positive pressures up to +40 cm were also often noted. Beyond mid-June a prolonged drying phase commenced creating a different hydrological situation so that any later rainfall events are not included in this analysis.

Tables III and IV show the respective Spearman's rank correlation coefficients and their levels of significance (Siegel, 1956). It will be noted that the analysis has been divided into "summer" and "winter" seasons. The basis of the separation were the histograms for the short-term maximum rainfall intensities. Fig.4 shows an example of the separation of the weekly maximum 6-min intensities into two populations. Other short-term intensity histograms showed the same distinct bimodal nature. It was remarkable that in

TABLE III

Weekly surface tipping-bucket volumes vs. rainfall variables correlation coefficients (r_s) for summer period

Weekly surface tipping-bucket volumes versus	Site *1a* n = 13		Site *1b* n = 13		Site *2* n = 9	
	r_s	significance level, p	r_s	significance level, p	r_s	significance level, p
Max. 6 min	0.62	0.05	0.64	0.05	0.43	not significant
Max. 12 min	0.62	0.05	0.72	0.01	0.85	0.01
Max. 18 min	0.63	0.05	0.82	0.01	0.90	0.01
Max. 24 min	0.58	0.05	0.76	0.01	0.82	0.01
Max. 30 min	0.54	0.05	0.71	0.01	0.80	0.01
Max. 1 h	0.42	not significant	0.51	0.05	0.85	0.01
Max. 2 h	0.57	0.05	0.67	0.01	0.85	0.01
Max. 3 h	0.60	0.05	0.73	0.01	0.90	0.01
Max. 4 h	0.61	0.05	0.76	0.01	0.93	0.01
Max. 5 h	0.49	0.05	0.74	0.01	0.87	0.01
Max. 6 h	0.56	0.05	0.80	0.01	0.92	0.01
Max. 12 h	0.44	not significant	0.63	0.05	0.81	0.01
Total weekly precipitation	0.85	0.01	0.82	0.01	0.73	0.05
Max. storm	0.54	0.05	0.71	0.01	0.88	0.01

The significance levels stated are for a one-tailed test as physically the area of interest is in in the positive spectrum, $r \geqslant 0$.

TABLE IV

Weekly surface tipping-bucket volumes versus rainfall variables correlation coefficients (r_s) for winter period

Weekly surface tipping-bucket volumes versus	Site *1a* $n = 8$		Site *1b* $n = 9$		Site *2* $n = 10$	
	r_s	significant level, p	r_s	significance level, p	r_s	significance level, p
Max. 6 min	0.20	not significant	0.43	not significant	0.46	not significant
Max. 12 min	0.46	not significant	0.66	0.05	0.71	0.05
Max. 18 min	0.40	not significant	0.58	not significant	0.67	0.05
Max. 24 min	0.99	0.01	0.96	0.01	0.92	0.01
Max. 30 min	0.97	0.01	0.92	0.01	0.85	0.01
Max. 1 h	0.92	0.01	0.94	0.01	0.83	0.01
Max. 2 h	0.90	0.01	0.90	0.01	0.80	0.01
Max. 3 h	0.95	0.01	0.94	0.01	0.91	0.01
Max. 4 h	0.95	0.01	0.94	0.01	0.87	0.01
Max. 5 h	0.98	0.01	0.95	0.01	0.88	0.01
Max. 6 h	0.98	0.01	0.97	0.01	0.90	0.01
Max. 12 h	0.65	0.05	0.60	0.05	0.76	0.01
Total weekly precipitation	0.89	0.01	0.93	0.01	0.95	0.01
Max. storm	0.97	0.01	0.98	0.01	0.91	0.01

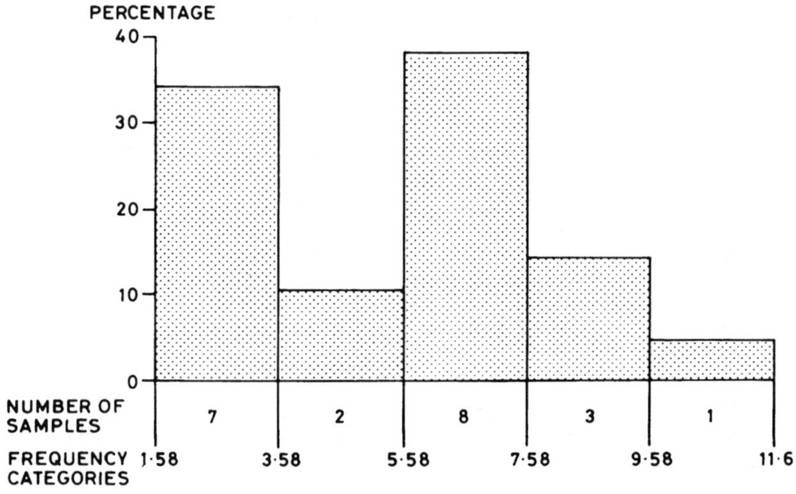

Fig. 4. Frequency diagram of maximum 6-min. rainfall intensity/week from 14 January—16 June 1976.

terms of time the first group incorporated the lower intensity "winter" rainfall dating from mid-April to early June. Conversely, the second group of higher intensity "summer" rainfall dated from the beginning of records in January to mid-April. It will be noted that the sample size for each site varies due to some loss in records.

Referring to Table II the most outstanding feature is that the majority of Spearman's r_s are significant for the summer period. The physical interpretation is probably that the short-term maximum intensities make a large contribution to the total weekly overland flow recorded despite their extremely short duration. Certainly, these intensities are far in excess of the k_s values below 10 cm and therefore capable of rapidly developing saturation overland flow. Furthermore, if they occur once the perched water table has emerged at the surface, which is commonly the case, then the catchement will be acting as an "impermeable" surface resulting in a very large proportion of this large short-duration input to move over the surface. The longer-term intensities, i.e. max. 2 h—max. 6 h, also include the lower 6-min intensities which as seen from Fig.2 are high enough to iniate overland flow. In addition, they will make a substantial contribution to saturation overland flow once the perched water table has emerged at the surface. The latter statement can also be applied to the significant r_s values corresponding to the maximum storm and the total weekly precipitation.

The small sample sizes in conjunction with random error effects make it difficult to evaluate major differences within and between the three study plots. In view of these statistical considerations any comparisons that are made must be regarded with caution. Nevertheless site *1a* consistently shows lower r_s values which, when significant, are at a lower level, than sites *1b* and *2*. Table II showed that site *1a* had the most subsurface flow. It is also the steepest and differs certainly from site *2* in that borings for piezometer installations detected a scattering of residual rocks in the clay just below the surface. No borings were made on site *1b* but the trough exposures suggested that it was intermediate in composition between the other two sites. The shallow-well pump-in tests showed that any above-average k_s values for the 20—100 cm zone were generally associated with rocky profiles. Consequently, this will not only lead to a much higher proportion of throughfall being transmitted either as subsurface flow or as deep vertical percolation but also will require much higher short-term rainfall intensities to cause the perched water table to emerge at the surface in such areas. In other words the effects of rainfall intensity do not always erase the influence of the lithological factor which probably contributes to the lower r_s values. This implies that overland flow is possibly only generated during higher rainfall intensity events which can exceed the rate of removal of stored water by subsurface flow. It is noteworthy that the steepest slopes do not necessarily give the highest overland flow amounts, c.f. Thomas (1973).

The winter situation presents interesting comparisons with the previous season (Table IV). The distribution of significant r_s values for all sites shows a much greater emphasis on the longer-duration maximum intensities. A notable feature is the sudden rise in the r_s between max. 18 min and max. 24 min. The initial three rainfall variables, i.e. max. 6 min to max. 18 min are either non-significant or only significant at the lower level. A further point of note is the highly significant r_s values for total weekly precipitation and maximum storm.

It is not surprising that the results show less emphasis on the ability of the short-term maximum intensities to generate substantial overland flow because at this time of the year intensities are much lower than during summer. For example, Fig. 4 shows max. 6-min values ranged from only 1.58 to 4 mm. The characteristics of the winter rainfall in 1976 being of lower intensity but longer duration probably accounts for the high r_s values between max. 24 min and max. 6 h. Nevertheless once the perched water table has emerged at the surface the higher-intensity parts of the storm, focusing on max. 24 min, seem to make a significant contribution to overland flow volumes. The longer-duration maximum intensities, as in the summer situation, will incorporate some of the lower short-term intensities which will also be capable of generating overland flow. The fact that max. 12-h has a much lower r_s at all sites can be attributed to some weeks not including a storm of that duration.

During the winter storms the r_s values for max. 12 min—max. 18 min were lower at site *1a* than at the other two sites. It is possible that the short-term intensities cannot make much impression on site *1a* because of the mosaic pattern of above average k_s created by subsurface rocks. Thus subsurface as well as overland flow is a major flow path. On the other hand, in the absence of such a factor (particularly at site *2*) a significant overland flow contribution can be made. Geological differences seem to be less important from max. 24 min onwards possibly because the long duration of the storm is capable of filling up even the larger voids thus enabling the perched water table to emerge at the surface. This statement could also explain why site *1a* shows such a high r_s value for the maximum winter storm compared with the maximum summer storm.

The results from this study are in marked contrast with findings in humid temperate areas. In such areas, if overland flow is to occur away from the shallow water table areas, it is likely to be associated with short periods when very high rainfall intensities are capable of exceeding the infiltration rate. Selby (1973) statistically demonstrated this fact in connection with ungrazed grass and scrub on pumice lithology in New Zealand. He noted that the most significant correlation coefficients corresponded with short-term maximum weekly rainfall variables. This contrasts with the high r_s values for both long- as well as short-term rainfall intensities in this work. Other statistical analyses by Hewlett et al. (1977), based on long-term records connected with a forested watershed in the southern Appalachians, U.S.A., indicated that maximum rainfall intensity (hourly and minutely) were of little practical value in predicting stormflow volume or peak discharge. Thus they concluded that rainfall intensity was not a dominant variable in stormflow generation in that environment.

CONCLUSION

This work has indicated that in this particular tropical rainforest environment the variable source area concept, associated with humid temperate areas, is not applicable due to the characteristics of rainfall intensity. Spearman's rank correlation analysis indicated that this applied to both seasons, despite some detailed differences, providing the assumption of a linear relation is correct. Thus the rainfall intensity is sufficient to exceed the average k_s for depths below 20 cm and therefore create a widespread perched water table, and corresponding subsurface flow, within the top 20 cm. Once this upper store has been filled, its emergence at the surface develops an "impermeable" surface and therefore leads to widespread saturation overland flow which can be tapped by a temporary dense drainage network. This phenomenon is thought to continue for the remainder of the storm. The only doubt raised by Spearman's correlation analysis was during the summer period when storms in 1976 were generally of shorter duration than normal and where a slope is underlain by residual rocks in the deeply weathered clays, e.g. site *1a.* It is hoped to examine this problem further with more sophisticated statistical analysis once continuous records become available.

The significance of subsurface flow contribution to the storm hydrograph, originating from the perched water table, becomes greater in the presence of residual rock. However, this does not preclude large volumes of overland flow occurring even on the steeper, well drained slopes in contrast with the situation in humid temperate areas.

ACKNOWLEDGEMENTS

The Australian Water Resources Council, Queensland Forestry Department and James Cook University of North Queensland are thanked for their continued financial support of this work. Mr. H.O. McColl is also thanked for his assistance in data collection under at times very trying conditions.

REFERENCES

Boersma, L., 1965. Field measurement of hydraulic conductivity above a water table. In: C.A. Black (Editor), Methods of Soil Analysis, Part 1, Agronomy No. 9, Am. Soc. Agron., Madison, Wisc., pp. 234—252.

Douglas, I., 1973. Rate of denudation in selected small catchments in eastern Australia. University of Hull, Occas. Pap. Geogr., No. 21, 127 pp. (quotation in text, p. 35).

Dunne, T. and Black, R.D., 1970a. An experimental investigation of runoff production in permeable soils. Water Resour. Res., 6: 179—191.

Dunne, T. and Black, R.D., 1970b. Partial area contributions to storm runoff in a small New England watershed. Water Resour. Res., 6: 1269—1311.

Dunne, T., Moore, T.R. and Taylor, C.H., 1975. Recognition and prediction of runoff-producing zones in humid regions. Hydrol. Sci. Bull., 20: 305—327.

Ernst, L.F., 1950. A new formula for the calculation of the permeability factor with the auger hole method. T.N.O., Groningen (reported by W.F.J. van Beers, 1958).

Freeze, R.A., 1972b. Role of subsurface flow in generating surface runoff, 2. Upstream source areas. Water Resour. Res., 8: 1272—1283.
Gilmour, D.A., 1975. Catchment water balance studies on the wet tropical coast of north Queensland. Ph.D. Thesis, James Cook University of North Queensland, Townsville, Qld. (unpublished).
Hewlett, J.D., 1961. Watershed management. In: U.S. For. Serv., Southeast. For. Exp. Stn. Rep., pp. 61—66.
Hewlett, J.D. and Hibbert, A.R., 1967. Factors affecting the response of small watersheds to precipitation in humid areas. In: W.E. Sopper and H.W. Lull (Editors), International Symposium on Forest Hydrology, Pergamon Press, Oxford, pp. 275—290.
Hewlett, J.D. and Troendle, C.A., 1975. Non-point and diffused water sources: a variable source area problem. In: Watershed Management, Symp., Comm. Watershed Manage., Irrig. Drain. Div. Am. Soc. Civ. Eng., Logan, Utah, pp. 21—46.
Hewlett, J.D., Fortson, J. and Cunningham, G., 1977. Effect of rainfall intensity on stormflow and peak discharge from forest land. Water Resour. Res., 13: 259—266.
Horton, R.E., 1933. The role of infiltration in the hydrological cycle. Trans. Am. Geophys. Union, 14: 446—460.
Horton, R.E., 1945. Erosional development of streams and their drainage basins: hydrological approach to quantitative morphology. Bull. Geol. Soc. Am., 56: 275—370.
Kirkham, D. and Van Bavel, C.H.M., 1948. Theory of seepage into auger holes. Soil Sci. Soc. Am., Proc., 13: 75—89.
Luthin, J.N. and Kirkham, D., 1949. A piezometer method for measuring permeability of soil in sites below a water table. Soil Sci., 68: 349—358.
Maki, T.E. and Hafley, W.L., 1972. Effects of land use on municipal watersheds. Water Resour. Res. Inst. Univ., North Carolina Report No. 71.
Northcote, K.H., 1971. A Factual Key for the Recognition of Australian Soils. Rellim Technical Publications, Glenside, S.A., 3rd ed., 122 pp.
Ruxton, B.P., 1967. Slopewash under Mature Primary Rainforest in Northern Papua. In: J.N. Jennings and J.A. Mabbutt (Editors), Landform Studies from Australia and New Guinea. Australian National University Press, Canberra, A.C.T., pp. 85—94 (quotation in text, p. 91).
Selby, M.J., 1973. An investigation into causes of runoff from a catchment of pumice lithology in New Zealand. Hydrol. Sci. Bull., 18: 255—280.
Siegel, S., 1956. Nonparametric Statistics for the Behavioural Sciences. McGraw-Hill, Tokyo, 312 pp.
Talsma, T., 1969. In situ measurement of sorptivity. Austr. J. Soil Res., 7: 269—276.
Talsma, T., 1974. Infiltration in field soils. Trans. 10th Int. Congr. on Soil Science, Moscow, 1: 68—74.
Thomas, M., 1973. Landforms in Equatorial Forest Areas. In: D. Brunsden and J.C. Doornkamp (Editors), The Unquiet Landscape, David and Charles, Newton Abbott, pp. 141—146 (quotation in text, p. 145).
Van Beers, W.F.J., 1958. The auger hole method, a field measurement of the hydraulic conductivity of soil below a water table. Int. Inst. Land Reclam. Improv., Wageningen, Bull. 1.
Ward, R.C., 1975. Principles of Hydrology. McGraw-Hill, London, 2nd ed., 367 pp. (quotation in text, p. 256).
Webb, L.J., 1969. A physiognomic classification of Australian rain forests. J. Ecol., 47: 551—570.
Whipkey, R.Z., 1965. Subsurface storm flow from forested slopes. Int. Assoc. Sci. Hydrol., bull., 10: 74—85.
Whipkey, R.Z., 1969. Storn runoff from forested catchments by subsurface routes. In: Floods and Their Computation, Studies and Reports in Hydrology, Leningrad, UNESCO—IASH—WMO, 3: 773—779.

Commentary on Paper 25

Lynch, J. A., Corbett, E. S. & Sopper, W. E. (1979) Effects of antecedent soil moisture on stormflow volumes and timing.

The experimental work on the Shale Hills Watersheds in Pennsylvania is another example of a streamflow generation study that has been largely forgotten but which is a classic of its type. The work was carried out by staff of the USDA Forest Service Northeastern Forest Resource Experimental Station and the School of Forest Resources at Penn State, led by Warren Sopper, Jim Lynch and Ed Corbett. There were three small experimental catchments at Shale Hills and three more not far away at Leading Ridge. Shale Hills Watershed 2, with an area of 22 acres (8.8 ha), was selected as a primary experimental site and was equipped with a rainfall simulation facility. The intention was to study the effects of antecedent conditions and storm size on runoff and hydrograph characteristics.

The rainfall simulator consisted of 520 rotating sprinklers on 5-foot (1.52 m) risers spaced on a 40-foot (12.1 m) grid. The water for the system was supplied by two pumps raising water from a small lake just downstream of the catchment outlet. The pumps were capable of supplying the equivalent of 0.24 ins/h (6.1 mm h^{-1}) and the sprinklers were linked by 5 miles (8 km) of aluminium piping. The system was designed so that either the whole catchment, or just the upper slopes, or mid-slope section, or lower riparian slopes, could be watered. Input rates to the slopes were monitored by 8 recording, 40 storage and 17 trough-like raingauges. Pictures of the experiments show the pumps glowing red when at full capacity and it is reported that the lake level would fall significantly after a full experiment. The experiments in which the upper parts of the hillslopes were sprinkled were particularly interesting, as the sound of air moving down the slope through macropores and escaping to the surface was apparently impressive (not that hydrological modellers normally worry about air pressure effects).

The response of the catchment was monitored at 31 soil moisture sites, 15 tensiometer sites, 52 shallow groundwater observation wells and three deep groundwater observation wells. The scale of this experiment was unprecedented and has scarcely been matched since (though there have been some remarkable sprinkling experiments at Coos Bay, Oregon, see, for example, Anderson *et al.*, 1997, and at the constructed HydroHill hillslope in China, Kendall *et al.*, 2001).

Huge amounts of data were collected in these experiments. Although the experiments were reported at several conferences, no journal publications were published and for many years the data (which would have been extremely useful for testing the distributed models that were being developed in the 1980s) remained virtually inaccessible. Thankfully, in the last few years Chris Duffy at Penn State has taken up the challenge of making all the data that could be recovered available (see http://cataract01.cee.psu.edu/ShaleHills/). Jim Lynch continues to work for the Forest Service and is involved in data collection programmes across the northeastern states.

References

Anderson, S. P., Dietrich, W. E., Montgomery, D. R., Torres, R., Conrad, M. E. & Loague, K. (1997) Subsurface flow paths in a steep, unchannelled catchment. *Water Resour. Res.* **33**(12), 2637–2653.

Kendall, C., McDonnell, J. J. & Gu, W. (2001) A look inside "black box" hydrograph separation models: a study at the Hydrohill catchments. *Hydrol. Processes* **15**, 1877–1902.

EFFECTS OF ANTECEDENT SOIL MOISTURE ON STORMFLOW VOLUMES AND TIMING

By

James A. Lynch, Assistant Professor of Forest Hydrology
School of Forest Resources
The Pennsylvania State University
University Park, PA 16802

Edward S. Corbett, Principal Hydrologist
U.S.D.A. - Forest Service
Northeastern Forest Experiment Station
Pennington, New Jersey 08534

and

William E. Sopper, Professor of Forest Hydrology
Institute for Research on Land and Water Resources
The Pennsylvania State University
University Park, PA 16802

Abstract

The effects of antecedent soil moisture content on stormflow amounts (quickflow, delayed flow, total storm runoff, and maximum peakflow) and timing (time to maximum peakflow, time of recession, and duration of quickflow) were studied by applying 38.1 mm of simulated rainfall at an intensity of 6.1 mm per hour to a 7.93 ha forested watershed. An irrigation system, which consisted of 8 kilometers of aluminum and plastic pipe and utilized 520 rotating sprinklers mounted on 1.5 m risers to distribute the water pumped from a nearby lake, was used to apply the simulated rainstorms. A 38.1 mm simulated rainstorm was replicated eight times at various antecedent soil moisture contents.

Quickflow, total storm runoff, maximum peakflow, time of recession, and duration of quickflow resulting from these artificial rainstorms were positively correlated with increasing antecedent soil moisture. Delayed flow and time to maximum peakflow were, for the most part, unaffected by changes in antecedent soil moisture. The percentage of rainfall converted to total storm runoff ranged from 12 to 86 percent, while antecedent soil moisture ranged from 87 to 99 percent of field capacity.

Introduction

One of the most important and fundamental needs in watershed research is to fully understand watershed behavior--that is to understand the entire hydrologic system which affects the disposition of precipitation within the drainage basin and its conversion to streamflow. This conversion is the result of complex interactions between hydrologic processes and numerous climatic and physiographic factors. One of these factors is antecedent soil moisture (ASM).

Although the importance of ASM in controlling storm-runoff relations has been realized, its actual affects on storm hydrograph characteristics and components have not been quantitatively defined. The primary problem encountered when utilizing natural rainfall-runoff events to evaluate the effects of ASM on stormflows is the variability of meteorological and antecedent watershed conditions. This variability makes it highly improbable that a precipitation event will be replicated on an experimental watershed at the same intensity and duration over a range of antecedent conditions. Physical laboratory models and computer models enable one to achieve the necessary replications. In addition, the utility of these models allows one to hold certain factors constant while at the same time varying others. However, one of the major disadvantages is the difficulty of transferring the results of model studies to natural watersheds (Black, 1970; Graveto and Eagleson, 1970; Roberts and Klingeman, 1970). Although both of these approaches have unquestionable value, their usefulness still depends upon a thorough knowledge of the hydrologic processes which can be extrapolated to the physical laboratory model or used to select parameters and linkages for conceptual models to characterize watershed behavior and response to rainfall.

The purpose of this study was to quantitatively evaluate the effects of antecedent soil moisture on stormflow amounts (quickflow, delayed flow, total storm runoff and maximum peakflow) and timing (time to maximum peakflow, time of recession and duration of quickflow). This was accomplished using an irrigation system installed on a 7.93 ha forested watershed in central Pennsylvania. The watershed is essentially elliptical, with symmetrical sides facing north and south. Average slideslope steepness is 30 percent. The rainfall simulation facility allows for control of watershed conditions prior to the application of simulated rainfall as well as rainfall intensity, duration, and consequently, amount.

The Rainfall Simulation System

The system consisted of approximately 8 kilometers of aluminum and plastic pipe. Two heavy-duty centrifugal pumps

90

mounted in tandem supplied water from a nearby lake. Five-hundred and twenty rotating sprinklers with 4 mm nozzles mounted on 1.5 m risers distributed the water over the basin. The sprinklers were located on the watershed in a 12 m grid spacing. Flow control valves on each sprinkler allowed a known quantity of simulated rainfall to be applied once the system was pressurized. The flow control valves had a maximum application rate of 6.1 mm per hour. For this study the irrigation system covered 7.59 ha of the 7.93 ha watershed; only the extreme upper portion of the rear half and several small areas on the ridge top were not irrigated.

The watershed was divided into sub-basins by three streamflow gaging stations on the perennial stream channel. Another streamflow gaging station monitored intermittent streamflow from a secondary channel in the upper portion of the watershed. The hydrologic status of the watershed was monitored at 31 soil moisture sites, 15 tensiometer sites, 52 shallow groundwater observation wells and 3 deep groundwater observation wells. Simulated and natural rainfall were measured by an extensive network of 8 recording, 40 standard, and 17 trough-type raingages.

The general shape of the hydrograph of a single storm occurring over the drainage area is shown in Fig. 1, which also shows the components of a storm hydrograph as used in this report. Separation of quickflow from delayed flow was based on the technique described by Hewlett and Hibbert (1967). A line was projected from the beginning of each storm hydrograph at a slope of 0.547 liters per second per square kilometer per hour until it intersected the recession side of the hydrograph. However, in our analysis, base flow volumes resulting from antecedent storms were excluded from delayed flow. This was accomplished by extending a horizontal line of constant discharge equal to antecedent flow rate from the beginning of storm runoff until it intersected the recession limb of the storm hydrograph.

Calibration of the Rainfall Simulation Facility

In order to quantitatively evaluate the effects of ASM on storm hydrographs resulting from simulated rainfall events, it was necessary to determine what, if any, difference in the storm hydrographs could be attributed solely to variations in the operation of the simulated rainfall facility.

The calibration involved the replication of a 38.1 mm simulated rainstorm at low antecedent soil moisture and a 25.4 mm simulated rainstorm at high antecedent moisture conditions. The two 38.1 mm storms were applied during August, while the 25.4 mm simulated rainstorm was replicated in October.

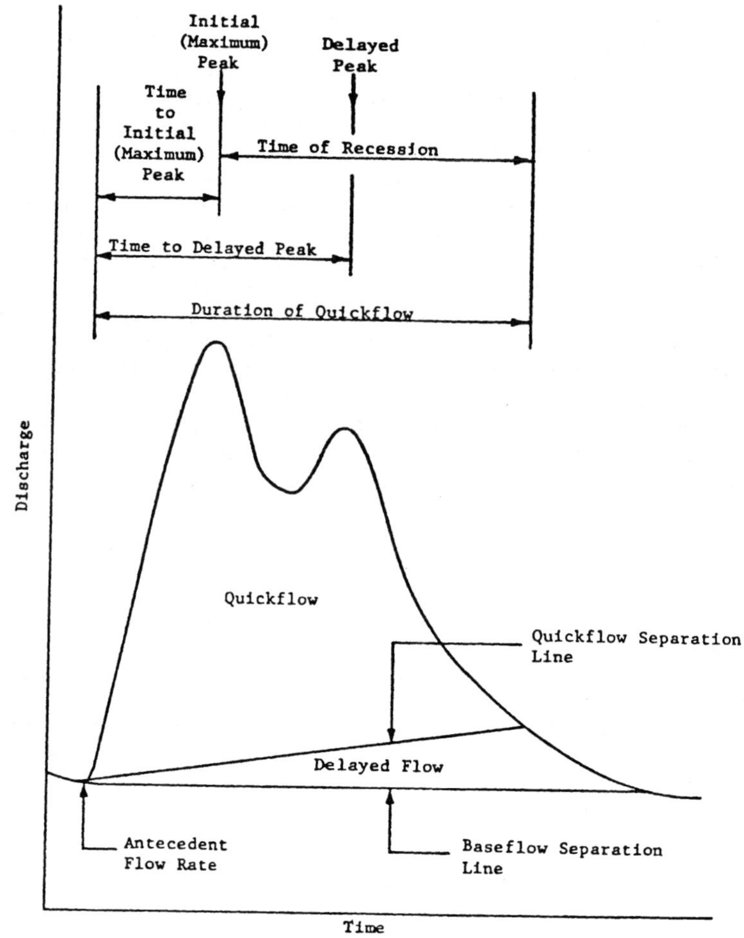

Fig. 1. Components and characteristics of a storm hydrograph resulting from a single storm occurring on the experimental watershed

The hydrographs resulting from the 38.1 mm rainstorm replications were almost identical with respect to rising limbs, peakflows, recession limbs, and storm runoff. Maximum peakflows (0.099 and 0.098 m^3s^{-1}/km^2), both of which occurred within minutes after cessation of rainfall, differed by only one percent. The times of recession and durations of stormflow differed by only 23 and 19 minutes, respectively. Total storm runoff from the replicated storms measured 4.5 and 4.4 mm, a variation of less than 2.2 percent.

92

The results of the 25.4 mm replications were very similar to those obtained from the 38.1 mm replications. The rising limbs, peakflows, and most of the recession limbs were almost identical for both storms. The only major discrepancy occurred during the first 13 hours of the second storm recession. During this period, a delayed peak occurred that was not evident following the first storm. Based on shallow groundwater well observations, the source of stormflow for this delayed peak was attributed to slightly wetter soil in the valley floor above the perennial stream at the time of the second storm. In terms of volume, the first simulated rainstorm produced a total storm runoff of 19.6 mm compared to 20.6 mm for the second rainstorm. The difference, 1.0 mm, represents only five percent of the total storm discharge.

Based on the results of both the 38.1 and 25.4 mm storm replications, very little difference in the size and shape of the storm hydrographs could be attributed to variations in the operation of the simulated rainfall facility.

Procedure

Soil moisture determination. Soils are mostly shaly silt loams and silt loams ranging in depth from 0.3 to 0.9 m on the upper side slopes and ridge top to 0.9 to 2.7 m on the lower side slopes and channel area. Soil infiltration and percolation rates are greater than the maximum expected rainfall intensity for the area. Field capacity for the entire soil mantle (average depth 1.42 m) has been estimated at 319.8 mm of water.

Using the neutron scattering technique, soil moisture was measured at 31 sites to monitor moisture fluctuations in the soil mantle. Both surface and depth probes were used. Thirteen of the soil moisture sites were located on a transect perpendicular to the stream channel at a point midway between two streamflow gaging stations. Each of the remaining soil moisture sites represented a physiographic unit of the watershed.

Effects of antecedent soil moisture on storm hydrographs. In order to quantitatively evaluate the effects of ASM on stormflow volumes and timing, eight 38.1 m simulated rainstorms were applied to the entire watershed under various ASM conditions. Rainfall intensity for all storms was 6.1 mm per hour; rainfall duration was 6.25 hours. The storms were applied between August 1 and September 19, 1974 under as similar climatic conditions as possible.

Quickflow, delayed flow, total storm runoff, maximum peakflow, time to maximum peakflow, time of recession and duration of quickflow from each simulated rainstorm were compared with ASM using correlation analysis. In addition, empirical relationships between the individual storm hydrograph parameters and ASM were derived using regression analysis.

Results and Discussion

Antecedent soil moisture measurements. Antecedent soil moisture on a total watershed basis reached its lowest level (278.1 mm) on 8/1/74 and progressively increased to its highest level (316.2 mm) on 9/19/74 (Table 1). A comparison of the watershed response to the simulated rainfall applications at the lowest and highest antecedent moisture contents is illustrated in Fig. 2. The relatively small difference in ASM (38.1 mm) between the applications of simulated rainfall under "wet" and "dry" conditions was due to the high ASM on 8/1/74. The high ASM on 8/1/74 was the result of 292.1 mm of rainfall (both natural and artificial) that fell on the watershed during the two months preceding this rainstorm. The abundance of rainfall during this period continually replenished soil moisture, but did not appreciably affect streamflow because of available moisture storage within the soil mantle. Consequently, streamflow prior to the 8/1/74 rainstorm was seasonally low at 0.0012 m^3s^{-1}/km^2. Although streamflows below 0.001 m^3s^{-1}/km^2 are not uncommon on the

Table 1. Stormflow volumes for eight 38.1-mm simulated rainfall events at different antecedent soil moisture contents

Date	Antecedent Soil Moisture	Total Storm Runoff	Quick-flow	Delayed Flow	Maximum Peakflow
	mm				(m^3s^{-1}/km^2)
8/01/74	278.1	4.55	2.64	1.90	0.099
8/07/74	287.5	12.19	8.36	3.84	0.201
8/14/74	290.1	15.52	11.94	3.58	0.253
8/19/74	295.1	15.90	12.27	3.63	0.239
8/23/74	304.8	20.75	17.02	3.73	0.304
8/27/74	310.4	23.34	19.58	3.76	0.328
9/16/74	313.2	29.74	25.78	3.96	0.526
9/19/74	316.2	32.66	28.88	3.78	0.628

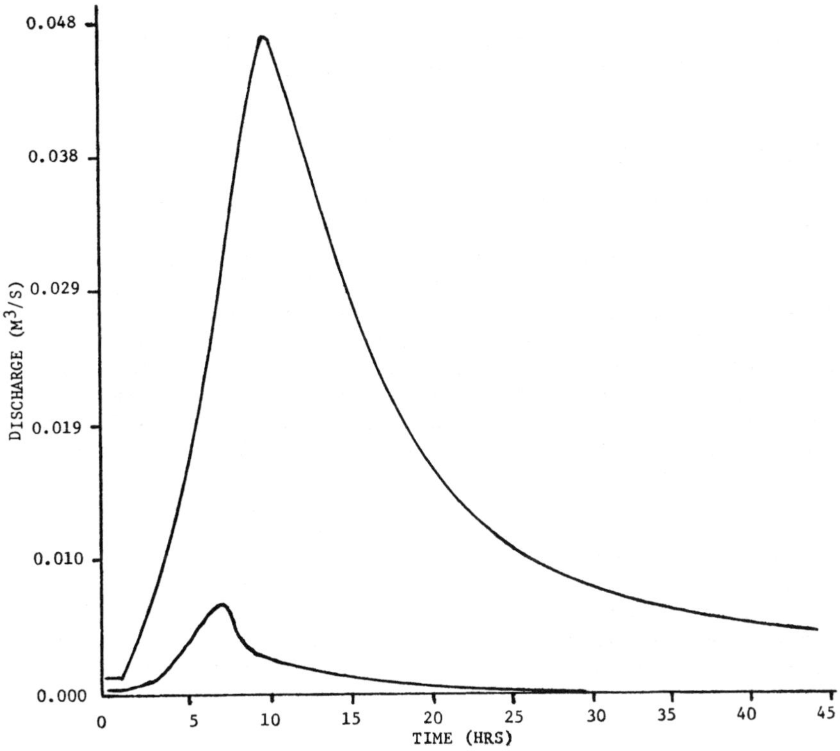

Fig. 2. Comparison of storm hydrographs from 38.1 mm simulated rainstorms at the lowest and highest measured antecedent moisture content

watershed, they occur only during long rainless periods with high evapotranspiration rates. Streamflow recession during such periods is very slow. As a result, soil moisture can decrease substantially with little change in stream discharge.

Stormflow volumes. Quickflow, delayed flow, total storm runoff and maximum peakflows from eight simulated 38.1 mm rainstorms at different ASM contents are given in Table 1. It is evident that ASM conditions of the basin were very important in determining the magnitude of storm discharge. Total storm runoff ranged from 4.55 to 32.66 mm. This increase in the efficiency of the watershed in converting precipitation to streamflow occurred over a relatively small range in ASM. The amount of moisture stored in the soil mantle at field capacity was estimated to be 319.8 mm. Based on this estimate, ASM on 8/1/74 and 9/19/74 represented 87 and 99 percent, respectively, of the total moisture stored at field capacity. This range represents a span of only 12 percent of the available moisture below field capacity. Storm runoff resulting from 38.1 mm rainstorms, occurring when ASM is at or below 87 percent of field capacity, can account for only 12 percent or less of the precipitation.

Quickflow volumes ranged from 2.64 to 28.88 mm and progressively increased with increasing ASM. Regardless of ASM, quickflow accounted for most (58 to 88 percent) of the total storm runoff from each simulated rainstorm. In fact, with the exception of the 8/1/74 and 8/7/74 storms, quickflow consistently accounted for 77 to 88 percent of storm runoff. Seven percent of the 8/1/74 rainstorm occurred as quickflow while 76 percent of the 9/19/74 rainfall was converted into quickflow.

Delayed flow contributions to storm runoff ranged from 1.90 to 3.96 mm. However, delayed flow did not progressively increase with increasing ASM, but remained fairly constant for all but the 8/1/74 storm. Delayed flow contributions to storm discharge in terms of percent of total runoff decreased from 42 percent at the lowest ASM to 12 percent when ASM was near field capacity.

Maximum peakflows ranged from 0.099 to 0.628 m^3s^{-1}/km^2. Although peakflows generally increased with ASM, the most substantial increases in terms of volume occurred when soil moisture contents approached field capacity.

Simple linear correlation coefficients between ASM and the various stormflow parameters ranged from 0.687 for delayed flow to 0.976 for total storm runoff. Linear correlation coefficients between ASM and quickflow and maximum peakflow were 0.975 and 0.903, respectively. All of the correlation coefficients, with the exception of delayed flow, were significant at the one percent level of confidence.

A good linear relationship existed between ASM and quickflow and total storm runoff, whereas no relationship was found between ASM and delayed flow. Almost 95 percent of the variation in quickflow and total storm runoff can be explained by the simple linear relationship $Y = b_o + b_1X$, where Y = quickflow or total storm runoff and X = ASM. These relationships apply only to the range of ASM conditions included in this study. However, it should be noted that the increase in ASM within this range resulted in a 74 percent increase in total storm runoff from a 38.1 mm rainfall.

Despite the good linear association between maximum peakflows and ASM, the mathematical function that best described this relationship was curvilinear. This curvilinear relationship is represented by the model $\ln Y = b_o + b_1X$, where Y = maximum peakflow and X = ASM. Using this model, approximately

89 percent of the variation in peak discharge could be explained on the basis of the soil moisture content prior to rainfall. A simple linear relationship could explain approximately 81 percent.

Stormflow timing. Time to maximum peakflow, time of recession and duration of quickflow are presented in Table 2. Maximum peakflows for storms applied at moisture contents of 304.8 mm or less occurred several minutes after cessation of rainfall. At ASM contents greater than 304.8 mm, maximum storm discharge occurred several hours after cessation of rainfall. This shift in the time of occurrence of peak discharge was caused by a change in the source-area controlling maximum peakflow as well as a decrease in the time for subsurface stormflow to reach the stream channel. Maximum peakflows at the lower ASM contents were regulated for the most part by the location and extent of the saturated soil zone along both the perennial and intermittent stream channel. This in turn determined the amount of direct channel interception of rainfall and consequently the magnitude and timing of the maximum peak. At higher ASM contents, the major source of stormflow contributing to peakflows was from subsurface storm water from the side slopes.

The duration of quickflow from the 8/1/74 simulated rainstorm lasted 25.43 hours. On 8/7/74 the duration of quickflow was 39.08 and on 8/27/74 storm runoff continued for 50.72

Table 2. Stormflow timing parameters for eight 38.1-mm simulated rainfall events at different antecedent soil moisture contents

Date	Antecedent Soil Moisture	Time to Maximum Peakflow	Time of Recession	Duration of Quickflow
	(mm)	-----------------	hrs.	-----------------
8/01/74	278.1	6.38	19.05	25.43
8/07/74	287.5	6.45	32.63	39.08
8/14/74	290.1	6.35	36.13	42.48
8/19/74	295.1	6.37	37.42	43.79
8/23/74	304.8	9.87	37.58	47.45
8/27/74	310.4	9.12	41.60	50.72
9/14/74	313.2	8.82	40.38	49.20
9/19/74	316.2	8.93	40.35	49.28

hours. Thereafter, the duration of quickflow decreased to 49.20 and 49.28 hours for the 9/16/74 and 9/19/74 storms, respectively.

The time of recession of the storm applied at the lowest ASM was only 19.05 hours. At the next lowest ASM, storm recession lasted 32.63 hours. Gradual increases in the length of storm recession occurred for each of the next four storms. The longest time of recession followed the 8/27/74 storm and was 41.60 hours. Thereafter, the duration of storm recession decreased to 40.38 and 40.35 hours for the 9/16/74 and 9/19/74 storms, respectively.

The linear correlation coefficient between ASM and time to maximum peakflow was 0.848. Although this correlation coefficient, which is significant at the one percent confidence level, indicated that ASM could explain 72 percent of the variation in the time to peakflow using a simple linear relationship, such a relationship is unrealistic in that it assumes a gradual transition over the range of ASM studied. This gradual transition was not supported by the data. In addition, no curvilinear relationship was evident between ASM and time to peakflow.

Despite the relatively good linear associations between ASM and time of recession and duration of quickflow ($r = 0.852$ and 0.905, respectively), a curvilinear function best described these relationships. This curvilinear function, a second-degree parabola, is represented by the model $Y = b_0 + b_1X + b_2X^2$, where Y = time of recession or duration of quickflow and X = ASM. Using this model, 97 and 90 percent of the variation in the duration of quickflow and time of recession, respectively, could be explained on the basis of variations in ASM conditions. These relationships were significantly better than straight-line linear relationships.

It should be emphasized that these curvilinear relationships apply only to the range of ASM studied. Both the duration of quickflow and the time of recession would become fairly constant as ASM approached field capacity as indicated by the data. This would not occur if the models are extended to or beyond field capacity.

Summary and Conclusions

Quickflow volume, total storm runoff, and maximum peakflow generally increased with increasing antecedent soil moisture, while delayed flow and time to maximum peakflow were, for the most part, unaffected. Time of recession and duration of quickflow progressively increased throughout the lower antecedent moisture contents, leveling-off as antecedent soil moisture approached field capacity. The changes in the size

98

and shape of the storm hydrographs occurred over a relatively small range in antecedent soil moisture conditions. This range represented a span of only 12 percent of the available moisture below field capacity. Consequently, it appears that specifying moisture contents of 20, 50 and 80 percent of field capacity to represent low, average, and high antecedent moisture conditions of a drainage basin, so common in stormflow modeling, is too insensitive to provide an adequate description of the affects of antecedent soil moisture on storm runoff.

Although significant linear or curvilinear relationships were found between antecedent soil moisture and quickflow volume, total storm runoff, maximum peakflow, time of recession and duration of quickflow, these relationships only apply to the range of antecedent soil moisture conditions included in this study.

Additional information is needed to assess the effects of seasonal changes on the relationships between these stormflow volumes and timing parameters and antecedent soil moisture.

Acknowledgment

Financial support was provided by the United States Department of the Interior, Office of Water Research and Technology; as authorized under the Water Resources Research Act of 1964; the Pennsylvania Science and Engineering Foundation of the Pennsylvania Department of Commerce; the Pennsylvania Department of Environmental Resources; and National Science Foundation Grant GA-22860.

References

[1] Black, P. E., 1970. "Runoff from Watershed Models," Water Resour. Res., 6(2):465-478.

[2] Graveto, V. M. and Eagleson, P. S.; 1970. "Scale Effects in the Physical Modeling of Surface Runoff," Hydrodynamics Laboratory Report No. 120, Mass. Inst. of Tech., Cambridge, Mass., 328 p.

[3] Hewlett, J. D. and Hibbert, A. R., 1967. "Factors Affecting the Response of Small Watersheds to Precipitation in Humid Areas," Internatl. Symposium on Forest Hydrology, pp. 275-290, Pergamon Press, N. Y.

[4] Roberts, M. C. and Klingeman, P. C., 1970. "The Influence of Land Form and Precipitation Parameters on Flood Hydrographs," Jour. of Hydrology, 11(4):393-411.

Commentary on Paper 26

Mosley, M. P. (1979) Streamflow generation in a forested watershed, New Zealand.

One of the better known field sites for field hydrology is the Maimai experimental catchment in New Zealand, due to the work there by (amongst others) Andy Pearce, Alistair McKercher, Paul Mosley, and more recently Jeff McDonnell and Ross Woods. The Maimai catchments are in a steep, wet area (mean annual rainfall 2610 mm; mean annual discharge 1500 mm), with highly permeable forested soils, a narrow riparian area, rapid peaky hydrograph responses and occasional debris flows (the original discharge gauge at the catchment outlet was completely destroyed by a debris flow in May 1988, see McDonnell, 1990b). An early summary of the work, and perceptual model of the response of the catchments, as dominated by subsurface stormflows, was given by Pearce & McKercher (1979). This was extended by Paul Mosley, then working with the Forest Research Institute in Christchurch, who investigated the importance of macropore flows in subsurface flows and stormflow generation based on experiments in one of the eight subcatchments (now known as M8; see also Pearce *et al.*, 1986; Sklash *et al.*, 1986; McDonnell, 1990a,b; McGlynn *et al.*, 2002, 2004; Freer *et al.*, 2004).

Paul Mosley addresses the problem of how the subsurface system can respond fast enough to explain the rapid storm flows in the catchment (stormflow is here defined relative to the "standard" separation line of Hewlett & Hibbert, Paper 9). Flow collected at a number of pits in the catchment respond rapidly to changes in rainfall intensity in a way that, he suggests, cannot be explained by flow through the soil matrix. Field observations suggested that significant subsurface flow might take place in pipes or macropores in the soil, so Mosley designed a series of dye tracing experiments to verify this explanation, looking at the flow, for one day, from 1–2 m upslope of an observation pit, at a number of different sites. Dye injection was also used to measure velocities of local surface runoff in valley bottom saturated areas. The surface velocities were faster than the macropore flow velocities which were 300 times faster than the measured saturated hydraulic conductivity of the soil. Mosley suggests that this means that there is no need to invoke a translatory flow hypothesis (see Paper 6) to explain the fast subsurface flow responses at Maimai. The inference is that the storm hydrograph could still be made up of rainwater flowing directly to stream channels from the soil surface through macropores.

This conclusion was later directly contradicted by Andy Pearce *et al.* (1986) and Mike Sklash *et al.* (1986) from the evidence of analyses of environmental tracers at M8. Their work suggested that pre-event stored water in the catchment dominates the composition of hydrograph peaks (see also Paper 27). They also provide evidence of rapid rises in near surface water tables and conclude: "*Since our isotopic analyses indicate that the major part of the new water contributions to quick flow can be generated by near-stream saturation overland flow, and since there have been only a few visual observations of macropore flow under natural rainfall, there is little need to subscribe to macropore flow or other rapid-flux concepts to explain current storm water contributions to storm runoff in the Maimai catchments*" (p.1282).

This controversy was, to some extent, resolved by an extended perceptual model of how Maimai works by Jeff McDonnell (1990a) building on his PhD work at the catchment (Fig. 26.1), and then by Brian McGlynn *et al.* (2002). They suggest that macropores may be involved in both transmitting event water into the profile, and in the transfers of water downslope, but in each case may result in the displacement of stored water into the collecting stream. A remaining intriguing question is how much of the flow in macropores is itself stored water rather than event water (see, for example, the results of Sklash *et al.* 1996, at Plynlimon, Wales). Some of the local information on hydrological responses was used in calibration of a version of TOPMODEL to M8 by Jim Freer *et al.* (2004). There have also been some interesting experimental studies of the spatial variability of subsurface runoff to the base of a hillslope by Ross Woods (Woods & Rowe, 1996; Woods *et al.*, 1997). McGlynn & McDonnell

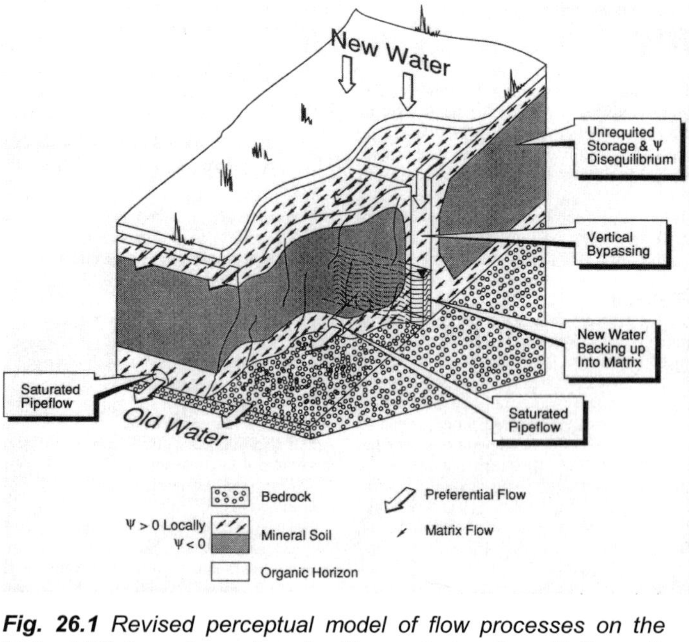

***Fig. 26.1** Revised perceptual model of flow processes on the Maimai hillslopes* (reproduced, with permission, from McDonnell, 1990a, © American Geophysical Union).

(2003) have studied the effect of the riparian zone in the catchment response and McGlynn *et al.* (2004) have looked at the scale dependence of the runoff response within the nested catchment network at Maimai.

Paul Mosley went on to work for the New Zealand Ministry of Works and Development, which was later reorganized into a new division of the Department of Scientific and Industrial Research. In 1992 the National Institute for Water and the Atmosphere (NIWA) was created from different government departments, including DSIR. He now works as a private consultant.

References

Freer, J., McMillan, H., McDonnell, J. J. & Beven, K. J. (2004) Constraining Dynamic TOPMODEL responses for imprecise water table information using fuzzy rule based performance measures. *J. Hydrol.* **291**, 254–277.

McDonnell, J. J. (1990a) A rationale for old water discharge through macropores in a steep humid catchment. *Water Resour. Res.* **26**, 2821–2831.

McDonnell, J. J. (1990b) The influence of macropores on debris flow initiation. *Quart. J. Eng. Geol.* **23**, 325–331.

McGlynn, B L. & McDonnell, J. J. (2003) Quantifying the relative contributions of riparian and hillslope zones to catchment runoff and composition. *Water Resour. Res.* **39**(11), doi: 10.1029/WR002091.

McGlynn, B. L., McDonnell, J. J. & Brammer, D. D. (2002) A review of the evolving perceptual model of hillslope flowpaths at the Maimai catchment. *J. Hydrol.* **257**, 1–26.

McGlynn, B. L., McDonnell, J. J., Seibert, J. & Kendall, C. (2004) Scale effects on headwater catchment runoff timing, flow sources and groundwater-streamflow relations. *Water Resour. Res.* **40**(7), W07504, doi: 10.1029/2003WR002494.

Pearce A. J. & McKercher, A. I. (1979) Upstream generation of storm runoff. In: *Physical Hydrology: New Zealand Experience* (ed. by D. L. Murray & P. Ackroyd), 165–192. NZ Hydrological Society, Wellington, New Zealand.

Pearce, A. J., Stewart, M. K. & Sklash, M. G. (1986) Storm runoff generation in humid headwater catchments. 1. Where does the water come from? *Water Resour. Res.* **22**(8), 1263–1272.

Sklash, M. G., Stewart, M. K. & Pearce, A. J. (1986) Storm runoff generation in humid headwater catchments. 2. A case study of hillslope and low-order stream response. *Water Resour. Res.* **22**(8), 1273–1282.

Sklash, M. G., Beven, K. J., Gilman, K. & Darling, W. G. (1996) Isotope studies of pipeflow at Plynlimon, Wales, UK. *Hydrol. Processes* **10**, 921–944.

Woods, R. A. & Rowe, L. K. (1996) The changing spatial variability of subsurface flow across a hillslope. *NZ J. Hydrol.* **35**, 51–86.

Woods, R. A., Sivapalan, M. & Robinson, J. S. (1997) Modeling the spatial variability of subsurface runoff using a topographic index. *Water Resour. Res.* **33**, 1061–1073.

VOL. 15, NO. 4 WATER RESOURCES RESEARCH AUGUST 1979

Streamflow Generation in a Forested Watershed, New Zealand

M. PAUL MOSLEY

Forest Research Institute, Christchurch, New Zealand

A 0.3028-ha watershed has been instrumented to monitor streamflow and subsurface flow through the soil mantle at a variety of topographic locations. The watershed is forested, with steep (35°) slopes and shallow (average 55 cm) soils on impermeable Old Man gravels. Data for a number of storms indicate that subsurface flow via 'macropores' (root channels, pipes) and seepage zones in the soil is the predominant mechanism of channel stormflow generation in storms with quickflows greater than about 1 mm. Subsurface flow from all parts of the watershed appears to contribute to stormflow even in very small storms (quickflow of the order of 3% of net precipitation). The saturated hydraulic conductivity of the soil matrix is not a limiting factor on the ability of subsurface flow to generate channel stormflow, because dye tracer experiments demonstrate that water may move through macropores (particularly root channels) at rates 2 orders of magnitude greater. However, subsurface flow from lower slope areas contributes to delayed flow; cessation of subsurface flow and streamflow after a drought period is roughly coincident in time. In the study area it appears that streamflow is at almost all times dominated by subsurface flow and that runoff from partial and variable source areas contributes significant quantities of streamflow only during the rising limb of small (less than 1 mm of quickflow) flood hydrographs.

INTRODUCTION

In recent years, considerable advances have been made in our understanding of the generation of streamflow and storm runoff, and a number of models are now available to describe mechanisms whereby precipitation appears as streamflow. Briefly, these mechanisms are (1) Hortonian overland flow, where rainfall rate exceeds the infiltration rate of the soil and the precipitation excess flows over the ground surface; (2) saturation overland flow from 'partial source areas,' in which high water tables cause saturation of small areas of a watershed, often near stream channels, and generate overland flow; (3) subsurface flow of infiltrated water moving laterally through the soil mantle toward the channel system; and (4) expansion of the channel system during storms to tap subsurface flow systems and permit overland flow from 'variable source areas.' *Dunne* [1978] and *Freeze* [1974] have provided thorough reviews of these mechanisms.

Much effort is currently going to define the situations in which each mechanism is important. It is now widely accepted that Hortonian overland flow is important primarily on unvegetated surfaces (desert environments, arable farmland, urban areas) where infiltration rates are low. Present thinking seems to emphasize the importance of stormflow generation on partial and variable source areas and to give subsurface stormflow a secondary role. *Freeze* [1974, p. 632] doubted that subsurface flow could deliver sufficient water to provide significant contributions to stormflow, citing the results of *Hewlett and Hibbert* [1967], *Weyman* [1970], and *Whipkey* [1965], which are frequently taken to support its importance. He concluded that 'only on convex hillslopes that feed deeply incised channels, and then only when the saturated hydraulic conductivities of the soil are very large, is subsurface storm flow a feasible mechanism.' More recent work by *Corbett et al.* [1975] and *Pilgrim et al.* [1978] does provide evidence for subsurface stormflow under less restrictive conditions, however.

As the partial and variable source area concepts become more widely accepted, water resource management decisions are increasingly being made in the light of their implications [*Dunne et al.*, 1975; *Engman*, 1974; *Hayward*, 1977; *Pearce and McKerchar*, 1979]. It is therefore necessary to define very closely the environments in which they are dominant and to test the predictions of theoretical studies such as those of *Freeze* [1972] against field data. The present study has monitored streamflow generation in a forested watershed to provide additional information on the relative importance of the stormflow-generating mechanisms.

STUDY AREA

Description

The study was located in one of eight experimental watersheds in Tawhai State Forest, near Reefton, New Zealand (Figure 1). The watersheds are underlain by moderately weathered, firmly compacted lower Pleistocene Old Man gravels, which are effectively impermeable. Soils are broadly classified as Blackball hill soils, shallow, strongly podzolized yellow-brown earths with a well-developed upper organic humus mantle (Figure 2). *Webster* [1977] found a substantial variation in soil depths and characteristics over very short distances (cf. Figure 3). The forest vegetation consists of undisturbed mixed beech-podocarp-hardwood forest [*Webster*, 1977] with a dense ground cover of ferns and small shrubs. Mean annual rainfall for the period 1963–1975 was 2610 mm [*Pearce et al.*, 1977], and annual runoff approximately 1500 mm.

Instrumentation

Gross precipitation is measured by a Lambrecht recording rain gauge (time resolution of 240 mm/d) just outside the main watershed (Figure 1). Throughfall through the forest canopy (net precipitation) is caught by 90 m of 10-cm-wide plastic guttering, leading to interconnected drums in which the water level is monitored by a Belfort FW1 water level recorder (time resolution of 45.5 mm/d) [*Rowe*, 1976].

Runoff from the main watershed, which has an area of 3.84 ha, is monitored by a 90° V-notch weir equipped with a Belfort FW1 water level recorder.

The hydrologic response of an 0.3028-ha subwatershed is monitored intensively by a variety of equipment (Figure 3). Streamflow in the permanent (but ephemeral) stream channel is monitored at sites C and D by 11° and 11°/90° compound V-notch weirs, respectively, each fitted with a Foxboro type

Paper number 9W0183.
0043-1397/79/009W-0183$01.00

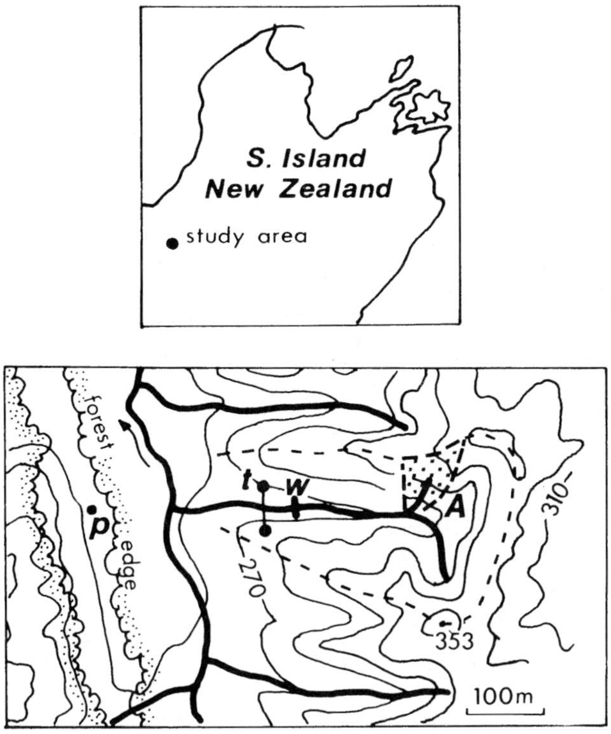

Fig. 1. Location of study area: p, rain gauge; t, throughfall measurement site; w, main weir; and A, experimental watershed (Figure 3). The contour interval is 20 m.

12R water level recorder (time resolution of 69.3 mm/d at zero flow).

Subsurface flow through the soil mantle is monitored at pits 1–6 and site A, which represent a wide range of topographic locations within the subwatershed. At each location a pit 2–3 m long was dug down through the soil into the underlying consolidated gravels, and a trough 1 m long (3.3 m at site A) was constructed at the base of the pit to intercept subsurface flow. By extending the pits beyond the collection troughs, convergence of flow lines toward the pits was minimized. Intercepted water is led from each trough to a tipping bucket; tips are recorded by an Esterline Angus eight-channel event recorder, running at 101.6 mm/hr. Discharge at a seepage site in the lower part of the watershed is also monitored by tipping bucket, and discharge from the seepage face at the head of the permanent channel by a 5½° V-notch weir equipped with a Foxboro 12R recorder.

At both seepage sites the collection troughs were placed at the base of preexisting vertical faces 1 m high cut into the Old Man gravels; seepage from within the soil mantle runs and drips down the face to the trough.

Data analysis

Because of equipment problems, particularly chart disengagement on the event recorder, completely reliable data for only 12 storms are presented (Table 1); fortunately, these cover a wide range of conditions. For each storm, total precipitation, net precipitation, total runoff at the main weir, and antecedent precipitation index API, defined by API = $\sum_{i=1}^{30} P_i/i$, where P_i is total gross precipitation on the ith day beforehand, were determined. For all sites, lag time T_{lag} and $T_{lag'}$, rise time T_{rise}, time to peak T_{peak}, and time to start of runoff T_q were computed (Figure 4 and Table 2). In addition, quickflow (as defined by *Hewlett and Hibbert* [1967, p. 280]) was computed for those sites (main weir, pit 1, site A, and site B) whose catchment areas could be confidently determined (Table 1). The separation line slope used by Hewlett and Hibbert (0.0055 l s^{-1} ha^{-1} hr^{-1}) was adopted for purposes of standardization and because no information is available to warrant adoption of a different value. Total runoff was taken to include quickflow and delayed flow only up to the time that quickflow ends.

It should be noted at this point that the time that runoff commences at the sites monitored by tipping buckets is not measured entirely accurately because of the time taken to fill the buckets and cause the first tip. At the very low discharge rates at the beginning of a storm, this time period may be quite long and lead to a significant overestimate of the time to start of runoff.

WATERSHED RESPONSE DURING A TYPICAL STORM EVENT

Subsurface Flow and Streamflow Hydrographs

At this stage the hydrologic response of the study watershed may be most clearly demonstrated by presenting data for a representative storm event. The storm of July 4–6, 1978, followed a prolonged dry period (API = 4.47 mm), and flow had ceased at all points within the catchment. The response to 90.6

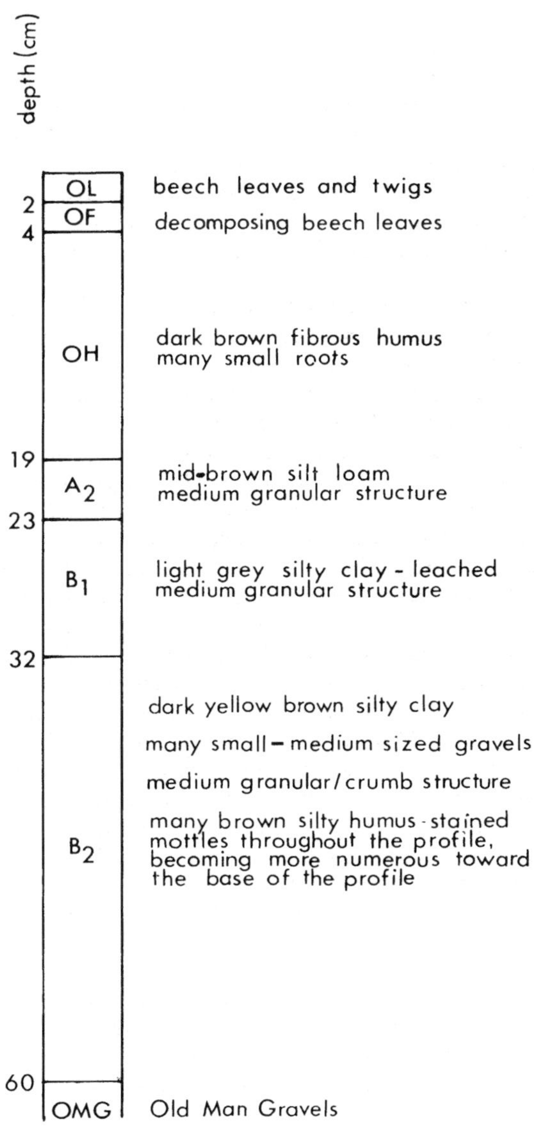

Fig. 2. Representative soil profile (sampling site 7, Table 3).

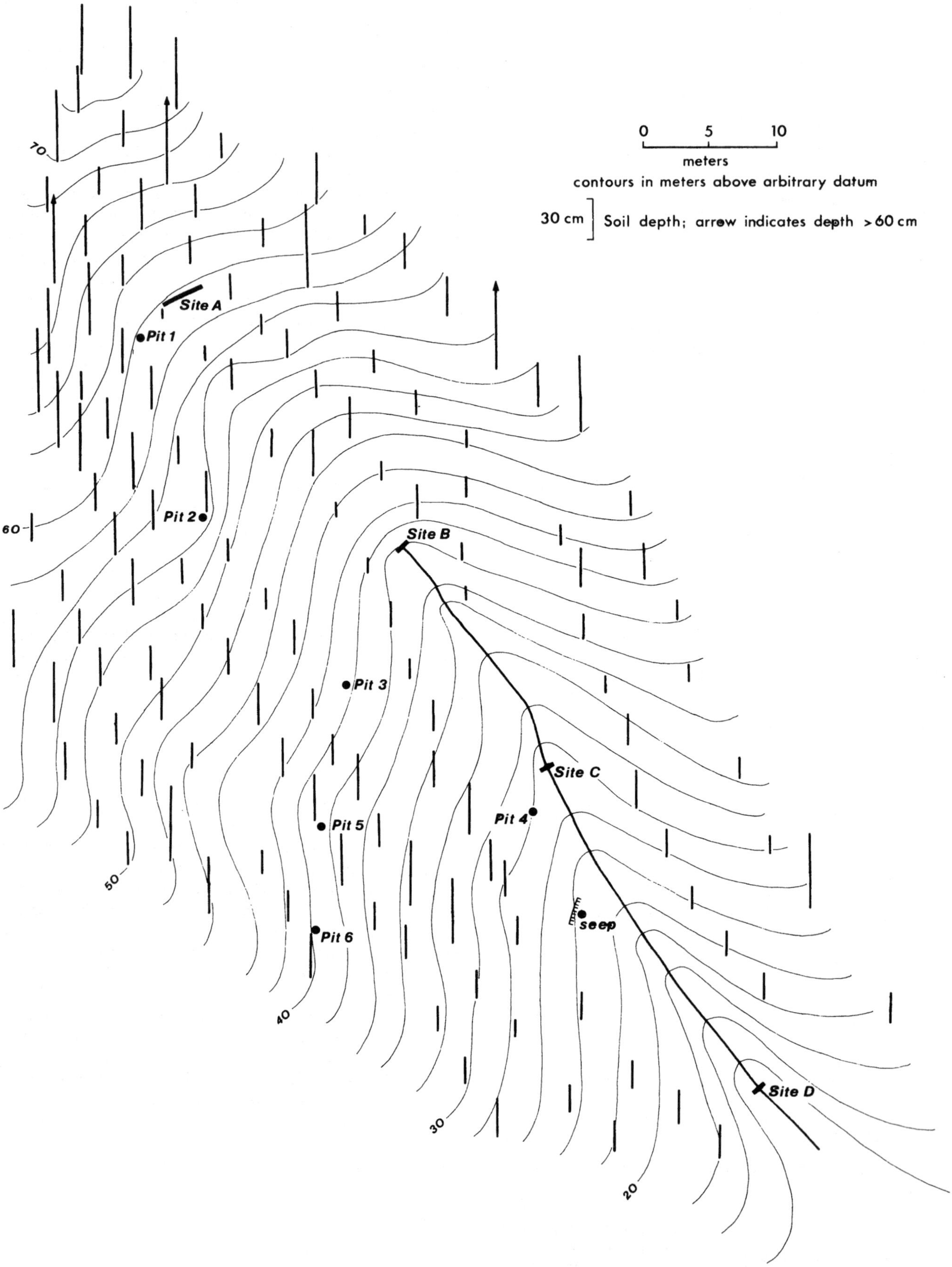

Fig. 3. Topographic map of experimental watershed; soil depths are shown (topographic survey by A. Watson and regolith survey by J. Novis).

TABLE 1. Throughflow and Streamflow During Selected Storms

Storm	Starting Date, 1978	Antecedent Precipitation Index	Total Precipitation	Net Precipitation	Total Runoff, Main Weir	Quickflow, mm				Peak Discharge, l/min												Quickflow Net Precipitation, %
						Main Weir	Pit 1	Site A	Site B	Pit 1	Pit 2	Pit 3	Pit 4	Pit 5	Pit 6	Seep	Site A	Site B	Site C	Site D	Main Weir	
1	May 6	5.58	2.6	0.40	0.07*	...†	0	0	0	0	0	0	0	0	0	0	0	0	>0.0	0.60	2.94	<<1.0
2	April 26	10.57	2.3	0.93	1.59	0.27	0	0	0.17	0	>0	0	0.013	0	0	0.096	0	0	1.44	1.44	10.2	29.0
3	April 29	6.55	7.8	5.1	0.25	0.05	0.07	0	0.01	0.0033	0	0.008	0.045	0	0	0.29	0	0.54	3.77	3.06	22.8	1.0
4	Aug. 23	26.55	7.7	5.3	1.05	0.35	0.32	0.11	0.14	0.01	0	0.008	0.044	0.031		0.47	0.125	2.40	9.00	16.2	77.4	6.6
5	April 30	11.24	8.9	9.1	1.75	1.31	1.4	0.1	1.3	0.067	0.20	0.13	0.24	0.041	0.005	1.47	0.042	2.34	16.8	30.0	200	14.4
6	July 27	41.40	18.4	14.2	9.6	5.3			7.7	0.13	0.12	0.13	0.21	0.88	0.047	1.37	0.85	7.20	16.8	33.6	332	37.5
7	May 31	12.54	23.6	17.3	2.5	1.9	2.6	0.2	2.2	0.049	0.05	0.056	0.17	0.017	>0	1.57	0.054	2.70	16.8	42.6	163	3.3
8	Aug. 24	30.00	33.5	21.4	14.7	9.8	17.1	9.7	10.7	0.17	0.16	0.12	0.24	0.75		1.59	0.90	10.2	29.4	48.6	422	45.9
										0.15	0.10	0.13	0.11	1.03		2.25	1.13	10.8	29.4	49.8	413	
9	Aug. 3	19.32	45.9	37.5	23.6	19.3	22.8	13.0	25.8	0.54	0.50	0.77	0.64	2.50		4.25	5.56	38.4	80.4	169	431	51.5
10	July 23	15.08	90.0	78.5	62.1	56.8	64.0	60.2	73.1	0.65	0.71	1.03	1.26		0.24	7.50	7.21	55.8	112	140	1830	7.20
11	Aug. 10	17.47	82.4	79.0	58.0	50.8	58.4	57.2	68.1	0.88	0.87	1.61	4.95	8.25		17.5	17.2	111.0	223	810	4040	64.3
12	July 4	4.47	90.6	80.6	46.6	43.1	45.5	45.8	59.8	1.82	2.18	2.99	7.85	19.7	1.44	23.9	34.4	115.2	257	468	4180	54.0
13	June 6	12.83	104.9	98.3	79.9	73.3				0.88	0.99	1.93	4.88	9.0	0.67	16.5	14.4	132.0	222	504	3390	74.6

Storm 8 is a double-peaked storm. Pit 6 was replaced by a time check device in July. Where there is no entry, there are no data.
*Over succeeding 24 hours.
†Slope of separation line exceeds slope of rising limb.

mm of gross precipitation (80.6 mm of net precipitation) at all the measurement sites is shown in Figures 5–8 (note the varying discharge scales). The most striking feature of the hydrographs is the close coincidence of peaks. Lag time varied between 1.0 and 1.9 hours (Table 2), with flow in pit 4, which is located in a swale, and from the seep peaking approximately 0.5 hours later than at the other sites. Although contributing areas for all sites cannot be precisely defined because of soil depth variability and contour curvature, there is a very clear increase in peak discharge and total flow volumes with distance from the watershed boundary. This indicates that flow is moving considerable distances through the soil (overland flow was observed only beside the channel between sites C and D) during the hydrograph rise. If flow at each site was derived only from the immediate locality, peak discharge and total flow volume would be independent of distance from the divide.

The subsurface flow in the subwatershed can account for the peak discharge of 69.7 l/s measured at the main weir. Within the subwatershed the peak discharge of 7.8 l/s at D can be accounted for by that at the seep (0.398 l/s), at site B (1.92 l/s), and at pit 4 (0.131 l/s per meter of valley wall). It is not possible to define precisely the contributing area at pit 4, but the pit is 44 m from the watershed divide. The mean distance from the right bank of the stream to the divide is 24 m, and from the left bank 14 m. Adjusting the pit 4 peak discharge for slope length, which is assumed to be roughly proportional to contributing area, the side slopes between B and D (48 m) would contribute 5.43 l/s, giving a total discharge at site D of 7.75 l/s, which is remarkably close to the actual measured peak discharge. The peak discharge at D (25.8 l/s/ha over the subwatershed area of 0.3028 ha) would give a peak discharge at the main weir of 99 l/s if it were supplied by the whole watershed, well in excess of the measured peak. The reasons that specific subwatershed discharge exceeds that for the whole watershed are discussed below.

Runoff Amounts

That subsurface flow can account for total streamflow throughout the storm hydrograph may be verified by computing cumulative runoff, in millimeters of water depth, for those sites for which contributing areas can be estimated (Figure 9). The cumulative runoff curves for the main weir, pit 1, and site A are quite similar; pit 1 in fact generated more runoff than the main weir during much of the storm. Site B generated substantially more runoff than the main weir during the recessions of the two hydrograph peaks, but not during the rising limb, while site D generated more runoff throughout the storm, probably because of runoff from the wet gully bottom (partial area contribution). Flow is sustained at these two locations in the subwatershed for longer and at higher levels than in the watershed as a whole; however, the close correspondence between the cumulative runoff curves for the main weir and pit 1 and site A suggest that the hydrologic performance of the whole watershed may be controlled to a large extent by subsurface flow. The sustained flow at sites B and D is presumably a function of their location in topographic hollows, which constitutes a relatively small proportion of the whole watershed.

Soil Moisture

Overland flow was not observed during the storm except in limited areas along the stream channel, so that subsurface stormflow was largely responsible for generation of the flow hydrograph. To monitor the hydrologic behavior of the soil

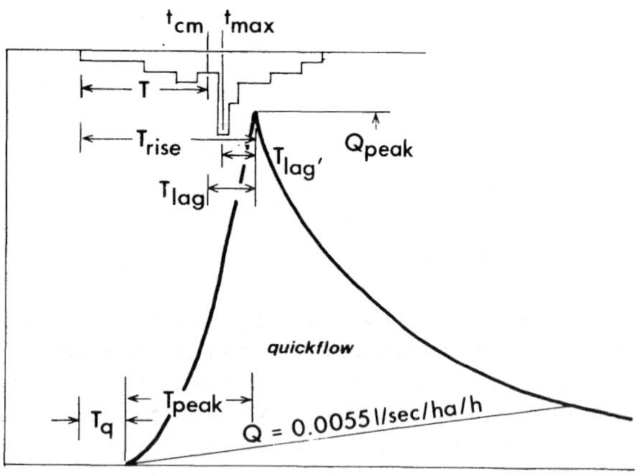

Fig. 4. Definition sketch showing hydrograph parameters.

mantle, a rectilinear slope profile stretching from valley bottom to ridge top was selected beside the throughfall measurement site, on the right bank of the main stream (t on Figure 1). A 60-cm auger was used to take soil samples in 20-cm increments from the top 60 cm of the soil at nine sites up the profile (Table 3) and at five points in time during the storm. Water contents were measured gravimetrically; because soil depths greater than 60 cm could not be sampled, total water content of the top 60 cm is presented in Figure 10, rather than percentages of the total water-holding capacity. *Webster* [1977] measured, using a laboratory permeameter, saturated water contents by volume of the mineral soil in the study area between 48% and 69% (mean, 57.6%; standard deviation s = 6.4%) and on the selected profile between 46% (28 cm of water in 60 cm of soil) at ridge top and over 70% (42 cm of water in 60 cm of soil) on the lower slopes.

Sites 1, 2, and 4 rapidly attained saturation throughout their depths, although overland flow was never observed. Free water rapidly appeared in the auger holes during sampling. Sites 1 and 2 are lower slope sites, adjacent to the valley floor, and site 4 is immediately upslope of an area of unusually thin soil (cf. site 3, Table 3). The midslope sites 3, 5, and 6 became saturated in the 20- to 40- and 40- to 60-cm depth increments; again, free water rapidly appeared in the auger holes. The upper slope sites 7, 8, and 9 approached saturation only in the 40- to 60-cm depth increment.

Samples were unfortunately not obtained before the start of the storm, but J. Webster (personal communication, 1978) considers that soil water content in the study area rarely declines below 10% of saturated water content. There was a slight tendency for water contents to decline during the hydrograph recession period; this was more marked for the upper slope sites. Broadly, the data suggest that there was rapid movement of water vertically in the soil profile and in the downslope direction, such that a saturated wedge, whose shape was modified by the shape of the bedrock surface, built up in the soil mantle. The wedge almost intersected the ground surface at the foot of the slope and above the zone of thin soil at site 4 and tapered off in the upslope direction, becoming restricted to only a thin layer above the bedrock surface at the top of the slope.

TABLE 2. Hydrograph Parameters for Selected Storms

	Pit							Site				Main Weir
	1	2	3	4	5	6	Seep	A	B	C	D	
	April 29, 1978; Qf = 1.0%; T = 6.1											
T_{lag}	4.6		4.6	3.6			3.6		2.4	2.4	1.6	3.1
$T_{lag'}$	4.2		4.2	3.2			3.2		2.0	2.0	1.2	2.7
T_{rise}	10.7		10.7	9.7			9.7		8.5	8.5	7.7	9.2
T_{peak}	1.5		0.6	5.4			5.4		1.5	1.5	1.7	3.0
T_{start}	9.2		10.1	4.3			4.3		7.0	7.0	6.0	6.2
	April 30, 1978; Qf = 14.4; T = 1.5											
T_{lag}	2.9	3.5	2.9	2.9	13.0	2.9	2.9	3.2	3.8	1.5	1.8	1.8
$T_{lag'}$	2.9	3.5	2.9	2.9	13.0	2.9	2.9	3.2	3.8	1.5	1.8	1.8
T_{rise}	4.4	5.0	4.4	4.4	14.5	4.4	4.4	4.7	5.3	3.0	3.3	3.3
T_{peak}	1.4	0.8	0.9	3.4	5.8	1.0	3.4	0.4	4.3	1.7	2.1	2.3
T_{start}	3.0	4.2	3.5	1.0	8.7	3.4	1.0	4.3	1.0	1.3	1.2	1.1
	July 27, 1978; Qf = 37.4%; T = 2.9											
T_{lag}	5.8	5.8	5.7	5.8	7.2	6.9	5.4	6.9	5.1	5.1	4.6	4.6
$T_{lag'}$	6.4	6.4	6.3	6.4	7.8	7.5	6.0	7.5	5.7	5.7	5.2	5.2
T_{rise}	8.7	8.7	8.6	8.7	10.1	9.8	8.3	9.8	8.0	8.0	7.5	7.5
T_{peak}	7.7	5.9	6.1	8.1	6.8	7.4	7.5	6.4	7.5	7.7	7.4	8.0
T_{start}	1.0	2.8	2.5	0.6	3.3	2.4	0.8	3.4	0.5	0.3	0.1	0
	July 4, 1978; Qf = 54.0%; T = 7.2											
T_{lag}	1.3	1.3	1.0	1.9	1.6	1.0	1.9	1.3	1.5	1.3	1.5	1.3
$T_{lag'}$	2.0	2.0	1.7	2.6	2.3	1.7	2.6	2.0	2.2	2.0	2.2	2.0
T_{rise}	8.5	8.5	8.2	9.1	8.8	8.2	9.1	8.5	8.7	8.5	8.7	8.5
T_{peak}	4.0	3.2	2.9	5.6	3.5	2.5	6.5	3.0	4.5	5.3	6.5	6.3
T_{start}	4.5	5.3	5.3	3.5	5.3	5.7	2.6	5.5	4.2	3.2	2.2	2.2
	June 6, 1978; Qf = 74.6; T = 16.3											
T_{lag}	2.0	3.1	2.5	3.0	2.5	2.5	3.0	2.5	0.2	0.2	0.4	1.7
$T_{lag'}$	2.8	3.9	3.3	3.8	3.3	3.3	3.8	3.3	1.0	1.0	1.2	2.5
T_{rise}	18.3	19.4	18.8	19.3	18.8	18.8	19.3	18.8	16.5	16.5	16.7	18.0
T_{peak}	12.1	9.8	11.9	13.8	5.7	6.1	14.1	7.3	18.0	18.0	18.2	18.7
T_{start}	6.2	9.6	6.9	5.5	13.1	12.7	5.2	11.5	−1.5	−1.5	−1.5	−0.7

Where there is no entry, there is no flow.

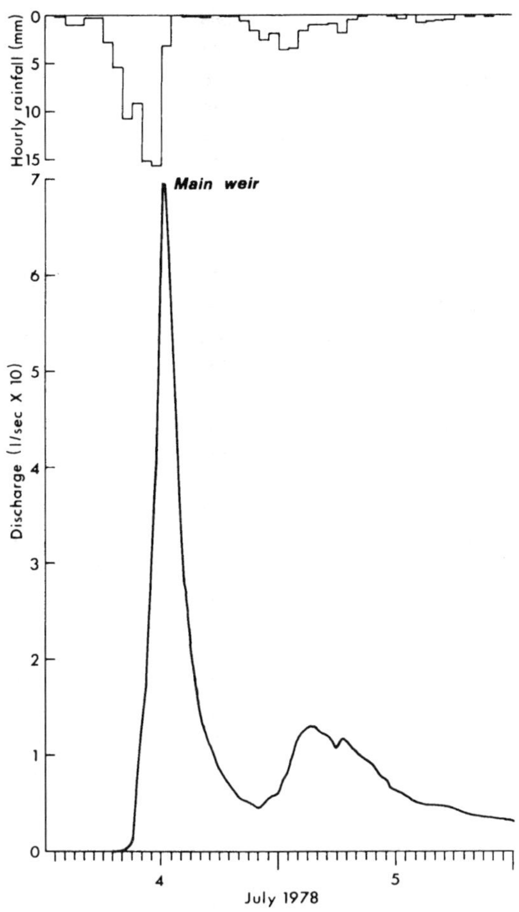

Fig. 5. Streamflow main weir, July 4–5, 1978.

Movement of Water Through the Soil

Figure 11 demonstrates the extreme sensitivity and rapidity with which the subsurface flow measurement sites respond to variations in precipitation intensity; lag times ($T_{lag'}$) are, as in the July 4–6 storm, about 2 hours.

Webster [1977] measured saturated hydraulic conductivities of the organic layer and mineral soil in the study area; mean values of 0.14 cm/s (s = 0.03 cm/s) and 0.007 cm/s (s = 0.001 cm/s), respectively, were obtained. For the representative profile in Figure 2, precipitated water transmitted vertically through the soil matrix at the saturated flow rate would require $(19/0.14) + (41/0.007) = 5993$ s, or 100 min, to reach the bedrock surface. With the wide range in soil depths in the study subwatershed (Figure 3), there would be a wide variation in this time requirement, but water moving at saturated flow rates vertically down through the soil profile then along a seepage zone at the base of the B horizon could cause a rapid response at the subsurface flow measurement sites.

However, because subsurface flow discharges increase in a downslope direction (Figures 6–8), it appears that there is downslope movement of water over distances of up to several tens of meters during the rising limb of the storm hydrograph; this could not be accomplished by saturated flow through the mineral soil. Pits dug in the study watershed for soil profile description and for installation of subsurface flow interception troughs displayed points of concentrated seepage, usually at the base of the B horizon, at which high rates of outflow were observed during storms. At one site on the 28-m contour between pit 5 and the seep (Figure 3), two pipes were discovered at the base of the B horizon through which water gushed at maximum rates of the order of 20 l/s. It is inferred that this water appeared at the seep.

Aubertin [1971], *Jones* [1978], and other workers have demonstrated that water may move through the soil mantle via 'macropores' at rates considerably greater than the saturated hydraulic conductivity of the soil, and a number of trials were carried out in the present study to evaluate the significance of this mechanism. The trials were conducted at the end of a storm that delivered 34 mm of rainfall in 48 hours. Dye was applied to the soil surface through a 15-cm-diameter cylinder; the cylinder was pressed into the humus layer to prevent leakage around the side. A mean depth of dye of 2 cm was maintained during the period of application; the dye soaked rapidly into the humus, so that a positive pressure head existed for only 2–4 min, and then only beneath the cylinder. Conditions were clearly artificial, but except in the immediate vicinity of the cylinder it is considered that movement of the dye was similar to that of water applied by an intense rainfall event.

Trial 1. Six liters of sodium dichromate was applied 1 m upslope from pit 6. The first traces of dye appeared through root holes in the B horizon 5.25 min after start of application. Depth of first appearance was 20.3 cm, well above the base of the B horizon. Dye velocity was 0.38 cm/s. Dye eventually seeped from the base of the B horizon, with a lateral spread across the face of the pit of 90 cm. No seepage from the top of the A horizon was observed. Maximum outflow rate occurred 12.5 min after start of application, at a rate of 0.0019 l/s; at least 3 l of dye flowed from the pit. Dye was still seeping from the pit 5 hours after the start of the trial.

Trial 2. Rhodamine B dye was injected 0.62 m upslope from pit 6, to the right of the trial 1 location. The first traces of dye appeared 1.5 min after first application at 15-cm depth, just below the base of the A_2 horizon; after 2 min a row of dye seeps, presumably at root channels, at 26-cm depth was visible. Maximum velocity was 0.81 cm/s. Dye also appeared at the base of a large root at the bottom of the humus layer; otherwise, no flow at the top of the mineral soil was observed. Dye eventually seeped out at a maximum depth of 35 cm, with a horizontal spread of 1 m.

Trial 3. Six liters of rhodamine B was applied 1 m upslope from pit 5 (depth, 1.6 m). Sodium dichromate was added after 20.5 min, at the same site. Rhodamine B was never observed in the pit, although clear fluid did percolate from a seep at the base of the B horizon. Sodium dichromate became visible from this seep 23 min after the start of its injection and after 30.5 min from smaller seeps at the base of the B horizon. The density of the dye color increased with time, the fluid initially being almost clear, probably because of adsorption of the dye by the soil. Dye appeared nowhere else in the profile. Maximum velocity from point of application to seepage point was 0.17 cm/s.

Trial 4. A new pit 90 cm deep was dug near sampling site 8 on the soil moisture sampling transect. At a point 4 m upslope from the pit applications of water were made as follows.

Elapsed time t, min	Volume, l
0–4	24
25–28	23
45–50	32

The first seep into the pit occurred at t = 54 min 30 s in the A_2 horizon, just below the base of the humus, and a second at t = 55 min 22 s. These two seeps appeared to be old root channels. Smaller seeps started to appear in the A_2 horizon

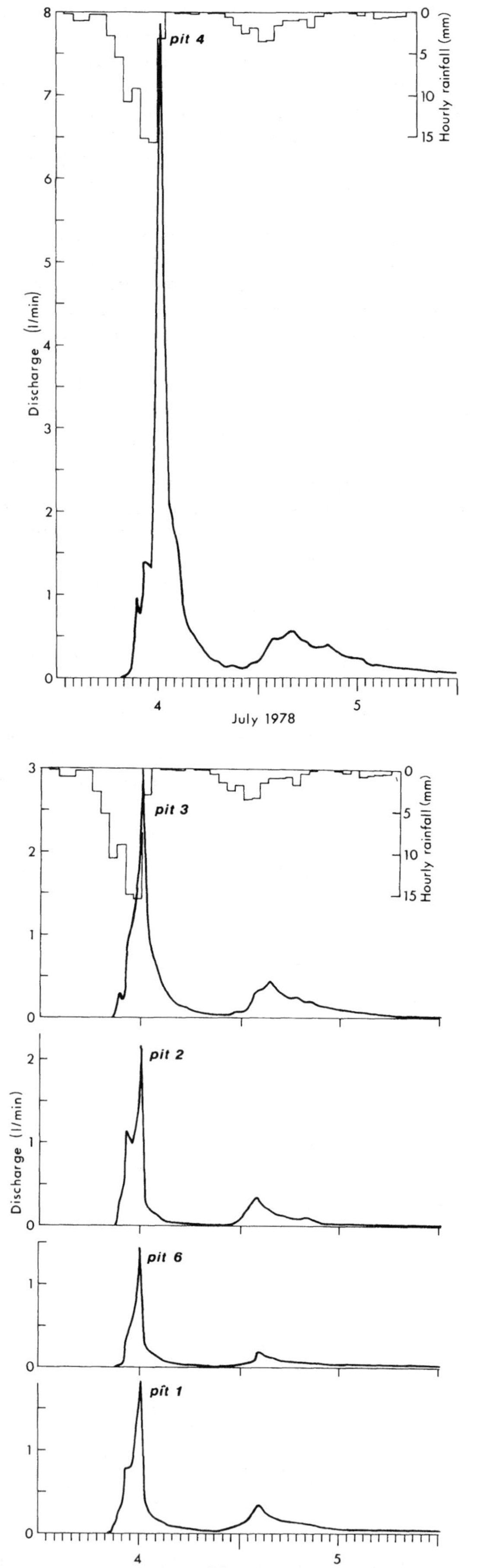

Fig. 6. Subsurface flow at side slope sites, July 4–5, 1978.

after 59 min; lateral spread from the injection point was 1 m. It is inferred that the first two applications of water (equivalent to approximately 15 mm of water over the area between the injection point and the pit) were completely soaked up by the soil and that the seepage observed came from the third application. A velocity of travel of the order of 0.7 cm/s is indicated if this inference is correct. Peak outflow occurred at t = 70 min 15 s.

At t = 105 min to 107 min 30s, 23 l of water were injected at a point 3 m upslope from the pit. A significant increase in outflow from the main seeps was observed at t = 108 min 15 s. A travel velocity of 0.54 cm/s is indicated. Peak flow occurred at t = 111 min.

At t = 145 min to 147 min 30 s, 10 l of water dyed with lissamine green dye was injected 2 m upslope from the pit. First seepage of dyed water was observed at t = 148 min; a travel velocity of 1.11 cm/s is indicated. Peak outflow occurred at t = 150 min. No seepage was observed at any point in the profile other than in the A_2 horizon; the B horizon has no observable root channels, unlike the other sites at which dye-tracing experiments were conducted. However, seepage from the base of the profile, at the soil-bedrock interface, was observed before the trial was started, and this appeared to be delayed seepage from a similar trial carried out in the pit 18 hours previously.

Additional tracer experiments were carried out toward the end of a storm which delivered 82 mm of precipitation in 26 hours. In each case, 1 l of rhodamine B dye was simply sprinkled onto the soil surface 1 m upslope from the pit; the dye soaked immediately into the humus layer.

Trial 5. The trial was conducted at pit 6. Outflow was first observed after 82 s in the B horizon, giving a maximum travel velocity of 1.2 cm/s, in comparison with a maximum velocity of 0.38 cm/s in trial 1 at the same site. Dye soon appeared at the row of seeps at 20- to 25-cm depth (cf. trial 2).

Trial 6. The trial was conducted at the trial 4 site. Outflow was observed after 60 s at the same seepage points observed in trial 4. Maximum travel velocity was 1.7 cm/s, in comparison with 0.54 cm/s in trial 4.

Trials 7 and 8. Similar trials at pit 2 and site A gave maximum travel velocities of 1.4 and 2.1 cm/s; at these sites, dye appeared at the base of the humus layer.

A search was made for locations at which overland flow velocities could also be measured. Four sites, three of which were in swales, were found at which water was moving downslope over the forest floor as unchanneled flow (although there was a tendency for flow to concentrate into an anastomosing system of small rivulets). Travel velocities measured by dye injection averaged 4.25 cm/s, in comparison with the mean of 1.6 cm/s for the four subsurface flow tests.

Maximum dye travel velocities in the subsurface flow trials were up to 300 times greater than the saturated hydraulic conductivity of 0.007 cm/s for mineral soil. At some locations, much of the fluid applied at the surface reappeared downslope at the base of the humus layer; at these sites a paucity of roots and root channels in the A and B horizons was observable. At other sites, much of the fluid reappeared from seeps, usually associated with root channels, or at the base of the B horizon, above the impermeable bedrock surface. At these latter sites, there appeared to be a greater density of roots and root channels in the mineral soil.

The dye-tracing experiments and other storm period observations permit a qualitative discussion of water movement through the soil mantle. Because of the considerable variability in soil and root network characteristics, topography, an-

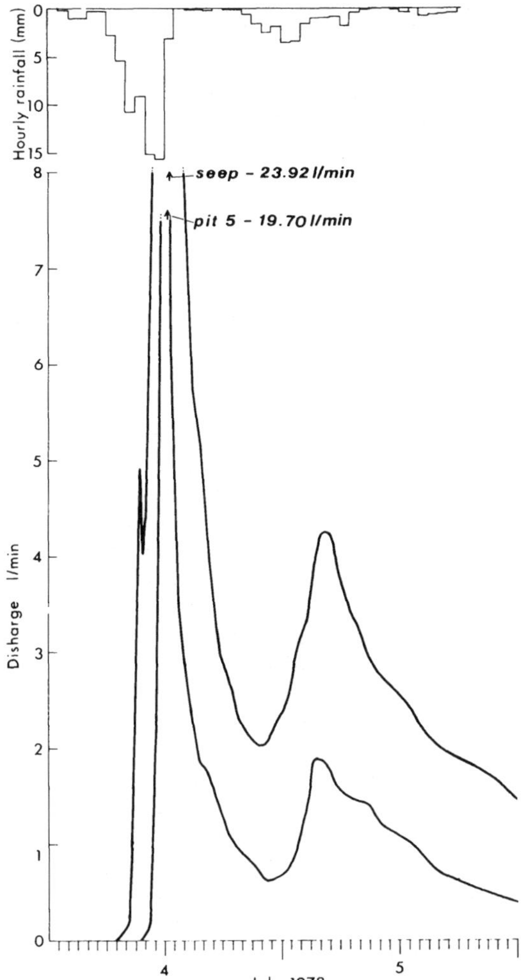

Fig. 7. Subsurface flow at swale and seep sites, July 4–5, 1978.

tecedent moisture conditions, and storm characteristics it is not feasible to apportion quantitatively amounts of water moving along different pathways, although further work is currently in progress to provide more detailed information. Water reaching the surface of the forest floor rapidly enters the organic layer, moving in a predominantly vertical direction through the highly porous humus (values for mean total porosity and macroporosity of 15 samples are 86% and 39% by volume, respectively [*Webster,* 1977]. Macroporosity is defined here as the percentage of the soil volume drained at a tension of 50 cm of water for 24 hours. It is distinct from the term macropore used elsewhere in the text, which refers qualitatively to such features as root channels, animal burrows, pipes, or cracks in the soil through which water may move. These will contribute to the macroporosity as defined above, but only partially.) A variable proportion of this water is held by the organic layer, depending on antecedent moisture conditions, precipitation intensity, and depth of the organic layer; J. Webster (personal communication, 1978) estimates that on average there would be 8 mm of unsatisfied water storage capacity in this layer. The remainder reaches the interface between the organic and mineral soils, at which the saturated hydraulic conductivity and porosity decline significantly.

Although the saturated hydraulic conductivity value of 0.007 cm/s for the mineral soil implies that up to 25 cm/hr of water could pass directly into the mineral soil, a large proportion in fact runs downslope above the surface of the A horizon and within its top 1–2 cm, which is more loosely aggregated and more porous than the rest of the mineral soil. The delayed seepage observed in trials 1 and 4 and apparent in subsurface flow hydrographs indicates that water enters and moves through the mineral soil at rates in line with the saturated hydraulic conductivity, but the rapidity of flow above the interface presumably prevents infiltration of all the water before it reaches the stream channels. A distinct saturated zone 1–2 cm deep in the base of the organic layer was frequently observed during storm conditions, and it is inferred that water moves downslope through this saturated, highly porous layer in a manner intermediate between free surface flow and flow through a porous medium (cf. the relative magnitudes of saturated hydraulic conductivity, subsurface flow velocities, and overland flow velocities).

Where roots, decayed root channels, and other macropores with diameters greater than about 3 mm (at which capillary forces become unimportant) provide pathways across the interface between the organic and mineral horizons, water moves rapidly into the mineral soil under the force of gravity and eventually reaches the interface between the B horizon and the impermeable Old Man gravels. A seepage zone has in most places been observed above this interface, and in some locations, pipes exist. The shallow, indefinite nature of the seepage zone has prevented measurement of characteristics such as porosity and hydraulic conductivity, but it seems probable that eluviation has increased hydraulic conductivities there.

That water may move rapidly along macropores and seepage zones rather than be drawn into the soil matrix is considered to be a function of the soil moisture conditions prevailing in the study area. As has already been noted, soil water contents rarely decline more than 10% below saturation. Hence

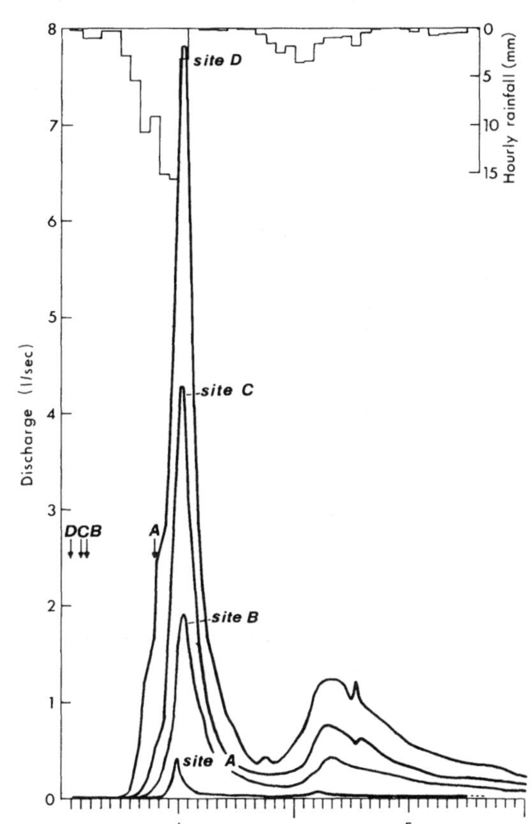

Fig. 8. Subsurface flow and streamflow up axis of experimental watershed, July 4–5, 1978.

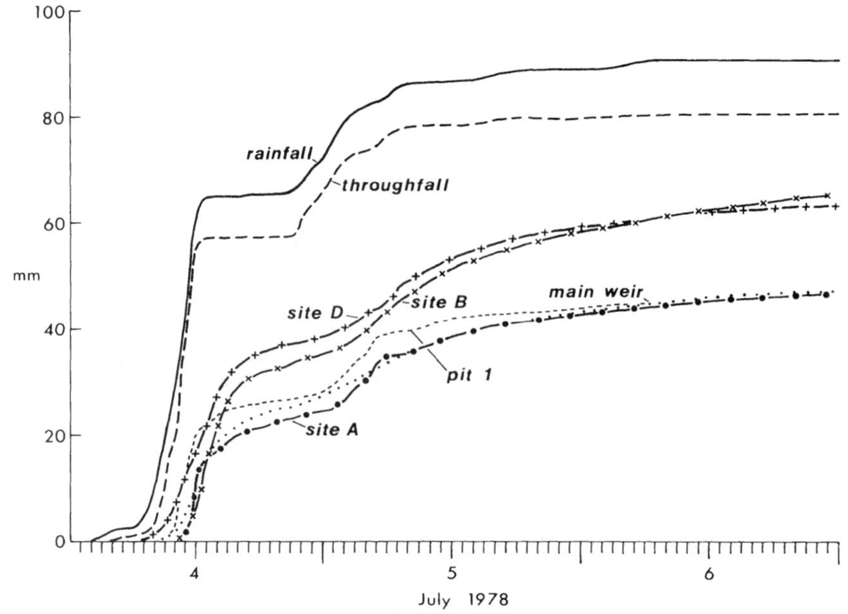

Fig. 9. Cumulative runoff at selected sites, July 4–6, 1978.

the mineral soil, especially below about 20-cm depth, may become saturated at an early stage of a storm and permit flow along macropores without losses to the soil matrix. Even where the soil matrix is unsaturated, water may not be drawn into the matrix rapidly enough to prevent flow reaching the saturated layers or the stream channel.

The processes of water movement through the soil in the study area require much more investigation, but even then they may prove so variable in response to variations in soil characteristics as to prevent precise description. Nevertheless, the observations reported herein demonstrate that subsurface flow through macropores and along discontinuities in the soil profile is capable of contributing to storm period streamflow and suggest that the translatory flow mechanism discussed by *Hewlett and Hibbert* [1967] may not be important in the study area.

CONTRIBUTION OF SUBSURFACE FLOW TO CHANNEL QUICKFLOW

The recent emphasis on the partial and variable source area concepts of streamflow generation has led to the view that there is a close relationship between the percentage of precipitation appearing as quickflow and the percentage of the total catchment area contributing to stormflow. In other words, as the percentage of precipitation appearing as quickflow increases, it is assumed that an increasing proportion of the watershed is directly contributing to storm runoff.

TABLE 3. Soil Moisture Sampling Site Characteristics

Sampling Site	Distance From Base of Slope, m	Depth of Organic Horizon, cm	Depth of Whole Profile, cm
1	0	3.0	>100
2	11.0	5.5	91.5
3	19.5	21.0	45.0
4	30.0	4.5	75.5
5	38.4	12.5	81.5
6	50.3	19.0	60.0
8	66.0	18.0	63.0
9	76.0 (ridge crest)	45.0	59.0

For the period December 1974 to February 1977, *Pearce and McKerchar* [1979] derived figures for precipitation, total runoff, and quickflow in the study watershed:

gross precipitation	5573 mm;
net precipitation	4403 mm;
total runoff	3275 mm;
quickflow	2201 mm.

Quickflow was 50% of net precipitation and 67% of total runoff. However, in 74 individual storms producing at least 0.5 mm of quickflow, quickflow as a proportion of net rainfall ranged from 4.3% to 99%, and Pearce and McKerchar concluded that 'the area of the catchment contributing to stormflow ranges from less than 5% to practically the whole catchment during some large storms with wet antecedent conditions.' They demonstrated that overland flow due to an excess of precipitation rate over infiltration rate or to runoff from permanently saturated wetlands (about 5% of total watershed area) could not alone generate storm runoff and sug-

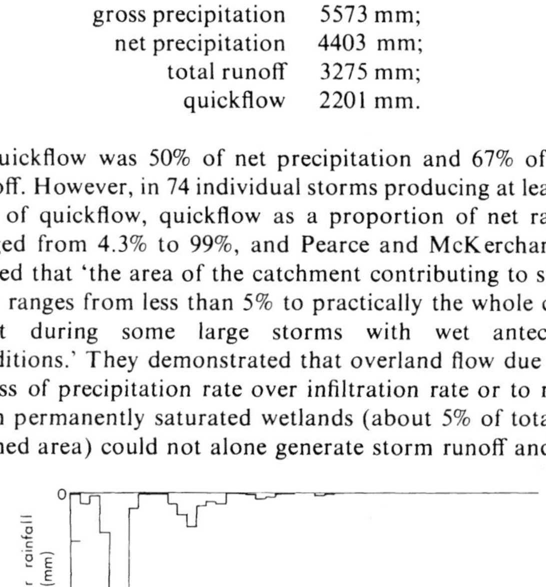

Fig. 10. Water contents of top 60 cm of soil, hillslope transect in experimental watershed, July 4–7, 1978. Note that sample site 3 has a soil depth of only 45 cm.

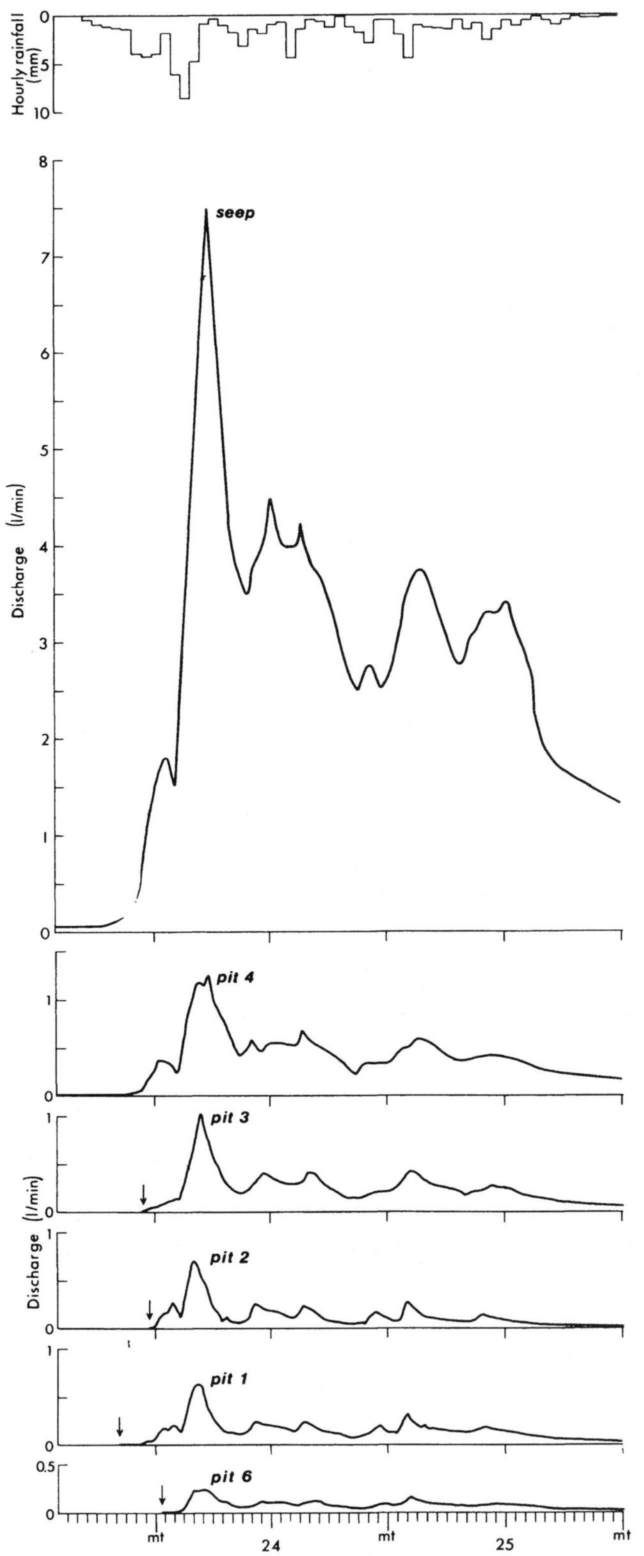

Fig. 11. Subsurface flow at side slope and seep sites, July 23–25, 1978.

gested that in events yielding more than 5 mm of quickflow the bulk of the storm runoff can be accounted for only by high rates of subsurface flow.

The data presented above for the July 4–6 storm (43 mm of quickflow) are in accord with Pearce and McKerchar's conclusions, but information for other storms indicate that they have underestimated the importance of subsurface flow in smaller events.

Thirteen storms have been ranked in Table 1 in terms of net precipitation; data are presented for peak discharges at all study sites and for quickflow at the sites whose contributing areas may be defined. Subsurface flow occurs in very small storm events in quantities sufficient to contribute a significant proportion of the runoff from the main weir. For example, in storm 7, quickflow at the main weir was 1.9 mm (3.3% of the net precipitation of 17.3 mm), at pit 1 it was 2.6 mm, and at site B 2.2 mm. The low figure for quickflow of 0.2 mm at site A is probably because the greater than average soil depths in its contributing area store a larger proportion of net precipitation in the early stages of a precipitation event; site A produces smaller quantities of quickflow than the main weir up to about 10 mm of quickflow. The large difference between the quickflows for pit 1 and site A, which are only 4 m apart, is indicative of the influence of slope shape and soil depth. Pit 1 is on a planar slope with a shallow (20 cm deep) soil, whereas the catchment area of site A is bowl shaped, and the soil at the site is 45 cm deep.

Subsurface flow does not occur throughout the study watershed until quickflow at the main weir exceeds 1.31 mm (storm 5, Table 1; net precipitation, 9.1 mm), but some sites respond when as little as 0.05 mm of quickflow is produced at the main weir (storm 3, Table 1; net precipitation, 5.1 mm). A plot of quickflow at the main weir against quickflow at the subsurface flow sites (Figure 12) indicates that quickflow at these sites is equal to or greater than that at the main weir at quickflows above 1 mm. Even below 1 mm, measurable quickflow at these sites may occur, but it tends to zero below this value, depending on antecedent precipitation.

The regression relationships between peak specific discharges in liters per second per hectare at the main weir and at pit 1 and sites A, B, and D for the storms in Table 1 (zero values omitted) are

$$\begin{aligned} Q_m &= 0.799Q_1^{0.82} & r^2 &= 0.92 \\ Q_m &= 1.98Q_A^{0.54} & r^2 &= 0.83 \\ Q_m &= 0.832Q_B^{0.87} & r^2 &= 0.93 \\ Q_m &= 0.532Q_D^{0.98} & r^2 &= 0.95 \end{aligned} \qquad (1)$$

Equality is attained at

$$\begin{aligned} Q_m &= Q_1 = 0.3 \text{ l/s/ha } (1.15 \text{ l/s}) \\ Q_m &= Q_A = 1.42 \text{ l/s/ha } (5.45 \text{ l/s}) \\ Q_m &= Q_B = 0.25 \text{ l/s/ha } (0.96 \text{ l/s}) \end{aligned} \qquad (2)$$

(where the values in parentheses are those at the main weir) and not at all for site D. Equations (1) and (2) indicate that in storms generating more than about 1 mm of quickflow, peak specific throughflow discharges exceed peak specific discharges in the channel and increase rather more quickly with storm size. This is in line with interpretation of Figure 12. The relationship for site A is strongly influenced by the lack of runoff during small storm events, which is probably caused by loss of precipitation to soil moisture storage. Peak specific discharge in the subwatershed above site D is double that for the main watershed as a whole, irrespective of storm size; precipitation into the channel and runoff from saturated areas along the channel may account for the difference in smaller storms but would rapidly become relatively unimportant with increasing storm size.

The study subwatershed is apparently not entirely representative of the whole watershed above the main weir, in that

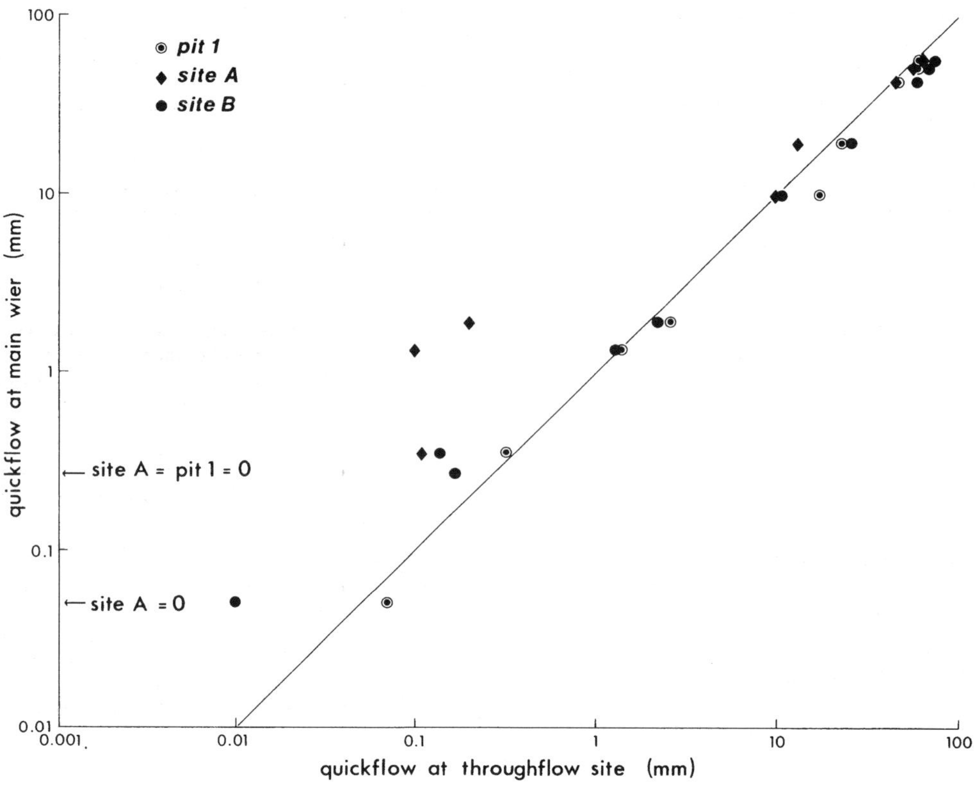

Fig. 12. Relation between quickflow at main weir site and quickflow at selected subsurface flow measurement sites.

subwatershed storm period specific discharges generally exceed those for the whole watershed. The subwatershed is a topographic hollow and hence might be expected to concentrate runoff to a greater extent than the side slopes and spurs that predominate in the main watershed. Furthermore, soils appear to be shallower than average.

Subsurface Flow Contributions to Low Streamflow

Quickflow in the storm of July 4–6, 1978, terminated at 1500 hours, July 6, at a discharge of 1.4 l/s at the main weir. Discharges at the throughflow sites are presented in Table 4; the final bucket tips in pits 2 and 6 were at 1942 and 1509 hours, respectively. Throughflow from site B, the seep, and pit 4 (adjusted as before) can account for discharge at site D, but specific discharge at site D (0.27 l/s/ha) is less than at the main weir (0.36 l/s/ha). This is consistent with the proportionately larger contribution to stormflow by the study subwatershed and suggests that other areas in the main watershed contribute proportionately less to stormflow and more to delayed flow. (Some areas with deep soils and a denser than average vegetation cover will presumably contribute less to both stormflow and delayed flow.)

Other storms amplify the relative importance of subsurface flow with respect to channel quickflow and delayed flow. For example, the storm event of July 4–6, discussed earlier, was preceded by 24 days without rain. The preceding storm, on June 7–9, produced 116.7 mm of precipitation; the precipitation center of mass was at 0400 hours on June 8. Quickflow at the main weir ceased at 1900 hours on June 10, and flow ceased altogether at 1200 hours on July 1. Table 5 shows the dates and times of the last bucket tip at each subsurface flow site; periods of flow may be underestimated because flow insufficient to fill the buckets may have continued. All sites continued to discharge after the end of channel quickflow, but the three upper slope sites did so for considerably shorter periods than the lower slope and swale sites. The time at which flow ceased at the lower slope sites coincides roughly with the

TABLE 4. Subsurface flow Discharges at End of Channel Quickflow Period, 1500 Hours, July 6, 1978

Site	Discharge at Subsurface Flow Site, l/min
Pit 1	0.009
Pit 2	0.001
Pit 3	0.013
Pit 4	0.048
Pit 5	0.189
Pit 6	0.02
Seep	0.79
Site A	0.39
Site B	3.6

TABLE 5. Periods of Flow Following Storm of June 7–9

Site	Last Bucket Tip	Period of Flow From Storm Center of Mass, hours
Pit 1	1409 hours, June 14	154
Pit 2	2315 hours, June 10	67
Pit 3	0157 hours, June 15	166
Pit 4	1749 hours, June 30	542
Pit 5	1425 hours, June 29	514
Pit 6	bucket faulty	
Seep	0214 hours, July 3	594
Site A	1106 hours, June 23	367
Site B	2400 hours, July 1	572

time that flow ceased at the main weir; subsurface seepage past the weir may explain why flow there ceased earlier.

Taken together, the data indicate that not only is subsurface flow important in generating stormflow in the channel system, but it also accounts for base flow, as was suggested by *Hewlett* [1961]. It is inferred that the contribution to stormflow is via macropores in the soil mantle and that slow release by saturated and unsaturated flow of the moisture that enters the soil matrix contributes to base flow discharge.

CONCLUSIONS

It is important to define the situations in which each streamflow-generating mechanism is important, so that implications for management can be confidently drawn. The present detailed study partially confirms the conclusions of *Freeze* [1972, 1974] and *Pearce and McKerchar* [1979] regarding the relative significance of subsurface stormflow as a contributor to streamflow but considerably extends their results.

In the study area it appears that direct runoff from precipitation on partial and variable source areas could be dominant only in those storms generating less than 1 mm of quickflow. In such small storms, however, overland flow is unlikely, and precipitation into the channel would be of more significance. Otherwise, subsurface flow is capable of accounting for the bulk of the channel quickflow. *Freeze* [1972, p. 1282] considered that a threshold value for saturated hydraulic conductivity of the order of 0.002 cm/s is necessary for subsurface stormflow to be significant, but in a soil that contains root channels, pipes, and seepage zones, saturated hydraulic conductivity of the soil matrix is not a limiting factor. Flow of dye tracer through macropores in the soil was observed at rates up to 3 orders of magnitude greater, and the sensitive and rapid response of subsurface flow to variations in precipitation suggests that flow through macropores rather than through the soil matrix contributes to channel stormflow. This flow was of 'new' water, and no evidence for translatory flow was observed. In addition, slow drainage of water from the soil mantle by unsaturated flow (and possibly saturated flow in a shallow layer above the soil-bedrock interface) appears to account for delayed flow throughout the hydrograph recession.

The notion that the proportion of precipitation appearing as quickflow is a good indicator of the total proportion of the watershed contributing to quickflow does not appear to apply to the study area. Subsurface flow was observed virtually throughout the study watershed in storms with as little as 3.3% of net precipitation appearing as quickflow, and subsurface flow from all parts of the subwatershed probably contributes to streamflow in storms with quickflows greater than 1 mm. The channel in the subwatershed is admittedly incised, so that the maximum area of near-channel wetlands is fairly restricted, but this does not fundamentally affect the conclusion that subsurface flow is the dominant mechanism of both stormflow and base flow generation in the study area.

Acknowledgments. Assistance in the field from C. Cooper, P. Hinchey, C. Smith, D. Tindale, and A. Watson, and particularly from D. Crequer and J. Webster, is gratefully acknowledged. C. O'Loughlin and J. Webster reviewed the manuscript, which also benefited considerably from the criticisms of the anonymous reviewers.

REFERENCES

Aubertin, G. M., Nature and extent of macropores in forest soils and their influence on subsurface water movement, *U.S. Forest Serv. Res. Pap., NE-192,* 33 pp, 1971.

Corbett, E. S., W. E. Sopper, H. W. Lull, Watershed response to partial area applications of simulated rainfall, *Int. Ass. Hydrol. Sci. Publ., 117,* 63–71, 1975.

Dunne, T., Integration of hillslope flow processes, in *Hillslope Hydrology,* edited by M. J. Kirkby, chap. 7, John Wiley, New York, 1978.

Dunne, T., T. R. Moore, and C. H. Taylor, Recognition and prediction of runoff-producing zones in humid regions, *Hydrol. Sci. Bull., 20,* 305–327, 1975.

Engman, E. T., Partial area hydrology and its application to water resources, *Water Res. Bull., 10,* 512–521, 1974.

Freeze, R. A., Role of subsurface flow in generating surface runoff, 2, Upstream source areas, *Water Resour. Res., 8,* 1272–1283, 1972.

Freeze, R. A., Streamflow generation, *Rev. Geophys. Space Phys., 12,* 627–647, 1974.

Hayward, J. A., Hydrology of a mountain catchment, and its implications for management, in *Proceedings of the 3rd Soil and Plant Water Symposium,* pp. 126–136, New Zealand Department of Scientific and Industrial Research, Wellington, 1977.

Hewlett, J. D., Soil moisture as a source of baseflow from steep mountain watersheds, *U.S. Forest Serv. Res. Pap., SE132,* 11 pp., 1961.

Hewlett, J. D., and Hibbert, A. R., Factors affecting the response of small watersheds to precipitation in humid areas, in *Forest Hydrology,* edited by W. E. Sopper and H. W. Lull, pp. 275–290, Pergamon, New York, 1967.

Jones, J. A. A., Soil pipe networks: Distribution and discharge, *Cambria, 5,* 1–21, 1978.

Pearce, A. J., and A. I. McKerchar, Upstream generation of storm runoff, in *Physical Hydrology—New Zealand Experience,* edited by D. L. Murray and P. Ackroyd, New Zealand Hydrological Society, Wellington, in press, 1979.

Pearce, A. J., C. L. O'Loughlin, and L. K. Rowe, Hydrologic regime of small, undisturbed beech forest catchments, North Westland, in *Proceedings of the 3rd Soil and Plant Water Symposium,* pp. 150–158, New Zealand Department of Scientific and Industrial Research, Wellington, 1977.

Pilgrim, D. H., D. D. Huff, and T. D. Steele, A field evaluation of subsurface and surface runoff, II, Runoff processes, *J. Hydrol., 38,* 319–342, 1978.

Rowe, L. K., Preliminary results from a rainfall interception study under a podocarp/beech/hardwood forest in North Westland, New Zealand, in *Proceedings of the New Zealand Hydrological Society Symposium,* pp. 151–164, New Zealand Hydrological Society, Wellington, New Zealand, 1976.

Webster, J., The hydrologic properties of the forest floor under beech/podocarp/hardwood forest, North Westland, M.Ag.Sci. thesis, 77 pp., Univ. of Canterbury, Canterbury, New Zealand, 1977.

Weyman, D. R., Throughflow on hillslopes and its relation to the stream hydrograph, *Bull. Int. Ass. Sci. Hydrol., 15,* 25–33, 1970.

Whipkey, R. Z., Subsurface stormflow from forested slopes, *Bull. Int. Ass. Sci. Hydrol., 10,* 74–85, 1965.

(Received October 11, 1978;
revised January 10, 1979;
accepted January 24, 1979.)

Commentary on Paper 27

Sklash, M. G. & Farvolden, R. N. (1979) The role of groundwater in storm runoff.

The study of Mike Sklash and Bob Farvolden of the universities of Windsor and Waterloo in Canada was not the first to use environmental isotopes or other chemical characteristics to separate the hydrograph into different source waters, but it was the first to get widely noticed and to use model results to explore the hypotheses suggested by the tracer measurements. Sklash & Farvolden themselves cite the earlier works of Dincer, Martinec, Fritz, Pinder & Jones and others. Others had used different chemical characteristics (e.g. Pinder & Jones, Paper 14; Pilgrim *et al.*, 1979) or were carrying out experiments with the application of artificial tracers; for example, Pilgrim (1966) and Pilgrim & Huff (1978), worked with radioactive tracers (in the latter study with the field assistance of the 1968 class in Nuclear Hydrology at Stanford University!).

The isotopes of hydrogen and oxygen are ideal water tracers in the sense that they are part of the water molecule. At the time, analyses of isotopic ratios were very expensive, and most of the early studies were based on relatively small numbers of measurements. Sklash & Farvolden also refer to equipment malfunctions reducing the number of storms that could be analysed, a problem that all field hydrologists will be familiar with. Their study, in the Ruisseau des Eaux Volées (Quebec, Canada) and Hillman Creek (Ontario, Canada) catchments, used a two component mixing theory to separate the hydrograph into event and pre-event (or "new" and "old") water sources. This simple form of mixing theory requires the assumption that the event and pre-event water sources are spatially and temporally constant in concentration. In addition, it is necessary to have a good separation between the concentrations of the different sources, something that is helped by a strong seasonal variation in the concentration in precipitation as a result of the temperature dependence of fractionation processes. In general, the separations showed that a major proportion of the fast response of the storm hydrograph in the two basins was made up of pre-event water.

One of the most interesting results in the Hillman Creek study was the grab sampling of overland flow water from three different sites during an April storm. Two of the samples showed isotope characteristics of event water, but the third had concentrations characteristic of groundwater that had returned to the surface. Sklash *et al.* (1996) later showed that very rapid flow responses in ephemeral subsurface pipes could also be dominated by old water. There have now been very many studies that have shown that old water dominates the hydrograph but it should not be assumed that this is always the case. There has to be storage of old water that can contribute to the hydrograph. This will not be the case where there is little storage and strong drying between rainfall events in arid and semiarid areas. Turner *et al.* (1991), for example, showed that for some pasture areas in Australia, event rainfalls made up some 65–75% of the storm hydrograph. There have also been studies to show that in some humid areas, such as the tropical forests of Australia, Puerto Rico and Ecuador, event water may contribute significantly to the hydrograph (Bonell *et al.*, 1998; Schellekens *et al.*, 2004; Goller *et al.*, 2005). It should also be expected that as the storm event gets larger, so will the contribution of event water to the hydrograph.

This study also threw some light on the "groundwater ridging" concept which had previously been proposed by Bob Ragan (Paper 11). The idea that a capillary fringe close to the surface at the base of a slope could, with small amounts of event water percolating down from the surface, be turned into a zone of saturation inducing both more rapid flow rates and steeper hydraulic gradients towards (and also in some circumstances away from) the stream channel, was proposed as one reason why old water could be displaced from storage during the time scale of the hydrograph as suggested by the isotope measurements. Numerical experiments with a two dimensional Darcy-Richards model of partially saturated flow on hypothetical hillslopes of different forms also demonstrated groundwater ridging close to the stream. The rapid response of a near-surface capillary fringe was also later revealed in the studies of Gillham (Paper 31) and Abdul & Gillham (1984), also at the University of Waterloo.

Sklash & Farvolden also give an illustration of the effect that a "vadose water" component, different in isotopic characteristics from either rainfall or groundwater, might have on the two component hydrograph separation (see Fig. 5, p.53). It is worth noting, however, that their method of distinguishing the vadose component is only valid under some special circumstances that do not necessarily hold. They suggest that a simple two-component mixing will result in a straight line on a concentration *versus* discharge plot. This is only true, however, if there is a linear change in the proportion of the two components between the start of the rising limb of the hydrograph (where it is assumed that discharge is 100% pre-event water) and the peak of the hydrograph (where only a proportion of the peak is pre-event water), and also on the falling limb. If there was a constant proportion of event and pre-event water then the resulting concentration would actually be constant through time. If the proportion of event water grows faster than linearly then the concentration *versus* discharge plot would deviate from a straight line in a way similar to that shown by Skalsh & Farvolden as indicating a contribution of vadose water, despite the fact that there is strictly two component mixing. There may, of course, be more than two components in reality. There may be a whole (possibly fractal) spectrum of residence times of water in a catchment (e.g. Kirchner *et al.*, 2001).

The Skalsh & Farvolden study resulted in a wide range of later studies using isotope separations. Mike Sklash later applied the techniques to the Maimai catchment in New Zealand (Sklash *et al.*, 1986; see also Paper 26) and the Plynlimon experimental catchments in Wales (Sklash *et al.*, 1996). DeWalle *et al.* (1988) introduced three component separations, Harris *et al.* (1995) provided a way of handling the variability in the isotopic concentrations of the input rainfall during storm events, while Bazemore *et al.* (1994), Genereux (1998), and Joerin *et al.* (2002) introduced different methods of examining the uncertainty in the hydrograph separations resulting from the uncertainty in characterizing the different water sources. Many of the early studies of tracer-based isotope separation are reviewed or presented in the chapter by Mike Sklash in the Anderson & Burt (1990) *Process Studies in Hillslope Hydrology* volume and Carol Kendall & Jeff McDonnell's (1998) *Isotope Tracers in Catchment Hydrology*.

Bob Farvolden remained at the University of Waterloo as Professor and Dean of Science for many years. He died in 1995. Mike Sklash left the University of Windsor to work as a private consultant in the 1990s.

References

Abdul, A. S. & Gillham, R. W. (1984) Laboratory studies of the effects of the capillary fringe on streamflow generation. *Water Resour. Res.* **20**(6), 691–698.

Anderson, M. G. & Burt, T. P. (eds) (1990) *Process Studies in Hillslope Hydrology*. Wiley, Chichester, UK.

Bazemore, D. E., Eshleman, K. N. & Hollenbeck, K. (1994) The role of soil water in stormflow generation in a forested headwater catchment: synthesis of natural tracer and hydrometric evidence. *J. Hydrol.* **162**, 47–75.

Bonell, M., Barnes, C. J., Grant, C. R., Howard, A. & Burns, J. (1998) Oxygen and hydrogen isotopes in rainfall–runoff studies. In: *Isotope Tracers in Catchment Hydrology* (ed. by C. Kendall & J. J. McDonnell), 347–390. Elsevier, Amsterdam, The Netherlands.

De Walle, D. R., Swistock, B. R. & Sharpe, W. E. (1988) Three-component tracer model for stormflow on a small Appalachian forested catchment. *J. Hydrol.* **104**, 30–40.

Genereux, D. (1998) Quantifying uncertainty in tracer-based hydrograph separations. *Water Resour. Res.* **34**,915–919.

Goller, R., Wilcke, W., Leng, M. J., Tobschall, H. J., Wagner, K., Valarezo, C. & Zech, W. (2005) Tracing water paths through small catchments under a tropical montane rain forest in south Ecuador by an oxygen isotope approach. *J. Hydrol.* **308**, 67–80.

Harris, D. M., McDonnell, J. J. & Rodhe, A. (1995) Hydrograph separation using continuous open system isotope mixing. *Water Resour. Res.* **31**(1), 157–172.

Joerin, C., Beven, K. J., Iorgulescu, I. & Musy, A. (2002) Uncertainty in hydrograph separations based on geochemical mixing models. *J. Hydrol.* **255**, 90–106.

Kendall, C. & McDonnell, J. J. (eds) (1998) *Isotope Tracers in Catchment Hydrology*. Elsevier, Amsterdam, The Netherlands.

Kirchner, J. W., Feng, X. & Neal, C. (2001) Catchment-scale advection and dispersion as a mechanism for fractal scaling in stream tracer concentrations. *J. Hydrol.* **254**, 82–101.

Pilgrim, D. H. (1966) Radioactive tracing of storm runoff on a small catchment. 1. Experimental technique, and 2. Discussion of results. *J. Hydrol.* **4**, 209–326.

Pilgrim, D. H. & Huff, D. D. (1978) A field evaluation of subsurface and surface runoff. 1. Tracer studies. *J. Hydrol.* **38**, 299–318.

Pilgrim, D. H., Huff, D. D. & Steele, T. D. (1979) Use of specific conductance and contact time relations for separating flow components in storm runoff. *Water Resour. Res.* **15**(2), 329–339.

Schellekens, J., Scatena, F. N., Bruijnzeel, L. A., van Dijk, A. I. J. M., Groen, M. M. A. & van Hogezand, R. J. P. (2004) Stormflow generation in a small rainforest catchment in the Luquillo experimental forest, Puerto Rico. *Hydrol. Processes* **18**, 505–530.

Sklash, M. G., Stewart, M. K. & Pearce, A. J. (1986) Storm runoff generation in humid headwater catchments. 2. A case study of hillslope and low-order stream response. *Water Resour. Res.* **22**, 1273–1282.

Sklash, M. G., Beven, K. J. Gilman, K. & Darling, W. G. (1996) Isotope studies of pipeflow at Plynlimon, Wales, UK. *Hydrol. Processes* **10**, 921–944.

Turner, J. V., Bradd, J. M. & Waite, T. D. (1991) The conjunctive use of isotopic techniques to elucidate solute concentration and flow processes in dryland salinized catchments. In: *International Symposium on the Use of Isotopes in Water Resources Development.* Int. Atomic Energy Auth., Vienna, Austria.

Journal of Hydrology, 43 (1979) 45—65 45
© Elsevier Scientific Publishing Company, Amsterdam — Printed in The Netherlands

[2]

THE ROLE OF GROUNDWATER IN STORM RUNOFF

MICHAEL G. SKLASH and ROBERT N. FARVOLDEN

Department of Geology, University of Windsor, Windsor, Ont. N9B 3P4 (Canada)
Department of Earth Sciences, University of Waterloo, Waterloo, Ont. N2L 3G1 (Canada)

(Accepted for publication April 25, 1979)

ABSTRACT

Sklash, M.G. and Farvolden, R.N., 1979. The role of groundwater in storm runoff. In: **W. Back and D.A. Stephenson (Guest-Editors), Contemporary Hydrogeology — The George Burke Maxey Memorial Volume. J. Hydrol., 43: 45—65.**

Groundwater plays a much more active, responsive and significant role in the generation, of storm and snow-melt runoff in streams than the recent literature on the subject suggests. Basin-wide tracer experiments using environmental isotopes (^{18}O, deuterium, tritium) and hydrometric studies carried out in hydrogeologically diverse watersheds, indicate that for all except the most intense rain storms and the most prolific melting days, groundwater dominates the runoff hydrographs in the study basins. The increased groundwater discharge during runoff events is apparently related to a rapid rise in hydraulic head along the perimeter of transient and perennial discharge areas. This groundwater ridging phenomenon probably arises from the almost instantaneous conversion of the near-surface tension-saturated capillary fringe into phreatic water. The ridging precedes, and is apparently independent of the response of the rest of the basin. In addition to its compatibility with several of the field observations commonly associated with contemporary concepts of runoff generation, the groundwater discharge theory explains some of the temporal variations in stream water chemistry which are not adequately accounted for by other theories.

INTRODUCTION

Most of the recent literature on storm runoff generation has overlooked true groundwater flow as a significant and active factor in the storm and snow-melt runoff process. Freeze (1974) summarized the hydrologic thought on the subject as:

> "True groundwater flow is seldom the cause of the major runoff during storms. Its primary role is in sustaining streams during low-flow periods between rainfall and snow-melt events . . ."

Indeed, it is difficult to conceptualize how slow-moving groundwater can respond rapidly enough to contribute to a storm or snow-melt runoff peak. In the past decade, however, basin-wide tracer experiments using environmental isotope techniques have demonstrated that groundwater often dominates snow-melt runoff in humid to sub-humid regions (e.g., Dincer et al., 1970; Martinec, 1975). Fritz et al. (1976) and Sklash (1978) have reported the occurrence of large groundwater components in both storm and snow-melt runoff in their environmental isotope studies of eight hydrogeologically diverse watersheds in Canada. The exact mechanism which enables groundwater to appear so quickly and in such large quantities in the stream during high runoff events, however, has yet to be established.

The purpose of this paper is two-fold. It presents further documentation for the active, responsive and significant role of groundwater in storm runoff. Secondly, it introduces a theory for storm (and snow-melt) runoff generation which, in addition to explaining how large quantities of groundwater can appear in the peak storm or snow-melt runoff, is compatible with some field phenomena which one might ascribe to other theories for runoff generation.

CONCEPTUAL MODELS FOR STORM RUNOFF GENERATION

The study of storm (and snow-melt) runoff generation can be approached in a variety of ways. For example, one may consider only what portion of the runoff water existed in the basin prior to the runoff event and what portion was added by the runoff-inducing event. Alternatively, one may choose only to examine the manner in which the runoff water travels over the last several tens of metres to the stream. A third approach could be concerned only with the history of the water from its arrival in the basin to its ultimate delivery to the stream. These three approaches, termed: time aspects, ultimate delivery mechanism aspects, and historical aspects, respectively, are outlined in Fig. 1.

Currently, the most widely accepted theories for storm (and snow-melt) runoff generation stem from the ultimate delivery mechanism approach to the subject. Each concept attempts to account for both the rapid response of the stream to runoff-inducing events and the observed increase in stream discharge. Freeze (1974) has summarized the most popular of these theories as: (1) partial area—overland flow; (2) variable source area—overland flow; and (3) variable source area—subsurface storm flow. Brief explanations of these mechanisms follow, however, the reader is referred to Freeze (1974) which provides an excellent review and reference list for each mechanism.

The partial area—overland flow concept suggests that runoff water is produced mainly from certain fixed portions of the watershed (usually controlled by soil characteristics) where the soil becomes saturated from above by the infiltrating water. After surface detention requirements have been satisfied, the excess water runs off rapidly to the stream as overland flow.

Perhaps the most widely accepted runoff generation mechanism concept is the variable source area—overland flow theory. In essence, runoff is generated from watershed areas (usually controlled by topography, geology and soil type) which have become saturated from below by a rising water table. The source areas, generally located near the stream, may expand and

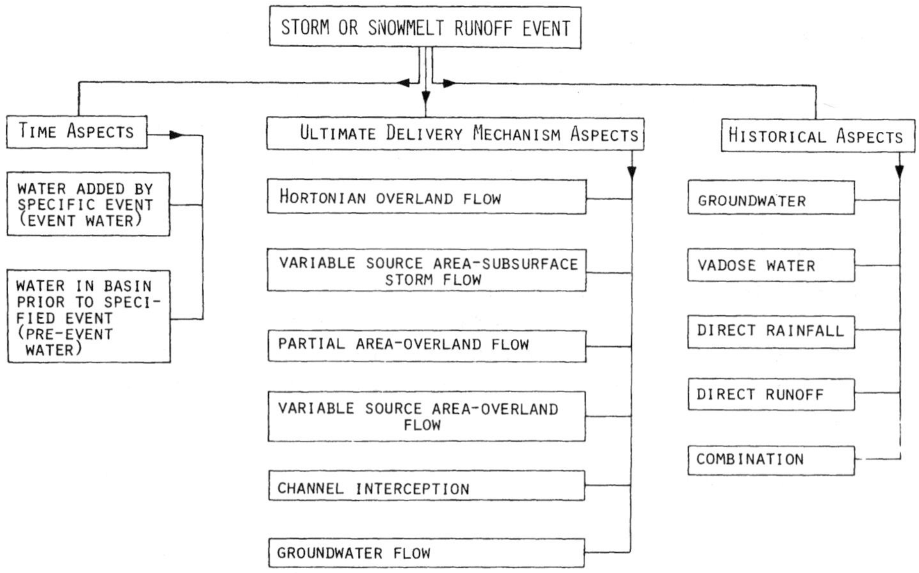

Fig.1. Classification chart for runoff generation terminology.

contract in response to climatic factors. The runoff from the variable source area, consisting of both rain and "return flow", runs off to the stream rapidly as overland flow.

The variable source area—subsurface storm flow concept states that the areas contributing to storm runoff expand and contract in response to climatic factors. Unlike the variable source area—overland flow theory, the transfer of water from the hillside to the stream is accomplished through subsurface routes. An expanding channel network and translatory flow (displacement or bumping of subsurface water toward the stream) allows the runoff water to reach the stream quickly. The subsurface storm flow may be either saturated—unsaturated Darcian flow through the porous soil matrix or turbulent flow through root channels, animal burrows and soilcracks (Cheng et al., 1975; Beasley, 1976).

Most of the research indicating groundwater as a significant factor in the storm (and snow-melt) runoff process can be classified under the heading of time aspect studies. Under this classification, the water which existed in the basin prior to a specified event is considered pre-event water and consists of groundwater, vadose water and surface storage. Water added to the basin by the specified runoff-inducing event, either a rain or snow-melt event, is called event water. Basin-wide tracer experiments involving such parameters as specific conductance, environmental isotopes, and conservative ions, have been used to distinguish pre-event water from event water (e.g., Pinder and Jones, 1969; Martinec, 1975). Groundwater contributions to peak storm and snow-melt runoff have been reported to exceed 50% in some instances.

The time aspect studies have all been stream-oriented; that is, their conclusions are based mainly on the examination of temporal variations of selected tracers in the stream. The ultimate delivery mechanism studies have all been slope-oriented; that is, their conclusions are based mainly on the examination of processes occurring somewhere above the stream. The conclusions from these two approaches have generally been divergent. Historical aspect studies attempt to identify changes in the runoff water chemistry while enroute to the stream in order to reconcile the slope- and stream-oriented results. Kennedy (1977) suggests that rain may infiltrate into micro-ridges (furrows), dissolve soluble materials, discharge from the furrows, and then run off to the stream as overland flow. In this manner, the chemistry of the runoff water may approach that of the groundwater and thereby confound any tracing attempts.

ENVIRONMENTAL ISOTOPE AND SPECIFIC CONDUCTANCE TECHNIQUES

Dincer et al. (1970), Martinec et al. (1974), Fritz et al. (1976), and others have demonstrated that environmental isotopes such as: ^{18}O, deuterium (D) and tritium (T), can be used to distinguish pre-event water from event water in the stream during runoff events. The storm (or snow-melt) runoff hydrograph can then be separated into its simple time components (pre-event and event water) by the simultaneous solution of the steady-state mass balance equations describing the fluxes of water and the tracer isotope in the stream. These equations are of the form:

$$Q_t = Q_p + Q_e \qquad \text{and} \qquad C_t Q_t = C_p Q_p + C_e Q_e \qquad (1),(2)$$

where Q is the discharge (L^3/T), C represents the tracer concentration, and subscripts t, p and e refer to the total stream discharge, pre-event component, and event component, respectively. Q_p and Q_e are unknowns which can be determined readily subject to the following criteria:

(1) The isotopic content (^{18}O, D, or T) of the event component is significantly different from that of the pre-event component.

(2) The event component maintains a constant isotopic content.

(3) The groundwater and vadose water are isotopically equivalent or vadose water contributions to runoff are negligible due to hydrogeologic constraints.

(4) Surface storage contributes minimally to the runoff event.

The environmental isotope technique for hydrograph separation is the dominant tool in this study. Isotopic, chemical and specific-conductance analyses of the various runoff components enhance the hydrograph interpretations.

HYDROMETRIC TECHNIQUES

The main hydrometric technique in this study involves the establishment of groundwater stage—groundwater discharge rating curves for hydrograph separation. According to this method, curves relating groundwater stage to stream discharge during baseflow periods (when stream discharge is assumed to be only groundwater) are assumed to hold true during high runoff episodes. By monitoring groundwater stage during runoff events, groundwater discharge is estimated by referring to the groundwater stage—groundwater discharge rating curve.

Unlike many of the previous studies (Schicht and Walton, 1961; Stevenson, 1967; Visocky, 1970) which used average groundwater stages for a number of index wells throughout the watershed, small-diameter recording wells installed in the stream bed and a few metres from the stream provide the groundwater stage data for one of the basins in this study. Although the individual wells may not represent the average response of the basin, they are indicative of the nature of the near-stream groundwater response.

Water level responses of wells and piezometers located farther from the stream offer some indication of the groundwater response of the remainder of the basin to runoff-inducing events.

WATERSHED MODELLING

The response of small hypothetical watersheds to runoff-inducing events is examined with the aid of a two-dimensional saturated—unsaturated transient finite-element flow model developed by Segol (1976). Observations are focussed on the groundwater response in the near-stream area when near-stream watershed relief and basin width are varied.

THE STUDY AREAS

Although several small watersheds have been examined during this study, only two, Ruisseau des Eaux Volées and Hillman Creek, are discussed here. The results from the other study basins, dealing mainly with snow-melt runoff, corroborate the findings reported in this paper.

The Ruisseau des Eaux Volées experimental watershed, situated at latitude 47°16′N and longitude 71°09′W, lies some 80 km north of Quebec City, Quebec, Canada. The basin is typical of many of the hanging valley tributaries found in the Laurentian Uplands of the Canadian Shield. Two subbasins, the west branch of Ruisseau des Aulnaies (subbasin *7A*) and the upper portion of Ruisseau des Eaux Volées (subbasin *6*), were monitored during the study (Fig.2). Subbasin *7A* occupies some 1.2 km^2 of heavily forested land between the 880- and 760-m elevation contours. Subbasin *6* covers 3.9 km^2 of heavily forested and experimentally logged land between 780 and 720 m in elevation. The climate of the area is considered moist cold temperate (Plamondon and Naud, 1975).

Surficial deposits, mainly of glacial origin, cover 80% of the basin to depths of 1—20 m. Jointed charnockitic gneiss, underlying the entire basin, outcrops only in the high areas (Rochette, 1971).

The groundwater system is considered unconfined with the bedrock groundwater demonstrating hydraulic connection to the surficial materials (Rochette, 1971). The water table, even in high areas, is close to ground surface.

Groundwater in the surficial materials is characterized by $\delta^{18}O$ and T values of approximately $-12.8^0/_{00}$ and 75 TU, respectively (all ^{18}O analyses are given relative to the SMOW standard). Bedrock groundwater has ^{18}O and T values of about $-11.9^0/_{00}$ and 175 TU, respectively. Baseflow ^{18}O values at the stream gauging stations for both subbasins range from $-12.6^0/_{00}$ in the winter to $-11.9^0/_{00}$ in the summer.

Suction lysimeters (soil moisture samplers) installed between depths of 0.26 and 0.47 m below ground surface were used to extract samples of vadose water from several sites (Fig.2). The ^{18}O analyses of the vadose water indicate some influence by the infiltrating rain but the values are consistently heavier than the groundwater by as much as $6^0/_{00}$.

The Hillman Creek study area, situated at latitude 42°04′N and longitude 82°39′W, is approximately 50 km SE of Windsor, Ontario, Canada. The drainage area above the stream gauging station is in the order of 1 km^2 which corresponds to a channel length of nearly 1.1 km (Fig.3). Relief over the basin is low with maximum and minimum elevations of approximately 212 and 202 m, respectively. Intensive agriculture in the area leaves only minimal tree cover and exposed soil conditions. An average of 740 mm of precipitation is distributed evenly throughout the year.

The uppermost unit in the basin, from 1 to 5 m of fine to coarse sand with a hydraulic conductivity in the order of 10^{-2} to 10^{-3} cm/s (Gillham et al.,

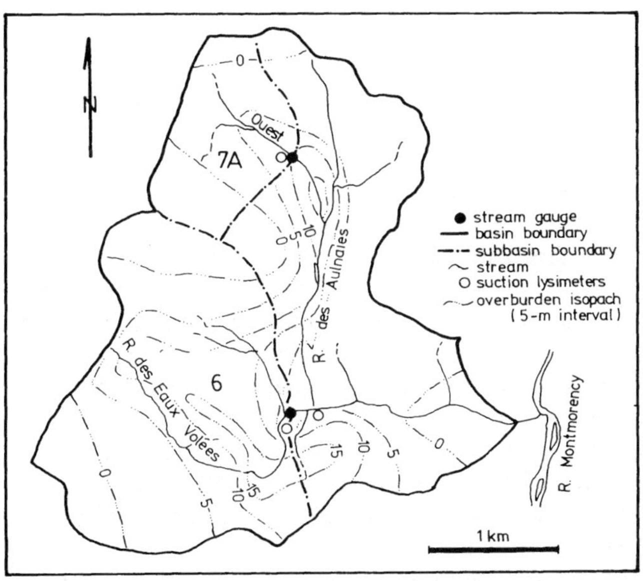

Fig.2. Instrumentation and depth of overburden in Ruisseau des Eaux Volées (after Rochette, 1971).

1978), overlies an impervious silty-clay unit. The water table, with a gradual slope from west to east, lies 1—4 m below ground surface. From water table maps (Gillham et al., 1978) and seepage meter data (Lee et al., 1977), it appears that groundwater discharge is low over the upper 600 m of the stream. From that point downstream beyond the gauging station, groundwater discharge is significant

The groundwater originates both from gravel pits along the western margin of the watershed and from local infiltration. The $\delta^{18}O$ values of the gravel pit groundwater are somewhat heavier than those of the locally infiltrated groundwater. Baseflow $\delta^{18}O$ values at the gauging station, ranging from −6.3 to -8.4$^0/_{00}$, indicate a mixture of the two groundwaters. The specific conductance of the baseflow is approximately 750 μS.

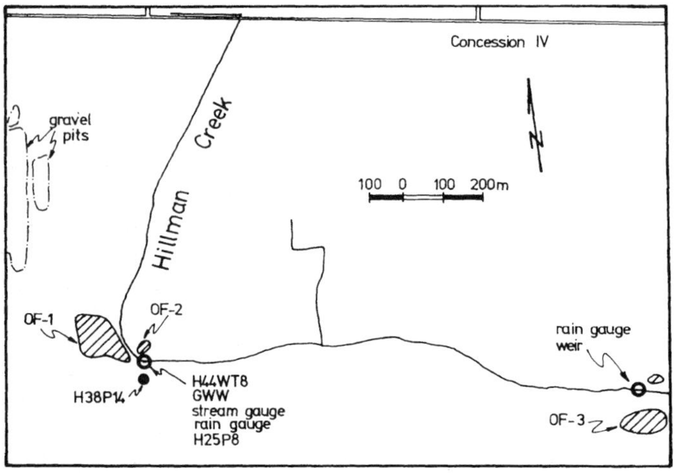

Fig.3. Instrumentation in the Hillman Creek study area.

RESULTS FROM RUISSEAU DES EAUX VOLÉES

Five rain events between August 5 and 25, 1976, resulted in stream discharge increases ranging from 45 to 820% in the two subbasins. Equipment malfunctions and storms unsuitable for the environmental isotope technique reduced the number of usable events which could be analyzed.

On August 5, a 32-mm rain event resulted in a discharge increase from subbasin *7A* of approximately 450% over its pre-storm baseflow discharge (Fig.4). The runoff hydrograph accounts for only 12% of the rainfall. Prior to the storm, baseflow $\delta^{18}O$ values were approximately -12.0$^0/_{00}$. Although the $\delta^{18}O$ value of the rain was -8.3$^0/_{00}$, the peak discharge $\delta^{18}O$ value of the stream only reached -10.8$^0/_{00}$ (Fig.4). Using eq. 1 and 2 and assuming that the vadose water contributions were negligible, it appears that groundwater contributed more than 65% of the peak discharge in the stream.

The validity of the assumption that vadose water contributions are negligible can be evaluated by examining a plot of stream $\delta^{18}O$ vs. stream discharge (Fig.5A). A simple two-component groundwater—rain mixture in the stream should result in essentially collinear $\delta^{18}O$-discharge data points for the rising and falling limbs of the hydrograph. As discussed previously, the vadose water in the basin is isotopically heavier than the groundwater.

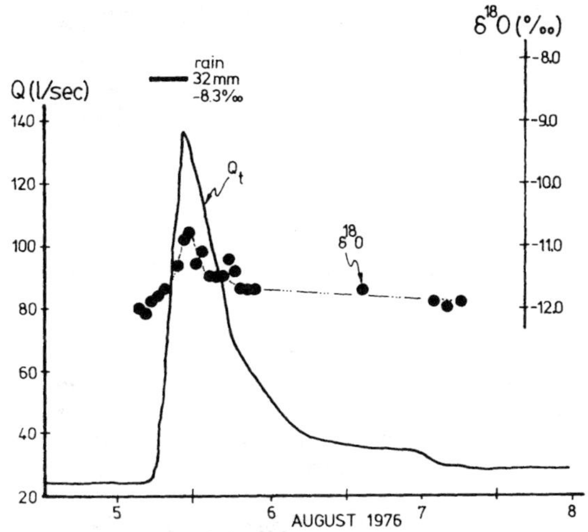

Fig.4. Storm runoff hydrograph and stream $\delta^{18}O$, subbasin *7A*, Ruisseau des Eaux Volées, August 5—8, 1976.

Significant vadose water contributions to storm runoff should cause a shift in the recession limb data to the isotopically heavier side of the groundwater —rain mixing line (which is defined by the data from the earlier part of the rising limb). If the shift exceeds the analytical error for ^{18}O analysis ($\pm 0.2^0/_{00}$), the assumption is questionable. Also, the magnitude of the shift should be directly related to the magnitude of the vadose water contribution.

Fig. 5B is a plot of the stream $\delta^{18}O$ vs. stream discharge for the August 5 storm in subbasin *7A*. The small shift between the rising and falling limb values of $\delta^{18}O$ supports the simple two-component groundwater—rain mixing assumption.

On August 12, a 35-mm rain event in subbasin *6* instigated a discharge increase of approximately 340% over the pre-storm baseflow discharge (Fig.6). The runoff hydrograph accounts for nearly 20% of the rainfall. Prior to the event, the baseflow $\delta^{18}O$ value was $-11.9^0/_{00}$. Although the rain $\delta^{18}O$ value was $-5.7^0/_{00}$, the stream attained a $\delta^{18}O$ value of only $-10.1^0/_{00}$ at peak discharge (Fig.6).

A plot of stream $\delta^{18}O$ vs. stream discharge for this event (Fig.7) suggests a significant vadose water contribution. Although a simple two-component hydrograph separation cannot be made in this case, it is apparent from the isotopic character of the various runoff components that pre-event water (groundwater and/or vadose water) dominated the storm runoff.

On August 22, a very brief but intense storm dropped 6 mm of rain on the basin in less than 15 min. Stream discharges from subbasins *6* and *7A* increased by 55 and 45%, respectively (Fig. 8A and B). Storm yields for both basins were about 6%.

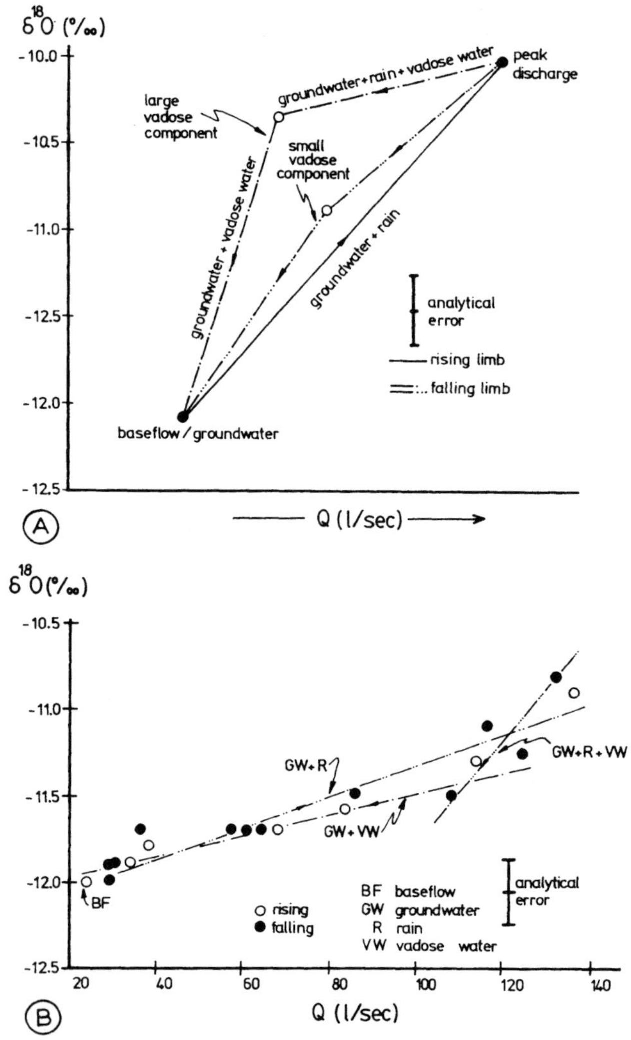

Fig.5. A. Model of $\delta^{18}O$ variations during storm runoff for the Ruisseau des Eaux Volées watershed.

B. $\delta^{18}O$ vs. discharge during storm runoff, subbasin *7A*, Ruisseau des Eaux Volées watershed, August 5—8, 1976.

The rain event was ideal for hydrograph separation by the isotope technique. The short duration and high intensity of the rain meant that the event water was added "instantaneously" to the watershed. Baseflow $\delta^{18}O$ prior to the event was -12.1°/oo in subbasin *6* and -11.9°/oo in subbasin *7A*. Although the rain $\delta^{18}O$ value was -6.6°/oo, peak discharge attained $\delta^{18}O$

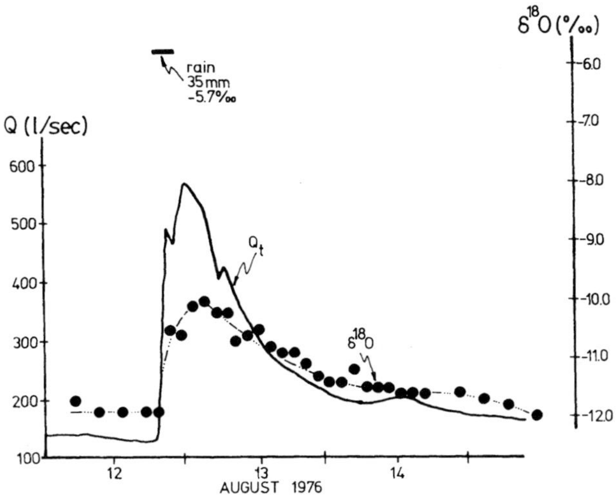

Fig.6. Storm runoff hydrograph and stream $\delta^{18}O$, subbasin *6*, Ruisseau des Eaux Volées watershed, August 12—15, 1976.

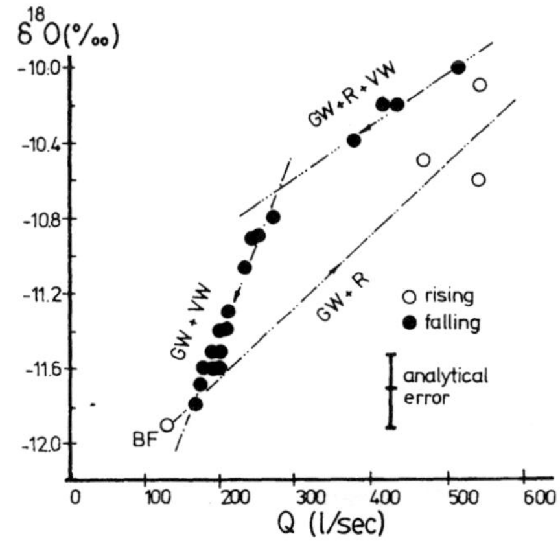

Fig.7. $\delta^{18}O$ vs. discharge during storm runoff event subbasin *6*, Ruisseau des Eaux Volées watershed, August 12—15, 1976.

values of -11.4‰ and -11.5‰ for subbasins *6* and *7A*, respectively (Fig.8). These correspond to groundwater components of more than 80% in peak discharge if vadose water contributions were negligible. A plot of stream $\delta^{18}O$ vs. stream discharge would support that assumption. Deuterium analyses for subbasin *6* gave similar groundwater contributions.

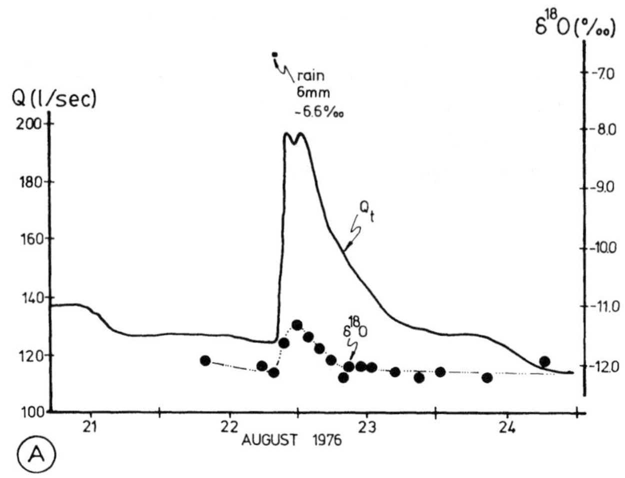

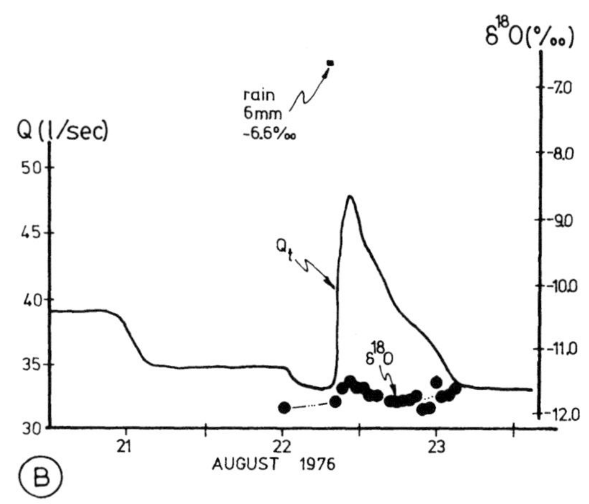

Fig.8.A. Storm runoff hydrograph and stream $\delta^{18}O$, subbasin *6*, Ruisseau des Eaux Volées watershed, August 21—24, 1976.
B. Storm runoff hydrograph and stream $\delta^{18}O$, subbasin *7A*, Ruisseau des Eaux Volées watershed, August 21—24, 1976.

RESULTS FROM HILLMAN CREEK

Groundwater stage—groundwater discharge rating curves for two near-stream observation wells were determined. H44WT8 is a 10-cm diameter, 2.44 m deep, recording water table well located approximately 3 m from the stream bank at the stream gauging site. GWW (Fig.9) is a 9-cm diameter recording piezometer open between 80 and 90 cm below the stream bed and adjacent to H44WT8. Vertical and horizontal hydraulic gradients during base-flow periods were approximately 0.12 and 0.03, respectively, between H44WT8 and GWW. These gradients are an order of magnitude greater than those given for the watershed average by Gillham et al. (1978).

Hydrogeologic constraints in the watershed; namely, low topographic relief and high hydraulic conductivities of the sand, preclude significant

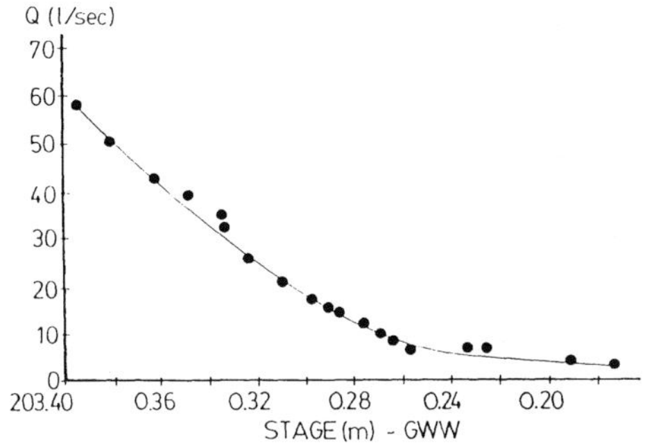

Fig.9. Groundwater stage—groundwater discharge rating curve for GWW, Hillman Creek study area.

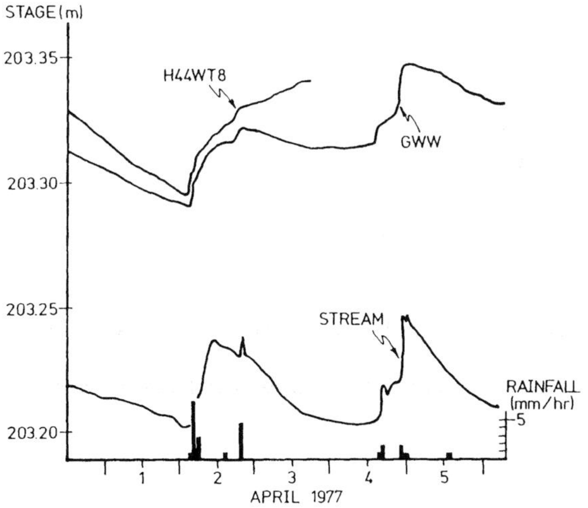

Fig.10. Stream and groundwater hydrographs for Hillman Creek study area, April 1—5, 1977.

vadose water contributions to storm runoff. Groundwater, therefore, is synonymous with pre-event water in this watershed.

Rain events of 19 mm on April 2 and 6 mm on April 4, 1977, caused two distinct runoff events at the Hillman Creek gauging site (Fig.10). Stream discharge increased from 15 l/s prior to the storm to 25 and 30 l/s at the first and second discharge peaks, respectively. The most noteworthy feature of these events is the rapid response of the near-stream groundwater exhibited by H44WT8 and GWW. The stream remained effluent throughout the events and although the hydraulic gradient to the stream decreased slightly during the first event, the gradient in the second event exceeded the pre-storm baseflow gradient. According to the groundwater stage—groundwater discharge rating curve technique, groundwater contributed more than 60 and 80% of the peak discharge on April 2 and 4, respectively. No isotope data are available for these events.

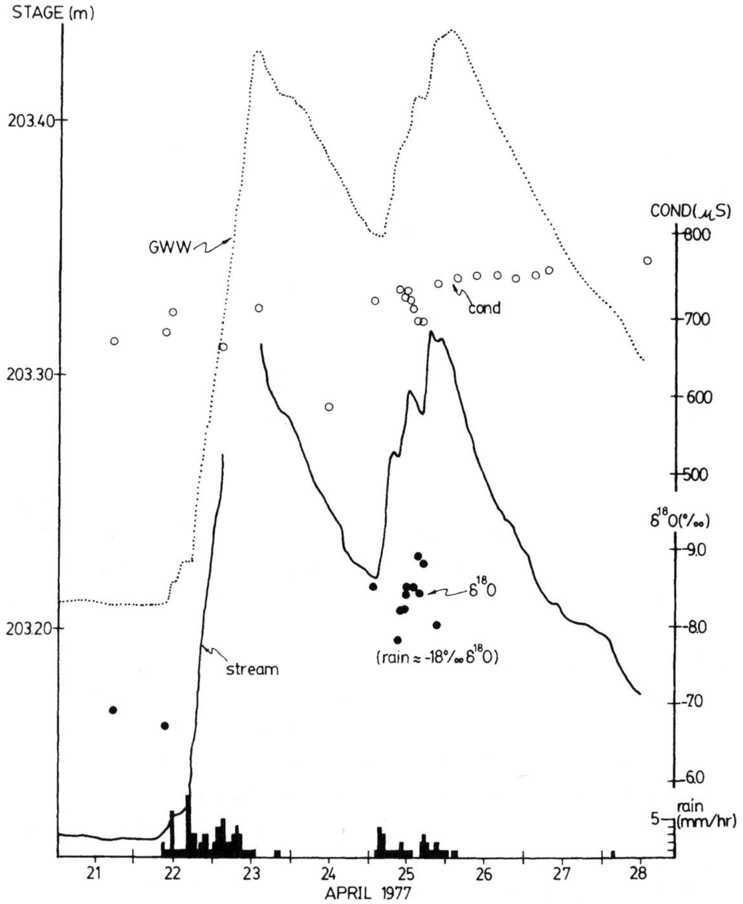

Fig.11. Stream and groundwater hydrographs, and stream $\delta^{18}O$ and specific conductance, Hillman Creek study area, April 21—28, 1977.

Rain events of 38, 35 and 31 mm occurring on April 22, 23 and 25, 1978, respectively, produced dramatic responses in both stream discharge and groundwater stage (Fig.11). Stream discharge increased from about 6 l/s on April 22 to peak discharge of at least 85 and 90 l/s on April 23 and 25, respectively. The length of the storms and instrument malfunctions curtailed some of the isotopic and specific conductance studies, however, many significant observations were made.

The stream remained effluent throughout the events even though stream discharge increased by more than an order of magnitude over the pre-storm discharge. The near-stream groundwater responded quickly to the rain events; however, the groundwater responsiveness apparently decreased away from the stream. Table I shows that over the duration of the study period, the stream discharge was more closely related to the near-stream groundwater stage than to the more remote groundwater. Although the near-stream hydraulic gradient toward the stream decreased slightly at the first discharge peak, the gradient at the second discharge peak was greater than the pre-storm base-flow gradient.

TABLE I

Relationship between stream discharge and depth to groundwater near H44WT8, Hillman Creek Watershed

Well or piezometer	Approximate lateral distance to stream (m)	Linear correlation coefficient between depth to water and stream stage
H38P14	65	-0.175
H25P8	5	-0.869
H44WT8	3	-0.954
GWW	0	-0.925

Visual observations were made and water samples were taken only on April 25. One area west of the stream gauging station (OF-1 on Fig.3) produced sufficient overland flow to erode a channel several centimetres deep to the stream. Two other overland flow producing areas, OF-2 and OF-3, produced much less overland flow.

The $\delta^{18}O$ values of the rain on April 25 ranged from -16.5 to $-20.0^0/_{00}$ with a weighted average of $-18.1^0/_{00}$. Since the baseflow prior to April 22 had $\delta^{18}O$ values of about $-6.9^0/_{00}$ and since the groundwater had $\delta^{18}O$ values of less than $-10.0^0/_{00}$ (Gillham et al., 1978), a very good distinction can be drawn between the groundwater/base flow (pre-event) and rain (event) components. Also, the specific conductance of the rain was less than 100 μS, whereas baseflow specific conductance is about 750 μS.

Fig.12A and B illustrates the temporal variations in $\delta^{18}O$ and specific conductance for the rain, overland flow and stream samples on April 25. The most prolific source of overland flow, OF-1, had $\delta^{18}O$ and specific conductance values indicative of large groundwater contributions. The two other observed sources, OF-2 and OF-3, had rain-like $\delta^{18}O$ and specific conductance. The stream $\delta^{18}O$ and specific conductance strongly suggest a large groundwater component. The groundwater stage—groundwater discharge rating curve technique supports these findings, with both methods giving more than 80% of the discharge as groundwater on April 25. Observed increases in the stream nitrate concentration on April 25 are apparently related to the high nitrate concentrations in the groundwater-dominant overland flow issuing from OF-1.

On May 4, 1977 a 1.7-mm rain event produced another storm runoff event in which the near-stream groundwater responded rapidly (Fig.13). Stream

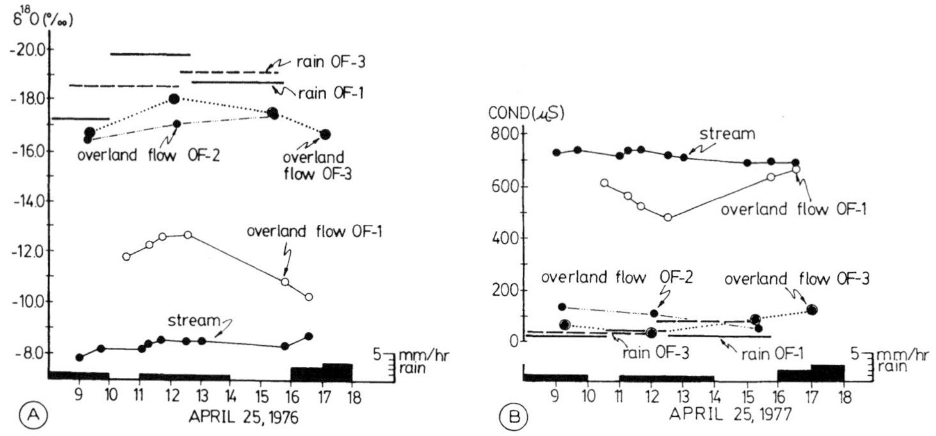

Fig.12. A. Temporal variations in $\delta^{18}O$ of overland flow, rain and the stream, Hillman Creek study area, April 25, 1977.
B. Temporal variations in specific conductance of overland flow, rain, and the stream, Hillman Creek study area, April 25, 1977.

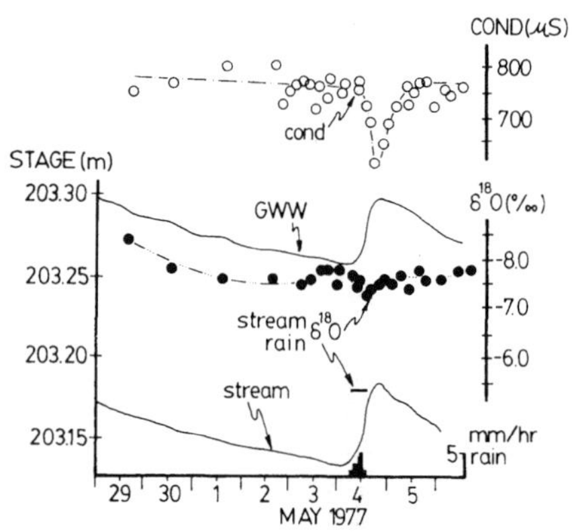

Fig.13. Stream and groundwater hydrograph, and stream $\delta^{18}O$ and specific conductance, Hillman Creek study area, April 29—May 5, 1977.

discharge increased from 9 to 22 l/s at peak discharge, yet the stream remained effluent. No overland flow was observed and baseflow $\delta^{18}O$ and specific conductance were diluted only slightly during the event. Both the groundwater stage--groundwater discharge rating curve and isotope techniques suggest that groundwater contributed over 80% of the peak discharge.

On June 6, 1977, following a month with only 20 mm of rain, a 36-mm rain event produced very prominent hydrographs in the stream and near-stream groundwater (Fig.14). Stream discharge increased from less than 5 l/s before the storm to over 200 l/s at peak discharge. A brief reversal of hy-

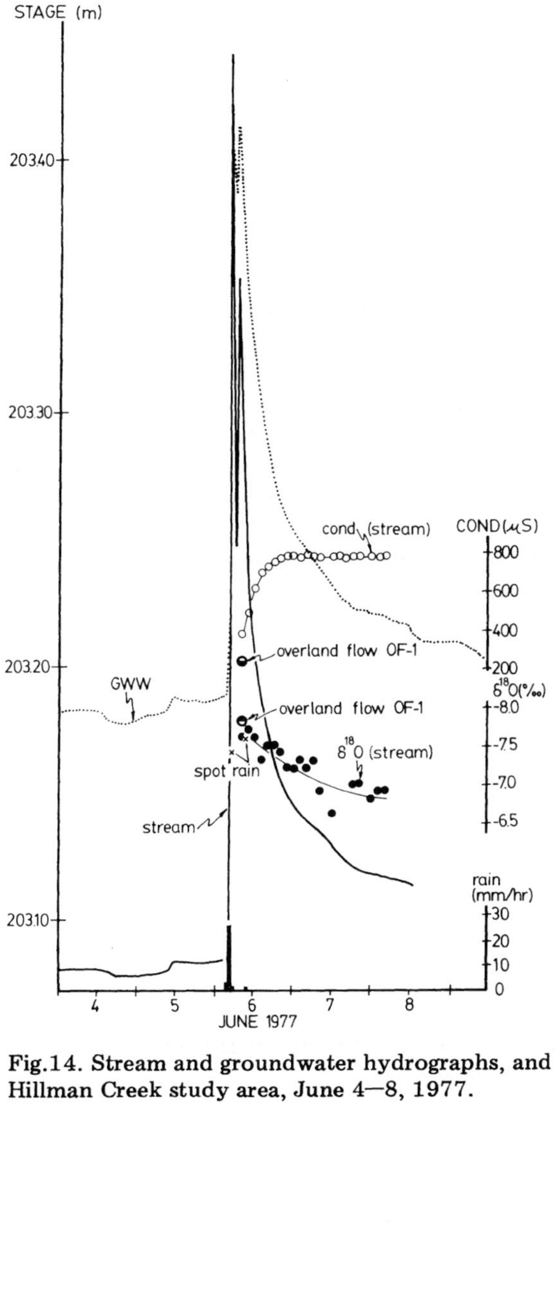

Fig.14. Stream and groundwater hydrographs, and stream $\delta^{18}O$ and specific conductance, Hillman Creek study area, June 4—8, 1977.

draulic gradient caused the stream to become influent for the only time during the study. The near-stream groundwater, however, did respond to the storm very rapidly. Significant amounts of overland flow emanated from OF-1, OF-2 and OF-3.

Fig.14 shows the temporal variations in stream $\delta^{18}O$ and specific conductance. Although the $\delta^{18}O$ value of the rain ($-7.8^0/_{00}$) was too close to the base-flow value ($-6.8^0/_{00}$) for very accurate separations, it is obvious that the stream runoff was dominated by rain runoff. The overland flow from OF-1, which was mostly groundwater during the April 25 storm, had rain-like $\delta^{18}O$ ($-7.8^0/_{00}$), and specific conductance (234 μS). The overland flow from OF-1 also had nitrate concentrations (4.5 mg/l) similar to those of peak discharge in the stream (4.2 mg/l). The stream nitrate concentrations increased to between 8 and 10 mg/l when the groundwater reasserted itself in the stream.

RESULTS FROM COMPUTER SIMULATIONS

Four small hypothetical watershed configurations (Fig.15) are examined to determine how near-stream watershed relief and basin width affect near-stream groundwater response. Watersheds *1*, *2* and *3* have comparatively low, medium and high near-stream topographic relief. Each configuration is 9 m from mid-stream to divide and varies in thickness above an impermeable base from 1 m at mid-stream to 2.9 m at the divide. The finite-element grid for each contains 464 nodes and 420 quadrilateral elements with the grids varying only in their near-stream topographic configuration. Watershed *4* is identical to watershed *2* in element configuration except that it is truncated 5 m from the stream leaving fewer nodes (400) and elements (360). The uppermost elements in the grids represent real thicknesses of no more than 3 cm.

The watersheds are homogeneous and isotropic. The saturated—unsaturated characteristics of the porous media are those of the Botany Sand as described by Watson (1967). The sand has a saturated hydraulic conductivity of 0.0186 cm/s, a porosity of 35%, and a capillary fringe of 39 cm. The initial conditions are static and the stream level is held constant throughout the simulation period. Seepage faces are allowed to form.

A rainfall rate equal to one-tenth of the saturated hydraulic conductivity (0.00186 cm/s) was applied to each watershed for 2.3 h after which drainage was allowed for at least 4.6 h. Computer output, consisting of total and pressure heads, for specified times during the event, are available for all the nodes.

Fig.16 compares the responses in total head for a common near-stream point *a* (Fig.15). Watershed *1*, with the lowest near-stream relief, exhibited the most rapid response to the storm. Watersheds *2* and *3* responded less rapidly but had higher peak stages. A comparison of the early responses of long watershed *2* and short watershed *4* reveals that the upland area has no effect on the early near-stream groundwater response.

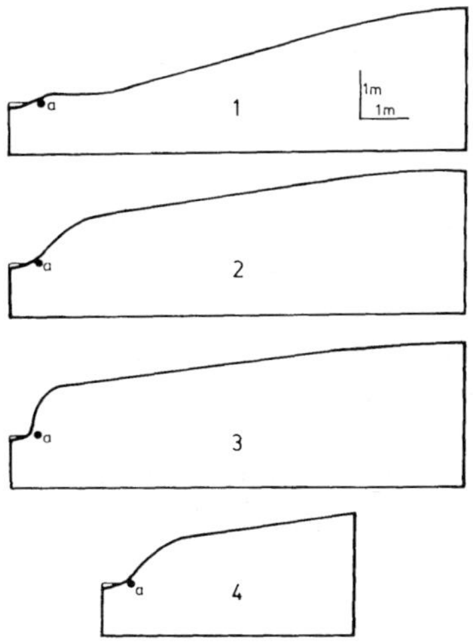

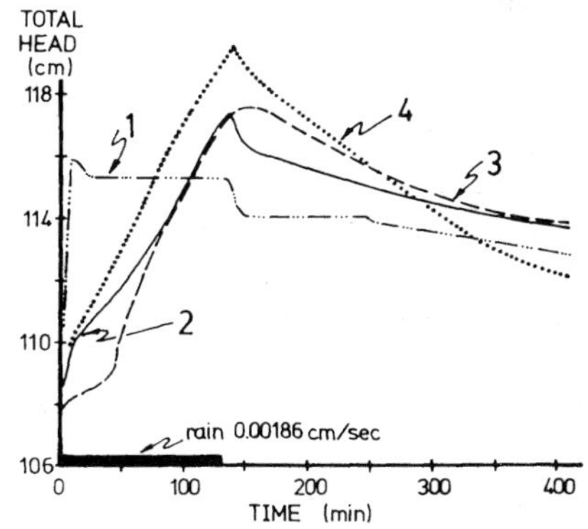

Fig.15. Watershed configurations used in mathematical simulations.

Fig.16. The effects of the near-stream unsaturated zone thickness and upland areas on early near-stream groundwater response to rain.

Fig.17 illustrates the temporal variations in the water table configuration for the rain event on watershed *2*. An apparent groundwater ridge, similar to the one observed by Ragan (1968), forms in the near-stream area prior to the response of the water table in the upland area. A seepage face forms in this case as it does in all the others. The discharges of the various runoff components in a 1-cm slice of watershed *2* are compared in Fig.18. Groundwater dominates the total discharge from the watershed.

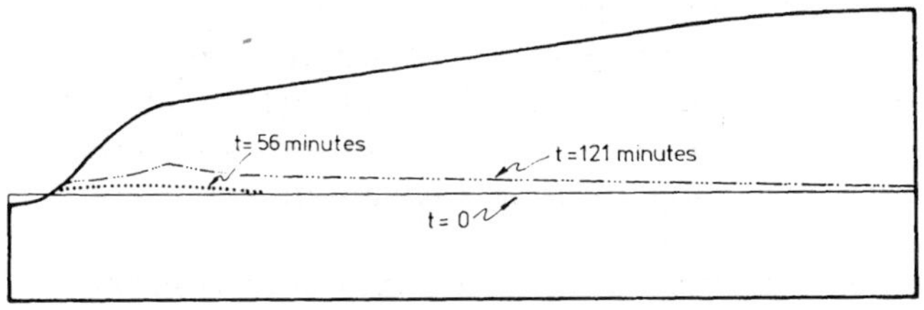

Fig.17. Formation of a near-stream "groundwater ridge" in response to a rain event.

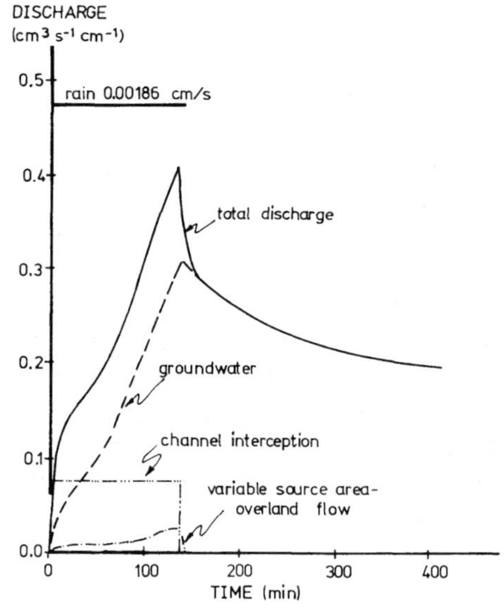

Fig.18. Discharge hydrograph from watershed *2*.

CONCLUSIONS

Groundwater was found to be a major component in virtually all of the runoff events documented in this study. Pinder and Jones (1969), Dincer et al. (1970), Martinec et al. (1974), Fritz et al. (1976), and others have arrived at the same conclusion, that is, groundwater is a significant and active factor in runoff generation. Martinec (1975), however, summarized the dilemma of how the groundwater appears in the stream so quickly as:

"The exact nature of the propagation is not known".

Several clues to the mechanism can be gathered from this study. Both field observations and computer simulations suggest that very large and rapid increases in the hydraulic head in the near-stream groundwater occur soon after the onset of rain. These responses precede, and are apparently independent of, the upland area response. The computer simulations also reveal the formation of a groundwater ridge adjacent to the stream in response to a rain event with the lag time brief and inversely related to the near-stream unsaturated zone thickness. Field observations indicate, however, that the near-stream hydraulic gradient may decrease slightly from its pre-storm value.

Isotopic and specific conductance data for overland flow in the Hillman Creek watershed imply that for one very prolific site (OF-1), overland flow is generated in more than one way. During a moderate intensity storm on a very wet basin, both the overland flow and stream flow were dominated by groundwater. In response to a very intense storm on a much drier basin, both the overland flow and stream were dominated by rain water.

The following theory outlines a physical mechanism which could explain the responsive, active and significant role of groundwater in storm and snow-melt runoff generation. Along the perimeter of transient and perennial discharge areas, the water table and its associated capillary fringe lie very close to the ground surface. Soon after a rain or snow-melt event begins, infiltrating water readily converts the near-surface tension-saturated capillary fringe into a pressure-saturated zone or groundwater ridge (Ragan, 1968). This groundwater ridge not only provides the early increased impetus for the displacement of the groundwater already in a discharge position, but it also results in an increase in the size of the groundwater discharge area which is essential in producing large groundwater contributions to the stream. The response of the upland area groundwater may become important at later times in the runoff event but has little influence in the early part of the runoff event.

The groundwater may discharge directly into the stream through the stream bed or it may issue from the growing near-stream and/or remote seep areas and flow as overland flow to the stream (as in the variable source area—overland flow theory). Following periods of drought during which the water table has fallen far below ground surface, intense storms may result in surface saturation from above and rain-like overland flow (partial area—overland flow) before the water table can emerge.

The recognition of groundwater as an important factor in storm and snow-melt runoff in humid to subhumid areas raises many questions regarding run-off events. What control does groundwater have on surface water quality during high runoff events? How do the increased heads in the near-stream groundwater affect the erodibility of soils? How can we improve predictive hydrologic models?

ACKNOWLEDGEMENTS

John Cherry, Emil Frind, Peter Fritz and Bob Gillham of the University of Waterloo provided many helpful comments during this study. Funds for this study were provided through National Research Council of Canada grants A7712, A6112 and A8368, and from Agriculture Canada and the Ontario Ministry of Agriculture and Food. The senior author wishes to acknowledge the NRCC for the support of his postgraduate education.

REFERENCES

Beasley, R.S., 1976. Contribution of subsurface flow from the upper slopes of forested watersheds to channel flow. Soil Sci. Soc. Am. J., 40: 955—957.

Dincer, T., Payne, B.R., Florkowski, T., Martinec, J. and Tongiorgi, E., 1970. Snow-melt runoff from measurements of tritium and oxygen-18. Water Resour. Res., 6: 110—124.

Cheng, J.D., Black, T.A. and Wellington, R.P., 1975. The generation of stormflow from small forested watersheds in the Coast Mountains of Southwestern British Columbia. Proc. Can. Hydrol. Symp., Winnipeg, Man., pp. 542—549.

Freeze, R.A., 1974. Streamflow generation. Rev. Geophys. Space Phys. 12: 627—647.

Fritz, P., Cherry, J.A., Weyer, K.V. and Sklash, M.G., 1976. Runoff analyses using environmental isotope and major ions. In Interpretation of Environmental Isotope and Hydrochemical Data in Groundwater Hydrology. I.A.E.A., Vienna, pp. 111—130.

Gillham, R.W., Hendry, M.J., Cherry, J.A., Frind, E.O. and Pucovsky, G.M., 1978. Studies of agricultural contribution to nitrate enrichment of groundwater and the subsequent nitrate loading to surface waters. Part I. Field investigation of the processes controlling the transport of nitrate in groundwater. Waterloo Res., Inst., Waterloo, Ont., 203 pp.

Kennedy, V.C., 1977. Hypothesis to explain the rapid release of solutes from soil during storm runoff. EOS (Trans. Am. Geophys. Union) 58: 386 (abstract).

Lee, D.R., Fritz, P. and Hynes, H.B.N., 1977. Groundwater flow into lakes and streams mass movement, transport of nutrients and organic matter. Waterloo Res. Inst., Waterloo, Ont., 57 pp.

Martinec, J., 1975. Subsurface flow from snow-melt traced by tritium. Water Resour. Res., 11: 496—497.

Martinec, J., Siegenthaler, H., Oescheger, H. and Tongiorgi, E., 1974. New insight into the runoff mechanism by environmental isotopes. Proc. Symp. on Isotope Techniques in Groundwater Hydrology. I.A.E.A., Vienna, 1: 129—143.

Pinder, G.F. and Jones, J.F., 1969. Determination of the groundwater component of peak discharge from the chemistry of total runoff. Water Resour. Res., 5: 438—445.

Plamondon, A.P. and Naud, R.C., 1975. The watershed management program at the Montmorency Experimental Forest. Symp. on Montmorency Experimental Forest, 27 pp.

Ragan, R.M., 1968. An experimental investigation of partial area contribution. Int. Assoc. Sci. Hydrol., Publ., 76: 241—249.

Rochette, F.J., 1971. Hydrogeological study of "Ruisseau des Eaux Volées" Experimental Basin. M.Sc. Thesis, University Western Ontario, London, Ont. (unpublished).

Schicht, R.J. and Walton, W.C., 1961. Hydrologic budgets for three small watersheds in Illinois. Ill., State Water Surv., Rep. Invest. No. 40, 40 pp.

Segol, G., 1976. A three-dimensional Galerkin finite element model for the analysis of contaminant transport in variably saturated porous media. Univ. Waterloo, Waterloo, Ont., Rep. (unpublished).

Sklash, M.G., 1978. The role of groundwater in storm and snowmelt runoff generation. Ph.D. Thesis, University of Waterloo, Waterloo, Ont. (unpublished).

Stevenson, D.R., 1967. Geological and groundwater investigations in the Marmot Creek Experimental Basin of Southwestern Alberta, Canada. M.Sc. Thesis, University of Alberta, Edmonton, Alta. (unpublished).

Visocky, A.P., 1970. Estimating the groundwater contribution to storm runoff by the electrical conductance method. Groundwater 8: 5—10.

Watson, K.K., 1967. Experimental and numerical study of column drainage. J. Hydraul. Div., Proc. Am. Soc. Civ. Eng., 93(HY2): 1—15.

Commentary on Paper 28

Sharma, M. L., Gander, G. A. & Hunt, S. G. (1980) Spatial variability of infiltration in a watershed.

This paper is important in that it was one of the first studies to examine the spatial variability of the infiltration characteristics in a small catchment, the R-5 catchment at Chickasha, Ohio, USA. Earlier studies by Don Nielsen and Jim Biggar and their students had established that heterogeneity in soil hydrological characteristics might be characterized by statistical distributions and the Miller & Miller (1956) scaling theory at the field scale (e.g. Nielsen *et al.*, 1973; Warrick *et al.*, 1977). Rogowski (1972) had already applied these concepts at the catchment scale. This paper has been chosen as a benchmark because of the later history of modelling studies at the R-5 catchment. The first studies were reported by Sharma & Luxmoore (1979), Luxmoore (1983) and Loague & Freeze (1985). The latter study involved applying different models to catchments of different types, including a distributed surface runoff model (QPBRRM). The R-5 catchment was perceived as one in which the dominant component of the storm hydrograph was due to overland flow but the results using only the measured infiltration parameters were not satisfactory.

Further modelling studies were reported by Keith Loague (1990, 1992). The 26 infiltration measurements reported in this paper were later extended by Loague & Gander (1990) to a further 247 measurements based on a grid of 25 m spacing with additional transects of 2 m and 5 m spacing. Geostatistical interpolation was used to provide estimates of soil parameters at 1 m spacing and Loague & Kyriakidis (1997) applied the model with 959 overland flow plane segments each with different infiltration characteristics. The results, however, were worse than in the original Loague & Freeze (1985) study. Adjusting the infiltration parameters to take account of temperature effects in the original measurements did result in some improvement, as did taking the ensemble average of a number of different stochastically generated parameter fields based on the geostatistical interpolations. The model significantly underestimated the volume of runoff, partly because overland flow generated on the upper slopes would, in the model, be re-infiltrated on higher infiltration areas downslope. Loague (1990) and Loague & Kyriakidis (1997) also suggested that subsurface runoff might be important at this site.

Keith Loague returned to the challenge of modelling the R-5 response in 2001 with Joel VanderKwaak, using the Integrated Hydrology Model (InHM) which includes a distributed variable saturation subsurface flow component based on the Darcy-Richards equation (VanderKwaak & Loague, 2001; Loague & VanderKwaak, 2004). They conclude that subsurface controls on the hydrograph may be important at the R-5 and that "*accurate reporting of soil water storage (e.g. the position of the water table and areas of high soil water content) may be more important than the exhaustive permeability estimates of Loague & Kyriakidis (1997) for simulating rainfall–runoff events*" (p.1010). A final (?) attempt at modelling R5 is reported in Loague *et al.* (2005) where, with a new mesh and taking account of evaporation, InHM was shown to produce better flood peak predictions than previously. The simulation results of the storm based simulations, however, remained sensitive to the initial conditions prior to each event.

References

Loague, K. M. (1990) R-5 revisited. 2. Re-evaluation of a quasi-physically based rainfall–runoff model with supplemental information. *Water Resour. Res.* **26**, 973–987.

Loague, K. M. (1992) Soil-water content at R-5. 2. Impact of antecedent conditions on rainfall-runoff simulations. *J. Hydrol.* **139**, 253–261.

Loague, K. M. & Freeze, R. A. (1985) A comparison of rainfall–runoff modeling techniques on small upland catchments. *Water Resour. Res.* **21**, 229–248.

Loague, K. M. & Gander, G. A. (1990) R-5 revisited. 1. Spatial variability of infiltration on a small rangeland catchment. *Water Resour. Res.* **26**, 957–971.

Loague, K. M. & Kyriakidis, P. C. (1997) Spatial and temporal variability in the R-5 infiltration data set: déjà vu and rainfall-runoff simulations. *Water Resour. Res.* **33**, 2883–2896.

Loague, K. & VanderKwaak, J. E. (2004) Physics-based hydrologic response simulation: platinum bridge, 1958 Edsel, or useful tool. *Hydrol. Processes* **18**(15), 2949–2956.

Loague, K., Heppner, C. S., Abrams, R. H., Carr, A. E., VanderKwaak, J. E. & Ebel, B. A. (2005) Further testing of the Integrated Hydrology Model (InHM): event-based simulations for a small rangeland catchment located near Chickasha, Oklahoma. *Hydrol. Processes* **19**(7), 1373–1398.

Luxmoore, R. J. (1983) Infiltration and runoff predictions for a grassland watershed. *J. Hydrol.* **65**, 271–278.

Miller. E. E. & Miller, R. D. (1956) Physical theory for capillary flow phenomena, *J. Appl. Phys.* **27**, 324–332.

Nielsen, D. R., Biggar, J. W. & Erh, K. T. (1973) Spatial variability of field measured soil-water properties. *Hilgardia* **42**, 215–259.

Rogowski, A. S. (1972) Watershed physics: soil variability criteria, *Water Resour. Res.* **8**, 1015–1023.

Sharma, M. L. & Luxmoore, R. J. (1979) Soil spatial variability and its consequences on simulated water balances. *Water Resour. Res.* **15**, 1567–1573.

VanderKwaak, J. E. & Loague, K. M. (2001) Hydrologic response simulations for the R5 catchment with a comprehensive physics-based model. *Water Resour. Res.* **37**, 999–1013.

Warrick, A. W., Miller, G. J. & Nielsen, D. R. (1977) Scaling field measured soil hydraulic properties using similar media concepts. *Water Resour. Res.* **13**, 355–362.

Journal of Hydrology, 45 (1980) 101—122 101

[4]

SPATIAL VARIABILITY OF INFILTRATION IN A WATERSHED

M.L. SHARMA[1], G.A. GANDER[2] and C.G. HUNT[2]

[1] *Division of Land Resources Management, CSIRO, Wembley, W.A. 6014 (Australia)*
[2] *USDA—Science and Education Administration, Chickasha, OK 73018 (U.S.A.)*

(Received February 19, 1979; revised and accepted June 13, 1979)

ABSTRACT

Sharma, M.L., Gander, G.A. and Hunt, C.G., 1980. Spatial variability of infiltration in a watershed. J. Hydrol., 45: 101—122.

Twenty-six infiltration tests were carried out on a 9.6-ha watershed, near Chickasha, Oklahoma. The data from each test were expressed in the form of Philip's two-parameter equation $I = St^{1/2} + At$, where I is cumulative infiltration; t is time; and S and A are the two parameters. In most cases this equation could be adequately matched to the data so that the variability of infiltration in the watershed could be expressed in terms of the parameters S and A.

There was no obvious pattern in the distribution of infiltration parameters with respect to soil type or position in the watershed. Due to large variability within soil units of the watershed, delineation of hydrologically homogeneous zones is not warranted. The range of values of S over the watershed was within an order of magnitude, while that of A was within two orders of magnitude. Parameters S and A were correlated ($r = 0.81$). The frequency distributons of S, A and I at a given time, were better approximated by log-normal than by normal distributions. However, assuming a normal distribution for S would not introduce as much error as it would for A.

The infiltration data from 26 sites, comprising of 618 observations, were scaled according to the concept of similar media, and one value of scaling factor was assigned to the infiltration data of each site. Scaling factors α_S and α_A were computed from the independently determined parameters, S and A. The reduction in error sums of squares of the scaled data was much greater with α_A than with α_S, although the scaling theory states that these factors should be identical. Remarkable improvement in scaling was achieved when the geometric (α_{SA}^{G}) or the harmonic (α_{SA}^{H}) mean of α_S and α_A was used. Results based on α_{SA}^{H} were almost identical to those calculated by statistical optimization. The frequency distribution of scaling factors in the watershed was log-normal.

INTRODUCTION

Modelling water balances of watersheds has been the concern of hydrologists for decades. Although theoretically possible, detailed waterflow modelling, based on basic soil hydraulic properties [the hydraulic conductivity—water content characteristic, $K(\theta)$; and the water potential—water content characteristic, $\psi(\theta)$], has limited application because of large field variability and hence the need for large amounts of data on hydraulic properties, and also by limited computer capacity (Amerman, 1973). Thus, purely physically-based water flow analysis may not be practical, and therefore use is often made of semi-empirical and empirical models.

Infiltration and evapotranspiration are the two most important processes which control the fate of water over large areas. Infiltration, although primarily dependent on soil properties, is also indirectly affected by evapotranspiration through its effect on the initial soil water status. Singh and Woolhiser (1976) argue that the rainfall excess (rainfall—infiltration) is the most sensitive parameter in many models for predicting streamflow.

One of the most serious drawbacks in hydrologic modelling has been the disregard of spatial and temporal heterogeneity; most models use an infiltration curve determined (or estimated) at one site in the watershed. The spatial variability of some agricultural lands, which have been evaluated in terms of basic hydraulic properties (e.g., Nielsen et al., 1973; Carvallo et al., 1976; Warrick et al., 1977a, b; Keisling et al., 1977) implies a large variability in infiltration behavior. Similar studies (Rogowski, 1972) could be made on watersheds and variability in infiltration could be assessed, but this is a lengthy process. We propose that the hydrologic variability of land surfaces be evaluated in terms of bulk parameters which can be measured easily and reliably.

Another problem is how to cope with the measured variability. One approach is to use Monte Carlo techniques (Warrick et al., 1977a; R.E. Smith, pers. commun., 1978) and another is to scale soil-water properties according to the concept of similar media (Warrick et al., 1977b). The scaling approach holds greater promise when realistic field processes are to be simulated (Peck et al., 1977).

The attractiveness of the scaling theory lies in its potential to express spatial variability in terms of a single physically-based parameter. Natural soils usually do not satisfy the strict similar-media requirements, and therefore the theory may have to be applied in an approximate form. Despite this, Warrick et al. (1977b) evaluated factors for scaling water retention and unsaturated hydraulic conductivity data collected from a large number of field sites. Reichardt et al. (1972) showed that laboratory-measured horizontal infiltration into soils having wide textural range could be scaled successfully and scaling factors could be determined from simple infiltration tests. This suggests a rather attractive way to evaluate scaling factors. However, no such study has been reported for field-measured infiltration data.

The objectives of this paper are: to evaluate surface hydrologic variability of a grassland watershed in terms of parameters of a physically-based infiltration equation developed by Philip (1957); and to test whether infiltration measured in the field at several sites could be scaled according to the similar-media concept so that the spatial variability could be characterized by the variability of a single parameter.

MATERIALS AND METHODS

Measurement and analyses of infiltration

Experimental design

The Washita River Basin, in the Southern Great Plains of the U.S.A. is currently the area of hydrologic research by the USDA—SEA staff at Chickasha, Oklahoma. In this basin, a grassland watershed (R-5), was chosen for this study. Some measurements on the spatial and temporal variability of soil moisture within this watershed have been made previously (G. Coleman, pers. commun., 1978).

R-5 is a native grassland watershed with a slope of about 3% and an area of 9.6 ha. Careful management has prevented erosion problems. The dominant vegetation consists of buffalo grass (*Buchloe dactyloides*), blue grama (*Bouteloua gracilis*) and bluestem grasses (*Andropogon* spp.). The Soil Conservation Service (SCS) recognises three soils in the watershed: Renfrow silt loam, Grant silt loam and Kingfisher silt loam, which respectively occupy 51, 43 and 6% of the area.

Twenty-six sites in the watershed were selected for infiltration measurement. These sites were basically on a grid pattern with additional measurements to characterize major soils. Soil types and surface contours are shown in Fig. 1.

Measuring procedure

Infiltration measurements were made with double-ring infiltrometers in September when the soil profile had dried out. A separate constant-head device supplied water to each ring. The inner ring was 45.7 cm in diameter and 15 cm high, while the outer one was 76.2 cm in diameter and 15 cm high. On the selected site, grass was cut down to 2 cm and rings were driven at least 5 cm into the ground. Special care was taken to ensure minimal disturbance within the inner ring.

6.5 cm of water was applied simultaneously to both inner and outer rings. Infiltration in the inner ring was recorded using a hook gauge at 30-s intervals for the first 3 min.; every minute from 3 to 15 min.; every 5 min. from 15 to 30 min.; every 15 min. until 60 min.; and subsequently every hour or so. All infiltration tests were run for at least 60 min. For the first 15 min. the water level in the outer ring was adjusted manually to that in the inner ring, and thereafter the inner and outer rings were connected to separate water supplies which maintained a constant water level 4 cm above the soil surface. The rings were covered with white plastic sheets to minimize evaporative losses.

During the period of measurement, the average depth of water penetration was 20 cm, which is less than half the radius of the buffer ring. Tests showed that increasing the size of the buffer ring did not affect the results. Thus, during the time interval of 0—60 min., the infiltration was essentially one dimensional.

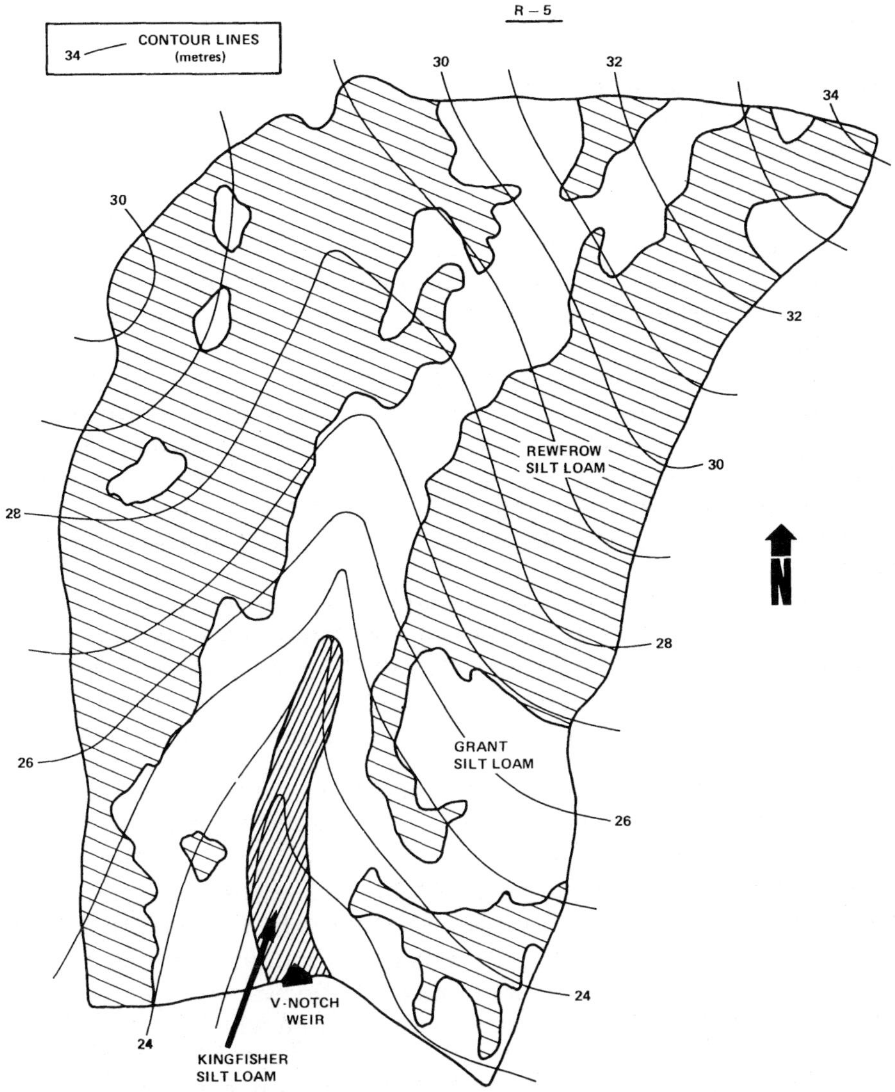

Fig. 1. Distribution of soils and topographic contours on the R-5 watershed, Chickasha, Oklahoma. Soil *1* = Renfrow silt loam; soil *2* = Grant silt loam; and soil *3* = Kingfisher silt loam.

Analysis of data

Several mathematical expressions describe infiltration into soils. Some are based on the physics of flow through porous media (e.g., Green and Ampt, 1911; Philip, 1957) while others are empirical (e.g., Kostiakov, 1932; Horton, 1940; Holtan, 1961; Smith, 1972). Among the physically based expressions, Philip's two-parameter equation is the most attractive for easiness of computation and also because time can be expressed as an explicit function of cumulative infiltration or vice versa (Swartzendruber and Youngs, 1974). This equation was applied to the data collected in the present study.

Philip (1957) showed that when water is ponded on a deep homogeneous soil with uniform initial water content θ_0, the cumulative infiltration I at time t is approximated by:

$$I = St^{1/2} + At \qquad (1)$$

where

$$S = \int_{\theta_0}^{\theta_s} \phi \, d\theta \qquad \text{and} \qquad A = \int_{\theta_0}^{\theta_s} \chi \, d\theta + K_0 \qquad (2), (3)$$

K_0 is the hydraulic conductivity at θ_0; and θ_s is the water content at saturation. ϕ and χ are functions of $K(\theta)$ and $\psi(\theta)$. K_0 usually makes a negligible contribution to A. Theoretical and experimental analyses (e.g., Youngs, 1968; Philip, 1969; Talsma, 1969; Brutsaert, 1976) suggest that the value of A lies between $\frac{2}{3} K_s$ and $\frac{1}{3} K_s$, but usually closer to $\frac{1}{3} K_s$, where K_s is the saturated hydraulic conductivity.

Values of ϕ and χ and therefore S and A can be calculated if the basic hydraulic properties, $K(\theta)$ and $\psi(\theta)$, are known for the soil. Experimental evaluation of these functional relationships for soils is time-consuming and expensive, particularly for large and variable land surfaces. However, recent work (e.g., Talsma, 1969; Talsma and Parlange, 1972; Smiles and Knight, 1976) suggests that S and A can be evaluated directly. In the present case, these parameters were calculated from infiltration data.

Determination of parameters

During the very early stages of infiltration when the contribution of gravity is small, the second term in eq. 1 is negligible relative to the first, and thus S can be calculated as the slope of an I vs. $t^{1/2}$ plot which should be linear. Values of S calculated by this method are termed S^*. On average, a water head of 6 cm was used for early infiltration measurements from which S^* was calculated, while for the rest of the infiltration a constant water head of 4 cm was used. This would cause over-estimation of sorptivity, since it has been reported that the initial infiltration rate increases with the increase in depth of water above the profile, although this effect decreases as $t \to \infty$ (Philip, 1957). Analyses (Talsma, 1969) suggested that the S^*'s were overestimated by about 15%, and therefore adjustment was made to calculate the corrected sorptivity S ($S = 0.85\, S^*$). We emphasize that this is only an approximate correction since the exact correction depends on the hydraulic properties of the soil at each measuring site.

Plots of I vs. t showed that in most cases this relationship was linear for $t > 45$ min., suggesting that a steady state had been reached. The steady infiltration rate (dI/dt) was calculated from the $I(t)$ curve at $t = 60$ min., and termed i_{60}. Independent measurements of K_s of some surface (0—15 cm) soil layers (C. Hunt, pers. commun., 1978) suggested that i_{60} was slightly higher than K_s, as has been observed by others. The coefficient A of eq. 1 was approximated by $\frac{1}{3}\ i_{60}$.

Scaling theory and its application

According to the concept introduced by Miller and Miller (1956), two porous media are similar if they are geometric scale models of one another in details, i.e. all the microscopic geometric details of one of the media could be multiplied by a constant to obtain the microscopic details of the second medium. Subsequent studies (Klute and Wilkinson, 1958; Elrick et al., 1959; Wilkinson and Klute, 1959) have experimentally tested the concept for idealized porous materials packed into laboratory columns. Philip (1967) applied the concept of similar media in an analysis of infiltration into heterogeneous media.

For similar media, the water potential ψ_j and hydraulic conductivity K_j of the jth soil can be expressed in terms of the water potential ψ_r and conductivity K_r of a reference soil at the same water content:

$$\psi_j\lambda_j = \psi_r\lambda_r \quad \text{and} \quad K_j/\lambda_j^2 = K_r/\lambda_r^2 \tag{4), (5}$$

where λ_j and λ_r are the microscopic characteristic lengths of the jth and reference soils, respectively.

Based on dimensional analysis (Philip, 1957, 1967) of the parameters of eq. 1, it can be shown that the sorptivity S and the parameter A can be scaled in the following manner:

$$S_j/\lambda_j^{1/2} = S_r/\lambda_r^{1/2} \quad \text{and} \quad A_j/\lambda_j^2 = A_r/\lambda_r^2 \tag{6), (7}$$

If similar-media conditions hold, then the λ_i's of eqs. 6 and 7 will have identical values.

For convenience, dimensionless scaling factors, α, can be defined as:

$$\alpha_j = \lambda_j/\lambda_r = \lambda_j/\bar{\lambda} \tag{8}$$

where $\bar{\lambda}$ is the characteristic length of a soil with average soil-water properties, and for such average soil the scaling factor would be equal to unity. Note that any soil in the set of geometrically similar soils can be selected as the reference soil; it is convenient but not necessary to use a soil with average soil-water properties for this purpose.

Based on n infiltration tests (n sets of S- and A-parameters), the dimensionless scaling factors, α_S and α_A, can be calculated as follows:

$$\alpha_{S_j} = (S_j/\bar{S})^2 \quad \text{and} \quad \alpha_{A_j} = (A_j/\bar{A})^{1/2} \tag{9), (10}$$

where

$$\bar{S} = \frac{1}{n} \sum_{j=1}^{n} S_j \qquad \text{and} \qquad \bar{A} = \frac{1}{n} \sum_{j=1}^{n} A_j \qquad (11), (12)$$

The measured infiltration $I(t)$ can be scaled using either α_S or α_A, according to the relationship:

$$I^{*} = \alpha I \qquad \text{and} \qquad t^{*} = \alpha^3 t \qquad (13), (14)$$

where I^{*} and t^{*} are the scaled cumulative infiltration and scaled time, respectively.

If the infiltration at each site can be adequately approximated by eq. 1, and if the similar-media conditions hold, then the infiltration of all sites in the scaled form should be described by:

$$I^{*} = \bar{S} t^{*\frac{1}{2}} + \bar{A} t^{*} \qquad (15)$$

Then the error sums of squares of the original data (where the average curve is described by $\bar{S}$ and $\bar{A}$) and the scaled data should be comparable, provided sets of scaling factors based on S and A (α_S, α_A) are identical. Least-squares fits of the form of eq. 1 to the original and the scaled data were also carried out and the error sums of squares determined. Least-squares fits, with additional terms, were tried but the reduction in the sums of squares was negligible.

Scaling factors were also calculated by statistical optimization. By using the relationships of eqs. 13 and 14, the optimum scaling factor (α_{opt}) for each infiltration run was selected by minimizing SS, the sum of squares of the differences between the scaled infiltration (I^{*}) and the average infiltration ($\bar{I}$ based on $\bar{S}$ and $\bar{A}$). Here SS is given by:

$$SS = \sum_{t^{*}_j} (I^{*}_{t^{*}_j} - \bar{I}_{t_j})^2 \qquad (16)$$

where the sum is taken over all values of t^{*}_j for the entire infiltration period of the run.

RESULTS AND DISCUSSION

Spatial variability and characterization of infiltration

Application of the Philip two-parameter equation

Calculation of S^* from plots of I vs. $t^{\frac{1}{2}}$ ($t \leqslant 4.0$ min.) was justified by the linearity (r always $\geqslant 0.98$, mostly $\geqslant 0.99$). Similarly, linearity in the latter part of the I vs. t plots justified calculation of i_{60}. In all cases a steady state was attained within 60 min.

We examined the infiltration data from each site to determine whether the infiltration behaviour over the full time range (0—60 min.) could be adequately described by using the parameters S and A, determined at the extremes of the range. Overall agreement between the measured and calculated infiltration was reasonable (the error was usually < 15%, but always < 25%).

In order to show how well eq. 1 describes all infiltration runs, the following dimensionless parameters are defined:

$$\beta = AI/S^2 \qquad \text{and} \qquad \tau = A^2 t/S^2 \qquad (17), (18)$$

In the dimensionless form eq. 1 becomes:

$$\beta = \tau^{1/2} + \tau \qquad (19)$$

If the infiltration process is adequately described by eq. 1, then the data from all of the infiltration tests should follow the single curve characterized by eq. 19. Variations in the original I vs. t data for the 26 infiltration runs are shown in Fig. 2A. The same data, converted into dimensionless form by eqs. 17 and 18, and presented in Fig. 2B, show that eq. 1 adequately describes the infiltration process.

In most field conditions S and K_s (and therefore A) can be measured easily and reliably (Talsma, 1969). Both the sorptivity and parameter A so evaluated have distinct physical meanings; sorptivity takes into account the hydraulic properties of the soil, $\psi(\theta)$ and $K(\theta)$, and the initial and final water content, while parameter A is related to steady-state infiltration or saturated hydraulic conductivity. Here, these parameters are called bulk properties, firstly because they are measured on a much larger volume of in situ soil compared to an "undisturbed" core sample usually used for characterization of hydraulic properties, and secondly because they encompass effects of many factors such as macropores, initial water content, soil texture and structure, etc., which influence water infiltration into field soils.

Eq. 1 has been reported to give reasonable predictions of infiltration in laboratory packed soil columns (Philip, 1957; Youngs, 1968; Talsma, 1969; Smiles and Knight, 1976), and also in the field (Watson, 1959; Talsma, 1969; Talsma and Parlange, 1972; Smiles and Knight, 1976). Difficulties in the use of this equation have also been experienced (Whisler and Bouwer, 1970; Collis-George, 1977). Skaggs et al. (1969), working with field data, reported a slightly better fit by using the Horton equation rather than eq. 1. Their data, however, were derived from rain infiltration—runoff studies where the initial ponding condition assumed by Philip did not occur. Our analyses of various algebraic equations to describe ponded field infiltration suggested a preference to eq. 1 (M.L. Sharma and G.A. Gander, unpublished data, 1978). Based on the evidence presented earlier in this section, field infiltration is adequately described by the Philip two-parameter equation and, therefore, we suggest that the surface hydrologic variability of the watershed in question can be expressed in terms of the parameters of this equation.

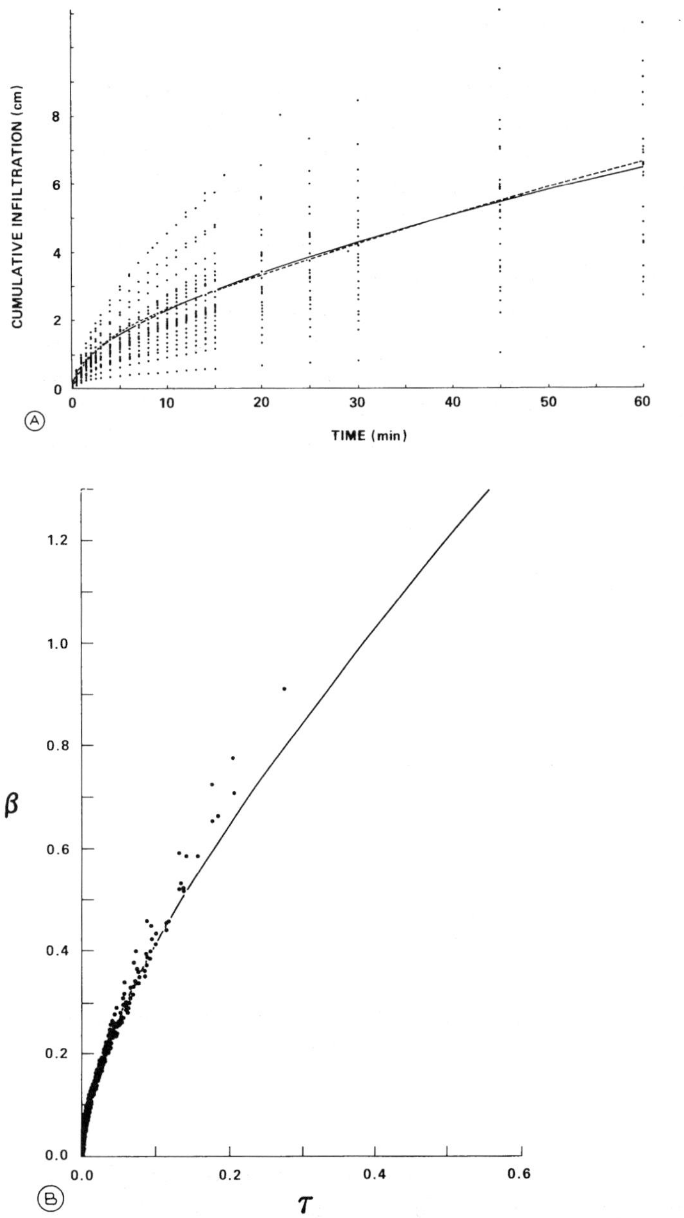

Fig. 2. A. Field-measured cumulative infiltration vs. time plots of 26 sites on a watershed. The *solid line* is an "average" curve computed from average infiltration parameters, $\bar{S}$ and $\bar{A}$, of the Philip equation, $I = St^{1/2} + At$ (eq. 1); while the *broken line* is the least-squares fit through the data. Total data points were 618, but only 616 appear on the graph, the remaining points fall beyond the selected graph scale.
B. Data of (A) in dimensionless form, where $\beta = AI/S^2$ and $\tau = A^2/S^2$.

Dependence of sorptivity on initial water content

Of the two parameters in eqs. 1—3, S is dependent on the initial water content, θ_0, but in sofar as A is a fixed fraction of i_{60} it may be considered independent of θ_0. Thus variation in S could be partly due to spatial variability in θ_0. Also, it is possible that the spatial variability in S may change if measured at different initial water contents. This was assessed by measuring S at exactly the same locations with three different initial water-content conditions on the R-5 watershed. Fig. 3 shows a linear dependence of S on θ_0. These data suggest that the variability in S was much higher than in θ_0.

Although S is dependent on θ_0, detailed analyses of variance suggested that the spatial variation in S at anyone set of initial water-content conditions was largely due to variation in inherent soil properties (pore size distribution) and the contribution due to variability in initial water content within a condition was rather small.

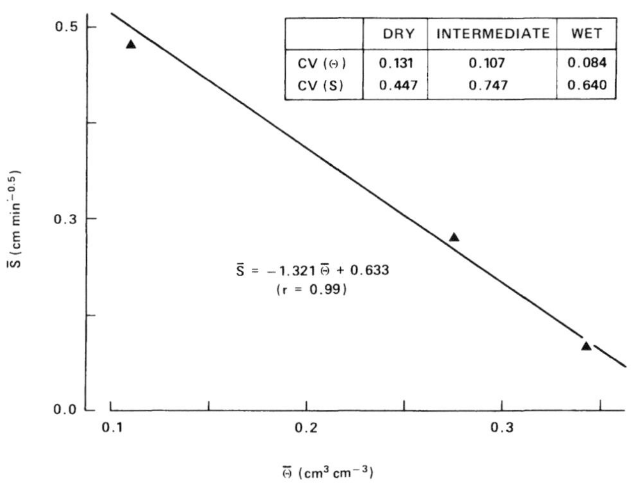

Fig. 3. Relationship between the average sorptivity $\bar{S}$ and the average initial water content $\bar{\theta}_0$ of the surface layer (0—15 cm) for the R-5 watershed (based on 15 observations).

Hydrologic variability and properties of parameters

The spatial variation of the parameters S and A over the watershed (Fig. 4) suggests no definite pattern. There may be a band of higher S and A values (higher-intake area) on the outer edges of the upper part of the watershed, but otherwise the distribution appears to be random, irrespective of soil type or topographic position. Variations in S over the watershed were within one order of magnitude; while variations in A were within two orders of magnitude. Similar variations in these or similar parameters have been previously reported (Talsma, 1969; Rogowski, 1972; Nielsen et al., 1973).

Now that hydrologic parameters can be averaged or lumped meaningfully,

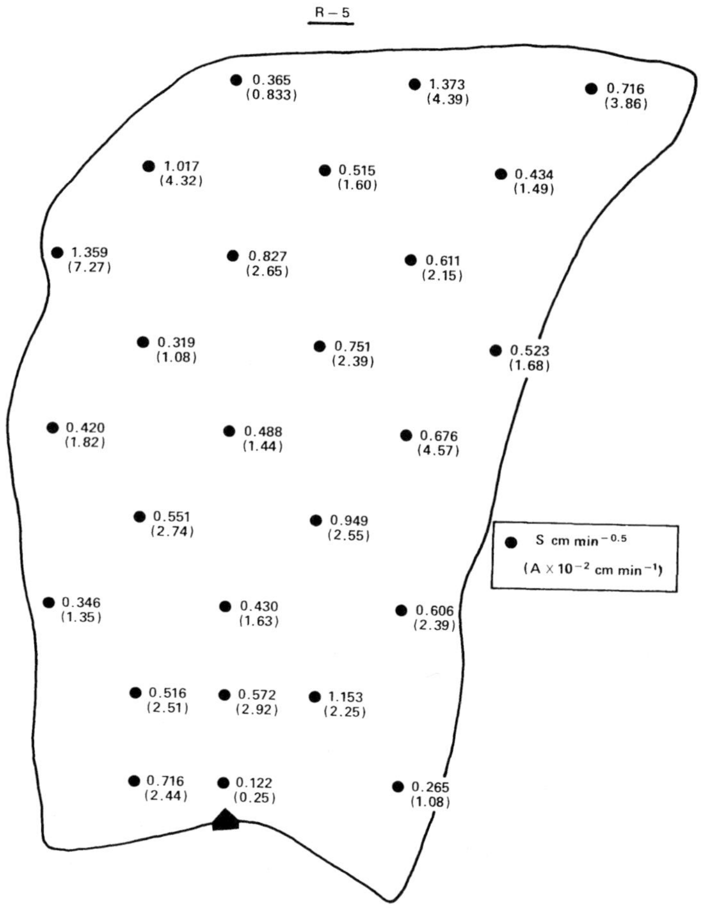

Fig. 4. Spatial distribution of the infiltration parameters, S and A, over the R-5 watershed.

it is appropriate to examine the properties of these parameters, regarding their relationships with each other and also their distribution functions. Fig. 5 shows a significant relationship between S and A. In general, there is a tendency for S to increase with increasing A. Talsma (1969) reported a poor relationship between sorptivity and saturated hydraulic conductivity.

Considering the watershed as one soil unit, the frequency distributions of S, A and measured I at several times were examined. These are presented as both normal and log-normal probability plots in Fig. 6. Approximate lines were drawn through all except one point. This point was ignored as it was associated with an exceptionally low infiltration rate. Examination revealed that this measuring site was unusual as it was very close to the streamflow-measuring station (Fig. 4) where clay materials have deposited over the years. This site represented an extremely small fraction of the watershed.

Fig. 6 suggests that both infiltration parameters, S and A, and I at t =

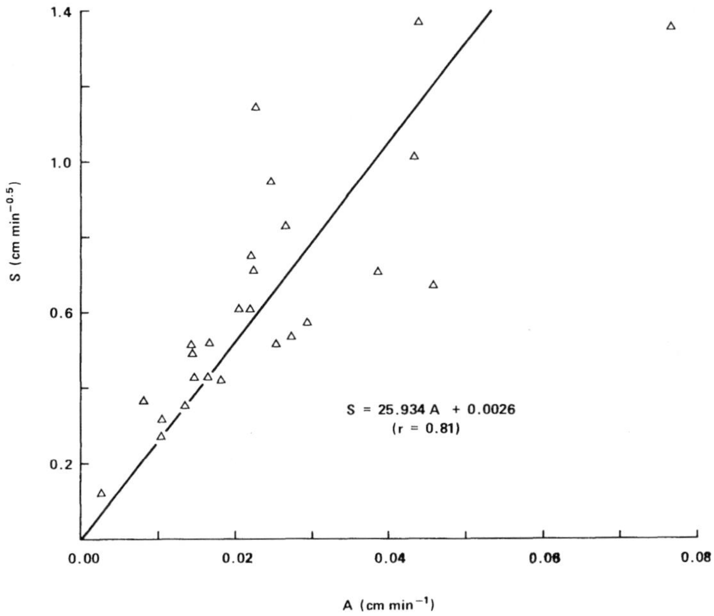

Fig. 5. Relationship between infiltration parameters S and A of the R-5 watershed.

30 min. are better approximated by log-normal rather than normal frequency distributions. Similar results were obtained for both infiltration rates and cumulative infiltrations examined at other times. Since A is closely related to the saturated hydraulic conductivity, its distribution function is in accordance with other observations (e.g., Rogowski, 1972; Nielsen et al., 1973). Although sorptivity is a much less variable parameter, its distribution is also closer to log-normal. A log-normal distribution for S, which has been suggested previously (Brutsaert, 1976), but has not been demonstrated with experimental data. However, it appears that assuming a normal distribution for S may not introduce as much error as the similar assumption for the hydraulic conductivity or A.

Calculating composite infiltration curves

Characterization of large watersheds in terms of hydrologic parameters should ultimately aim at identifying relatively homogeneous areas so that some sort of averaging or lumping could be used to obtain a composite $I(t)$ function, which should be defineable with some degree of confidence in terms of the scatter of expected values. The degree of lumping would be dictated by the extent of variability, desired accuracy or level of error tolerance, and the need for which the information would be used.

We assume there are n infiltrometers distributed randomly on the watershed and infiltration in each of the infiltrometers behaves independently. If cumulative infiltration as a function of time $I(t)$ at n locations is characterized by n

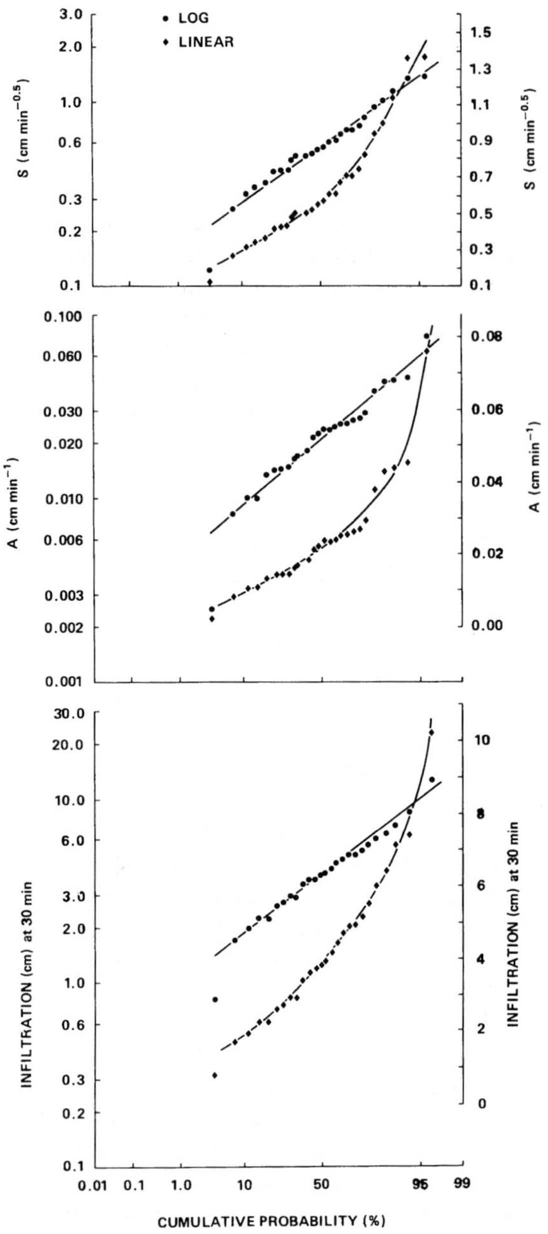

Fig. 6. The frequency distributions of S, A and I at t = 30 min. for observations on the R-5 watershed.

sets of S- and A-parameters, then the composite infiltration response of the watershed $I_w(t)$ will be given by:

$$I_w(t) = \frac{1}{n} \sum_{j=1}^{n} I_j(t) \qquad (20)$$

This is simply described by $\overline{S}_w$- and $\overline{A}_w$-parameters. Thus, the composite response of a watershed could be characterized by the arithmetic averages of the parameters S and A. However, studies based on Monte Carlo simulations suggest that the extent of variability of parameters and their frequency distributions affect the shape of the composite infiltration curve (R.E. Smith, personal commun., 1978; E. Seely and M.L. Sharma, unpublished data, 1978). Thus, for a watershed with great soil spatial variability, an infiltration curve based on average parameters can not be sufficient to describe the composite response.

Furthermore, vertical water fluxes at n locations, supposedly representing behaviour of homogeneous areas, may not behave independently. It is conceivable that during early stages of infiltration, the fluxes from spatially varying soil would be essentially one-dimensional and independent. For this period, arithmetic means of the parameters would be appropriate. However, as the wetting front depth increases, it is likely that the flow becomes multidimensional, particularly from areas having high infiltration rates. Under these situations, an areal flux would be between their arithmetic and harmonic means (Bouwer, 1969) and for this period an infiltration curve based on the geometric means of the parameters could be appropriate.

Composite $I(t)$ functions based on 26 infiltration tests for the whole watershed and separately for the three soils, calculating of which is based on arithmethic as well as geometric means of S and A, are presented in Fig. 7. The approximately 95% confidence limits, calculated as twice the standard errors of the means of measured I at selected times are presented in Table I. In our case, the composite curves based on arithmetic and geometric means of the parameters are not significantly different.

The infiltration behaviour of soils *1* and *2* are very similar and their infiltration parameters are higher than those of soil *3* (Fig. 7). However, soil *3* occupies only a small proportion of the watershed (Fig. 1), where only three measurements were made, one of which had an exceptionally low infiltration rate and represented negligible fraction of the main type of soil (as discussed earlier). Considering the variability (Table I), differences between the soils appear to be marginal and therefore for hydrologic considerations, the whole watershed can be assumed to have one soil type. This does not conflict with the SCS classification which identifies these soils, because the classification is not primarily based on surface hydrologic considerations. However, these conclusions should be treated with caution, since they are based on an assumption that measured infiltration behaviour within and between soils does not inter-

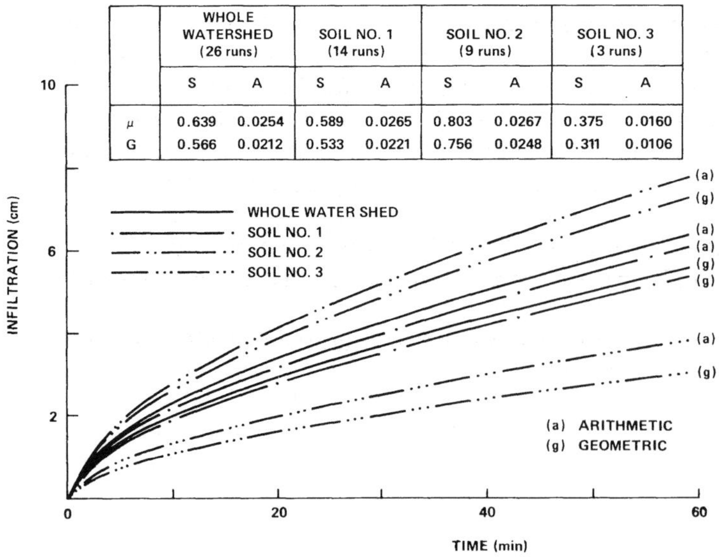

Fig. 7. Composite infiltration curves of the whole R-5 watershed and the three soils *1*, *2* and *3* calculated by using arithmetic and geometric means of the parameters.

TABLE I

Arithmetic means of measured I, $\mu(I)$, and standard errors, SE, at selected times for the whole watershed and also separately for the three soils of the R-5 watershed

t (min.)	Whole watershed		Soil *1*		Soil *2*		Soil *3*	
	$\mu(I)$ (cm)	2SE	$\mu(I)$ (cm)	2SE	$\mu(I)$ (cm)	2SE	$\mu(I)$ (cm)	2SE
2	0.849	0.134	0.776	0.168	1.055	0.216	0.572	0.305
5	1.477	0.258	1.353	0.332	1.856	0.412	0.920	0.534
10	2.215	0.410	2.033	0.534	2.792	0.648	1.335	0.858
15	2.818	0.534	2.600	0.708	3.535	0.828	1.684	1.136
30	4.271	0.840	4.054	1.216	5.176	1.196	2.562	1.828
60	6.624	1.377	6.514	2.110	7.654	1.832	4.050	3.152

act. Perhaps, the significance of identifying several hydrologically different areas can be assessed better by looking at its effect on the hydrologic response of the watershed.

Scaling of infiltration

From 26 sets of S- and A-values, α_S and α_A were calculated by using eqs. 9 and 10 and the measured infiltration I at time t, and the data were scaled according to eqs. 13 and 14. Comparison of measured infiltration data (Fig. 2A) with the scaled data (Figs. 8 and 9) suggests that the scatter of data around a "best-fit" curve is much less in the scaled form than for the original data. It is also clear that the "average" curve based on $\bar{S}$ and $\bar{A}$ does not describe the average of the scaled infiltration in both cases. Scaled-data points based on α_A fall above the "average" curve while those based on α_S usually fall below it. The error sums of squares from "average" and best-fit curves, presented in Table II, suggest that better scaling is achieved with α_A. The scaling factors, α_S and α_A are significantly different, although they are correlated to some extent (Fig. 10).

The conditions of the similar-media concept are that soils at different loca-

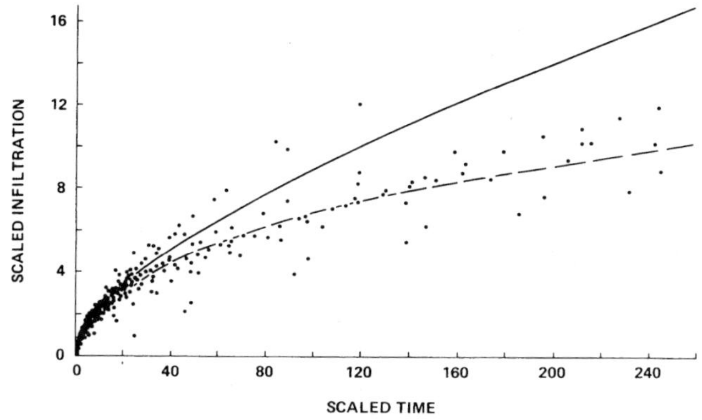

Fig. 8. Infiltration data from 26 sites scaled by α_S. Only 570 out of 618 data points appear on the graph, the rest fall beyond the selected scale. The *solid line* is computed from $\bar{S}$ and $\bar{A}$, while the broken line is the least-squares fit through the scaled data.

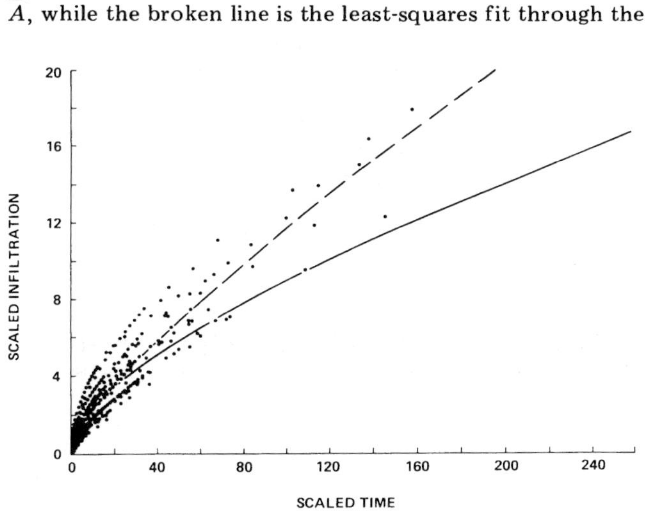

Fig. 9. Infiltration data from 26 sites scaled by α_A. Number of data points appearing on the graph is 617. The *solid line* is computed from $\bar{S}$ and $\bar{A}$, while the *broken line* is the least-squares fit through the scaled data.

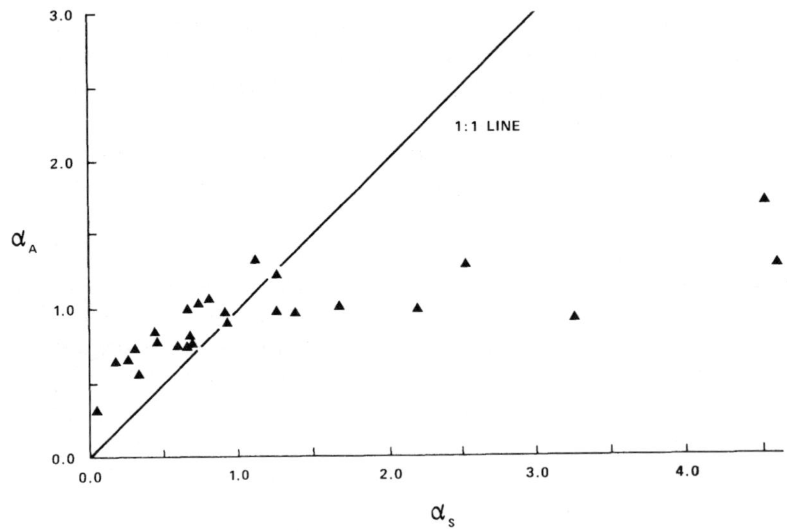

Fig. 10. Relationship between α_S and α_A for the 26 infiltration sites.

TABLE II

Scaling factors calculated by different methods and used to scale field infiltration data sets of 26 tests, each of which can be adequately described by the two-parameter Philip equation, $I = St^{1/2} + At$ (eq. 1)

Scaling factors used	*SS* from the average curve $\overline{SS}$ = 0.639 cm min.$^{-1/2}$ *AA* = 0.0254 cm min.$^{-1}$	*SS* from the best fitted	Scaled infiltration parameters	
			S^*	A^*
α_S	86,984.1	1,069.7	0.604	0.0036
α_A	677.4	279.4	0.491	0.0681
α_{SA}	2,986.0	297.4	0.677	0.0109
α_{SA}^H	83.2	78.5	0.676	0.0241
α_{SA}^G	471.2	131.8	0.682	0.0166
α_{opt}	67.1	65.8	0.656	0.0247
Unscaled	1,085	1,079	0.555	0.0399

Average parameters, $\overline{S}$ = 0.639 cm min.$^{-1/2}$, $\overline{A}$ = 0.0254 cm min.$^{-1}$, and the "average" infiltration curve described by these parameters were assigned a scaling factor of 1.00. The error sums of squares of the scaled data were calculated as deviations from this "average" curve; and also from statistically best-fit curve for the scaled data. A total of 618 points were used in the analysis.

tions in the watershed must have identical porosity and the same relative pore-size distributions and that these should not change with time or degree of saturation (Miller and Miller, 1956). However, in practice soils do vary with respect to porosity and exhibit some shrinking or swelling with varying degree of saturation. Therefore, it is unlikely that any field soil will ever fully satisfy the conditions required by the similar-media theory.

Theoretically, either of the infiltration parameters S or A could be used in calculating scaling factors and these should be identical if all the demands of similar media are satisfied. However, in this study the scaling factors α_S and α_A are significantly different (Fig. 10). This difference may be due to changes in porosity and pore-size distribution with saturation as infiltration proceeds, and also to inherent variation in initial soil-water content which affects S and not A. Apparently, scaling of infiltration based on α_A would be preferred over that based on α_S.

The infiltration data scaled by using α_{opt} show a considerable reduction in the scatter of points around the "average" and best-fit curves, and a similar reduction in the error sums of squares (Fig. 11, Table II). Comparisons suggest

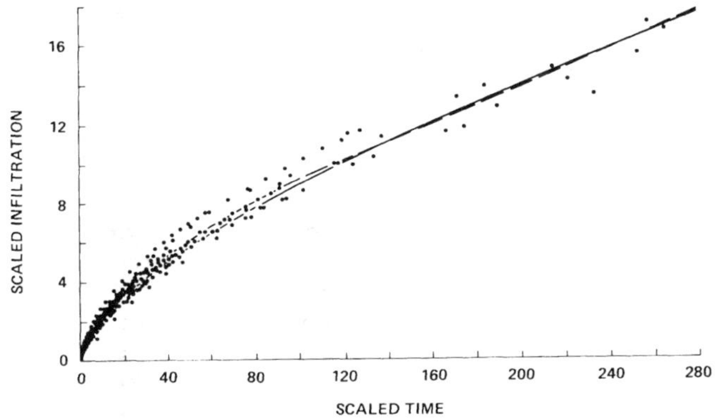

Fig. 11. Infiltration data from 26 sites scaled by α_{opt}. Number of data points appearing on the graph is 611. The *solid line* is computed from $\overline{S}$ and $\overline{A}$, while the *broken line* is the least-squares fit through the scaled data.

that α_{opt} values fall between those of α_S and α_A. Therefore, infiltration data were scaled by using arithmetic ($\overline{\alpha}_{SA}$), geometric (α_{SA}^G) and harmonic (α_{SA}^H) means of these scaling factors. Use of $\overline{\alpha}_{SA}$ resulted in considerable reduction in the error sums of squares compared with that with α_S, but the result was almost identical to that of α_A (Table II). A remarkable reduction in the error sums of squares was achieved both by α_{SA}^G as well as α_{SA}^H, the best results being obtained with α_{SA}^H. All scaling factors are fairly well correlated with α_{opt}, the strongest correlation being with α_{SA}^H (Table III). Similarities between α_{opt} and α_{SA}^H are evident by their almost identical means and standard deviations (Table III) and almost 1:1 relationship between the two (Fig. 12). This indicates that both early- and late-stage infiltration parameters must be considered and that scaling by only one of these parameters may give poor results. The fact that the geometric (α_{SA}^G) and harmonic (α_{SA}^H) means were better than the arithmetic mean ($\overline{\alpha}_{SA}$), suggests that equal weighting is not required for early- and late-stage parameters. More research is required to fully answer the reasons behind this observation.

TABLE III

Arithmetic mean, μ; geometric mean, G; standard deviation, σ; and coefficient of variation, CV of 26 scaling factors calculated by different methods and their correlation (r = correlation coefficient) with α_{opt}

Scaling factors	μ	σ	CV	G	r
α_S	1.244	1.233	0.991	0.793	0.927
α_A	0.942	0.288	0.306	0.897	0.913
$\bar{\alpha}_{SA}$	1.093	0.732	0.669	0.904	0.960
α_{SA}^H	0.963	0.544	0.565	0.790	0.991
α_{SA}^G	1.014	0.638	0.629	0.832	0.984
α_{opt}	0.965	0.584	0.605	0.766	1.000

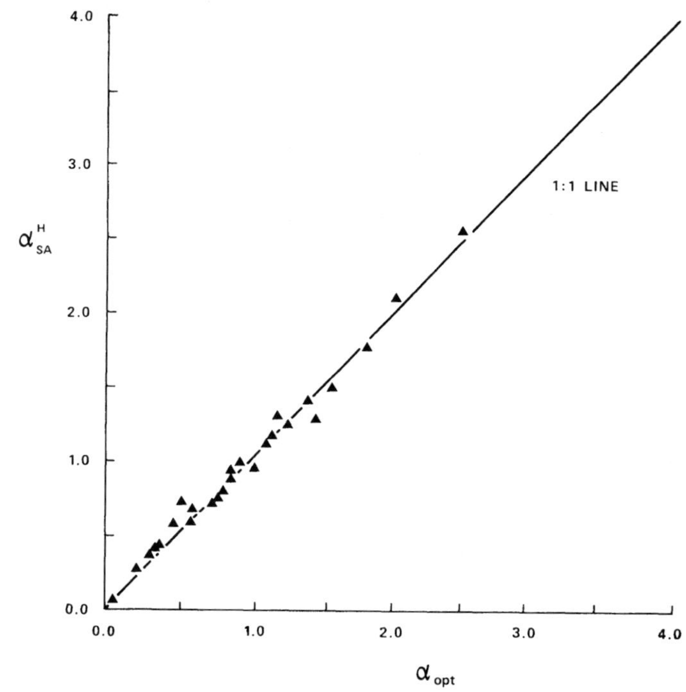

Fig. 12. Relationship between α_{opt} and α_{SA}^H for the 26 infiltration tests.

In a recent analysis, Warrick et al. (1977b) reported that the factors calculated to scale the $\psi(\theta)$ data of field soils were not the same as those calculated to scale the $K(\theta)$ data of the same set of soils. They found that scaling factors calculated from $\psi(\theta)$ were more effective in scaling the unsaturated hydraulic conductivity data than vice versa. Considerably larger variability in the scaling factors from $K(\theta)$ compared to those from $\psi(\theta)$ was attributed to the fact that K is more sensitive to variation in θ than is ψ, and also because the experimental techniques for measuring K in the field are not as well developed as those for ψ. In our case, scaling factors calculated from S exhibited much larger variability than those calculated from A. This may be primarily because S is much more sensitive to variable initial soil-water conditions than is A, although the magnitude of spatial variability in A is usually much larger than in S.

Scaling factors calculated by different methods show considerable deviation from the mean and large skewness (Table III). Frequency distribution plots of α_{opt} (Fig. 13) suggest that, if we ignore the lowest scaling factor which is associated with exceptional soil of rare occurrence (as discussed earlier), the distribution is better approximated by a log-normal rather than a normal function. Similar plots of all the other sets of scaling factors, not presented here, suggested a general preference to a log-normal distribution. Log-normal frequency distributions for scaling factors have also been reported by Warrick et al. (1977b).

The observed coefficient of variation of scaling factors of about 0.60 (Table III) was well within the range (0.48—1.70) of the coefficient of variations for scaling factors reported by Warrick et al. (1977b). However, it was considerably higher than that (0.25) assumed by Peck et al. (1977) in a recent simulation study. By assuming a normal frequency distribution, Peck et al. concluded that the soil spatial variability had a rather small effect on the components of water balance. However, further studies (Sharma and Luxmoore, 1979) suggest that their conclusions may not hold if more realistic values of

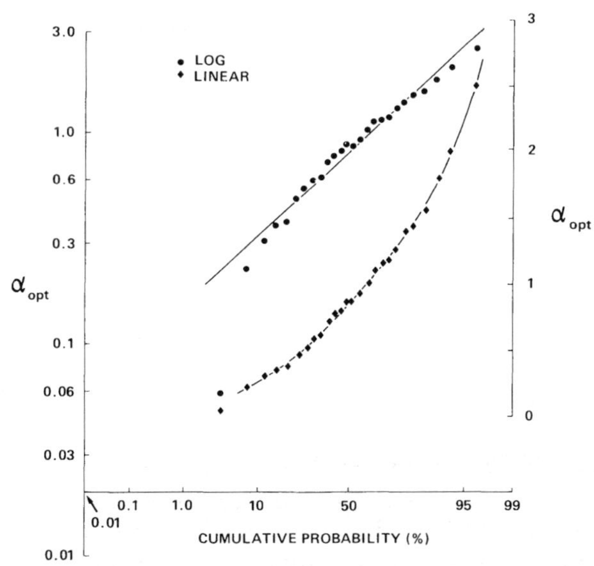

Fig. 13. The frequency distribution of α_{opt}.

coefficients of variation of scaling factors are used and when frequency distribution for scaling factors is considered log-normal.

Our studies demonstrate that, with some approximations, the similar-media concept can be used to scale field-measured infiltration data. Further work may be required to test whether the same scaling factors could be applied in scaling other soil-water properties, such as water retention, $\psi(\theta)$, and unsaturated hydraulic conductivity, $K(\theta)$. If the same factors apply, then calculation of scaling factors from simple infiltration tests would offer a rather convenient method for generating basic soil hydraulic properties, $\psi(\theta)$ and $K(\theta)$, the determination of which is otherwise both time-consuming and tedious.

ACKNOWLEDGEMENT

We are thankful to Dr. A.J. Peck for helpful comments on the manuscript.

REFERENCES

Amerman, C.R., 1973. Hydrology and soil science. In: Field Soil Water Regime, Soil Sci. Soc. Am. Spec. Publ. No. 5, pp. 167—180.

Bouwer, H., 1969. Planning and interpreting soil permeability measurements. J. Irrig. Drain Div., Proc. Am. Soc. Civ. Eng., 95: 391—402.

Brutsaert, W., 1976. The concise formulation of diffusive sorption of water in a dry soil. Water Resour. Res., 12: 1118—1124.

Carvallo, H.O., Cassel, D.K., Hammond, J. and Bauer, A., 1976. Spatial variability of in situ unsaturated hydraulic conductivity of Maddock sandy loam. Soil Sci., 121: 1—8.

Collis-George, N., 1977. Infiltration equations for simple soil systems. Water Resour. Res., 13: 395—403.

Elrick, D.E., Scandrett, J.H. and Miller, E.E., 1959. Tests of capillary flow scaling. Soil Sci. Soc. Am. Proc., 23: 329—332.

Green, W.H. and Ampt, G.A., 1911. Studies on soil physics, 1. The flow of air and water through soils. J. Agric. Sci., 4: 1—24.

Holtan, H.N., 1961. A concept for infiltration estimates in watershed engineering. USDA, Washington, D.C., Agric. Res. Ser. Pap., 41: 1—51.

Horton, R.E., 1940. An approach toward a physical interpretation of infiltration capacity. Soil Sci. Soc. Am. Proc., 5: 399—417.

Keisling, T.C., Davidson, J.M., Weeks, D.L. and Morrison, R.D., 1977. Precision with which selected soil parameters can be estimated. Soil Sci., 124: 241—248.

Klute, A. and Wilkinson, G.E., 1958. Some tests of the similar media concept of capillary flow, I. Reduced capillary conductivity and moisture characteristic data. Soil Sci. Soc. Am. Proc., 19: 267—270.

Kostiakov, A.N., 1932. On the dynamics of the coefficient of water percolation in soils and the necessity for studying it from a dynamic point of view of purposes of ameliorations. Trans. 6th Comm. Int. Soil Sci. Soc., Russian Part A, pp. 17—21.

Miller, E.E. and Miller, R.D., 1956. Physical theory for capillary flow phenomena. J. Appl. Phys., 27: 324—332.

Nielsen, D.R., Biggar, J.W. and Erh, K.T., 1973. Spatial variability of field measured soil-water properties. Hilgardia, 42: 215—259.

Peck, A.J., Luxmoore, R.J. and Stolzy, J.L., 1977. Effects of spatial variability of soil hydraulic properties in water budget modeling. Water Resour. Res., 13: 348—354.

Philip, J.R., 1957. The theory of infiltration, 4. Sorptivity and algebraic infiltration equations. Soil Sci., 84: 257—264.
Philip, J.R., 1967. Sorption and infiltration in heterogeneous media. Aust. J. Soil Res., 5: 1—10.
Philip, J.R., 1969. Theory of infiltration. Adv. Hydrosci., 5: 215—296.
Reichardt, K., Nielsen, D.R. and Bigger, J.W., 1972. Scaling of horizontal infiltration into homogeneous soils. Soil Sci. Soc. Am. Proc., 36: 241—245.
Rogowski, A.S., 1972. Watershed physics: soil variability criteria. Water Resour. Res., 8: 1015—1023.
Sharma, M.L. and Luxmoore, R.J., 1979. Soil spatial variability and its consequences on simulated water balance. Water Resour. Res. (in press).
Singh, V.P. and Woolhiser, D.A., 1976. Sensitivity of linear and non-linear surface runoff models to input errors. J. Hydrol., 29: 243—249.
Skaggs, R.W., Huggins, L.E., Monke, E.J. and Foster, E.R., 1969. Experimental evaluations of infiltration equations. Trans. Am. Soc. Agric. Eng., 12: 822—828.
Smiles, D.E. and Knight, J.H., 1976. A note on the use of the Philip infiltration equation. Aust. J. Soil Res., 14: 103—108.
Smith, R.E., 1972. The infiltration envelope: results from a theoretical infiltrometer. J. Hydrol., 17: 1—21.
Swartzendruber, D. and Youngs, E.G., 1974. A comparison of physically-based infiltration equations. Soil Sci., 117: 165—167.
Talsma, T., 1969. In situ measurement of sorptivity. Aust. J. Soil Res., 7: 269—276.
Talsma, T. and Parlange, J.Y., 1972. One-dimensional vertical infiltration. Aust. J. Soil Res., 10: 143—150.
Warrick, A.W., Mullen, G.J. and Nielsen, D.R., 1977a. Prediction of the soil-water flux based upon field-measured soil-water properties. Soil Sci. Soc. Am. J., 41: 4—19.
Warrick, A.W., Mullen, G.J. and Nielsen, D.R., 1977b. Scaling field-measured soil hydraulic properties using similar-media concept. Water Resour. Res., 13: 355—362.
Watson, K.K., 1959. A note on the field use of a theoretically derived infiltration equation. J. Geophys. Res., 64: 1611—1615.
Whisler, F.D. and Bouwer, H., 1970. Comparison of methods for calculating vertical drainage and infiltration for soils. J. Hydrol., 10: 1—19.
Wilkinson, G.W. and Klute, A., 1959. Some tests of the similar media concept of capillary flow, II. Flow system data. Soil Sci. Soc. Am. Proc., 23: 434—437.
Youngs, E.G., 1968. An estimation of sorptivity for infiltration studies from moisture movement consideration. Soil Sci., 106: 157—163.

Commentary on Paper 29

Huff, D. D., O'Neill, R. V., Emmanuel, W. R., Elwood, J. W. & Newbold, J. D. (1982) Flow variability and hillslope hydrology.

Here is another paper that has not been widely cited but which deserves to have had a greater impact on hydrological thought and practice. This may be because the results presented (as also in Paper 30, by Hjelmfelt & Burwell) have disturbing implications that hydrologists, and particularly catchment hydrological modellers, would prefer not to think about. It is only a short paper, effectively a technical note, about the technique of using a tracer to determine the incremental discharges downstream in the west branch of the Walker Branch experimental catchment in Tennessee, USA. This was the catchment that was used by the Environmental Studies Division of the Oak Ridge National Laboratory. Dale Huff and the other authors worked at Oak Ridge at the time.

The important result of the tracing experiments was the very large variability in hillslope inputs to the channel, and the change in the relative magnitude of the inputs under different conditions. Similar variability had been previously suggested by Anderson & Burt (Paper 23) and Beven (1977, 1978) in relation to the variation in topography (spurs and hollows) contributing to the stream channel. At Walker Branch, the soils have a relatively deep regolith on the divides, with shallow depths in the valley bottom overlying a dipping dolomite bedrock. Huff *et al.* speculate that it is the geological structure rather than surface topography that controls the variability in the inputs to the channel, such that when one of the dipping bedrock layers intersects the stream channel, water is forced back into the channel. The problem then, in characterizing the system for a hydrological model, is in having a sufficiently accurate conceptual model of the subsurface geological structures and soil horizons, in order to be able to specify appropriate model parameters. A similar problem is encountered in groundwater modelling problems, where the results of models are often strongly dependent on the specification of the geological conceptual model. But if the effects on the hillslope contributions to the stream are so strong, how will it be possible to characterize the system without extensive subsurface geophysics across the catchment? And, if this is just one small catchment area, how will it be possible to do so across larger scale catchments?

There is an analogy here, with the more recent study of Freer *et al.* (2002) on the hillslopes of the Panola catchment in Georgia. There, they showed that the predictions of the spatial patterns of downslope flows on a small hillslope plot could be greatly improved if flows over the undulations of the bedrock surface were considered explicitly. If this were generally true (even if only for areas of similar bedrock in that region), how would it be possible to obtain information on bedrock topography everywhere? This problem of characterization of local effects that might be important in governing the responses of hillslopes within a region has been called the problem of "uniqueness-of-place" by Beven (2000). The Huff *et al.* study contributed to the long-term thinking that led to that paper.

The tracing experiments in Walker Branch were later extended by Genereux *et al.* (1993) who used a combination of naturally occurring radon and calcium in spatial flow separations, and Mulholland (1993) who used calcium and sulphate in three component mixing of storm responses at subcatchment and catchment scales.

We need more studies of the incremental discharges into stream channels, so that we are encouraged to explore the reasons for the heterogeneity of inputs found. This is, however, difficult to achieve. Gauging structures are expensive, may affect the flow pathways, and may not capture all the down-valley subsurface flows. Current meter gaugings at successive cross-sections may not provide sufficient accuracy or time resolution to allow an accurate estimation of the discharge increments. Experimental tracers may be affected by sorption or other biogeochemical effects, or may be lost into the subsurface, or may not be permitted by regulatory agencies. Naturally occurring tracers may be open to interpretation, but as shown by Genereux *et al.* (1993), might be useful in certain circumstances).

Studies in this catchment by Bob Luxmoore were also influential in suggesting that preferential flow pathways in macropores and mesopores might be important in soil hydrology (e.g. Luxmoore, 1981; Luxmoore *et al.*, 1990). Other interesting work on the hillslope hydrology of Walker Branch included an elaborate study of the flows from a hillslope hollow (e.g. Luxmoore *et al.*, 1990; Wilson & Luxmoore, 1988; Wilson *et al.*, 1989, 1991; Mulholland *et al.*, 1990) where tensiometer measurements sometimes showed wetting at depth in the soil before a response at higher levels, revealing the importance of bypassing. The profiles were intended to be used for the calculation of field soil moisture characteristic functions, but such bypassing effects would, of course, result in the calculation of negative hydraulic conductivities!!

References

Beven, K. J. (1977) Hillslope hydrographs by the finite element method. *Earth Surface Processes* **2**, 13–28.

Beven, K. J. (1978) The hydrological response of headwater and slideslopes areas. *Hydrol. Sci. Bull.* **23**(4), 419–437.

Freer, J., McDonnell, J. J., Beven, K. J., Peters, N. E., Burns, D., Hooper, R. P., Aulenbach, B. & Kendal, C. (2002) The role of bedrock topography on subsurface stormflow. *Water Resour. Res.* **38**(12), doi: 10.1029/2001WR000872.

Genereux, D. P., Hemond, H. F. & Mulholland, P. J. (1993) Spatial and temporal variability in streamflow generation on the west fork of Walker Branch watershed. *J. Hydrol.* **142**, 137–166.

Luxmoore, R. J. (1981) Micro-, meso- and macroporosity of soil. *Soil Sci. Soc. Am. J.* **45**, 6771.

Luxmoore, R. J., Jardine, P. M., Wilson, G. V., Jones, J. R. & Zelazny, L. W. (1990) Physical and chemical controls on preferred flow path through a forested hillslope. *Geoderma* **46**, 139–154.

Mulholland, P. J. (1993) Hydrometric and stream chemistry evidence of 3 storm flow paths in Walker Branch Watershed. *J. Hydrol.* **151**, 291–316.

Mulholland, P. J., Wilson, G. V. & Jardine, P. M. (1990) Hydrogeochemical response of a forested watershed to storms: effects of preferential flow along shallow and deep pathways. *Water Resour. Res.* **26**, 3021–3036.

Wilson, G. V. & Luxmoore, R. J. (1988) Infiltration, macroporosity and mesoporosity distributions on two forested watershed. *Soil Sci. Soc. Am. J.* **52**, 329–335.

Wilson, G. V., Alfonsi, J. M. & Jardine, P. M. (1989) Spatial variability of saturated hydraulic conductivity of the subsoil of two forested watersheds. *Soil Sci. Soc. Am. J.* **53**, 679–685.

Wilson, G. V., Jardine, P. M., Luxmoore, R. J., Zelazny, L. W., Tood, D. E. & Lietzke, D. A. (1991) Subsurface transport from an upper subcatchment of Walker Branch Watershed during storm events. 1. Hydrologic transport processes, 2. Solute transport. *J. Hydrol.* **123**, 297–336.

EARTH SURFACE PROCESSES AND LANDFORMS, VOL. 7, 91–94 (1982)

FLOW VARIABILITY AND HILLSLOPE HYDROLOGY*

D. D. HUFF, R. V. O'NEILL, W. R. EMANUEL, J. W. ELWOOD AND J. D. NEWBOLD
Environmental Sciences Division, Oak Ridge National Laboratory, Oak Ridge, TN 37830, U.S.A.

Received 4 February 1981
Revised 8 May 1981

ABSTRACT

Examination of spatial variability of streamflow in headwater areas can provide important insight about factors that influence hillslope hydrology. Detailed observations of variations in stream channel input, based on a tracer experiment, indicate that topography alone cannot explain flow variability. However, determination of changes in channel input on a small spatial scale can provide valuable clues to factors, such as structural geology that control subsurface flows.

KEY WORDS Hillslope hydrology Small catchments Flow variability Tracer

INTRODUCTION

As interest in hillslope hydrology has grown, it has become more obvious that certain types of field data are seldom collected. Even though there are excellent studies that report the spatial distribution of soil moisture in relation to slope position (e.g., Anderson and Burt, 1978; Kirkby and Chorley, 1967; Dunne and Black, 1970; and Beven, 1978, only one study (Anderson and Burt, 1978) presents data to indicate the spatial variability of stream channel inputs at a fixed point in time. Most work on hillslope processes has instead focused on the temporal aspects of runoff at a few specific sites on a hillslope. Both points of view are useful, and we suggest that there is much to be learned about factors that influence hillslope hydrology by examining the spatial variability of streamflow in first and second order streams. Our objective here is to indicate briefly how detailed hydrologic mechanisms can be discovered through intensive simultaneous measurements along a stream reach.

STUDY SITE AND METHODS

The study site is located on a first-order reach of the West Fork of Walker Branch, a 38·4 ha watershed on the U.S. National Environmental Research Park near Oak Ridge, Tennessee at latitude 35°58′N and longitude 84°17′W. The lithologic unit underlying the catchment is the Knox Group, which is of late Cambrian and Early Ordovician age. The Knox Group is composed primarily of massive, siliceous dolomite, which weathers rapidly to form a deep residual mantle with abundant chert on the surface (McMaster, 1963). Soils are generally well drained with saturated hydraulic conductivities near the surface ranging from 15 to greater than 140 cm/hr (Peters *et al.*, 1970). Depth of the weathered zone ranges from a metre or less next to the stream to about 30 m at the basin divides. Most of the basin has slope angles between 12 per cent and 20 per cent but the full range is 5–30 per cent. The catchment vegetation

* Research sponsored jointly by the National Science Foundation's Ecosystem Studies Program under Interagency Agreement No. DEB-78-03012 with the U.S. Department of Energy under contract W-7405-eng-26 with Union Carbide Corporation. Publication No. 1819, Environmental Sciences Division, Oak Ridge National Laboratory, Oak Ridge, Tennessee 37830, U.S.A.

0197-9337/82/010092-04$01.00

is predominantly oak–hickory (Grigal and Goldstein, 1971). The stream averages about 0·05 m in depth, 2·1 m in width and 4·5 per cent in slope with considerable variability among pools and riffles.

In conjunction with a comprehensive ecological study (Elwood *et al.*, in press) of the stream, 9·75 mCi of ^3H, as tritiated water, were released to the study site over a 30-minute period on 5 July, 1978. The diluted isotope was held in a large carboy and released through a flow-meter to ensure a constant release rate. The tracer release was made 186 m upstream from a permanent weir, which was established in 1968 and serves as a reference point for the stream channel survey (Curlin and Nelson, 1968). At the time of tracer release, the hydrologic conditions could be characterized as stable low flow. The previous major peak flow at the weir had been on 8 June, when mean daily flow was 75 l sec^{-1} in response to 81·4 mm of rainfall. On 2 July, a rainfall of 8·5 mm occurred and streamflow showed a slight response that peaked on 3 July when mean daily flow was 6·63 l sec^{-1}. The flow of the weir had dropped to 6·18 l sec^{-1} by 5 July, when the experiment was initiated. Flow decreased about 4·6 per cent over the following three days.

For a period of eight hours after the ^3H release, water samples were taken simultaneously by a team of 40 investigators at 11 sampling stations distributed along the 120 m reach below the point of release. Samples were taken every six minutes, and the frequency was increased to every three minutes during the period of highest ^3H concentrations. Water samples were collected in 25 ml polyethylene bottles from which 1 ml was transferred to liquid scintillation vials containing 10 ml of Insta-gel for counting on a liquid scintillation spectrometer. All samples were corrected for decay and counting efficiency.

Flow rates (Table I) were calculated by the standard method of taking the ratio of the amount of introduced isotope to the concentration at each sampling station integrated over time (Clayton and Smith, 1963). The integration was performed by plotting the concentration data versus time on large sheets of graph paper and determining the area under the curve with a planimeter. Velocities reported in Table I were derived by nonlinear least-squares fits to a multiple compartment model of the stream (O'Neill *et al.*, 1979).

RESULTS

Mean flow rate over the reach was 3·93 l sec^{-1} with a coefficient of variation (CV = SD × 100/Mean) of 11 per cent. The flow rate increased over this short distance by 43 per cent. Mean water velocity over the reach was 2·2 m min^{-1} with a CV of 25 per cent.

Figure 1 shows the relationship between distance along the stream channel, flow increases per unit length of channel, and physical attributes of the channel and surrounding hillslopes. An earlier stream

Table I. Variability in hydrological properties over a short (120 m) reach of first-order woodland stream. Analysis is based on 3- to 6-minute measurements following a 30-minute release of ^3H_2O

Sampling station	Distance below release (m)	Flow rates (l s^{-1})	Change in flow (l s^{-1})	Water velocity (m min^{-1})
1	5·23	3·18		
2	15·23	3·44	0·26	2·6
3	31·38	3·57	0·14	2·4
4	47·12	3·55		2·6
5	60·78	3·75	0·19	2·9
6	76·02	4·12	0·37	2·4
7	84·55	4·16		2·0
8	101·62	4·32	0·16	2·5
9	107·72	4·32		2·2
10	110·77	4·31		0·9
11	120·06	4·55	0·24	1·8

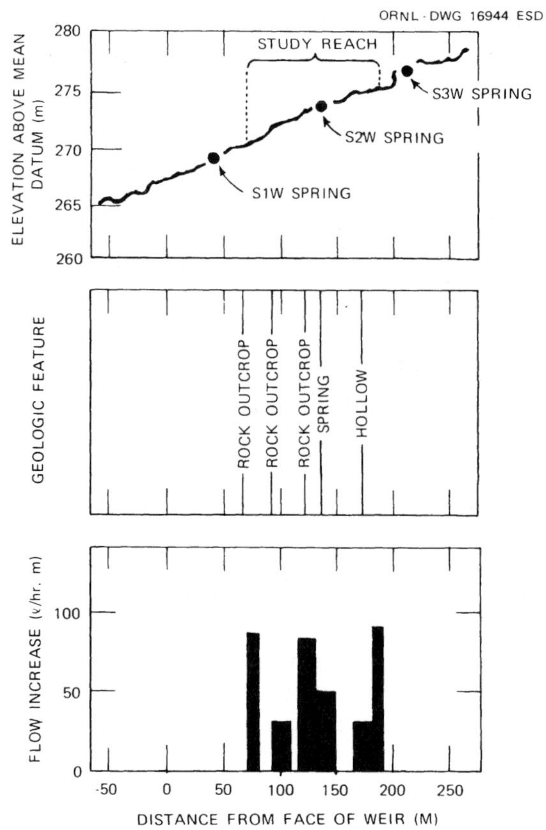

Figure 1. Variability of streamflow increases along a study reach of Walker Branch watershed

survey (Curlin and Nelson, 1968) showed the presence of spring S2W within the study reach. It was described as an intermittent spring that is active during wet periods. As one might anticipate, flow in the stream segment containing the spring showed an appreciable gain.

There was a total of six sampling stations that showed flow increases (Table I). Two other stations (4 and 10) showed decreases of 0·01 and 0·02 l sec^{-1}. These decreases are about an order of magnitude smaller than the increases and can be attributed to experimental error. The presence of the flow increases along this short (120 m) reach of the stream prompted a field inspection to identify significant features associated with them.

The field inspection revealed a correlation between flow and hillside topography. The side slopes along both sides of the channel were quite uniform, with the exception of one hollow that extended laterally from the left edge of the channel (facing downstream) at a distance of about 168 m upstream from the permanent weir. There appears to be a significant input of flow into the channel just opposite the hollow, which would tend to support the contention that hollows are more significant contributors of subsurface flow than spurs or uniform slopes (Anderson and Burt, 1978). However, examination of Figure 1 also shows three additional segments of channel that have significant subsurface inputs. All three of these segments have massive beds of siliceous dolomite that cut across them. The dolomite beds dip to the southeast at about 35°, and strike northeast at about 56°, which is nearly normal to the channel. One possible explanation of the increase in flows associated with these outcrops is that subsurface drainage that is moving down the stream valley collects at the prominent bedding planes and moves laterally along strike to the stream channel. This explanation strongly suggests that factors other than surface topography play a significant role in determining hillslope hydrology at Walker Branch. However, the important point is that determination of flow variations along the stream channel can provide valuable clues concerning factors controlling subsurface flows. Intensive simultaneous sampling appears to be an excellent approach to elucidating the detailed mechanisms of water movement on the watershed.

REFERENCES

Anderson, M. G., and Burt, T. P. 1978. 'The role of topography in controlling throughflow generation', *Earth Surface Processes*, **3**, 331–344.

Beven, K. 1978. 'The hydrological response of headwater and sideslope areas', *Hydrological Sciences Bulletin*, **23**, 419–437.

Clayton, C. G., and Smith, D. B. 1963. 'Radioisotopes in hydrology', *International Atomic Energy Agency, Vienna*, 1–24.

Curlin, J. W., and Nelson, D. J. 1968. 'Walker Branch watershed projects: objectives, facilities, and ecological characteristics', *Oak Ridge National Laboratory Report ORNL–TM–2271.*

Dunne, T., and Black, R. D. 1970. 'An experimental investigation of runoff production in permeable soils', *Water Resources Research*, **6**, 478–490.

Elwood, J. W., Newbold, J. D., O'Neill, R. V., and Van Winkle, W. in press. 'Resource spiralling: An operational paradigm for analyzing lotic ecosystems', in Fontaine, T. D. III, and Bartell, S. M. (Eds.), *The Dynamics of Lotic Ecosytems*, SREL Symposium.

Grigall, D. F., and Goldstein, R. A. 1971, 'An integrated ordination–classification analysis of an intensively sampled oak–hickory forest', *Journal of Ecology*, **59**, 481–492.

Kirkby, M. J., and Chorley, R. J. 1967. 'Throughflow, overland flow and erosion', *International Association for Scientific Hydrology Bulletin*, **12**, 5–21.

McMaster, W. M. 1963. 'Geologic map of the Oak Ridge Reservation, Tennessee', *Oak Ridge National Laboratory Report ORNL/TM–713.*

O'Neill, R. V., Emanuel, W. R., Newbold, J. D., and Elwood, J. W. 1979. 'Simulation in the service of environmental research', *Summer Computer Simulation Conference, 1979, Simulation Councils, La Jolla, CA*, p. 401–405.

Peters, L. N., Grigal, D. F., Curlin, J. W., Selvidge, W. J. 1970. 'Walker Branch Watershed Project: chemical, physical and morphological properties of the soils of Walker Branch Watershed', *Oak Ridge National Laboratory Report ORNL/TM–2968.*

Commentary on Paper 30

Hjelmfelt, A. T. & Burwell, R. T. (1984) Spatial variability of runoff.

This fascinating paper reports on the measurements of surface runoff from 40 identical plots at the Claypan Experimental Watershed, Kingdom City, Missouri, USA. This was a sufficient sample to enable a statistical analysis of the distribution of responses during each storm, and comparisons between storms. However, Hjelmfelt & Burwell note that the plots (each 27.4 m by 3.2 m) remain a small sample relative to a 40 ha field. They found that there was little consistency in the responses, either across treatments of the plots, or in ranking of the measured outputs of the different plots for different storms (e.g. their Fig. 5). The experiment had been designed with a view to quantifying the controls on surface runoff in this environment. This proved to be impossible on the basis of the data collected.

They also considered the implications of the results for hydrological modelling. They demonstrated how, for a representative model using SCS curve numbers, spatial variability of curve numbers could be used in prediction of responses but note that: "*Use of the curve number runoff model was for illustration only. It is not adequate for estimation of the mean runoff quantity with consistency. A much more detailed model that incorporates effect of cultivation on surface roughness, soil porosity, development of a surface seal, and time distribution and duration of rainfall would be required to achieve truly adequate representation*" (p.53).

We now know, of course, that incorporation of all these additional processes, and all the additional model parameters associated with them, does not necessarily make it easier to understand and predict responses. It had also been recognized previously that subsurface flow processes might also contribute to storm hydrographs in claypan soils (e.g. Minshall & Jamison, 1965). The discussion of Paper 28 in this volume tells the story of the attempts to predict the response of the R5 catchment at Chickasha, Ohio, by incorporating more and more information about soil heterogeneity and process representations.

Allen Hjelmfelt later became Professor in the Department of Biological Engineering at the University of Missouri, Columbia.

Reference

Minshall, N. E. & Jamison, V. C. (1965) Interflow in claypan soils. *Water Resour. Res.* **1**(3), 381–390.

Spatial Variability of Runoff

By Allen T. Hjelmfelt, Jr.,[1] M. ASCE and Robert E. Burwell[2]

Abstract: Forty contiguous plots (each 27.5 m × 3.2 m) were maintained in a cultivated fallow condition during spring and summer 1981. Twenty-six runoff events were measured during that period. The runoff variability as expressed by differences between plots was analyzed. The coefficient of variation for individual events in which all plots exhibited runoff ranged from 0.65–0.07. Individual plots did not react consistently throughout the series of storms, but were seen to shift between producing more and less than the average volume of runoff. The implication of these results to rainfall-runoff modeling is discussed.

Introduction

During the spring and summer of 1981, 40 adjacent runoff-erosion plots were maintained in a fallow condition. Twenty-six rainfall events occurred that produced runoff during this period. The array of plots simulate an intensively sampled watershed of field size, so the results are analyzed to describe spatial variation of runoff and water retention. Uniformity in applied management was maintained among the 40 plots, so that in the context of field-size watershed modeling even comprehensive mathematical models would treat the plots as identical. The variability among plots then is a measure of the precision that can be expected using mathematical models.

Spatial variability of cover and soil properties has long been recognized and its influence on infiltration noted. Vieira, et al. (10), Sisson and Wierenga (8), and Grah, et al. (3) have all published observations of infiltration rates exhibiting spatial variation and have used various techniques to identify the scale of spatial dependence. The influence of this spatial variability on runoff has been investigated through numerical experiments by Smith and Hebbert (9) and by Hawkins (4). This study describes a field study of the influence of the soil and surface variability on the runoff from 27.5 m × 3.2 m erosion plots.

Experimental System

The series of tests to determine spatial variation of water retention and runoff was conducted at the Claypan Experimental Watershed, operated by the U.S.D.A., Agricultural Research Service and the University of Missouri-Columbia near Kingdom City, Missouri. The test facility consisted of 40 plots (each 27.4 m × 3.2 m) arrayed as shown in Fig. 1. The soil is classified as Mexico silt loam, a member of the fine, montmoril-

[1]Hydr. Engr., Watershed Research Unit, Agricultural Research Service, U.S. Dept. of Agr., 207 Business Loop 70 E, Columbia, Mo. 65201.

[2]Soil Scientist, Retired, Watershed Research Unit, Agricultural Research Service, U.S. Dept. of Agr., 207 Business Loop 70 E, Columbia, Mo. 65201.

Note.—Discussion open until August 1, 1984. To extend the closing date one month, a written request must be filed with the ASCE Manager of Technical and Professional Publications. The manuscript for this paper was submitted for review and possible publication on May 4, 1983. This paper is part of the ***Journal of Irrigation and Drainage Engineering***, Vol. 110, No. 1, March, 1984. ©ASCE, ISSN 0733-9437/84/0001-0046/$01.00. Paper No. 18661.

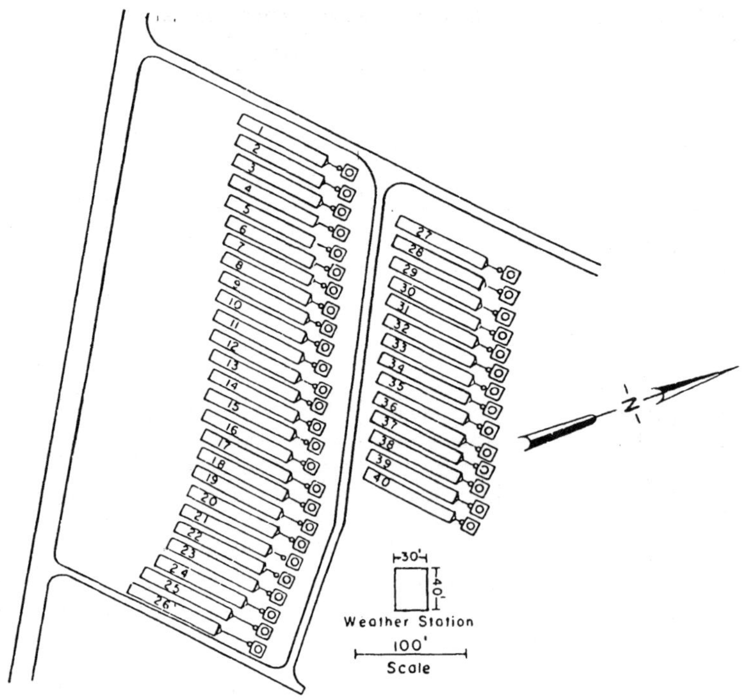

FIG. 1.—Arrangement of Plots and Weather Station at Claypan Experimental Watershed, Kingdom City, Mo.

lonitic, mesic Udollic Ochraqualfs. The soil was derived from loess, but a high content of colloidal material in the claypan subsoil that occurs 200–280 mm below the surface retards deep percolation of water. Slopes of the plots range from 3–3.5%.

Total runoff volume was measured in tanks located at the outlet of each plot. The volume of captured water was measured, samples for sediment and chemical analyses were taken, and the tanks were emptied and cleaned to prepare for the next event. To accomplish this, it is necessary to have several daylight hours without rainfall. A runoff event is defined then as rainfall periods separated by a working day of no rainfall. This definition occasionally leads to difficulties. An unusually violent storm developed without warning on July 20, 1981. Tanks were being cleaned when the storm struck. As a result, some plots had been completely processed, and others were being processed. The high winds accompanying the event blew tank covers awry or completely off. As a result, event number 18 (Table 1) includes less than 40 plot measurements to eliminate confounded data. Rainfall was recorded using a single recording raingage located as shown in Fig. 1.

The facility was constructed in 1937 and used for a variety of tests through 1977. In 1978, soil observations and measurements were made and the slopes were reshaped. A more detailed description of the facility was given by Jamison et al. (6), Saxton and Whitaker (7), and Burwell

47

TABLE 1.—Summary of Rainfall and Runoff for Spring and Summer 1981, by Event

Event		Precipitation, in millimeters	Runoff, in Millimeters						
Number (1)	Date (2)	(3)	Number[a] of zero's (4)	Mean (M), in millimeters (5)	Standard deviation (S), in millimeters (6)	Coefficient of variation, S/M (7)	Minimum, in millimeters (8)	Median, in millimeters (9)	Maximum, in millimeters (10)
Turn plowed and disked	3/24–27								
1	4/11	38	trace	0.43	0.28	0.651	trace	0.2	1.3
2	4/14	8	1	0.22	0.20	0.870	0.0	0.2	0.8
3	4/19	13	1	0.91	0.81	0.890	0.0	0.5	3.6
4	4/21–22	43	0	12.75	3.30	0.259	6.6	12.7	19.6
Harrowed	4/28								
5	5/9–10	47	16	1.98	2.16	1.091	0.0	1.0	7.1
6	5/17–19	96	0	47.52	12.70	0.267	18.0	48.0	71.9
7	5/23–24	16	0	2.44	0.91	0.375	0.3	2.5	4.1
Cultivation	5/29								
8	6/1	27	5	1.90	1.07	0.563	0.0	2.3	4.6
9	6/4	9	13	0.41	0.20	0.488	0.0	0.3	1.0
10	6/5	6	17	0.36	0.18	0.500	0.0	0.3	0.8
Cultivation	6/9								
11	6/16	20	0	3.23	0.86	0.266	0.8	3.3	4.8
12	6/20	21	0	6.40	1.30	0.203	2.3	6.4	11.7
13	6/22	70	0	50.11	9.12	0.182	19.8	51.8	63.5
14	7/1	31	0	9.96	1.30	0.131	7.6	9.7	13.5
15	7/2	27	0	22.02	2.34	0.106	17.0	21.8	29.2
16	7/5	23	0	18.62	1.32	0.071	15.5	18.8	21.1
Cultivation	7/7								
17	7/18	95	0	56.65	10.08	0.178	30.7	57.4	76.2
18	7/20	21	0	16.43	3.56	0.217	5.3	16.5	24.6
19	7/23	96	0	78.56	13.28	0.169	54.9	79.8	101.6
20	7/25	39	0	28.75	2.67	0.093	21.3	28.7	34.0
21	7/26	5	3	2.49	1.09	0.438	0.0	2.5	5.1
22	7/27	32	0	22.20	4.80	0.216	8.6	22.1	30.7
23	7/28	12	0	7.47	1.50	0.201	1.5	7.6	9.9
Cultivation	7/31								
24	8/25	26	0	1.27	0.58	0.457	0.5	1.0	2.8
25	8/31	25	0	8.13	1.45	0.178	5.8	7.9	13.2
26	9/13	18	0	4.14	0.84	0.203	2.8	3.8	6.6

[a]Number of plots without measurable runoff out of 40 plot total.

and Kramer (1). In 1979 and 1980, all plots were uniformly cropped to soybeans. All plots were moldboard plowed and disked between March 24 and 27, 1981, and maintained in a fallow condition throughout the summer. Cultivation with a chisel plow was used to control weeds and to break up soil crust.

During spring and summer of 1981, 26 rainfall events occurred that produced 984 measurable plot runoff events of a possible 1,040 (95%). The results are summarized in Table 1. A best summary of results was not obvious, so two systems were used. First, the data were segregated into plots producing runoff and plots that did not produce runoff. For each event, the number of plots that did not produce runoff is tabulated. Using measurements from only those plots producing runoff, each event was characterized by measures of central tendency and dispersion. The arithmetic mean and median are tabulated as measures of central ten-

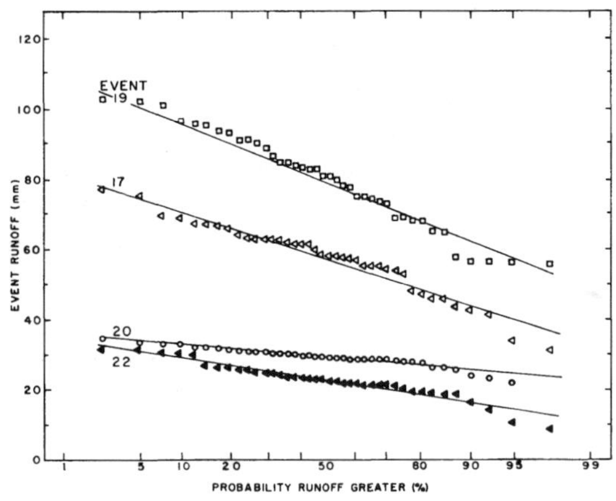

FIG. 2.—Frequency Distribution of Runoff Volumes of Forty Plots for Events 17, 19, 20 and 22. Straight Lines Represent Gaussian Distributions Using Mean and Standard Deviations Given in Table 1

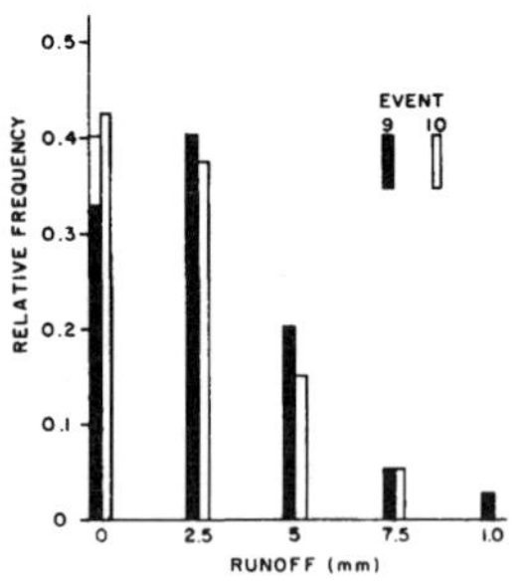

FIG. 3.—Relative Frequency of Runoff Quantities for Events 9 and 10 in which 33% and 43% of Plots Produced No Runoff

dency, and standard deviation, coefficient of variation and range are tabulated as measures of dispersion.

A more detailed display of the within-event variation is shown in Figs. 2 and 3. Runoff from four of the events producing, on the average, more than 12.5 mm of runoff are represented in Fig. 2. Small events, those with several plots producing no runoff, are illustrated by relative frequency diagrams for two events in Fig. 3.

For events in which all plots produced runoff, the mean and median are in substantial agreement, suggesting a Gaussian distribution for the runoff variability. The Gaussian distribution is unbounded so it must be truncated at zero runoff and at runoff equal to rainfall. The influence of this truncation becomes significant for the smaller events, so these must be treated using a mixed distribution. Fitting of such a distribution was not explored, as for most studies these small events are not of interest.

The dispersion, as measured by standard deviation or by coefficient of variation, is not constant. One expects this to vary with degree of surface roughness and sealing. The surface roughness was observed but not measured, and appeared to decrease with each event after a tillage operation. Events 11–16 show a progressive decrease in coefficient of variation from a high of 0.266 for the first event after cultivation to a low of 0.071 for the last event before the next cultivation. This progressive decrease is not confirmed, however, by the next series of events, 17–23.

Variability of total runoff for the 26 events is displayed in Fig. 4. The mean total runoff was 457 mm with a coefficient of variation of 0.101. This coefficient of variation is less than the average for the nine large events (average coefficient of variation = 0.17). In fact, only two individual events showed coefficients of variation less than 0.10. The ag-

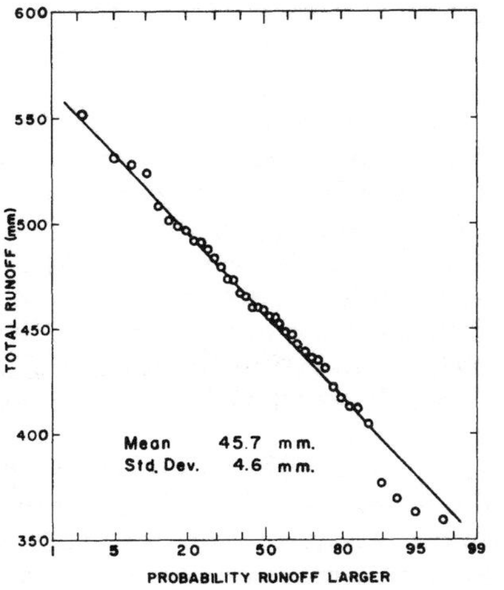

FIG. 4.—Frequency Distribution for Total Runoff by Plot for Spring and Summer, 1981

gregate then shows less variability than the individual events. This implies that individual plots are not consistent in their relative runoff production, and this was observed in the data.

To investigate consistency of the individual plots, standard normal deviations from the mean runoff for each plot were graphed in Fig. 5 for events 17 and 19. Event 17 represents conditions after cultivation. Surface characteristics produced by previous events should have been elim-

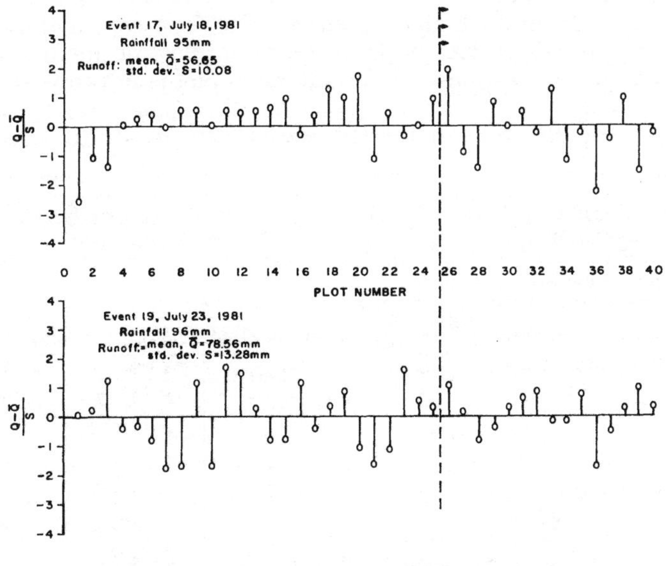

FIG. 5.—Runoff from Individual Plots for Events 17 and 19

50

inated giving a new base condition. Event 18, which is not displayed, is expected to "age" the surface. Event 19, then, is shown for comparison with the base. Plot 36 is significantly low in both events. It is also the second from the lowest in total runoff for the twenty-six events. The plot with the lowest total runoff for the twenty-six events, number 27, was above average for event 19, though below average for event 17. Plot 26 was significantly above average for both events. Two plots, 12 and 13, were higher than 26 for the 26 event total. Plots 12 and 13 produced above average runoff for both events, but only plot 12 produced runoff more than one standard deviation from the mean. That occurred on event 19. In general, the plots were not consistent in their runoff production. Less than half, 18 of 40, deviated in the same direction from the mean for both of the events shown.

Consistency of individual plots was also investigated by considering the response of six selected plots to the nine events producing more than 12.5 mm of runoff (event 18 was not used due to measurement problems). Average conditions were investigated by selecting two plots whose accumulated runoff was nearest the 50% value of Fig. 4. High runoff conditions were investigated using the two plots nearest the 50% plus one standard deviation value and low runoff by the two plots nearest the 50% minus one standard deviation value. The 40-plot array of runoff values for each of these events was divided into quartiles and the quartile position of each of the six selected plots was recorded. The results are shown in Table 2.

A high runoff plot must have more events occurring in the higher quartiles than in the lower quartiles, whereas a low runoff plot must have more events in the lower quartiles. The distribution of occurrences for a plot then must be skewed toward the character of the 26-event total for that plot. This necessary skew is evident in Table 2. The low runoff plots have more events in the lowest quartile, whereas high runoff plots have more events in the highest quartile. Note, however, that most plots showed runoff values that fell in all four quartiles, no matter what the 26 event total. The response of an individual plot is not consistent from event to event.

Quantifying the reason for the variability in runoff was a goal

TABLE 2.—Quartile Ranking of Runoff Values by Plot for Nine Largest Events and for Plots Representing Low, Median and High Runoff Production

Annual Total Runoff Ranking			Large Event Quartile Ranking of Runoff			
Plot number (1)		Total seasonal runoff, in millimeters (2)	1 (lowest) (3)	2 (4)	3 (5)	4 (highest) (6)
Low	3	405.6	4	2	1	2
	11	412.0	4	2	2	1
Median	5	456.2	1	3	3	2
	40	458.7	0	3	6	0
High	20	502.2	1	2	2	4
	26	508.8	1	1	2	5

throughout this study. Carryover effects resulting from antecedent tillage practices were tested by comparing the mean of accumulated runoff from 13 plots that had been cropped to corn with conventional tillage, with the same characteristic for 14 plots previously cropped to corn using no-till practice. These antecedent practices were in effect from 1971–1977. Equivalence of the means could not be rejected at the 5% level for these sets of 1981 runoff data. In addition, one plot that had been in meadow from 1968–1977 would be expected to show considerable difference when compared to tilled plots if carryover effects were present. Its accumulated runoff was, however, very close to the 40-plot median for the 26 event total. This indicates that the monoculture treatment of soybean cropping and chisel plow tillage from 1978–1980 masked any previous tillage or crop carryover effect on run-off production. This observation suggests that residual crop and tillage management practices that affect runoff are relatively short-termed and can be altered within a 3-yr period after changing crop or tillage management, or both.

Machine cultivation leads to some unavoidable and unaccountable differences in surface conditions among the plots. These differences are visible in the form of surface depressions and location of clods relative to tool marks, micro changes in slope and minor variation in size distribution of clods and continuity of tool marks. Weathering of these variations in surface condition progresses at different rates. An attempt was made to carefully observe and record, in a qualitative fashion, these differences between plots, and so be able to predict which should be high and low runoff plots on the next event. This effort was stymied because the variation within an individual plot was at least as large as the variation among plots.

All variations within and among plots were relatively minor and were considered to represent a typical range for the specific management used. The variations in surface conditions appear to be, for all practical purposes, randomly distributed within and among plots.

Soil properties can vary greatly across any field. The 40 plots have been extensively sampled, so soil characteristics are quite well defined. If one is seeking a field with a uniform soil, these plots represent a field with more uniformity than most. A comparison of accumulated runoff, with average clay content in the top 15 cm, showed no correlation. Similar results were found when comparing to depth to claypan (depth to 40% clay).

These observations indicate that the runoff variability must be attributed to relatively subtle variations within plots, which though similar among the plots, may be located at different positions on the various plots. The elements making up the observable differences are small, but the response of the plot represents an aggregate of these differences. Different aggregations lead to a total difference that is measurable.

Implication to Runoff Modeling

A primary element of detailed hydrologic models for field-size areas is the overland flow plane, for which these plots are physical models. In principle, one could make detailed topographic and soil studies of each plot to express differences. In practice, however, this would de-

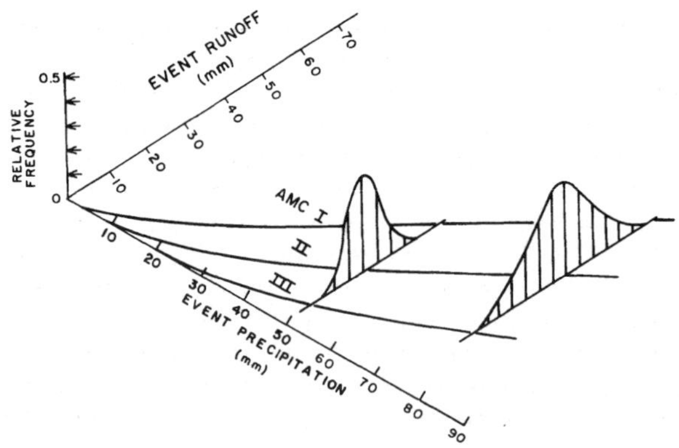

FIG. 6.—Runoff from Rainfall Predicted Using SCS Curve Number Method Including Measure of Variability Based on Kingdom City Observations

stroy the concept of the overland flow plane. Too, it is not obvious how one would incorporate this detail into the model. In the case of modeling a field-size watershed, it would certainly be impractical to approach the detail that would distinguish between the forty plots. It takes over 4,000 plots of this size to make a 40 ha field. In a watershed modeling sense, then, the plots are usually treated as identical.

One criterion for a valid rainfall-runoff model, in the face of the observed variability, is that it predicts the mean runoff for each event. If such a model were constructed for this set of plots and the estimates compared with the runoff from a single plot, the predicted and measured values would seldom agree. The Curve Number Runoff Equation of the Soil Conservation Service (2) is a mathematical model dealing with only the quantities measured, so is appropriate for illustration.

A previous study by Hjelmfelt, et al. (5) showed that a Curve Number of approximately 88 is appropriate for fallow plots at Kingdom City. This value gives the rainfall-runoff curve labeled AMC-II on Fig. 6. Runoff values given by this curve were taken to be representative of the mean value. Using the observation that the runoff variation from plot to plot seems to be normally distributed, with coefficient of variation 0.17 (based on nine events with mean runoff greater than 12.5 mm), permits expressing the distribution of observed runoffs that can be expected. This is shown for event rainfalls of 50 and 75 mm on Fig. 6. This distribution indicates that 80% of the observations should lie between $7.2\bar{Q} < Q < 12.8\bar{Q}$ mm. The runoff predicted, using the Curve Number transformation to AMC-I and AMC-III, is displayed in Fig. 6 for reference.

Use of the Curve Number runoff model was for illustration only. It is not adequate for estimation of the mean runoff quantity with consistency. A much more detailed model that incorporates effect of cultivation on surface roughness, soil porosity, development of a surface seal, and time distribution and duration of rainfall would be required to achieve truly adequate representation.

Conclusions

The results of the spring and summer runoff study using the 40 plots at Kingdom City, Mo., describe the spatial variability of runoff. The coefficient of variation for events with an average plot runoff of 1.25 cm or more was 0.17 with a range among events of 0.093–0.267. This variation could not be attributed to carryover effects, quantifiable surface variations, or to spatial variation in soil texture. Individual plots did not respond consistently through the sequence of 26 events.

In the usual descriptions applied by current mathematical models of small watersheds, each plot was identical, so these results contribute to defining confidence limits that can be applied to model results, i.e., if a mathematical model were constructed that predicted the 40-plot mean for each event, and that model were compared with the observed values for any one plot, there would be differences between predicted and observed results. This study helps to establish the significance of that difference.

These results are representative of a claypan soil in the climatic region of central Missouri. Transferability of these results to other soils and climates is not obvious. We hope that other test facilities will be used to expand the region of validity.

Appendix I.—References

1. Burwell, R. E., and Kramer, L. A., "Long-term Annual Runoff and Soil Loss from Conventional and Conservation Tillage of Corn," *Journal of Soil and Water Conservation*, Vol. 38, No. 3, 1983, pp. 315–319.
2. "Estimation of Direct Runoff from Storm Rainfall," *National Engineering Handbook*, Soil Conservation Service, Section 4, Hydrology, pp. 10-1–10-24.
3. Grah, O. J., Hawkins, R. H., and Cundy, T. W., "Distribution of Infiltration on a Small Watershed," *Advances in Irrigation and Drainage*, ASCE, July, 1983, pp. 44–54.
4. Hawkins, R. H., "Interpretation of Source Area Variability in Rainfall-Runoff Relations," *Rainfall-Runoff Relationship*, Water Resources Publications, Littleton, Colo., 1982, pp. 303–324.
5. Hjelmfelt, A. T., Kramer, L. A., and Burwell, R. E., "Curve Numbers as Random Variables," *Rainfall-Runoff Relationship*, Water Resources Publications, Littleton, Colo., 1982, pp. 365–370.
6. Jamison, V. C., Smith, D. C., and Thornton, J. F., "Soil and Water Research on a Claypan Soil," *Technical Bulletin 1379*, Agricultural Research Service, United States Department of Agriculture, 1968.
7. Saxton, K. E., and Whitaker, F. D., "Hydrology of a Claypan Watershed," *Research Bulletin 974*, College of Agriculture, University of Missouri, July, 1970.
8. Sisson, J. B., and Wierenga, P. J., "Spatial Variability of Steady-State Infiltration Rates as a Stochastic Process," *Soil Science Society of America Journal*, Vol. 45, 1981, pp. 699–704.
9. Smith, R. E., and Hebbert, R. H. B., "A Monte Carlo Analysis of the Hydrologic Effects of Spatial Variability of Infiltration," *Water Resources Research*, Vol. 15, No. 2, Apr., 1979, pp. 419–429.
10. Vieira, S. R., Nielsen, D. R., and Biggar, J. W., "Spatial Variability of Field Measured Infiltration Rates," *Soil Science Society of America Journal*, Vol. 45, 1981, pp. 1040–1048.

Commentary on Paper 31

Gillham, R. W. (1984) The capillary fringe and its effect on water-table response.

This paper relied on the special characteristics of a particular field site (visited by generations of students on the Masters course in Hydrogeology at the University of Waterloo, Ontario, Canada) to make an extremely effective demonstration of the rapid response of the water table under conditions of a deep capillary fringe. At the site, there is a shallow water table and the capillary fringe reaches almost to the soil surface. Addition of only a very small volume of water (a cup full) close to an observation well results in a rise in the water table that is both fast and dramatic. If no more water is added, of course the water table falls again (albeit more slowly) as the small amount of water is dispersed away from the observation well site. The mechanism was also demonstrated in the laboratory studies reported by Abdul & Gillham (1984).

The significance of this paper lies in its implications for the response of shallow water tables under natural rainfall conditions. This can be demonstrated by a simple kinematic analysis. The Darcian velocity, v, for downslope saturated flow through a unit area, assuming that flow is parallel to the slope is given by:

$$v = K_s \sin\beta$$

where K_s is the saturated hydraulic conductivity of the slope and β is the slope angle.

The mean pore water velocity, v_p, at saturation is given by:

$$v_p = K_s \sin\beta / \theta_s$$

where θ_s is the porosity of the soil.

The wave velocity or celerity, c, is the speed at which disturbances to the flow will be propagated downslope. It is a function of the amount of pore space that must be filled for the water table to rise. In a kinematic analysis it is given simply by:

$$c = K_s \sin\beta / \theta_e$$

where θ_e is the effective porosity of the soil. In the capillary fringe, of course, the soil will be very close to saturation and so the effective porosity will be very small. Assume it is 1% of the porosity. That means that the celerity will be 100 times faster than the mean pore water velocity. Thus the effect of a disturbance to the subsurface flow, such as the first recharge arriving at the top of the capillary fringe, will propagate downslope very much faster than the water can flow. This is why tracer experiments often show a lag between the initial discharge response to an addition of tracer and the arrival of the first tracer tagged water. This, effectively, is an explanation of the "translatory flow" concept of Hewlett and others, and a reason why tracer experiments often reveal significant percentages of pre-event "old" water in hydrographs. This water is being displaced out of the soil by the disturbance of recharges to the water table. There will be a rapid response all the while that the effective porosity, θ_e, remains much smaller, particularly where the water table initially rises into the capillary fringe.

In a more complete (dispersion equation) analysis, the celerity depends on the magnitude of the disturbance and changes in effective hydraulic gradient close to the stream, as well as on the effective porosity, but the same effect holds. Continued addition of water as recharge to the saturated zone may mean, of course, that the effective porosity might increase (and the celerity decrease) if the water table rises faster than the unsaturated flow can replenish the capillary fringe, but this effect might also be offset by increases in saturated hydraulic conductivity as the water table rises.

The speed of response also depends on water being transmitted through the unsaturated zone to the capillary fringe (see Beven, 1982a,b, for a more complete kinematic analysis including unsaturated flows). This will be a function of the depth to the capillary fringe (which will tend to be smaller close to stream channels and in hillslope hollows because of convergence of flow) and the pathways by which water flows through the unsaturated zone. The fact that the unsaturated zone is generally not so deep close to the stream means that the water table might rise first in such areas (see Paper 11), with the possibility of groundwater ridging leading to rapid subsurface contributions to streamflow (Paper 27). These pathways may involve preferential flows, either due to macropores or to heterogeneities in the soil characteristics (e.g. Beven & Germann, 1982; Moore *et al.*, 1986; Flury *et al.*, 1994).

The remaining question is then: How widespread are the conditions for which the effective porosity is much less than the porosity of the soil? Abdul & Gillham (1989), Novakowski & Gillham (1988) and Blowes & Gillham (1988) all report similar types of rapid near surface response in the field. Zaltsburg (1986) and McDonnell & Buttle (1998), on the other hand, both suggest that conditions where the mechanism might be important will be found rather rarely in the field and that mechanisms other than groundwater ridging are needed to account for large proportions of pre-event water in the storm hydrograph.

What can we conclude? What can be said is that the effective porosity is necessarily always less than the porosity, including cases where the water table might rise without saturating all the soil, so that the celerity is always greater than the mean pore water velocity. However, the celerity may not be necessarily faster than the fastest velocities at the upper tail of the velocity distribution if some of the downslope flow is through some preferential flow pathways. Thus, it may be important to be aware of the possibility of capillary fringe responses and groundwater ridging in generating old water inputs to the stream when the water table is close to the surface, but this does not mean that the mechanism is important everywhere.

Bob Gillham is still a Professor in the Department of Earth Sciences at the University of Waterloo.

References

Abdul, A. S. & Gillham, R. W. (1984) Laboratory studies of the effects of the capillary fringe on streamflow generation. *Water Resour. Res.* **20**(6), 691–698.

Abdul, A. S. & Gillham, R. W. (1989) Field studies of the effects of the capillary-fringe on streamflow generation. *J. Hydrol.* **112**, 1–8.

Beven, K. J. (1982a) On subsurface stormflow: predictions with simple kinematic theory for saturated and unsaturated flows. *Water Resour. Res.* **18**(6), 1627–1633.

Beven, K. J. (1982b) On subsurface stormflow: an analysis of response times. *Hydrol. Sci. J.* **27**(4), 505–521.

Beven, K. J. & Germann, P. F. (1982) Macropores and water flow in soils. *Water Resour. Res.* **18**(5), 303–325.

Blowes, D. & Gillham, R. W. (1988) The generation and quality of streamflow on inactive uranium mine tailings near Elliott Lake, Ontario. *J. Hydrol.* **97**, 1–22.

Flury, M., Flühler, H., Jury, W. A. & Leuenberger, J. (1994) Susceptibility of soils to preferential flow of water: a field study. *Water Resour. Res.* **30**(7), 1945–1954.

McDonnell, J. J. & Buttle, J. M. (1998) Comment on "A deterministic-empirical model of the effect of the capillary-fringe on near-stream area runoff. 1. Description of the model" by Jayatilaka, C. J. & Gillham, R. W. *J. Hydrol.* **207**, 280–285.

Moore, I. D., Burch, G. J. & Wallbrink, P. J. (1986) Preferential flow and hydraulic conductivity of soils. *Soil Sci. Soc. Am. J.* **50**, 876–881.

Novakowski, K. S. & Gillham, R. W. (1988) Field investigation of the nature of water-table response to precipitation in shallow water table environments. *J. Hydrol.* **97**, 23–32.

Zaltsberg, E. (1986) Comment on "Laboratory studies of the effects of the capillary fringe on streamflow generation", by A. S. Abdul & R. W. Gillham. *Water Resour. Res.* **22**(5), 837–838.

Journal of Hydrology, 67 (1984) 307—324
Elsevier Science Publishers B.V., Amsterdam — Printed in The Netherlands

[4]

THE CAPILLARY FRINGE AND ITS EFFECT ON WATER-TABLE RESPONSE

R.W. GILLHAM

Department of Earth Sciences, University of Waterloo, Waterloo, Ont. N2L 3G1 (Canada)

(Received May 6, 1982; accepted for publication April 13, 1983)

ABSTRACT

Gillham, R.W., 1984. The capillary fringe and its effect on water-table response. J. Hydrol., 67: 307—324.

Shallow water tables are often observed to respond in a highly disproportionate manner to precipitation events. That is, the magnitude of the response is often much greater than would be predicted on the basis of the specific yield of the geologic material and the amount of rainfall. Physical arguments are presented to show that if the capillary fringe (the zone of tension saturation) extends to ground surface, then the addition of a very small amount of water can result in an immediate and large rise in the water table. The arguments are supported by a field experiment in which the addition of 0.3 cm of water caused the water table to rise 30 cm in 0.25 min., thus demonstrating the large and highly transient influence of the capillary fringe on the position of the water table. The evidence in this study, as well as recent work of others, suggests that the capillary-fringe effect could have a major influence on the process of stream-flow generation, contaminant transport to surface waters and contaminant transport in hydrogeologic regimes having shallow water tables. It is further shown that calculations of groundwater recharge and consumptive use, based on water-table response and an assumed specific yield, could be substantially in error.

INTRODUCTION

The water table is generally considered to be the upper boundary of saturated groundwater flow, and as such, it assumes an important role in many groundwater investigations. In particular, the water-table configuration has a pronounced effect on the direction and rate of groundwater flow, the rate of decline of the water table is often used as an indication of the effects of prolonged pumping of unconfined aquifers, and variations in the water-table elevation are often used to estimate rates of groundwater recharge and discharge or for evaluating consumptive use by phreatophytes. An appreciation of the physical factors associated with the water-table response is therefore a necessary prerequisite for the interpretation of many hydrogeologic phenomena.

The specific yield (Sy) of an unconfined aquifer has been defined as the

0022-1694/84/$03.00

volume of water released from storage per unit cross-sectional area of the aquifer, per unit decline in the elevation of the water table (Freeze and Cherry, 1979) and is approximately equal to the difference between the saturated volumetric water content and field capacity. Assuming a typical value for specific yield in a well-sorted sand or gravel of 0.3, the infiltration of 1.0 cm of water would be expected to result in a water-table rise of 3.3 cm. However, as observed by Meyboom (1967), shallow water tables frequently show a highly disproportionate response to infiltration. That is, the rise in the water table is frequently much greater than would be predicted on the basis of calculations using specific yield.

The disproportionate rise has been attributed to an increase in the gas-phase pressure in the unsaturated zone above the water table during the infiltration process. This has been termed the Lisse effect by Thal Larsen in 1932 (Hooghoudt, 1947), and is described in greater detail in Meyboom (1967).

A second explanation of the response considers the specific yield above a shallow water table to be variable, with values decreasing as the water table approaches ground surface. This phenomenon is related to the shape of the water-content profile above the water table and its potential significance is determined by the water content—pressure head relation for the particular geologic material. The variable specific yield concept is discussed further in Childs (1960), Luthin (1960), Duke (1972) and McWhorter and Sunada (1978).

This paper considers a special case of the variable specific yield concept; the case where the zone of tension saturation (the capillary fringe) extends from the water table to ground surface. In this situation, the effective specific yield is near zero, and the application of a small amount of water should result in a rapid and large rise in the water table. Hooghoudt (1947) referred to the rapid decline of a water table across the zone of tension saturation as the Wieringermeer effect, while the rapid rise was referred to as the reversed Wieringermeer effect.

Although the variable specific yield associated with shallow water tables, and the special case of the capillary fringe effect (Wieringermeer effect) have been recognised for many years, the phenomenon has generally been neglected in hydrogeological investigations, and little attention has been given to the potential significance in the interpretation of hydrologic processes. In this paper, the physical principles associated with the capillary-fringe phenomenon are discussed, field evidence of the phenomenon is presented, and possible implications with respect to several hydrogeologic processes are discussed.

PHYSICAL BASIS OF THE CAPILLARY-FRINGE PHENOMENON

The principles presented in this section can be found in most books

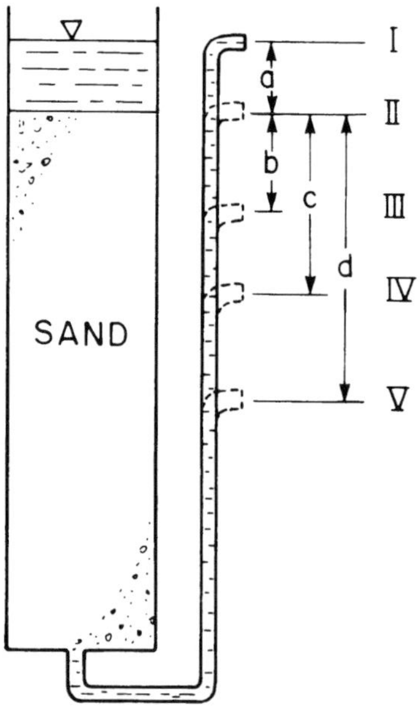

Fig. 1. Sand column showing different positions of the outflow tube.

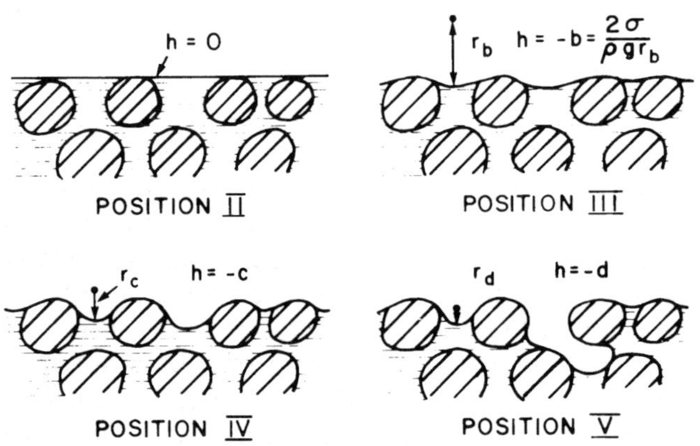

Fig. 2. Schematic of the air—water interface for different positions of the outflow tube of Fig. 1.

dealing with unsaturated flow processes. The purpose here is to examine those principles in a hydrogeological context, and in particular, to show their potential influence on the response of shallow water tables.

Fig. 1 shows a column of uniformly packed sand with a water-filled outflow tube connected at the bottom. Initially, the column is completely saturated, with free water standing on the surface to a depth of *a*. The capillary model of a porous medium will be used to examine the effect of lowering the outflow tube on the hydraulic conditions at the sand surface. Fig. 2 is a conceptual, magnified view of the soil surface, while

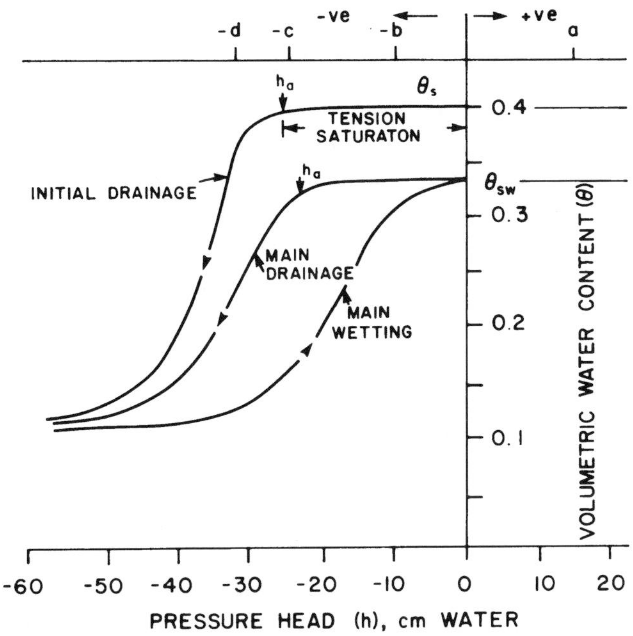

Fig. 3. Water content—pressure head relations for a medium- to fine-grained sand.

Fig. 3 shows the effect of the changing pressure on the volumetric water content of the upper sand layer.

With the outflow tube at position I, the sand is saturated and the surface of the sand is at a pressure head of $h = +a$. When lowered to position II and the free water has drained away, the pressure in the water at the soil surface is atmospheric and thus the pressure head is $h = 0$. Because the pressure in the water is atmospheric, there is no pressure difference across the air—water interface and thus the interface is flat and horizontal as shown in Fig. 2, position II. Referring only to the initial drainage curve of Fig. 3, this corresponds to the point $h = 0$ and $\theta = \theta_s$ on the water content—pressure head relation. Lowering the outflow to position III results in a negative pressure head ($h = -b$), or suction being applied to the water in the upper layer of soil. As water is drawn into the pores, a curved interface forms, giving rise to a pressure difference across the air—water interface. The water will be drawn into the pores until the radius of curvature is sufficiently small to cause a pressure difference across the interface that will balance the applied negative pressure head. The pressure difference across the interface is given by the capillary-rise formula. Expressing the pressure difference in units of head, and assuming the angle between the meniscus and the solid phase to be zero, the capillary-rise formula can be written:

$$h = 2\sigma/\rho g r \tag{1}$$

where σ is the surface tension of the water: ρ is the density of water; g is the accelaration due to gravity; and r is the radius of curvature of the meniscus.

In lowering the outflow tube to position III, the only water lost from the sand would be a very small amount caused by the withdrawal of the menisci into the pores. As a result, as shown in Fig. 3, the surface soil remains very near saturation, but at a negative pressure head of $h = -b$. Position IV is similar to III except the menisci (Fig. 2) have been drawn further into the pores, resulting in a smaller radius of curvature that is capable of balancing the greater applied negative pressure. At position V, the applied pressure is $h = -d$. At this pressure the radius of the largest pore is too great to maintain the radius of the meniscus necessary to balance the applied negative pressure, and thus the pore drains (Fig. 2). This results in the volumetric water content decreasing at a pressure head of $-d$ (Fig. 3). As the pressure head becomes progressively more negative, progressively smaller pores drain, giving the initial drainage curve of Fig. 3.

The pressure head at which the first pore drains is commonly referred to as the air-entry value (h_a) for the porous medium, and the region between $h = 0$ and $h = h_a$ is referred to as the zone of tension saturation.

The curves of Fig. 3 are taken from Gillham et al. (1976) and are for a medium- to fine-grained dune sand material. Upon rewetting, the saturated water content (θ_{sw}) at $h = 0$ is less by $\sim 8\%$ than the fully saturated water content (θ_s). This is the result of air entrapment in the pores during the wetting process and could result in the effective specific yield of the material being less than that anticipated on the basis of the fully saturated water content.

Reducing the pressure head from the point ($h = 0, \theta = \theta_{sw}$) generates the main drainage curve. The shapes of the three curves of Fig. 3 differ markedly, with the zone of tension saturation for the initial drainage, main drainage and main wetting curves being ~ 26, ~ 20 and ~ 5 cm, respectively. Thus it is apparent, that for a particular soil, the zone of tension saturation is, to some extent, dependent upon the previous wetting and drying history of the sample.

The extent of the tension-saturated zone is highly dependent upon the texture and structural characteristics of the soil material. Fig. 4 shows the height of rise of water in a capillary vs. the radius of the capillary calculated using eq. 1. If the radius of Fig. 4 is considered to be the largest pore in a soil material, then height of rise (l), can be interpreted as the air-entry value ($h_a = -l$) for the medium. If it is further assumed that the pore radius is equal to the grain diameter, then Fig. 5 indicates the extent of the zone of tension saturation that could be expected for materials of different texture. For example, in coarse sand, the zone of tension saturation would extend to a pressure head of only -0.5 cm, while in fine silt, it could extend to a pressure head of $-1.5 \cdot 10^3$ cm. Assuming the radius of the largest pore to be equal to the grain diameter is, of course, a questionable approximation; nevertheless, based on Fig. 4, it is safe to conclude that in the absence of secondary porosity caused by structuring or aggregation, medium- to fine-textured materials can remain saturated under quite substantial negative pressures.

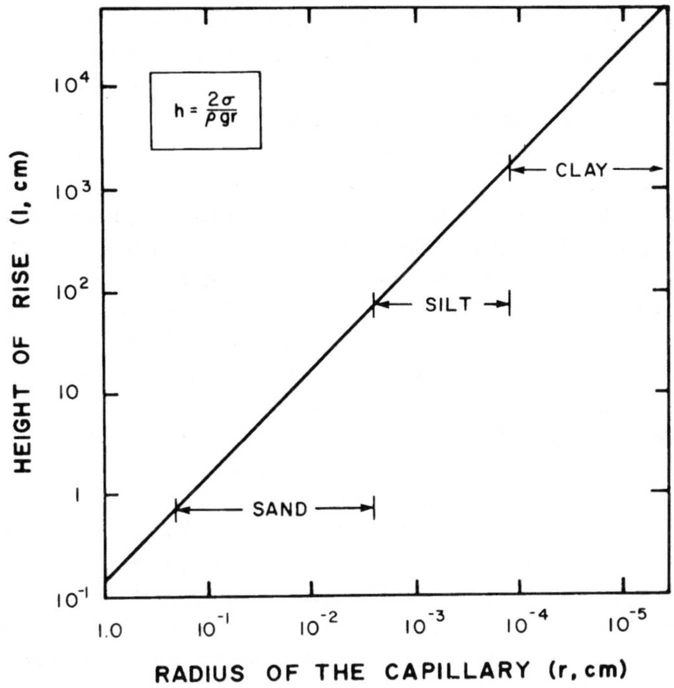

Fig. 4. Extent of the capillary fringe for different soil materials as estimated from the capillary-rise formula.

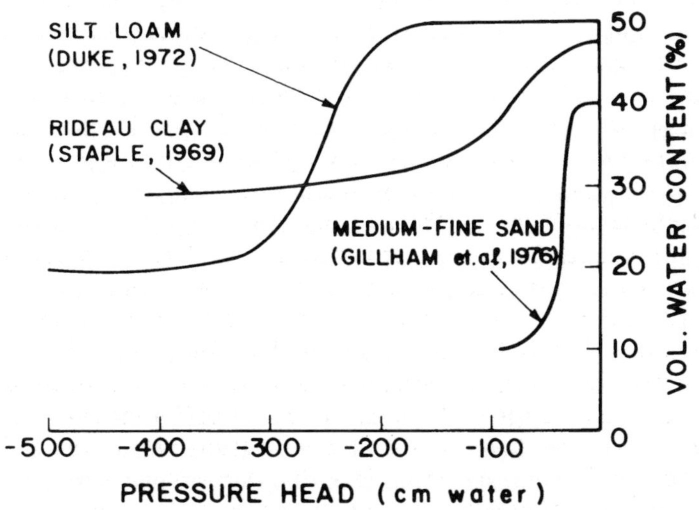

Fig. 5. Drainage curves for soil materials of different texture.

Fig. 5 shows drainage curves for three soil materials. The greater zone of tension saturation in the silt loam, relative to the sand, reflects the smaller grain size and thus the smaller pore size in the silt. Although the clay material has a substantially smaller grain size than the silt, the sample was

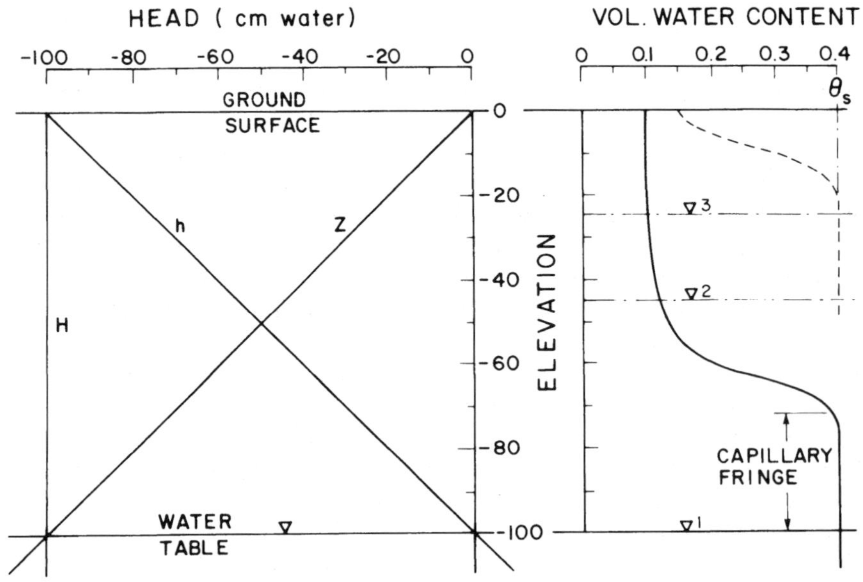

Fig. 6. Schematic demonstrating the effect of the capillary fringe on specific yield.

highly aggregated. Thus the low air-entry value for the clay can be attributed to drainage of the relatively large pore spaces between aggregates rather than to intergranular drainage.

Clay materials that are not structured generally show a gradual decline in water content as the pressure head progresses in the negative direction from zero. In these materials the loss of water that is observed for the first several hundred centimetres of negative pressure is the result of sample consolidation rather than air entry. Although these materials generally do not exhibit the horizontal tension-saturated zones characteristic of incompressible materials such as the silt loam and sand of Fig. 5, it should nevertheless be noted that small changes in water content will result in very substantial changes in pressure head. Thus grain size, structuring and compressibility can all have a substantial effect on the shape of the drainage curve and thus on the extent of the tension-saturated zone.

The above concepts will now be considered in a hydrogeological context. Fig. 6 shows a cross-section through a saturated—unsaturated zone. Initially, the water table is situated at a depth of 100 cm below ground surface, and the hydraulic head (H), pressure head (h) and gravitational head (z) are plotted as functions of depth. In constructing this diagram, it was assumed that static equilibrium conditions prevailed.

If the geologic material is assumed to be the dune sand as represented in Fig. 3, then the water-content profile can be constructed using the pressure-head profile of Fig. 6 and the water content—pressure head relation of Fig. 3. Because the pressure head is equal to the negative of the elevation above the water table, the water-content profile is simply that of Fig. 3 rotated clockwise through 90°. In this case the initial drainage curve of Fig. 3 was used.

Although defined as the locus of points having zero pressure head, the water table is frequently and erroneously referred to as the boundary between the saturated and unsaturated zones. As shown in Fig. 6, the medium remains saturated for a distance above the water table that is equal to the tension-saturated zone of Fig. 3. This zone of saturation above the water table is referred to as the capillary fringe. Based on the previous discussions, the capillary fringe may extend for less than a centimetre above the water table in very coarse sand; however, fine-textured materials may remain saturated for several metres above the water table.

The water-content profile is also shown for two additional positions of the water table. In the second position, ~ 45 cm below ground surface, the rapidly changing portion of the water-content profile intersects the ground surface. In this region the effective specific yield is highly variable, depending upon the elevation of the water table (Childs, 1960; Duke 1972).

Of particular relevance to this paper is the third water-table position. In this position, or any shallower position, the capillary fringe extends to ground surface. The geologic material is therefore saturated to ground surface and has an effective specific yield very close to zero. The upper surface of the water would have a configuration similar to that shown in Fig. 2, III or IV. The addition of a very small amount of water would relieve the tension in the capillaries, resulting in a horizontal water surface (Fig. 2, II). The pressure head at the surface would therefore be zero, i.e. the water table would rise to ground surface. If the system is totally incompressible and there is no downward flux of water, the water-table response would be instantaneous and would occur with the application of perhaps less than 1.0 mm of water. For the situation of Fig. 6, the water table would rise ~ 25 cm, while for finer-textured materials, even larger responses could be expected. It should also be noted that with the loss of a small amount of water, either by downward drainage or by evapotranspiration, the water table would return very quickly to its original position. The effect is therefore highly transient. Even if the conditions of incompressibility and no downward flow are not met, the response should be very rapid and highly disproportionate to the amount of water added.

FIELD EVALUATION

A field test was conducted to evaluate the principles presented in the previous section. A study site was chosen in a flat sandy area of the Canadian Forces Base at Borden, Ontario. The upper metre or two of sand had been removed and used as cover material for an adjacent sanitary landfill. Although the area had been re-vegetated with red-pine, there was little soil development and no significant accumulation of organic matter in the upper soil

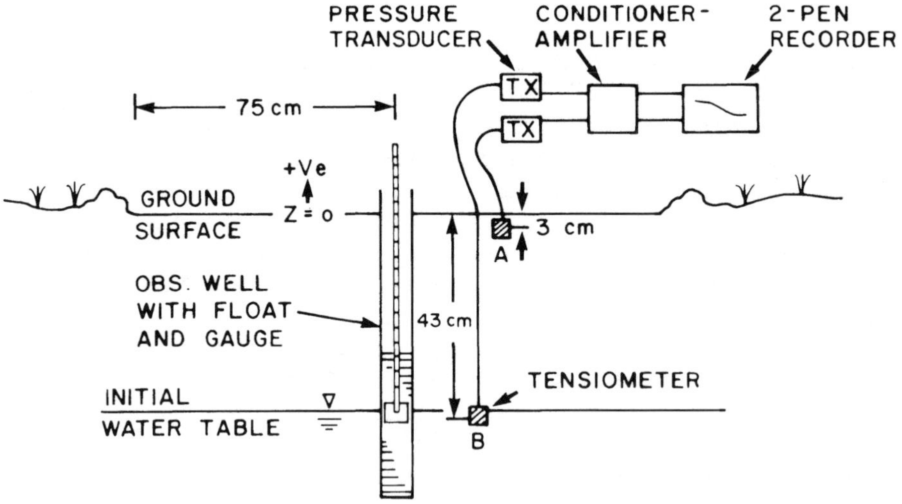

Fig. 7. Schematic of the study site and instrumentation at Borden, Ontario.

horizon was evident. A detailed description of the hydrogeology of the region associated with the landfill is given in MacFarlane et al. (1983).

A small pit dug 5 m from the experimental site, showed that the sand was very homogeneous to a depth of 0.5 m and that the water table was located at a depth of 40 cm below ground surface. Although a grain-size analysis was not conducted, visual comparison showed the sand to be somewhat finer than that represented in Fig. 3. The surface of the soil appeared moist, but was not obviously saturated; however, by compacting the surface slightly free water would appear, indicating that the soil was indeed saturated, or very nearly saturated, to ground surface.

A circular area 0.75 m in radius was leveled, and a low berm was constructed around the perimeter. A 3.8-cm I.D. water-table observation well was installed in the middle of the site, with the screened interval extending from 0.3 to 0.6 m below ground surface. A Styrofoam® float with a stick attached was used as a water-level gauge. In addition, two tensiometers were installed. The intake zone of one tensiometer extended from 1.0 to 4.0 cm below ground surface while that of the second extended from 41 to 45 cm below ground surface. The intake zone of the tensiometers was 4.0 cm in length and 3.5 cm in diameter and was constructed of porous plastic having a bubbling pressure of ~ 300 cm of water. Each tensiometer was connected to a pressure transducer which was connected to a two-pen chart recorder through an amplifier-and-transducer conditioning unit. A cross-section through the experimental site is shown in Fig. 7.

After assuring that all air was removed from the hydraulic system, the transducers were calibrated and an initial set of hydraulic-head readings was obtained. A quantity of water (5.3 l) equivalent to a 0.3 cm application

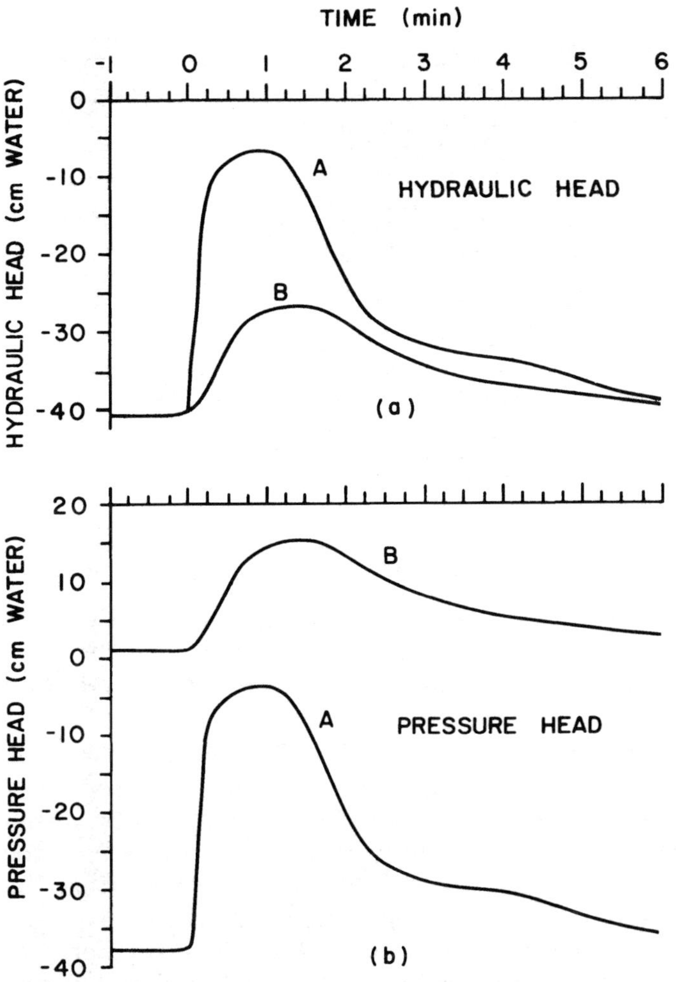

Fig. 8. Hydraulic head and pressure head vs. time in response to the surface application of 3 mm of water.

was spread over the site as quickly and evenly as possible. Because of small irregularities in the surface, and the small amount of water applied, the application was not entirely uniform. All free water had disappeared ~ 1 min. after the application.

For the purpose of calculating the hydraulic and pressure heads, the gravitational head was considered to be zero at ground surface and positive in the upward direction. Fig. 8a and b shows the hydraulic head and pressure head, respectively, at tensiometers A and B, vs. time. Zero time was taken as the time at which water was first applied.

At times less than zero, the hydraulic head was the same for both tensiometers, indicating no vertical gradients and thus static equilibrium conditions in the vertical direction. The pressure-head values differed by an amount

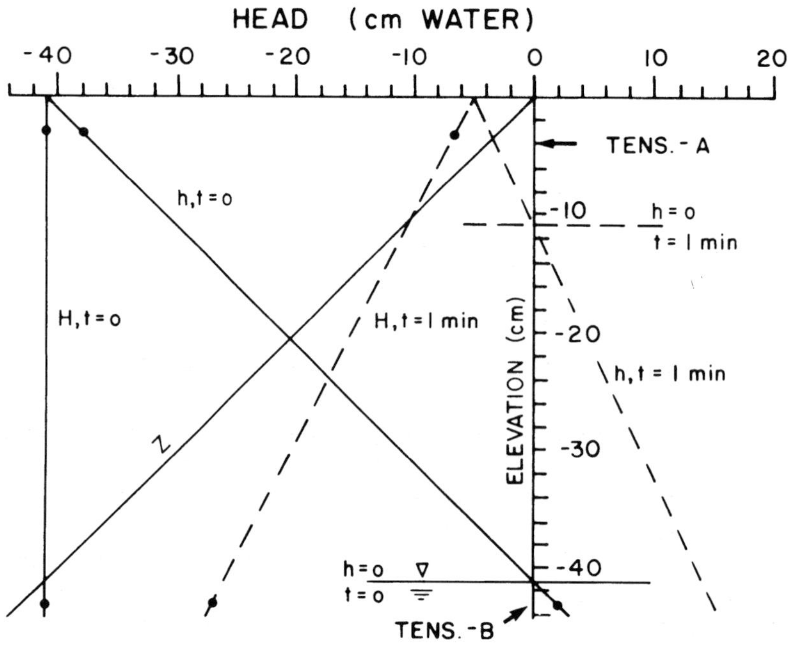

Fig. 9. Vertical head distributions prior to the application of water and 1 min. after the application (H = hydraulic head; h = pressure head)

equal to the difference in elevation between the tensiometers. With the application of water, both tensiometers responded immediately. The response of A was particularly rapid, with a change in head of ~ 30 cm occurring within 15 s. The top of the curve remained relatively flat from ~ 15 to 75 s, the approximate time at which free water disappeared from the soil surface. The flat top suggests that a response of almost equal magnitude could have been obtained with the application of considerably less water. After 75 s there was a rapid decline in head, with the head distribution returning almost to the initial conditions 6 min. after the water was applied. The reduced magnitude in the response at B indicates that the water application generated strong downward hydraulic gradients.

Although the head response near the surface was large and very rapid, it was not instantaneous as might have been expected from the discussion of the previous section. The slight delay in response may have been caused by the fact that the water was not applied instantaneously. However, it is more likely that the non-instantaneous response was due to the compressibility of isolated air bubbles trapped within the porous medium.

The effect of the water application on the head distribution is demonstrated further in Fig. 9. This diagram shows the initial head distributions and the distributions 1.0 min. after the water was applied. In preparing this diagram, it was assumed that the medium was saturated and uniform and thus the hydraulic gradients were considered to be linear between values

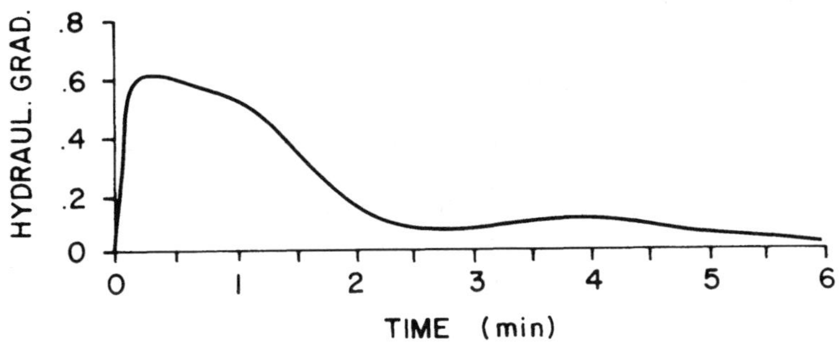

Fig. 10. Vertical hydraulic gradient vs. time in response to the water application.

measured by the tensiometers. The strong downward hydraulic-head gradient (~ 0.38) at $t = 1.0$ min. indicates the change from static equilibrium to downward flow conditions. By inspecting the pressure-head profile at $t = 1$ min., it can be seen that the water table ($h = 0$) was located 11 cm below ground surface. Thus the water table rose 30 cm in response to the 0.3 cm of applied water, giving an effective specific yield of 0.01. Thus the presence of the capillary fringe resulted in a rapid and highly disproportionate rise in the water table.

The magnitude of the hydraulic gradient between tensiometers A and B, as determined from Fig. 8, was plotted vs. time (Fig. 10). The amount of water passing vertically through the site was calculated by integrating graphically the area under the curve and multiplying by the hydraulic conductivity. The hydraulic conductivity of similar sands near the test site is $\sim 5 \cdot 10^{-3}$ cm s^{-1} (MacFarlane et al., 1983). The calculation indicated that 0.27 cm of water had passed through the site; a value remarkably similar to the 0.3 cm applied. Considering the uncertainty in the hydraulic conductivity, the good agreement is considered to be somewhat fortuitous. Nevertheless, the result does provide an additional degree of confidence in the hydraulic head data and the interpretations of the data.

The results observed in the field test are consistent with the principles discussed previously and confirm that the capillary fringe can indeed have a dramatic effect on the response of shallow water tables to applied water. Furthermore, from the discussions of the previous section, one would anticipate that in fine-grained materials an even larger response could be observed.

The response of the observation well installed in the middle of the site was not used in the above analysis. The water level rose ~ 20 cm but the response lagged behind the pressure transducers by several seconds. Furthermore, because of the long screened interval, the water level indicates an averaged hydraulic head rather than the actual position of the water table. Although not considered suitable for detailed analysis, the well can nevertheless serve as a very simple and effective demonstration of the phenomenon.

HYDROLOGICAL IMPLICATIONS

Most approaches to regional groundwater-flow problems assume steady-state flow conditions. It is apparent that in areas with shallow water tables, the flow could be very highly transient. The significance of these transients would depend upon the nature of the particular problem and, in all probability, on the duration of the transient behavior. The following discussions present some specific situations where the capillary-fringe effect could have a significant influence on hydrologic processes.

Recharge and consumptive use calculations

The specific yield of the sand at the study site described above was ~ 0.3. Multiplying this value by the water-table response (30 cm) would give a predicted recharge value of 9 cm, or ~ 30 times the volume actually applied. It is apparent that in situations with shallow water tables, this approach to determining groundwater recharge is highly unreliable and could lead to significant errors when used in water-budget calculations. Similarly, calculations of consumptive use by phreatophytes in which a constant value for specific yield is used could be substantially in error. The use of a specific yield function that depends on the shape of the water content—pressure head relation and the depth to the water table, such as that proposed by Duke (1972), would increase the computational difficulties, but should result in improved estimates. The results would nevertheless be quite sensitive to the degree to which the θ—h relation used in the calculations actually represents the in situ field conditions, and could also be influenced by hysteresis in the θ—h relation.

Streamflow generation

Understanding the processes that lead to the generation of stream flow is important to the development of water-management programs and in flood forecasting. It is therefore not surprising that streamflow generation has been a major area of research interest for several decades. A useful description and summary of the recognized mechanisms is given in Freeze (1974) and Freeze and Cherry (1979).

Using environmental isotopes as a means of separating the components of storm hydrographs, previous studies (Martinec, 1975; Fritz et al., 1976; Sklash and Farvolden, 1979) have shown that groundwater constitutes a much larger proportion of storm runoff than would be predicted using graphical methods of hydrograph separation. It has not been clear however, how groundwater flow systems, which are generally viewed as being sluggish, could respond in such a rapid manner to precipitation events. In a recent paper, Sklash and Farvolden (1979) used the capillary-fringe effect to account for the rapid groundwater response in explaining measured runoff events.

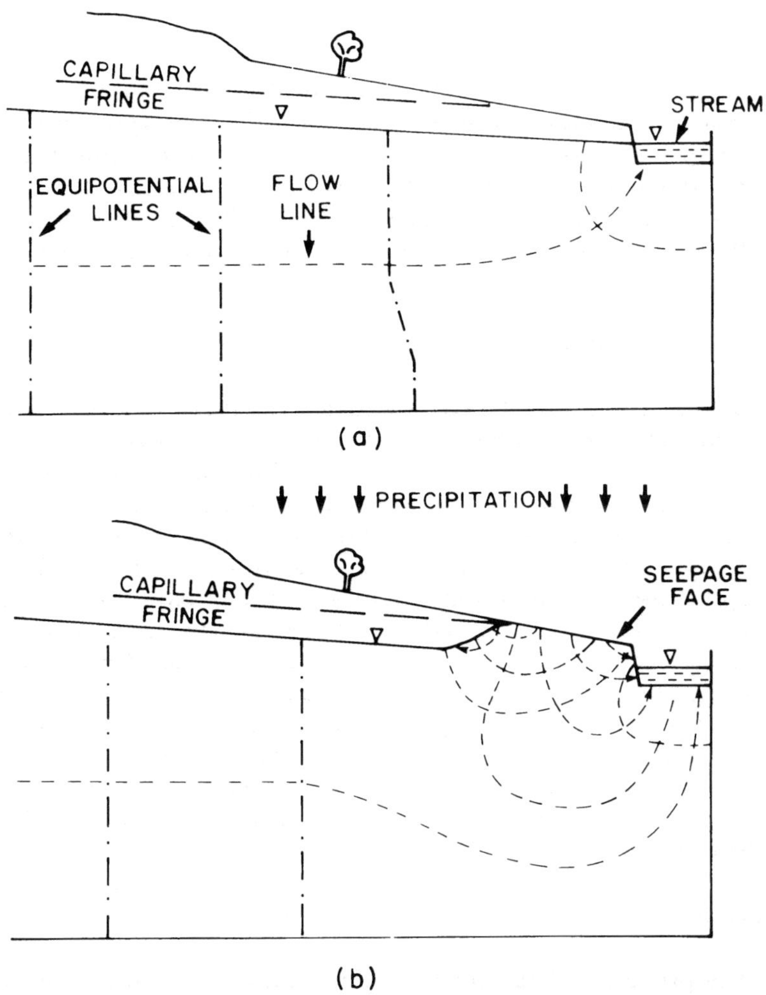

Fig. 11. Schematic showing the proposed effect of the capillary fringe on the discharge of groundwater to surface waters: (a) prior to rainfall; and (b) shortly after the onset of rainfall.

Fig. 11a shows a hypothetical cross-section through a stream, with a uniformly sloping water table, and the capillary fringe intersecting the ground surface at some distance to the left of the stream. These conditions could be common, particularly in floodplain areas during wet periods of the year. Fig. 11b shows the hypothetical response of the system to a precipitation event. In principle, the rise of the water table across the capillary fringe, in the vicinity of the stream could occur very quickly in response to the start of the rain event. Although the flow nets of Fig. 11 are highly schematic, they show a greater concentration of stream tubes entering the stream as a result of the water-table rise, and thus a very pronounced and rapid response of the groundwater system to the precipitation event.

Contaminant migration to surface waters

For the purpose of flood forecasting and water management, understanding the mechanisms of streamflow generation may be of greater scientific interest than of practical importance. However, in the development of models for predicting streamflow quality, quantifying the respective sources is essential. Thus an understanding of the role of the capillary fringe in streamflow generation may be essential to the development of reliable runoff-quality models.

As a specific example, consider the management of mine tailings or other extensive deposits of waste materials containing toxic and soluble constituents. Particularly in humid climate regions, these deposits will be traversed by perennial or intermittent streams, that for at least a part of the year, will be in hydraulic connection with contaminated groundwater in the waste material. In the absence of the capillary-fringe effect, contaminants immediately adjacent to the stream could be leached out over a relatively short period of time, with the only remaining contribution being that caused by the regional groundwater flow system. Thus a significantly improved stream-water quality could be expected within a reasonable period of time. However, if the capillary fringe contributes significantly to flow in the stream, then the deep circulation of flow lines indicated in Fig. 11b would cause contaminants to be leached for a considerable distance around the stream. As a result, the water quality in the stream could remain poor for an extremely long period of time. This could be of great practical importance in developing abandonment or long-term management requirements for the site.

Contaminant transport in shallow water-table regimes

Current methods of predicting the migration of contaminants in hydrogeologic environments generally involve a solution of the advection—dispersion equation, with the flow field assumed to be steady. Contaminants in groundwater are of concern primarily when they impinge on water supplies, or when they are returned to the biosphere. Return to the biosphere generally implies that they are carried to a groundwater discharge zone. Because discharge zones are often characterized by shallow water tables, from our previous discussions, there is good reason to expect the hydraulic characteristics of many of these zones to be highly transient. Thus, for perhaps the most critical region in the analysis, predictions based on the steady-flow assumption could be substantially in error.

Fig. 12 is a cross-section through a hypothetical groundwater discharge zone showing a shallow water table and an undulating topography. In the initial condition (Fig. 12a) the water table is horizontal, with the capillary fringe intersecting the bottom of the depressional area. A distinct unsaturated zone exists between the capillary fringe and the tops of the higher areas. Fig. 12b shows the anticipated response to rainfall. The water table would rise

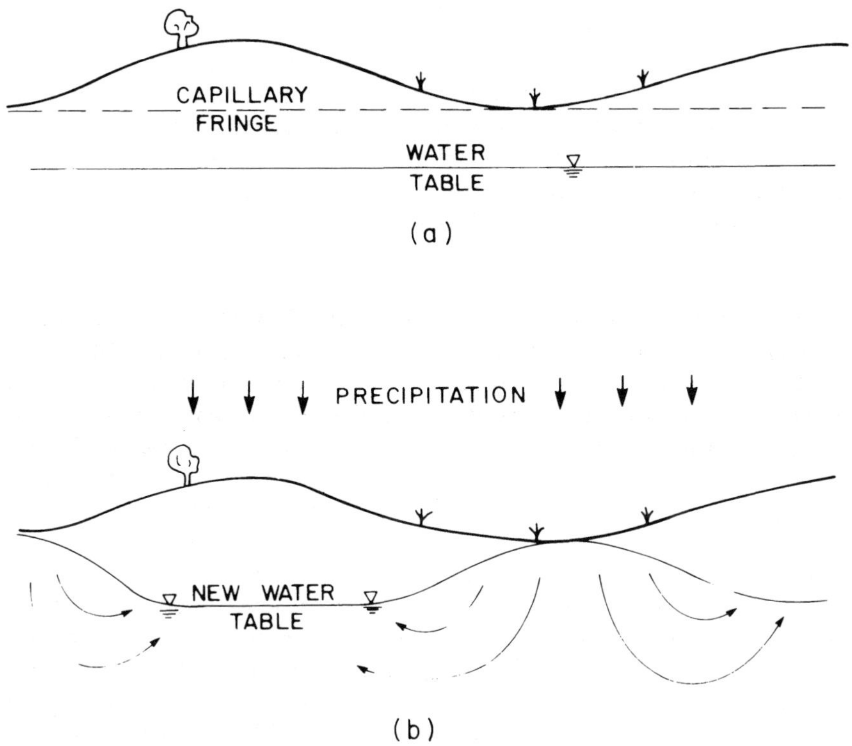

Fig. 12. Schematic showing the effect of the capillary fringe on the water-table response in an area with irregular topography.

almost immediately to the bottom of the depressional areas while the response beneath the higher areas would be substantially delayed. The result would be a highly complex flow pattern that would tend to cause contaminants to move in directions that would not be expected under the assumption of steady-state flow conditions. The net effect would be to spread contaminants over a larger area than expected, but to reduce the concentrations to values that would be lower than expected. This phenomenon would not be accounted for in existing contaminant transport models.

CONCLUSIONS

The large water-table response to precipitation that can be expected in situations where the capillary fringe extends to ground surface has been recognised for many years. Nevertheless, in most hydrogeological investigations, the effect has been ignored, or discounted as being of little significance, and few detailed field investigations of the phenomenon have been reported.

The physical arguments presented in this paper indicate that in fine-grained geologic materials, the capillary fringe, and thus the zone of saturation, may extend for several metres above the water table. In addition, the results of the field study indicated that the water-table response under natural conditions was consistent with the physical basis of the capillary-fringe phenomenon. It therefore appears that the capillary fringe can indeed have a major influence on the response of shallow water tables to precipitation.

The capillary fringe could be an important consideration in the analysis of several hydrologic processes. Examples include recharge and consumptive use calculations, stream-flow generation, migration of contaminants to surface waters and contaminant transport in shallow water-table regimes. Undoubtedly, there are others. Additional studies are required in order to evaluate the effect of the capillary fringe on the water-table response in a wider range of geologic materials, and to evaluate its importance in the analysis of hydrologic processes associated with shallow water-table regimes.

ACKNOWLEDGEMENTS

Constructive comments on this manuscript were provided by Arnold Klute and Dave McWhorter of Colorado State University. The author also wishes to acknowledge the unselfish support of his field assistant, David Gillham. Financial support for this study was obtained through an operating grant of the National Science and Engineering Research Council of Canada.

REFERENCES

Childs, E.C., 1960. The nonsteady state of the water table in drained land. J. Geophys. Res., 65: 780—782.

Duke, H.R., 1972. Capillary properties of soils — influence upon specific yield. Trans. Am. Soc. Agric. Eng., 15: 688—691.

Freeze, R.A., 1974. Streamflow generation. Rev. Geophys. Space Phys., 12: 627—647.

Freeze, R.A. and Cherry, J.A., 1979. Groundwater. Prentice-Hall, Englewood Cliffs, N.J., 604 pp.

Fritz, P., Cherry, J.A., Weyer, K.V. and Sklash, M.E., 1976. Runoff analyses using environmental isotope and major ions. In: Interpretation of Environmental Isotope and Hydrochemical Data in Groundwater Hydrology. Int. At. Energy Agency (I.A.E.A.), Vienna, pp. 111—130.

Gillham, R.W., Klute, A. and Hermann, D.F., 1976. Hydraulic properties of a porous medium: measurement and empirical representation. Soil Sci. Soc. Am. Proc., 40: 203—207.

Hooghoudt, S.B., 1947. Waarnemingen van grondwaterstanden voor de landbouw. Comm. Hydrol. Onderz., Tech. Natuurw. Onderz. (T.N.O.), The Hague, Versl. Tech. Bijeenk. *1—6*, pp. 94—110 (published in 1952).

Luthin, J.N., 1960. Discussion of *The nonsteady state of the water table in drained land*, by E.C. Childs. J. Geophys. Res., 65: 4221—4222.

MacFarlane, D.S., Cherry, J.A., Gillham, R.W. and Sudicky, E.A., 1983. Migration of contaminants in groundwater at a landfill: a case study, 1. Groundwater flow and plume delineation. In: J.A. Cherry (Guest-Editor), Migration of Contaminants in Groundwater at a Landfill: A Case Study. J. Hydrol., 63: 1—29 (special issue).

Martinec, J., 1975. Subsurface flow from snow-melt traced by tritium. Water Resour. Res., 11: 496—497.

McWhorter, D. and Sunada, D.K., 1977. Ground-water Hydrology and Hydraulics. Water Resources Publications, Fort Collins, Colo., 290 pp.

Meyboom, P., 1967. Groundwater studies in the Assiniboine River drainage basin, Part II. Hydrologic characteristics of phreatophytic vegetation in south-central Saskatchewan. Geol. Surv. Can., Bull. No. 139, pp. 18—23.

Sklash, M.G. and Farvolden, R.N., 1979. The role of groundwater in storm runoff. J. Hydrol., 43: 45—65.

Staple, W.I., 1969. Comparison of computed and measured moisture redistribution following infiltration. Soil Soc. Am. Proc., 33: 840—847.

IAHS, the International Association of Hydrological Sciences

IAHS is a nonprofit-making nongovernmental scientific organization committed to serving the science of hydrology and the worldwide community of hydrologists. IAHS, established in 1922, now has more than 4300 members from 124 countries.

The IAHS objective is to promote hydrology as part of the geosciences and as the basis for water management applications. The primary aims are to:

- initiate, facilitate, and coordinate research into hydrological problems requiring international cooperation
- provide for international discussion, comparison, and publication of hydrological research
- support hydrological sciences in developing countries
- bridge the gap between science and practice, including providing training opportunities and materials

International scientific commissions deal with surface water, groundwater, snow and ice hydrology, erosion and sedimentation, water quality, water resources, land–atmosphere systems and specific techniques, while working groups address particular issues. The Predictions in Ungauged Basins (PUB) Decade, 2003–2012, is a major initiative.

IAHS fulfils its mission of worldwide dissemination of research results and practice, and raises funds, by publishing and selling *Hydrological Sciences Journal* and books. Details of all publications are available at the web site, including abstracts of published papers. IAHS supports hydrological science in financially disadvantaged countries by: providing publications free of charge to libraries in those countries; and, offering discounts on publications purchased by members and contributing towards their costs when attending IAHS meetings.

Membership of IAHS is *free of charge* and open to anyone wishing to participate in IAHS activities, such as IAHS symposia and workshops.

IAHS Ltd is a UK registered charity (no. 1078635); all revenues are used to support the aims of the Association. Also, IAHS is one of the seven associations that comprise the International Union of Geodesy and Geophysics (IUGG, *www.iugg.org*).

IAHS AISH

www.iahs.info

Other publications from IAHS

Predictions in Ungauged Basins: International Perspectives on the State of the Art and Pathways Forward

edited by Stewart W. Franks, Murugesu Sivapalan, Kuniyoshi Takeuchi & Yasuto Tachikawa

The Australia–Japan PUB Workshop at Perth, Western Australia, in 2004, considered current hydrological prediction capabilities and future research directions to achieve future improvements, given the decline of hydrological gauging networks, natural variability, and long-term climate and land-use changes. This book combines chapters elaborating innovative theoretical and practical possibilities of different approaches for prediction, with contributions describing the specific needs in Australia and Japan. Topics covered include:

PUB Perspectives; *Floods, drought, sediment and water quality*; *New data sources and modelling*; *Approaches for process understanding and integration; Uncertainty estimation and model identification*

Publ. 301 (*2005*) ISBN 1-901502-38-4, ~350 pp. £60.00

Bringing Groundwater Quality Research to the Watershed Scale

edited by Neil R. Thomson

Sustainable economic growth requires plentiful supplies of high-quality water. This has placed pressure on groundwater resources globally and forced the international groundwater science, engineering and regulatory community to recognize the current limitations of research and resource management. Integrated and technically feasible approaches tackling local issues and watershed-scale concerns concurrently are required. These were considered by GQ2004 (Waterloo, Canada) and this collection of papers comprises a reviewed selection from the conference.

The introductory contributions include global and several national perspectives, followed by sections dealing with:

- Contaminant input processes
- Site characterization
- Management and decision making
- Natural attenuation processes and applications
- *In situ* remediation
- Flow and transport modelling at the national, the watershed and smaller scales.

Publ. 297 (*2005*) ISBN 1-901502-18-X, 576 + xiv pp. £85.00

Permeable Reactive Barriers

edited by Genevieve A. Boshoff & Brian D. Bone

In the foreword Bob Gillham concludes that the contributions in this volume, and the conference from which they were derived, provide considerable insight concerning the evolution of PRB technology. With the advancement of the technology, it appears certain that PRBs will continue as a significant remediation method and that their stature and acceptance is likely to increase in the future.

Publ. 298 (*2005*) ISBN 1-901502-23-6, 176 + viii pp. £38.00

IAHS AISH *More details at www.iahs.info*

Sediment Budgets 1

edited by Des E. Walling & Arthur J. Horowitz

Publ. 291 (*2005*) ISBN 1-901502-87-2, 372 + xii pp., £63.00

Sediment Budgets 2

edited by Arthur J. Horowitz & Des E. Walling

Publ. 292 (*2005*) ISBN 1-901502-92-9, 358 + xii pp., £63.00

A sediment budget provides an integrated view of the sediment sources, transfers, sinks and outputs of a drainage basin, and so draws together the many different aspects of erosion and sediment mobilization, sediment transport, sediment storage, and sediment yield. Key developments have influenced recent sediment budget investigations:

- increasing awareness of the environmental implications of sediment loads in rivers and of the need for effective sediment management strategies
- advances in the techniques available including use of ^{137}Cs, ^{210}Pb and ^{7}Be as sediment tracers; continuous recording of suspended sediment fluxes; application of distributed modelling, remote sensing and GIS
- recognition of the need to quantify sediment-associated chemical fluxes.

The papers (81 selected from >150 submissions) reviewed and edited for these volumes, coming from many countries, exemplify these developments internationally.

Sediment Budgets 1 focuses on the main components of the sediment budget: *Sediment mobilization and sources*, *Sediment transport and transfer*, *Sediment storage* and *Sediment yields*. The papers highlight important advances in understanding of soil and sediment redistribution in small catchments, the major developments in sediment source tracing, key improvements in monitoring techniques, the role and behaviour of sediment sinks, particularly river flood plains, and the increasing availability of reliable data on sediment yields from different areas of the world.

Sediment Budgets 2 emphasizes the value of sediment budgets in providing an integrating framework, for both scientific investigations and environmental management. *Sediment budgets* reports results of studies in a wide range of environments. *Modelling sediment budgets and their components* demonstrates the range of approaches and the potential for applying existing models. *Human impact on sediment budgets* emphasizes human modifications to individual components of, as well as the overall sediment budget, including the impact of dams. *Sediment problems and sediment management strategies* illustrates how the sediment budget concept can be integrated into management strategies.

The International Commission on Continental Erosion (ICCE) and UNESCO organized the symposium as a contribution to the International Sedimentation Initiative (ISI) of the UNESCO International Hydrological Programme (IHP VI).

IAHS AISH

More details at www.iahs.info